ROUTLEDGE LIBRARY EDITIONS:
AGRIBUSINESS AND LAND USE

Volume 24

ENVIRONMENT AND LAND USE IN AFRICA

ENVIRONMENT AND LAND USE IN AFRICA

Edited by
M. F. THOMAS
AND
G. W. WHITTINGTON

Routledge
Taylor & Francis Group

LONDON AND NEW YORK

First published in 1969 by Methuen & Co. Ltd

This edition first published in 2024
by Routledge
4 Park Square, Milton Park, Abingdon, Oxon OX14 4RN

and by Routledge
605 Third Avenue, New York, NY 10158

Routledge is an imprint of the Taylor & Francis Group, an informa business

© 1969 Methuen & Co. Ltd

British Library Cataloguing in Publication Data
A catalogue record for this book is available from the British Library

ISBN: 978-1-032-48321-4 (Set)
ISBN: 978-1-032-46943-0 (Volume 24) (hbk)
ISBN: 978-1-032-46953-9 (Volume 24) (pbk)
ISBN: 978-1-003-38392-5 (Volume 24) (ebk)

DOI: 10.4324/9781003383925

Publisher's Note
The publisher has gone to great lengths to ensure the quality of this reprint but points out that some imperfections in the original copies may be apparent.

Disclaimer
The publisher has made every effort to trace copyright holders and would welcome correspondence from those they have been unable to trace.

Environment and Land Use in Africa

Edited by
M. F. THOMAS and
G. W. WHITTINGTON

London
Methuen & Co Ltd
11 New Fetter Lane EC4

First published 1969
© 1969 Methuen & Co Ltd
Printed and bound in Great Britain
by Richard Clay (The Chaucer Press) Ltd,
Bungay, Suffolk
SBN 416 10840 7

Distributed in the U.S.A.
by Barnes and Noble, Inc.

Contents

Figures

Both in the maps and in the text the spelling of place names follows the *Oxford Atlas*, 1963.

Plates

1 Introduction

THE EDITORS

The upsurge of interest in Africa

In the last twenty years interest in Africa, especially in the essential Africa lying to the south of the Sahara, has increased enormously. This growth of interest, not always complemented by an increase in the external understanding of Africa, has led to a great flood of writing which in most instances has added to our knowledge of that continent. Information has stemmed from two main sources. In the first instance there is the work undertaken by a growing group of African writers. Secondly, there has been the contribution made by expatriates of the various nations intimately involved in the legacy of the colonial era of the late nineteenth century. This has been supplemented increasingly in recent years by American workers and others of various nationalities working under the aegis of such bodies as FAO or I.B.R.D. Information in any one particular branch of learning reveals a very uneven depth of study and areal spread. This results from the great differences in the numbers of educated Africans from country to country and also, where the expatriates are concerned, from the different emphases placed upon trade, education, government and extension services in the former colonial territories.

These factors are illustrated very clearly in geographical writings. Until very recently African writers, excepting those from the West African zone and the Republic of South Africa, have made only a very small contribution to geography. Most of the information has come from the expatriate group, and here differences in depth of and approach to study are well illustrated. Perhaps nowhere is this clearer than in the British and French contributions. Differences in methods of rule are partly at the root of this contrast. Also the existence in the French territory of the Institut Français de l'Afrique Noire (IFAN) was not paralleled in the British zone – the only comparable centre, the Rhodes–Livingstone Institute, although of great importance, hardly compares in its total contribution. The British territories did, however, possess an advantage in the greater number of universities and centres of higher education. Even in this realm, however, an uneven spread is again encountered, a factor which accounts for the domination of the geographical scene by the West African geographers, both native and expatriate, especially those from Nigeria and Ghana.

The dissemination of knowledge

Many of the difficulties associated with the undoubted existence of the general lack of a clear understanding of and insight into African problems and conditions are due to the inadequacies of communication of available information. The work achieved by geographers in Africa has appeared in different forms, ranging from papers in journals, through monographs and memoirs to books organized on either a continental or a regional/national scale.

The paper on the whole allows a quick dissemination of information, usually on a particular or circumscribed topic and usually aimed at a specialized readership. Difficulties of space in general and the high cost of printing limit the potential of this vehicle of communication. Furthermore, geographical papers tend to be widely dispersed through a variety of journals reflecting the many approaches the geographers have made to their subject. Partly in an attempt to bring some kind of order into this latter situation, partly to have more space in which to present facts and arguments and partly to provide material for the educationalist, certain authors, mainly by means of a systematizing of existing material, have been tempted to expand their writings into the book form.

The book form has been exploited in different ways which in terms of approach can most succinctly be entitled *The Conventional Text* and *The Thematic Text*.

1. *The Conventional Text:* A large number of books have come into existence in a desire to provide an orderly and regional view of Africa: one which in most instances confuses region with nation and which conforms to the 'conventional' geographical text. A review is made of physical conditions, usually of the most inadequate nature, and sometimes of over-all economic conditions. Then the area under review is split up into a number of 'regions' in which attempts are made to relate physical and human factors into a coherent entity. Such books have enjoyed varied amounts of success. From the outset, however, writers engaged in such a task suffer from distinct disadvantages, especially in the case of the single-author approach. Despite the greater availability of material on African geography today, even leaving out of consideration the difficulty of locating it, great problems still face the author, for there are enormous gaps in the necessary basic information. Any *comprehensive* geographical treatment of Africa is therefore bound to suffer from the great contrasts in the amount of available information that exist – compare for example that for Nigeria with that for Moçambique. Furthermore, many of the books in this category are 'dead'. This stems largely from the sheer size of the task of writing a book on such a large area as Africa, even at the sub-Saharan scale, and thus from a con-

comitant lack of intimate knowledge of the enormously varied and complex African scene. Such books all too often become a catalogue of facts, historical before they are published and in many instances before they are written. There is in this type of book a great danger of over-generalization and over-simplification, which all too often leads to the distorted view of Africa and African conditions that exists in the minds of all too many. How often has the statement been made that African soils are infertile? This is just one example of the many hundreds of similar over-generalizations that could be given. The uniformity of the human scene in much of Africa conveyed by many texts is quite false. Indeed, the localized physical, economic and social framework within which most African communities function has led to great variety in custom and practice over very short distances.

In a variant of this type of book an attempt has been made to get away from the comprehensive approach of the complete synthesis of physical and human parts. Hance (1964) achieved this to a certain extent by slanting his book to the economic aspects. He provided a well-written compendium of facts which by the very nature of the approach was difficult to make stimulating. De Blij (1962) used a varied approach to the area with which he dealt; the method had great potential but the result was largely disappointing because of a lack of intimate knowledge of much of the area he considered. This was a disadvantage of which Hance was acutely aware but which some writers of the type of book discussed above have unfortunately ignored. Books in which the regional approach to African geography has been employed have been of very limited success. This is partly due to the size of the area being examined. Perhaps more realistically, however, it is because the time is not yet arrived for such adventures. Even in the more successful books of this nature, for example Paterson (1960) on North America, the criteria used for the basic regional division prove somewhat illusory on close examination. Indeed, perhaps the approach is doomed to failure from the start judging from the disagreement among geographers as a whole on the regional issue.

2. *The Thematic Text:* Far more successful has been the class of books which have had a more circumscribed and easily handled approach: books which have a realistic field of concern and which have given a lead in the approaches which can be made in the furthering of African geography. Unfortunately such books are none too common and in many instances have been written by non-geographers. Works in this category which spring readily to mind are by such authors as Allan (1965), Biebuyck (1963), Bohannan and Dalton (1962), King (1963), Phillips (1961) and Yudelman (1964). The geographer's contribution has been by such authors as Hance (1958), Fair and Green (1962), Barbour and Prothero (1961) and O'Connor

(1966). It is noticeable that in many instances such books as are highly
successful result either from a large number of authors with detailed areal
experience collaborating under an over-all editorship, or from a lifetime of
intensive and areally wide experience in administration, teaching or extension service by one man. It is in the former approach that the greatest
hope lies if coherent, stimulating studies of African geography on, for
example, the sub-Saharan scale are to be attempted. The growing number
of worthwhile books appearing in this form in other areas of geographical
study shows the success of the method (Chorley and Haggett, 1965, 1967;
Eyre and Jones, 1966; Hoffman, 1961; Thomas, 1956).

The scope and aims of the present volume
The justification for a volume of separate though related essays by individual specialists thus rests upon the realization that Africa is too large and
too diverse an area to be covered in a serious and comprehensive manner
within the covers of a single book, and, furthermore, that specialization
within the general field of geography has now reached, in common with
other scientific disciplines, such a level that one author is unlikely to
achieve equal depth in a wide range of geographical studies. Although this
volume has a dominant theme, that of the relationships between environment and land use within the African continent, it has a wide compass of
physical, biological and social studies, and as wide a geographical spread
as it has proved editorially possible to achieve. It differs from most other
books of similar structure in that it is not designed as a systematic text in
the style of Watson and Sissons (1964), nor is it regionally organized in the
manner of Hodder and Harris (1967). It is more widely based than Barbour
and Prothero (1961) and would appear to fill a serious gap in the books on
African studies in that it brings together studies of both the natural environment and the social environment, as they appear relevant to land use.
Apart from the detailed study of Ghana edited by Wills (1962) there have
been few other attempts in this field.

It is a part of the geographer's training to recognize the need for complementary studies in a diversity of fields, and in this instance the need is
not entirely, or indeed primarily, academic. Regional economic development in Africa hinges upon rural land use, and the present patterns and
future policies for change must be understood and formulated with a
realization that neither natural nor social conditions can be ignored or
lightly considered.

It is not, however, possible within a work of this kind to weave the
threads together to produce a complete design; this must be foregone in
order to make gains in detail and depth of individual studies. It is hoped,
none the less, that a volume of this kind will assist future assessments of

the geography of Africa by other writers, and will stimulate those concerned with the practical aspects of regional development to make as thorough and as comprehensive an assessment of environment as is possible, before action to alter the present situation in an area is undertaken.

Inevitably, shortcomings of the volume as a whole are evident to the editors. The uneven geographical spread with its concentration upon the English-speaking areas of Africa and further bias towards West African conditions, reflects the experience of English-speaking geographers in African studies.

The division of the book into three parts illustrates its theme and purpose. In Parts I and II general principles and methodology in the study of elements of African environments are emphasized along with reviews of factual and conceptual material relating to man's use of his environment. In Part III case studies of different kinds illustrate the geographer's approach to the study of environment and land use in specific areas. Since many of the concepts and methods used, particularly in Part I, are recent and in certain instances being presented in this form for the first time, it is not possible to point in each case to their application to a particular area or problem in Part III. In any individual case it may not be possible, at the present time, to employ sophisticated methods of recording and analysis in the study of each and every element of the environment of land use. More important, it is a corollary of the multiple authorship of this volume that comprehensive study in a given area should be undertaken by a team of specialists. Our contributors have for the most part had to work as individuals.

Only the development agencies of the United Nations and other government-aided bodies have been able to afford teams of specialists to work on particular problems or in special areas. On the other hand these teams have to work according to a tight schedule and cannot often take a detached and critical view of the concepts and methods which they employ. This, however, is what the present authors have attempted to do. Thus the studies are academic to the extent that they attempt to arrive at more useful methods of analysing and mapping man's environment and his patterns of land use. But they are applied inasmuch as they concern areas of study that are important to African rural land use. Furthermore, they have implications for those concerned with the recording and analysis of data and with mapping and reconnaissance surveys for development purposes.

The aims of the present volume can therefore be summarized: to bring together the results of original geographical research in the fields of environmental and land-use study in Africa, and as a result, further the understanding of some of the basic spatial patterns of the continent and of the functional relationships which exist between them. It is addressed to all serious students of Africa and it is hoped that it will be appreciated by a wide range of specialists, not only geographers. The book has been offered to its contributors as a forum for individual views and no attempt is made to reconcile these where they diverge or conflict.

Future approaches to the geographical study of Africa

No attempt is made here to forecast developments in the study of African geography. The following remarks are offered rather as observations upon the present state of our knowledge of the continent, and as reflections on the content of the present volume.

The urgency of the need for regional economic development in Africa has focused attention upon fields of geographical study that are commonly neglected by geographers who work in technologically advanced and economically highly specialized communities. The concern with pro- grammes of basic mapping and compilation of data regarding man's environment, and with the relationship of the agriculturalist to his environ- ment, are but a few of these. Africa presents a challenge to those who are geographically trained, to exercise the basic principles of their craft. At the same time there is a need for the application of new concepts and quantita- tive methods to the study of African problems. Many of the generalizations concerning African conditions are explicitly or implicitly challenged by the studies in this volume, as they have been in other fields by recent publica- tions. Yet there are subjects of common concern throughout wide areas of the continent, and there remain broad similarities of conditions, both natural and human, within these areas. These similarities identify our image of Africa and range from the rhythms of climate, the lithology of the rocks and character of land forms to the intricate relationship of the African peasant with his environment, as expressed through the delicate vegetation/land-use associations evident throughout so much of Africa. While these features exist, simplification of them on the basis of our present knowledge is likely to distort our view of the continent and its great variety.

The geographer must sooner or later emulate the team organization of the development agencies, if his work is not to become, or remain super- ficial. This may involve groups of geographers with a diversity of interests, or, equally, teams of specialists drawn from various disciplines. The alternative is the repetition of isolated studies of particular elements of the landscape. These of course will, in any case, continue to be made and will enlarge our understanding in specific fields. But within a given area a com- prehensive study of its character can hardly be achieved by individual work, unless it is prolonged over a period of many years; perhaps not even then. The furtherance of our understanding of Africa as a whole can only come from the detailed study of many localized areas within which the diversity of specialism represented by the authors of this volume – together with others not here included – is applied to the equal if not much greater diversity of the natural and human scene.

References

ALLAN, W. 1965. *The African Husbandman*, Edinburgh.

BARBOUR, K. M. and PROTHERO, R. M. (eds.). 1961. *Essays on African Population*, London.

BIEBUYCK, D. (ed.). 1963. *African Agrarian Systems*, Leopoldville.

BOHANNAN, P. and DALTON, G. (eds.). 1962. *Markets in Africa*, Evanston.

CHORLEY, R. J. and HAGGETT, P. (eds.). 1965. *Frontiers in Geographical Teaching*, London.

  1967. *Models in Geography*, London.

DE BLIJ, H. J. 1962. *Africa South*, Evanston.

EYRE, S. R. and JONES, G. R. J. (eds.). 1966. *Geography as Human Ecology*, London.

GREEN, L. P. and FAIR, T. J. D. 1962. *Development in Africa*, Johannesburg.

HANCE, W. A. 1958. *African Economic Development*, New York.

  1964. *The Geography of Modern Africa*, New York.

HODDER, B. W. and HARRIS, D. R. (eds.). 1967. *Africa in Transition*, London.

HOFFMAN, G. W. (ed.). 1961. *A Geography of Europe*, London.

KING, L. C. 1963. *South African Scenery*, Edinburgh.

O'CONNOR, A. M. 1966. *An Economic Geography of East Africa*, London.

PATERSON, J. H., 1960. *North America*, London.

PHILLIPS, J. 1961. *The Development of Agriculture and Forestry in the Tropics*, London.

THOMAS, W. L. (ed.). 1956. *Man's Role in Changing the Face of the Earth*, Chicago.

WATSON, J. W. and SISSONS, J. B. (eds.). 1964. *The British Isles: A Systematic Geography*, London.

WILLS, J. B. (ed.). 1962. *Agriculture and Land Use in Ghana*, London.

YUDELMAN, M. 1964. *Africans on the Land*, Cambridge, Mass.

2 Agricultural geography in tropical Africa

J. T. COPPOCK

Since agriculture is the most common and widespread of the ways in which man gets his living and since geographers are primarily concerned with his varied impact on the earth's surface, it might be supposed that a considerable proportion of geographical research would be devoted to the study of agricultural geography. Yet this is not the case and, while it is true that geographical journals contain many descriptive articles and that descriptions of agriculture often bulk large in elementary textbooks, few substantive works have appeared in this field and no adequate methodology or body of theory has yet been devised. In developed countries some justification for such neglect might be sought in the small and declining contribution of agriculture to both national income and employment and in the urban background of most geographers, but the situation is not greatly different in developing countries where agriculture may account for two-thirds of the national income and occupy four-fiths of the population and where many of the growing number of university-trained geographers are the children of farmers or were brought up in rural communities. It seems likely that a major cause of this neglect is the difficulty, which stems largely from the nature of available data, of undertaking worthwhile and intellectually rewarding investigations in this field by comparison with those which are possible in other branches of geography, a view which is supported by the fact that the proportion of agricultural articles in geographical journals is declining. This explanation seems even more plausible in tropical Africa where many of the investigations which have been undertaken by geographers have either been in physical geography, where the ground itself provides all necessary data, or in the study of communications, industries and settlement, for which material is both more accessible and more manageable. Even in agricultural geography there has been some tendency to concentrate on those topics, like settlement schemes, which are not representative of agriculture but which are comparatively well documented and it is not surprising that the only monograph to appear in English on the agricultural geography of any country in tropical Africa, D. McMaster's study of subsistence crops in Uganda, should concern a territory with unusually rich sources of statistical data (McMaster, 1962).

Despite these problems, a strong case that geographers should pay more attention to tropical agriculture can be made, on both practical and academic grounds. In the foreseeable future it is likely that agriculture will continue to occupy a leading place in the economies of African countries and, both out of academic self-interest and as responsible citizens, geographers ought to play a part in economic development by employing their skills in describing and understanding the existing use of agricultural resources and so contributing to the planning of their more effective exploitation; for the geographer's concern with total environment, both physical and human, and his training in methods of reconnaissance survey are very appropriate in countries in tropical Africa, whose problems are often too urgent to wait on detailed inquiries by many specialists. It would, of course, be wrong to allow such immediate practical objectives wholly to dictate academic priorities, but geographers may find it easier to obtain data for strictly academic investigations when they have demonstrated their capacity to make practical contributions. In any case, the distinction between academic and practical should not be over-stressed; a better theoretical understanding of the location of agricultural production may be expected to increase geographers' ability to predict the likely results of proposed changes and so make a more effective contribution to the planning process and, conversely, the investigation of the agricultural consequences of, say, the construction of new communications may lead to improvements in theory. In this respect, the tropics offer opportunities which are not so readily available in the developed world; the existence of large numbers of peasant farmers, operating on a small scale in an environment with few external links and a relatively simple economic climate and yet subject to radical and rapid change, offers the nearest approach to laboratory conditions that the social scientist is likely to find.

The central aim of agricultural geography as an academic discipline is an understanding of the complex variety of agricultural patterns on the earth's surface. These may be examined at all scales, from the world and continental views characteristic of general atlases and textbooks to the ultimate unit of decision-making, the individual farm. In practice, the minimum area is more likely to be a group of farms or the territory of a village or other similar rural community, since the geographer, like the economist and the sociologist, is generally more concerned with the trends and tendencies which are the resultant of many individual actions, although he recognizes that individual decisions and idiosyncrasies may be important where the unit of decision-making is large. For the agricultural geographer working in tropical Africa no one scale is intrinsically more appropriate than another and both large- and small-scale investigations present difficulties. The choice of the scale will depend upon the nature of the problem being investigated and upon the resources available for its solution. Where farms are very small, as in most of tropical Africa, the territory occupied by a single rural community is likely to be the largest area which can be studied comprehensively in the field by an individual investigator, especially if he has to rely on field observation and inquiry for all his data. Much larger areas can, of course, be studied more superficially on a reconnaissance basis, by sampling, by reliance on secondary sources, by teamwork, or by limiting the aspects to be investigated, and all these approaches have a part to play.

Ideally, each investigation in agricultural geography should involve four stages: the identification of a problem; the formulation of a hypothesis or hypotheses; the collection of the relevant data; and the testing and modification of the hypothesis to provide an adequate explanation. Such a sequence presupposes that the problem can be readily identified, that sufficient is known to permit the formulation of hypotheses and that appropriate data can be obtained, but it will rarely be possible for all of these conditions to be satisfied in the present state of knowledge of tropical Africa and objectives must consequently be more limited. Like the fly on the newspaper photograph, the earthbound geographer can observe at any one time only the details of the picture and one of his first tasks must be frequently to construct a reconnaissance agricultural geography, giving a general conspectus of the main agricultural features of the country in which he is working from such data as are available, perhaps in the form of an agricultural atlas. Such a reconnaissance study is unlikely to be intellectually rewarding, but it will help to define problems and to provide the context in which specific investigations can be set. It will also serve a useful purpose by influencing the mental maps which all administrators and politicians have of the territories for which they are responsible and whose form is indicated by P. Gould's (1966) pioneer study in Ghana. Such 'maps' must play an important part in decisions about where agricultural and other developments will occur and the nearer they approach reality, the better the prospect that decisions will match the actual needs of the country.

Such a reconnaissance study will, of course, be less necessary in territories where the World Bank, FAO or similar bodies have made general surveys, although the published reports often provide only a very sketchy geographical appraisal. It can be argued that to attempt such a broad survey is to reverse priorities, in that it ought to await the completion of many detailed investigations, yet as K. M. Buchanan and J. C. Pugh have convincingly shown, it is possible to provide a surprisingly full picture of the agricultural geography of a territory, for there is often much information, whether statistical data, reports of administrators, agricultural officers, soil surveyors and visiting experts, or investigations by research workers in a wide range of disciplines, which, however unsatisfactory or meagre its contribution may be in isolation, can be usefully brought together in such a synthesis (Buchanan and Pugh, 1958). A valuable first step would thus be a thorough bibliographic survey, without regard for subject boundaries, of all books, papers and published and unpublished reports relating to the agriculture of each territory.

To complement this work there is also an urgent need for detailed studies of individual agricultural communities; in many areas too little is known about the organization of agriculture, especially in relation to local environment, for adequate hypotheses to be made about such topics as the choice of particular crops or particular agricultural methods. Consequently, in advance of, or parallel to, more problem-orientated investigations, studies of the agricultural organization of selected communities covering a range of environments, should be undertaken in depth. Anthropologists have long realized the importance of such investigations, which presuppose a willingness to undergo long residence in a rural community, and have accumulated a wealth of information, much of it germane to an understanding of the use of agricultural resources. Agricultural economists and sociologists, too, have come to appreciate the necessity for such studies in depth and there is increasing need for geographers to follow the example of J. Hunter (1961), R. M. Prothero (1957) and others in undertaking surveys of small areas. Such tasks are particularly appropriate for African geographers, unhampered by linguistic or cultural difficulties.

Between these two extremes, of broad appraisal of the whole agricultural economy and the detailed study of small areas, the range of possible problems which might be investigated is very large, although most will involve either the study of a single agricultural enterprise, such as cocoa-growing or yam-production, throughout a territory, or the investigation of a more limited problem area, particularly from the viewpoint of agricultural development. In practice, the choice will tend to be limited by the information that is available or that can be collected and by the skills of individual geographers; a major obstacle here is the lack of formal training in agriculture provided in most courses in geography.

The nature of available data is of critical importance. Written records, of the kind found in the journals of learned societies, in government archives and libraries, and in the publications of official and semi-official bodies, can be valuable in suggesting possible hypotheses and giving an insight into information gleaned from other sources; but, while they may sometimes contain the results of special surveys, they will not usually yield sufficiently specifically located data to form the basis of a satisfactory geographical analysis. Even so, in the absence of suitable statistical data, written records, especially reports on local administrative areas like the agricultural notebooks prepared by district agricultural officers in Northern Nigeria (Nigeria, 1957) or the agricultural surveys in Uganda (Uganda, 1960), can sometimes provide sufficient qualitative data about the occurrence of particular crops and perhaps even their relative importance to allow the preparation of distribution maps showing the approximate limits of their cultivation and the areas of greatest importance. Although such maps must be treated with great care, they will be particularly useful where a distribution has changed markedly over the course of time and they can be used also as a check on statistical data.

Agricultural geographers in developed countries frequently complain about the inadequacy of the statistical data available to them, and it is true that the more interesting the data, the more likely they are to be regarded as confidential and hence to be highly generalized and difficult to locate exactly; but for many aspects there is an embarrassment of material rather than a lack and the principal difficulty, now more readily soluble with the advent of digital computers, has been how to handle it. In Africa the situation is very different, for the major problem is the paucity of statistics of any kind. In many ways the problems of the statistician in tropical Africa resemble those of the agricultural geographer in that he must attempt to describe, with quite inadequate resources, the characteristics of large areas about which comparatively little is known. For some territories there are still no available statistics which can be used to establish the regional character of agriculture, and for others there are only rough estimates collected by administrative officers and other officials. In an increasing number of states, statistics are derived from small samples, designed to produce acceptable national totals rather than the detailed breakdown by areas desired by the geographer.

Where statistical material exists the agricultural geographer can use it to provide, at least in outline, the main spatial characteristics of his problems, but his first task must be to evaluate its accuracy, both by inquiry into the way in which it has been collected and, where possible, by comparing it with other sources, e.g. the records of purchases by marketing boards or by commercial companies. Statistics derived from the informed guesses of district officers or collected as a by-product of taxation returns are not likely to have a high degree of accuracy, but, unless there is reason to suppose that there are marked variations in the scale and direction of error between one administrative area and the next, even these can be used for the production of qualitative maps of distribution and perhaps of relative importance by arraying crops or livestock in rank order (International Bank, 1955; Coppock, 1965). It may also be possible to use indirect measures of the distribution of different crops or livestock by plotting features such as cotton-buying stations, rice mills or veterinary inspection

posts whose location is often accurately known and can be presumed to
bear some relation to the particular agricultural enterprise being investigated (Johnsrud, 1960).

Where statistics are derived from samples they are potentially more useful, provided that the sample has been properly drawn and the sample areas
accurately measured and recorded (Holleman, 1964). While the basis of
sampling is usually made explicit in official reports it must be admitted
that, in view of the scarcity of skilled staff, the absence of large-scale maps,
the small size of most farms and the complexities of inter-cropping and
successional cropping, the latter are big assumptions. Nevertheless, the use
of sampling does conserve scarce resources of skill and makes it possible
for survey teams to spend some considerable time in each sample area and
to measure plots by chain and compass, so that the results are likely to be
more accurate than any other available data (Sierra Leone, 1967). If it is
possible to have access to the sample data for each village or other enumera-
tion area, if these can be correctly located and if the crop or livestock being
considered is of major importance and not too highly localized, it is probably better to treat each as a point sample rather than to use the estimates
which have been computed for the administrative areas in which they lie;
for not only are these often very unsatisfactory areal units, but the factors
used in raising the sample, e.g. tax lists, are likely to be less reliable than
the sample data themselves. An increasing number of statistical procedures
is available for the analysis of sample data of this kind such as trend–surface
analysis (Chorley and Haggett, 1965) and they are also amenable to systems
of computer mapping like Symap, in which isopleths may be interpolated
between sample points. The use of sampling also offers other potential
benefits, for if the records of the sample villages are accessible (and this is
likely to be a less serious problem than in developed countries where rules
on disclosure are very restrictive), they can provide a wealth of detail about
the agricultural structure of individual communities. Such data have
already been used to good effect by B. Floyd (1965) in his study of terrace
agriculture, but their potential has hardly begun to be exploited.

Data on yields are sometimes available and, where based on actual
measurements of sample plots, as in Sierra Leone where produce from
randomly chosen plots 22 ft square is weighed, are likely to be more reliable
than the generality of estimates of yields (Sierra Leone, 1967). Economic
data, on the other hand, are very scarce, although the use of indirect
measures, like standard labour requirements, is probably more justifiable
than in temperate latitudes where livestock play a more important role and
there are wide variations in the use made of machinery. Nor is it generally
possible to obtain satisfactory data about the seasonal aspects of agriculture
which are of such importance throughout tropical Africa or about the

seasonal movements of crops, livestock and people which link the agricultural economies of widely separated areas, although there are sample
ergographs and records of veterinary check posts which have proved valuable in analysing movements of trade cattle (Fricke, 1964; Morgan, 1959a).

Little if any of the statistical information available will permit detailed
comparisons over time, although some indications of major change can be
obtained by subjective visual comparison of maps of different periods
(Mabogunje, 1959). No direct comparisons will be possible because the
bases of collection have generally changed and, whatever the accuracy of
recent statistics, it is probably much higher than that of any earlier estimates. Plotting indirect indices, like the location of buying stations for
different periods can also provide a useful indication of changes in distribution, although the data are essentially qualitative and must be handled with
caution (Coppock, 1965).

Potentially one of the most valuable sources of data are air photographs,
which are increasingly available in many parts of Africa, although inter-
cropping makes crop interpretation difficult and the large-scale photographs necessary would demand resources beyond the capabilities of the
individual research worker if any considerable area was to be covered.
Where photographs are already available their most immediate application
is in the identification and measurement of the major categories of land use,
especially the extent of land actually under cultivation and available for
cultivation, as in the maps of the Gambia prepared by the Directorate of
Overseas Surveys, and in the preparation of accurate base maps for the
study of sample areas (Gambia, 1958–9). Even colour photographs taken
from the air with a hand-held 35-mm camera can throw useful light on
land use.

For the investigation of any individual country or of any large part of it
the agricultural geographer will have to depend on data such as these which
have been collected by others, often for quite different purposes. At best,
he can hope to supplement the information they provide by field inquiries
of his own and, given the resources he can command, this too will generally
involve sampling. The value of the impressionistic reconnaissance survey
must not, of course, be discounted; the experienced geographer with a good
eye for country can glean much valuable information by a carefully planned
reconnaissance survey of an area, recording absence or presence and
relative frequency of different crops and classes of livestock.

The possibility of using the rapidly growing number of secondary school
pupils to undertake qualitative surveys should also be borne in mind; for,
while the nature of land use in Africa and lack of large-scale maps makes
land-use surveys similar to those carried out by schoolchildren in Great
Britain too slow and too difficult, it is quite possible to record the number

and relative abundance of crops, the nature and timing of events in the agricultural calendar and the like.

Nevertheless, these impressions will have to be checked by sample surveys, for which tropical Africa is well suited, with its relatively small variation in size of farm and traditional agricultural practices over large areas. There is much scope for field experimentation in different methods of sampling, but the primary rural community is probably the most useful sampling unit, partly because the universe from which the sample is to be drawn can be readily identified from census records and often from medium-scale maps and partly because the community frequently plays an important part in decision-making and is therefore a meaningful agricultural unit in a way that the parish or village in a developed country is not. Resources of manpower are likely to require some sub-sampling of farmers within the sample areas and some form of areal stratification will also be desirable, although the theoretical advantages of random sampling will thereby be lost. Line samples have been used elsewhere in sample work, as in the Soil Survey of the Nigerian Cocoa Belt, although difficulties of access are likely to be a major obstacle in wooded country (Vine et al., 1954). Traces cut for overhead electricity cables offer one possibility for fairly rapid traverses across wooded country and, while they have the disadvantage of being neither random nor parallel, they are not tied to existing tracks and roads which are themselves likely to modify the pattern of land use. Traverses by air have obvious advantages in this respect and have been successfully used in counts of cattle in Northern Nigeria (Fricke, 1965).

The geographer's contribution to agricultural development can be made in three ways: by the provision of broad studies of the major crops and livestock which will facilitate the evaluation of existing resources; by reconnaissance studies to identify and clarify the nature of particular problems and problem areas; and by service as a member of a team on particular development projects. Contributions of the first type, which serve both practical and academic ends, can be made in advance of any specific need, although most national development plans in fact call for extensions of the acreages of the leading crops, both to provide additional foreign exchange through increased production of export crops and to feed rapidly growing populations. The second type of contribution can also be made either as a by-product of academic inquiry or for specific purposes. The geographer's training well qualifies him for the task of identifying the main features of a problem area; he is accustomed to working with data of doubtful accuracy on problems involving many variables and his tradition of exploration (in the widest sense of the term) helps to avoid some of the pitfalls which arise from excessive specialization.

These same qualities can also be valuable in development projects which can originate either independently or as a result of such reconnaissances; for once the problem has been identified in outline it must be the subject of more detailed surveys, so that solutions can be devised and plans prepared to put them into effect. Resources of time and money will generally be inadequate for prolonged investigations and the geographer's familiarity with both physical and human aspects can make him a valuable member of a small team. As yet, there has been comparatively little opportunity to test this assertion, although geographers like N. Hilton have been employed by FAO on work of this kind, sometimes under designations other than geographer. It is a matter of regret that the valuable experience of such teamwork in Ashanti should not have led to many similar investigations elsewhere (Fortes et al., 1948).

Although such work is important, the attraction of tropical Africa for the agricultural geographer lies also in its possible contribution to the study of agricultural geography as an academic discipline. The spatial arrangement of agriculture around settlements is of particular interest, but it is difficult to evaluate in developed countries where holdings are often so large that individual preferences can greatly modify the zonation of land uses, where both the availability of mechanical transport and pricing policy have made location of declining importance and where the hurdle of disclosure of confidential data has to be surmounted successfully before any satisfactory analysis can be undertaken (Chisholm, 1962; Prothero, 1957). None of these problems is a serious obstacle in tropical Africa. Holdings are small, so that individual decisions matter less, the great bulk of local transport and travel is undertaken on foot, so that effort is directly proportional to distance, and the question of disclosure of confidential information does not generally arise. It should be possible to discover settlements in which the surrounding land was sufficiently homogeneous for land quality to be discounted as a variable so that the effects of distance alone could more easily be measured; and, although it is true that the motivation of African farmers is only to a limited extent economic and that systems of land tenure and traditional ways of working the land are likely to modify the theoretical arrangement of land uses, some of the more serious distortions can be avoided by the careful selection of examples. Indeed, it ought to be possible to find a sequence of settlements which illustrates the effects of a progressive relaxation of the idealized conditions assumed in theories of land-use zonation. Alternatively, the rapidity of change in some areas makes it possible to achieve the same end by studying land uses in selected settlements over a period of years.

There are other situations in which practice and theory might be studied with equal profit, such as the areal diffusion of new crops and new techniques, changes in the values of a single variable and the introduction of some new factor. The tropics have certain advantages for the study of the spread of new agricultural enterprises and techniques, partly because channels of communication are relatively few, partly because innovations are generally so different from existing practices that they can be fairly easily mapped by observation, and partly because it may frequently be possible to determine fairly accurately the date and place of more recent introductions, since these have usually been made through official channels. The effects of price changes are also more easily studied in areas where a single cash crop is produced and marketed through official buying points and where marketing boards and governments can cause prices to vary widely and uniformly: thus a study of the distribution of export crops since the Second World War might be rewarding in this respect, especially where, as in Nigeria, marketing arrangements have changed considerably. The introduction of some quite new factor is also more likely than in developed countries. There are numerous instances of situations which have been radically transformed by the building of new roads and even of new railway lines, or by the erection of new processing plant. Where firm proposals for new communications are known sufficiently far in advance, as with the Bauchi–Maiduguri railway, it should be possible to study selective aspects of the agricultural geography before and after the opening of the new link: the effects of the creation of a new rubber factory or palm-oil mill could be studied in a similar way. Such situations would have the great merit of allowing the construction of hypotheses which could then be tested by the actual course of events (Barbour, 1967).

Other topics within the field of agricultural geography will also repay study. Many of these are primarily of intellectual interest, such as the investigation of a highly localized and distinctive type of agriculture or the study of regional variations in methods of grain storage or in the plants used to surround enclosures, although each of these has practical applications in countries where soil erosion is a hazard, where as much as a quarter of harvested crops may be lost in storage and where the integration of crops and livestock and the individualization of land tenure are likely to require the use of some cheap and effective hedge material, like the euphorbia used by the pagan peoples of the Jos Plateau or the Mauritius thorn used by the Kipsigis (Morgan, 1959b).

In all these investigations the geographer's main contribution must be to identify the spatial characteristics of the topic he is investigating and to place them firmly in their human and physical context. Admittedly this will not be easy in countries where large-scale maps are rare and where so much of the work of environmental stocktaking has yet to be done. It is for this latter reason that much of the work done by geographers in tropical

Africa has been concerned with inventories of physical resources rather than with their agricultural utilization. The geographer's contribution here is likely to be especially valuable in those fields where geography and the systematic sciences overlap, notably in biogeography, climatology, hydrology and geomorphology, and it is no accident that some of the most noteworthy practical contributions have been made in these branches, such as the reconnaissance studies of soil erosion in Nigeria undertaken by A. T. Grove (1951).

The context in which agricultural features are placed must not only be local. Comparative studies have an important place in academic geography and there is much to be gained by comparing homologous situations not only in other countries in tropical Africa but also in Asia and Latin America. Conditions will never be identical, but comparison will focus attention on both similarities and differences and the identification of these is likely to yield fresh insights and to draw attention to features which might otherwise escape notice; the appearance of a crop atlas of West Africa is especially welcome in this connection (Papadakis, 1965). This comparative approach will be particularly valuable in the study of the major cash crops, for example, the cultivation of oil palm and rubber in West Africa and in the East Indies; but it will also be appropriate in studies in land settlement, land consolidation and the like.

In the short run, at least, the most valuable service that the few agricultural geographers working in tropical Africa can perform is essentially educational, the bringing together of a range of material from a wide variety of sources to provide a new perspective and possibly even a new dimension for the technical expert and for the administrator. The volume of relevant material which can be used to construct at least an outline geography of agriculture in an African territory is very considerable and new sources are continually coming to light. There is also increasing awareness of the need for agricultural surveys as a basis for planning research and development and hence an opportunity for the geographer with some training in agriculture both to provide what is currently available in an appropriate form and to employ his skills in the filling of major gaps. The intellectual harvest from work in agricultural geography in tropical Africa also offers exciting prospects but it, too, will require much patient recording and investigation; for, as J. C. Weaver (1958) has noted 'Identification and description may not be our ultimate aim, but they do constitute the first order of geographic business.'

References

BARBOUR, K. M. 1967. A survey of the Bornu railway extension in Nigeria – a geographical audit. *Niger. geogr. J.* **10,** 11–28.

BUCHANAN, K. M. and PUGH, J. C. 1958. *Land and People in Nigeria*, London.

CHISHOLM, M. 1962. *Land Use and Rural Settlement*, London.

CHORLEY, R. J. and HAGGETT, P. 1965. Trend surface mapping in geographical research. *Trans. Inst. Br. Geogr.* **37**, 47–67.

COPPOCK, J. T. 1965. Agricultural geography in Nigeria. *Niger. geogr. J.* **7**, 67–90.

FLOYD, B. 1965. Terrace agriculture in Eastern Nigeria: the case of Maku. *Niger. geogr. J.* **7**, 91–108.

FORTES, M., STEEL, R. W. and ADY, P. 1948. Ashanti Survey 1945–6: an experiment in social research. *Geogr. J.* **110**, 149–79.

FRICKE, W. 1964. *The Cattle and Meat Industry in Northern Nigeria*, Chapter A, unpublished report, Frankfurt-am-Main.

    1965. Herdenzählung mit Hilfe von Luftbildern. *Die Erde* **96**, 206–23.

GAMBIA. 1958–9. *Land Use, 1 : 25,000*, Directorate of Overseas Surveys.

GOULD, P. 1966. *On Mental Maps*, Discussion Paper No. 9, Michigan Inter-University Community of Mathematical Geographers.

GROVE, A. T. 1951. *Land Use and Soil Conservation in Parts of Onitsha and Owerri Provinces*, Bulletin 21, Geological Survey of Nigeria.

HOLLEMAN, J. F. (ed.). 1964. *Experiment in Swaziland*, Cape Town.

HUNTER, J. M. 1961. Akotuakrom: a devastated cocoa village in Ghana. *Trans. Inst. Br. Geogr.* **29**, 161–86.

INTERNATIONAL BANK. 1955. *Report on the Economic Development of Nigeria*, Map 2.

JOHNSRUD, R. O. 1960. A decade of Nigerian cotton. *Niger. geogr. J.* **3**, Fig. 3.

MABOGUNJE, A. L. 1959. Rice cultivation in Southern Nigeria. *Niger. geogr. J.* **2**, 59–69.

MCMASTER, D. N. 1962. *A Subsistence Crop Geography of Uganda*, World Land Use Survey, Occasional Papers, No. 2.

MORGAN, W. B. 1959a. Agriculture in Southern Nigeria (excluding the Cameroons). *Econ. Geogr.* **35**, Fig. 2.

    1959b. The distribution of food crop storage methods in Nigeria. *J. trop. Geogr.* **13**, 58–64.

NIGERIA. 1957 onwards. *Ministry of Agriculture*, Kaduna.

PAPADAKIS, J. 1965. *Crop Ecologic Survey in West Africa*, 2 vols, F.A.O.

PROTHERO, R. M. 1957. Land use at Soba, Zaria Province, Northern Nigeria. *Econ. Geogr.* **33**, 72–86.

SIERRA LEONE. 1967. *Agricultural Statistical Survey of Sierra Leone 1965–6*, Appendix, Form AS9, Freetown.

UGANDA. 1960. *The Systems of Agriculture Practised in Uganda*, Memoirs of the Research Division, 5 vols, Kampala.

VINE, H., WESTON, V. J., MONTGOMERY, R. F., SMYTH, A. J. and MOSS, R. P. 1954. Progress of soil surveys in South Western Nigeria. *Proc. 2nd Inter. Afr. Soils Conf.* **1**, 211–36.

WEAVER, J. C. 1958. A design for research in the geography of agriculture. *Prof. Geogr.* **10**, 5.

PART I

Studies of the natural environment

3 A simple energy balance approach to the moisture balance climatology of Africa

J. A. DAVIES *and* P. J. ROBINSON

Two of the essential needs for the survival and growth of plants are energy and moisture and these must be satisfied by climate. It is one of the tasks of the climatologist to define these parameters for vegetation communities and crops. In the context of this book it must be emphasized that the vegetation or land-use patterns of Africa cannot be considered as simple functions of climate. Both vegetation and land use are functions of physical and human environments which are essentially multivariate. Each of the large number of plant species involved in Africa may be a law unto itself with respect to response to environmental conditions, and this will increase the complexity (Fogg, 1966). If any climatic parameters are ecologically significant they must be the energy and moisture balances (Gates, 1962; Monteith, 1965). This chapter will attempt to approximate the important terms of these balances and demonstrate the significance of the findings to the vegetation pattern.

Radiation and evaporation

Plants obtain their energy from the sun through the photochemical process of photosynthesis. In the tropical world there is no period during the year when solar energy is limiting to the process: the main causes of below 'potential' rates of photosynthesis will be the rate of carbon dioxide diffusion into chloroplasts and plant crowding, which limits maximum leaf exposure. Radiant energy, however, does have an indirect limiting effect upon plant growth. Because leaf stomata are open for photosynthesis, plants transpire at a rate which is governed mainly by the available radiant energy. Through its control on evapotranspiration, radiant energy also controls the water balance of a plant community. Measurements of soil moisture changes, weight changes of a vegetated lysimeter[1] or input-output differences of precipitation and evapotranspiration could define the balance, but, at present, these are largely inaccurate and available at only

---

[1] A column of soil isolated in a container but still mounted within the main soil body.

a few stations in Africa. Evapotranspiration has therefore to be calculated, and it is through the available radiant energy that this is best done.

At any surface the available radiant energy is the balance between the incoming and outgoing fluxes[1] of both solar (short-wave: 0·3–3·0 microns) and terrestrial (long-wave: > 3·0 microns) radiation. Solar radiation consists of a direct component (Q) and a diffuse component (q). A fraction of the total amount is reflected (α). Long-wave radiation is emitted from the surface (L_O) as a function of surface temperature, and a smaller quantity is usually radiated to the earth mainly from clouds and atmospheric moisture (L_I). The available radiant energy at a surface is known as the net radiation (R_N), which, in terms of the defined energy fluxes, is written:

$$R_N = (1 - \alpha)\,(Q + q) + (L_I - L_O). \tag{1}$$

Another source of energy, advection (thermal energy that is transported over the surface by winds), is present if the vertical temperature gradient is directed towards the surface (i.e. overlying warmer air). This will occur when the surface is downwind from a source of warm air such as a desert or warm ocean current. If it is assumed that the surface lies within an infinite field of the same thermal character there will be no energy advected. Then the net radiation will define the available energy at the surface.

The net radiation is expended in evaporation (E), which includes evapotranspiration, sensible heat flux (H) or convection, soil heat flux (G) and photosynthesis (Ph):

$$R_N = E + H + G + Ph. \tag{2}$$

Photosynthesis, important as it is to the plant, utilizes only a small percentage of R_N (Monteith, 1965) and can be ignored in our task to calculate E. The soil heat flux is usually ignored for the same reason unless there are very marked seasonal changes in soil temperature. Work at Dakar (Salvador, 1964) has shown that the heat flux for a sand surface at maximum solar irradiation was always < 5 per cent of R_N. Equation (2) reduces to:

$$R_N = E + H, \tag{3}$$

and the problem becomes that of defining the ratio $H/E = \beta$ (Bowen, 1926). If β (the Bowen ratio) is known, evaporation can be obtained from:

$$E = R_N/(1 + \beta.) \tag{4}$$

The evaluation of β requires gradient measurement of temperature and humidity which are difficult to obtain. Penman (1948) eliminated the gradient parameters algebraically by introducing, and combining with equation (3), equations for the fluxes of heat and water vapour. A simple

---

[1] Energy per unit area per unit time.

solution is preferred here. Measurements of evapotranspiration, at stations where a high level of accuracy can be expected, have shown that for freely transpiring, well-covered surfaces with no shortage of water in the root zone (field capacity conditions), evapotranspiration utilizes most of the net radiation and $\beta \rightarrow 0$. This evapotranspiration rate can be equated with Thornthwaite's (1948) potential evapotranspiration (PE) or Penman's potential transpiration. Penman (1956) defines it as 'The amount of water transpired in unit time by a short green crop, completely shading the ground, of uniform height and never short of water.'

Under these conditions:

$$PE \simeq R_N/L, \tag{5}$$

where L is the latent heat of vaporization (59 gcal of radiation=1 mm of water). The use of equation (5) has been discussed by House, Rider and Tugwell (1960), Thornthwaite and Hare (1965), Monteith (1965), Blackwell and Tyldesley (1965) and Davies (1965a).

If advection exists, the conditions for the potential evapotranspiration model are not fulfilled, and evapotranspiration may well exceed the limit set by the net radiation. This is known as the 'oasis' effect, whereby a moist area is surrounded by or adjacent to drier land, and is probably very strong in Africa where a severe dry season is experienced. However, the concept of potential evapotranspiration is significant even in the dry season because it is the hypothetical water need for vegetation and crops where water supply is not a limiting factor for growth. It is therefore a measure of irrigation need when compared with rainfall. Where rainfall exceeds potential evapotranspiration the moisture régime is more than adequate for growth, but where rainfall falls short of potential evapotranspiration irrigation is required. Maps of the difference between these two values are important in defining areas of water need and may serve as first estimates for irrigation. In the future it may also be possible to predict yields for different crops from maps of potential evapotranspiration alone. Penman (1962) and Monteith (1965) have shown for parts of the temperate world that cumulative yield is linearly related to cumulative potential evapotranspiration. Before this advance can be made for Africa there is need for experimental work to determine yield-potential evapotranspiration relationships for individual crops.

To distinguish the water equivalent of the net radiation from potential evapotranspiration obtained by empirical formulae, such as those of Penman and Thornthwaite, the term 'potential water loss' will be used and designated E_R. It has the important advantage over the other formulae that it does not depend upon measurements that are themselves influenced by the nature of the surface. For example, temperature, which is the sole

climatic parameter in Thornthwaite's formula, will be affected by the
β value at the point of measurement which may be very different from
β → o that is essential to the potential evapotranspiration model. The
Penman formula requires sunshine, temperature, wind run and humidity
measurements which makes it unsuitable for large areas of Africa where
only temperature is recorded. In addition, the use of humidity in an
evaporation formula can be queried since it is the result of evaporation
rather than a contributory factor (Deacon, Priestley and Swinbank, 1958).
Nevertheless, the Thornthwaite and Penman formulae have been widely
used in Africa. Thornthwaite's method has proved to be quite inaccurate
in the high areas of East Africa where lower temperatures belie the high
solar radiation input and high rates of evapotranspiration (Blackie 1965).
In West Africa, on the other hand, the formula over-estimates considerably
(Davies, 1965b), although Garnier (1956) thought it underestimated in dry
areas, and produced a variant by introducing the saturation deficit of the
air (i.e. the difference between the actual and saturation vapour pressure
at air temperature). More recent work suggests that this modification is
insecurely based (Davies, 1965b). The Penman formula has been more
successful and in East Africa there have been interesting attempts to provide more suitable terms (McCulloch, 1963). However, it tends to give a
value of evaporation too close to evaporation from pans, and these are noted
for their exaggerated rates due to absorption of radiation through their
sides, shallow depth, rim effects upon turbulence and, above all, 'oasis'
effects.

   In general, the literature suggests that the radiation and moisture balances have not received very much attention. Radiation in particular, seems
to have been neglected. Solar radiation studies, have, in the main, concentrated upon the empirical estimation of radiation from sunshine
(Masson, 1954; Glover and McCulloch, 1958a, b; Page, 1961; Davies,
1965a). However, solar radiation maps, based upon available measurements,
were published for the whole of Africa in 1965 in *The Climate of Africa*
(Thompson, 1965). Net radiation data studies are rare, although several
research institutes are known to use net radiometers. Apart from data at
Dakar (Salvador, 1964) the only data that were available for this study were
collected from Benin and Ibadan in Nigeria (Davies, 1965a, c).

   Maps of the water balance of Africa have been published by the Thornthwaite Laboratory of Climatology (Carter, 1954), and Garnier (1956)
produced maps for West Africa based upon his particular variant of the
Thornthwaite formula. Maps are at present being prepared for East Africa
by the East Africa Agriculture and Forestry Research Organization. This
organization has also been prominent in conducting catchment experiments
(Pereira *et al.*, 1962), to study changes in the water balance as forest is

cleared and the effect of cultivation on the hydrology of steep valleys. A review of this work has been given by Penman (1963).

The Nigeria model

The initial work which produced a formula for potential water loss to be used in this study was carried out in Nigeria (Davies, 1965b, 1967). Since this method will be applied to the whole of tropical Africa it is relevant at this point to give an account of the Nigerian study.

At the time the study was made only three stations in Nigeria had a usable solar radiation record and only two had a record of net radiation. Therefore, it was necessary to approximate these parameters from more readily available climatic data. The derivation of the formula had three stages.

Firstly, to approximate solar radiation, sunshine was selected as a predictor. Cloudiness and sunshine have been widely used for this purpose (e.g. Budyko, 1956; Black, 1956; Ångström, 1956; Glover and McCulloch, 1958a, b; Page, 1961). However, cloudiness was eliminated because there were fewer recording stations and because the method of cloud-cover assessment is visual and the assessments are carried out, except at airports, at intervals as great as or even greater than three hours. At climatic stations and other non-synoptic stations, only one estimate a day is made. Objective sunshine measurements by card-burning recorders (Campbell-Stokes) were preferred. Linear correlations between radiation and sunshine were obtained using the equation:

$$Q + q = Q_A (a + b \, n/N) \tag{6}$$

where Q_A is the radiation at the outside of the atmosphere, n is measured sunshine (hours), N is the possible number of sunshine hours and a and b are respectively the regression constant and coefficient. Only three stations (Benin, Ibadan and Kano) with concurrent records of solar radiation and sunshine could be mustered from Nigeria but the number was increased to six by including Accra, Fort Lamy and Niamey from neighbouring countries. These six stations gave quite a good representation of the climatic range in West Africa. Satisfactory correlations were obtained using all six sets of data in monthly groupings. Grouping by month was necessary to accommodate the effect of the latitudinal movement of the Inter-Tropical Convergence (I.T.C.). Marked variations in the a and b values in equation (6) were attributed to cloudiness and turbidity patterns each side of the I.T.C. (Davies, 1965a). Correlation coefficients, in the main, were higher than 0·90. Monthly solar radiation maps were then constructed using data for all stations (39) in Nigeria which had records of sunshine.

Secondly, the lack of information concerning reflection coefficient (α)

and the net long-wave radiation $(L_I - L_O)$ had to be overcome. This was achieved by relating daytime positive net radiation (the portion of the 24-hour value that is active in evaporation and evapotranspiration) directly to solar radiation using the linear formula:

$$R_N = b(Q + q) + a. \tag{7}$$

The constant, a, represents the mean net radiation (1) between sunrise and the onset of positive net radiation, (2) between the ending of positive net radiation and sunset. The coefficient, b, must then account for the absorbed solar radiation $(1 - \alpha)(Q + q)$ and the heating or cooling of the surface which determines in the main the net long-wave exchange. Following Monteith and Szeicz (1961, 1962), a 'heating coefficient' (ε) can be introduced where:

$$\varepsilon = -dL_N/dR_N. \tag{8}$$

Here $$L_N = (L_I - L_O).$$

For b we write: $$b = \frac{1 - \alpha}{1 + \varepsilon}.$$

Separate analyses of data for Benin, Ibadan and Dakar and of data from various parts of the world representing tropical, temperate, sub-Arctic and Arctic climates showed that one regression line fitted all the data satisfactorily (Davies, 1967). The line is:

$$R_N = 0.617\,(Q + q) - 24, \quad r = 0.98. \tag{9}$$

Since most of the reflection coefficients for these data ranged between 0.20 and 0.30, the mean of 0.25 gave $\varepsilon = 0.22$, which is within the range found by Monteith and Szeicz in temperate latitudes. From this relationship monthly maps of net radiation were drawn.

Thirdly, potential water loss calculated as the water equivalent of the net radiation showed a very good relationship with measured potential evapotranspiration in Nigeria at Benin, Ibadan and Samaru during the wet season when 'oasis' effects are minimal (Davies, 1966; see Fig. 3.1).

Although it was not possible to check the validity of the net solar radiation relationship elsewhere in Africa because of the lack of data, potential water-loss calculations at Kimakia and Kericho in Kenya showed excellent agreement with Penman calculations to indicate that even in upland areas of tropical Africa the regression relationship to obtain net radiation is valid. A comparison with Penman values for the coastal station, Mombasa, was not so good due to the important contribution of aero-dynamic factors to the Penman evaluation (Fig. 3.2). For these three stations potential water loss was calculated from solar radiation data alone. This suggests that the wind and humidity terms in Penman's equation can be dispensed with so that the problem of calculating evaporation becomes that of the calculation of solar radiation.

Evaporation pan data in Nigeria showed that the potential water-loss calculations were in good agreement during the wet season when 'oasis' effects were reduced (Fig. 3.3) and suitably lower during the dry season when 'oasis' effects strongly influence pan values.

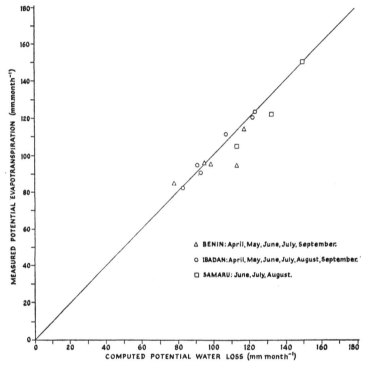

Figure 3.1 *Comparison of measured potential evapotranspiration and computed potential water loss for Benin, Ibadan and Samaru. The solid line on the graph represents the 1:1 line.*

The complete equation for potential water loss (mm per day) in Nigeria, using equations (5), (6) and (9) is:

$$E_R = \frac{R_N}{L} = 0.017 \left\{ 0.617 Q_A \left(a + b\, n/N \right) - 24 \right\} \qquad (10)$$

(0.017 is the reciprocal of 59, the latent heat of vaporization).

It was found that potential water loss also had some considerable ecological significance. By comparing E_R with rainfall P (by simple difference) wet season surpluses ($P - E_R > 0$) were obtained. Plotting these two sets of values against latitude for all stations and distinguishing savanna stations

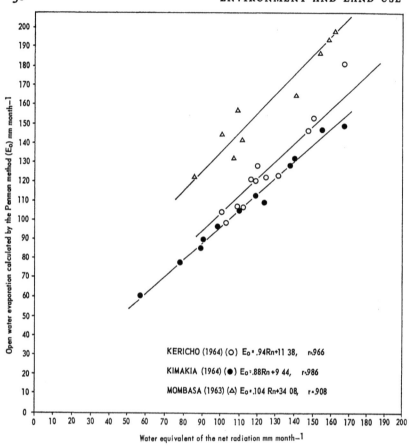

Fig. 3.2 *Comparison of open water evaporation, computed by the Penman method and computed potential water loss for Kimakia, Kericho and Mombasa.*

from forest stations, it was clearly shown that the forest–savanna boundary could be set where wet season surplus = dry season deficit or where, on an annual basis, $P - E_R = 0$ (Fig. 3.4). A plot of these data on to a vegetation map of Nigeria (Keay, 1959a) showed that the forest–savanna boundary and the $P - E_R = 0$ contour were not as clearly associated as Figure 3.4 suggested (Fig. 3.5). If one assumed that forest would grow in areas of negative $P - E_R$ values, since wet season storage in the soil could be utilized in the dry season, a better spatial fit is given by the −200 mm per year contour. Other vegetation boundaries could also be fitted. This finding is significant for it suggests a rational climatic explanation for the forest–savanna boundary, although the consensus of ecological opinion would suggest that the activities of man have been more important.

The assumption is now made that since the climatic variation encountered between north and south in Nigeria covers much of the variation within tropical Africa, this climatic index can be expected to apply to the whole

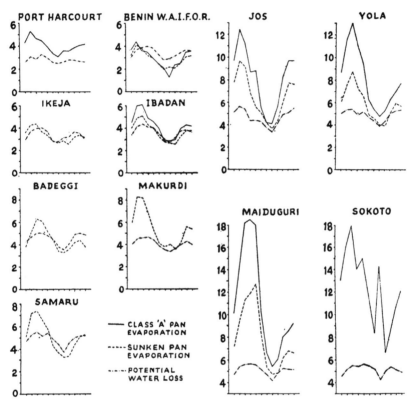

Figure 3.3 *Comparison of measured evaporation from pans and computed water loss in Nigeria. Divisions of the x-axes are months of the year (January to December). Values on the y-axes represent the evaporation rate (mm day⁻¹).*

of the area under study. There is already some confirmation of this from East Africa, which falls outside the range of conditions encountered in Nigeria.

The data

Since the scope of this study is so large, it was found that the solar radiation data published in the Quarterly Radiation Bulletins of the Union of South Africa Weather Bureau, supplemented by additional data collected in West Africa, were sufficient. The distribution of the stations used in the study is shown in Figure 3.6. Although the coverage is poor in some areas (e.g. Tanzania and Ethiopia), no attempt has been made at empirical calculations since Page (1961) has shown a great variation from station to station in the regression relationship between radiation and sunshine.

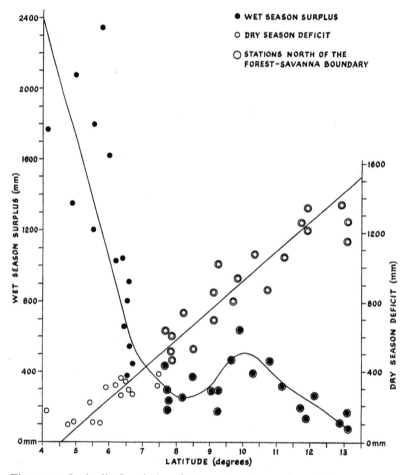

Figure 3.4 *Latitudinal variation of wet season water surplus and dry season water deficit in Nigeria.*

Rainfall values were taken from *The Average Climatic Water Balance of the Continents, Part I: Africa* (Laboratory of Climatology, 1962), and from the most recent summary of climatic data for Africa published by the British Meteorological Office (1958). It was not possible to use a standard period for either radiation or rainfall since a great variety of periods are represented (see Appendix 2). Some of the radiation data apply to only three years or less. Fortunately, radiation is a more conservative parameter than rainfall and short-period data are therefore more representative.

Maps of solar and net radiation, potential water loss and rainfall minus potential water loss have been prepared. Only the last two sets are relevant here and they are shown in Figures 3.7 and 3.8.

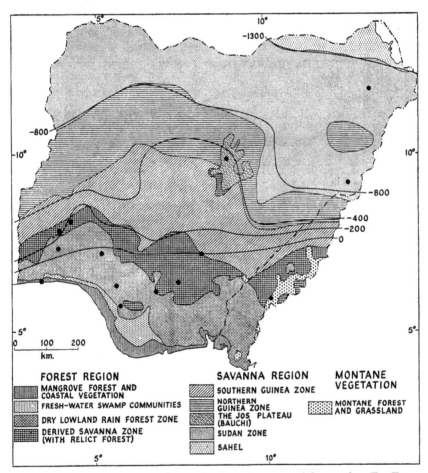

Figure 3.5 *Spatial pattern of precipitation minus potential water loss* $(P - E_R)$ *contours in relation to the vegetation distribution in Nigeria.* (Vegetation Zones after Keay, 1959).

Maps of potential water loss (E_R)

The annual map shows the main features of the potential water-loss distribution which are also found on the monthly maps. These are:

1. Lowest values in the cloudy equatorial zone.

2. Highest values (within tropical Africa) in the uplands of north central Kenya.

3. A marked N–S gradient over West Africa associated with a strong cloudiness gradient across the I.T.C. where dry northerly flow and moist southerly flow converge.

The annual map on p. 41 shows considerable uniformity over the Saharan region and the portion of Africa south of the Congo and Tanzania, except

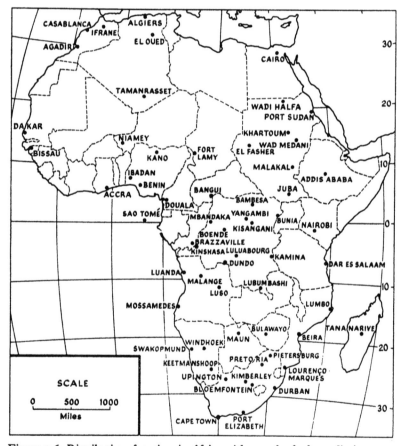

Figure 3.6 *Distribution of stations in Africa with records of solar radiation.*

for an intense cell over the Kalahari. This can be rationally attributed to the great uniformity of cloudiness conditions within the sub-tropical high-pressure cells but the sparse network prevents any valid generalizations.

Figure 3.7 *Maps of potential water loss* (E_R).

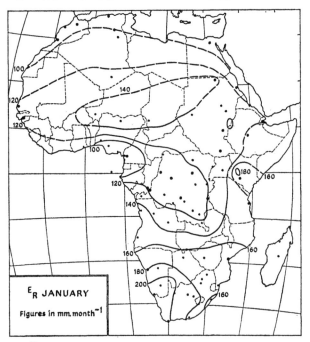

3.7 (i)

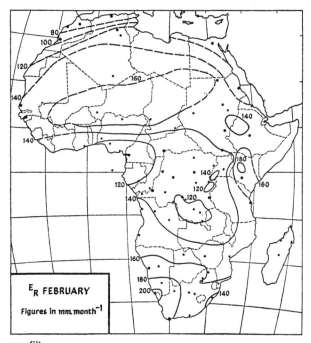

3.7 (ii)

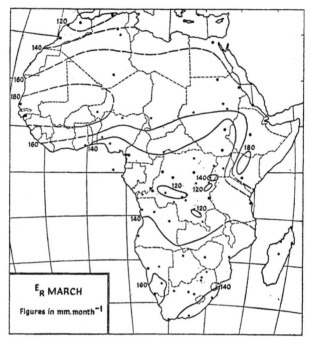

E_R MARCH

Figures in mm.month^{-1}

3·7 (iii)

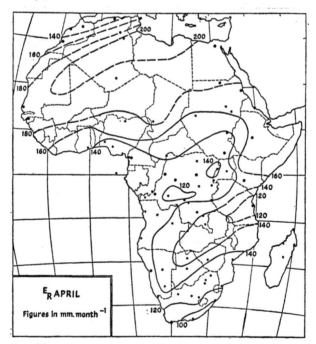

E_R APRIL

Figures in mm.month^{-1}

3·7 (iv)

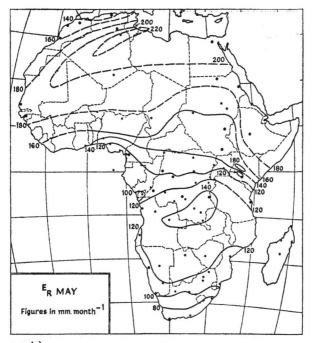

E_R MAY

Figures in mm. month⁻¹

3.7 (v)

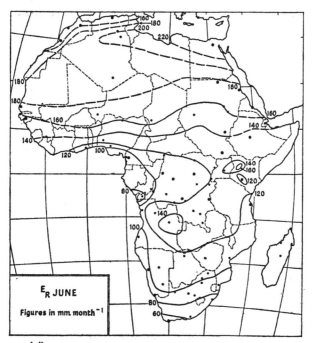

E_R JUNE

Figures in mm. month⁻¹

3.7 (vi)

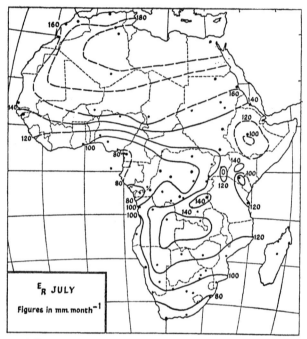

E_R JULY

Figures in mm. month⁻¹

3·7 (vii)

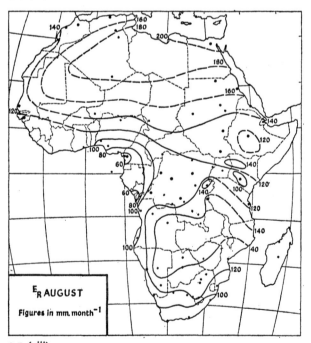

E_R AUGUST

Figures in mm. month⁻¹

3·7 (viii)

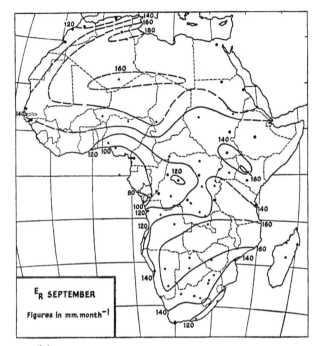

E_R SEPTEMBER

Figures in mm. month^{-1}

3·7 (ix)

E_R OCTOBER

Figures in mm. month^{-1}

3·7 (x)

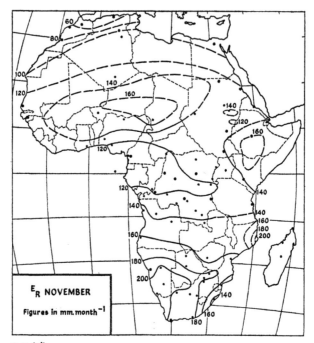

E$_R$ NOVEMBER

Figures in mm.month^{-1}

3·7 (xi)

E$_R$ DECEMBER

Figures in mm.month^{-1}

3·7 (xii)

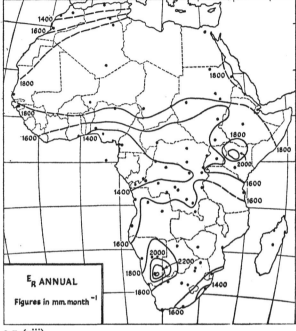

3.7 (xiii)

The monthly charts show:

1. That values in the equatorial belt remain consistently below 140 mm per month.

2. That areas of lowest potential are found on the west side of the continent, extending northwards and westwards into Cameroon and southern Nigeria in January and southwards around the Congo (Kinshasha)–Angola border in May to October.

3. That considerable complexity in upland areas of East Africa changes to more open patterns during April–May and October–December – the periods of the rains. On the May chart the marked latitudinal gradient may reflect the position of the I.T.C.

Maps of the difference between precipitation and potential evapotranspiration ($P - E_R$)

Only a small percentage of the continent in any one month has positive values. The area enclosed by the zero value moves northwards from its southernmost position in January to reach northern areas of West Africa in July and August. After June it tends to concentrate towards the west, attaining the highest values of +600 mm in Cameroon. Most of this positive area has values throughout the year which are less than +500 mm. The strongest gradients appear over Cameroon, southern Nigeria, and farther west in July, August and September. Therefore, in all months, only a small fraction of the continent has adequate rainfall for optimum growth (i.e. to satisfy E_R levels). These fractions for each month, as obtained by planimetry, are listed in Table 3.1.

TABLE 3.1 *Percentage of Africa, excluding Malagasy (Madagascar) with precipitation exceeding potential water loss*

Month	%
January	13·6
February	13·8
March	17·6
April	16·0
May	12·8
June	11·9
July	21·9
August	23·5
September	16·2
October	13·9
November	14·9
December	17·9
Year	10·3

The annual map shows only 10 per cent of Africa with positive values. The significance of the delimited area is emphasized by comparison with the forest-savanna boundary as taken from the best available vegetation map (Keay, 1959b). The two delimited areas are in good agreement. A perfect agreement could not be expected because, firstly, the vegetation boundary is imprecise, and secondly, there is insufficient station control to allow the zero contour to be accurately drawn. It would seem that the West African agreement between this vegetation boundary and climate is confirmed. Although climate can be only one of many variables determining the boundaries of vegetation zones, the coincidence of the forest-savanna boundary and a rational climatic index cannot be ignored. The question must now be asked: is perhaps the influence of man's cultivation and burning not overstated by ecologists? (Richards, 1952).

It is hoped that these maps will be of use to ecologists and agriculturists. Unfortunately, the climatologist is restricted to this scale at present because of the sparse radiation network. The spatial and temporal variability of the $P - E_R$ parameter is another problem which is of considerable importance

Figure 3.8 *Maps of difference between precipitation and potential evapotranspiration* $(P - E_R)$.

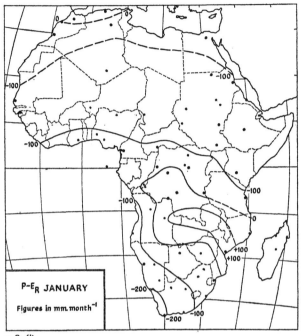

3.8 (i)

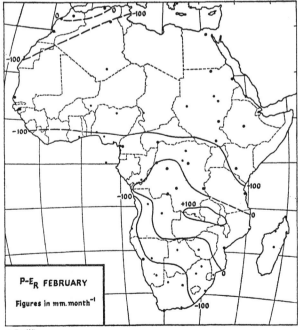

3.8 (ii)

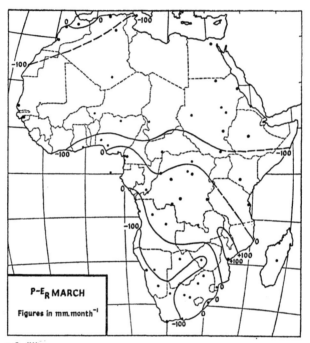

P-E_R MARCH

Figures in mm.month^{-1}

3.8 (iii)

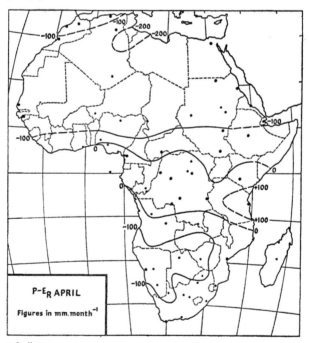

P-E_R APRIL

Figures in mm.month^{-1}

3.8 (iv)

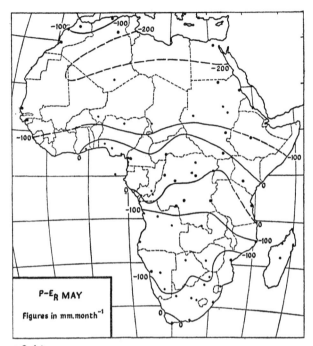

P-E$_R$ MAY

Figures in mm. month^{-1}

3.8 (v)

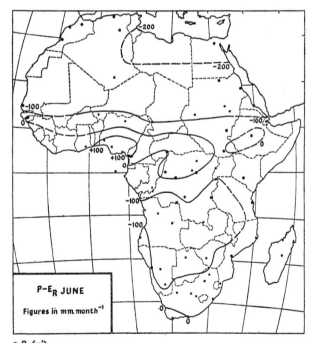

P-E$_R$ JUNE

Figures in mm. month^{-1}

3.8 (vi)

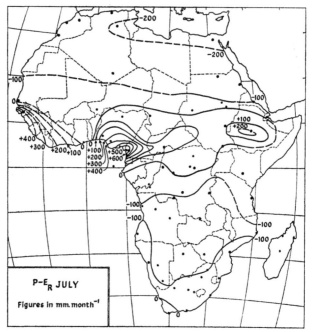

3.8 (vii)

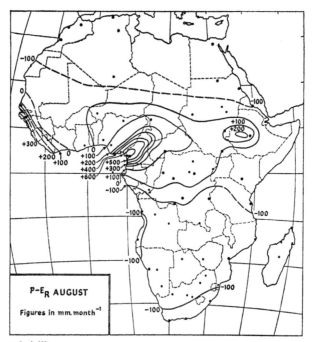

3.8 (viii)

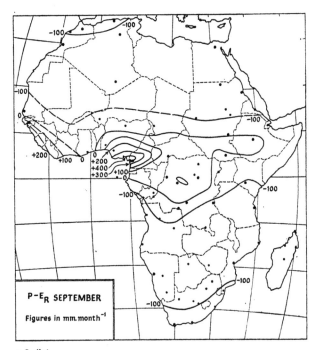

P-E$_R$ SEPTEMBER

Figures in mm.month^{-1}

3.8 (ix)

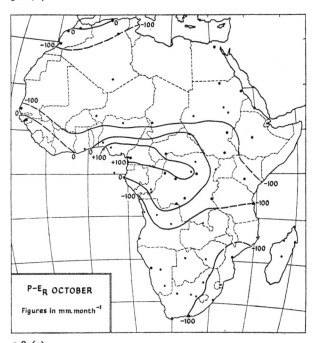

P-E$_R$ OCTOBER

Figures in mm.month^{-1}

3.8 (x)

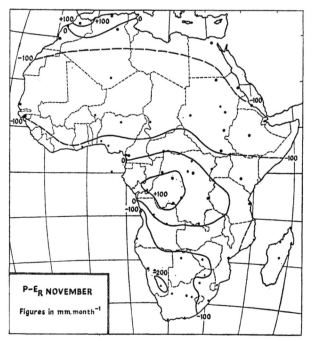

3.8 (xi)

3.8 (xii)

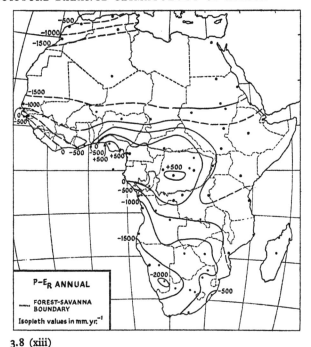

3.8 (xiii)

to ecology and agriculture, but with such a short series of data on hand it cannot yet be tackled.

Comparison with estimates from general circulation studies

Although it is encouraging to find that a climatic index has a significant spatial expression on the ground, further work is needed to arrive at actual water loss (actual evapotranspiration) by evaluating departures of soil moisture from field capacity and advected energy. An aid to this end is provided by the approach and results of a dynamic climatological study by Peixoto and Obasi (1965). Using data accumulated during the International Geophysical Year the water-vapour flux fields over Africa were computed. From these and a study of the wind field they determined the net flux divergence.[1] The divergence is equated with the difference between precipitation and evaporation at the ground surface. Where divergence is positive $P < E$. Where divergence is negative (i.e. convergence) $P > E$. The computed values of the divergence are spatial means for each area defined by a $5°$ longitude $\times$ $5°$ latitude grid. Peixoto and Obasi claim that the calculations are representative of the gross features of the general

---

[1] The difference between the outflow and inflow of moisture into a column of the atmosphere above a station.

circulation, and that the results should be helpful for the evaluation of water resources over parts of Africa as large as say West Africa or East Africa, but the size of the grid limits the size of the area that can be represented accurately. They maintain that: 'The excess precipitation in regions of strong convergence could be tapped for irrigational purposes in adjacent

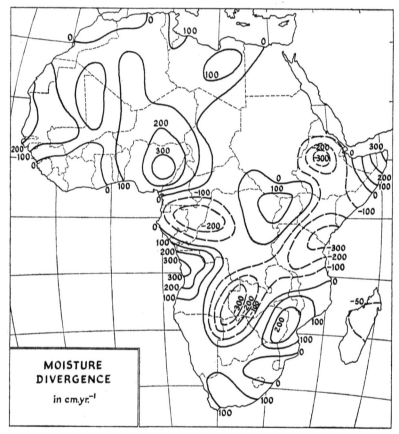

Figure 3.9 *Piexoto and Obasi's map of the moisture divergence field over Africa.*

regions where there is a water deficit.' A simplified version of their annual map is shown in Figure 3.9. Divergence is indicated by positive values and convergence by negative values. The grid and average values given on their map have been removed for clarity.

Strongest divergence is found over the southern half of Nigeria, the west coast of Saharan Africa, the headwaters of the White Nile, Moçambique, and the Angolan Coast. Somewhat surprisingly the southern parts of

Nigeria, including the Niger delta area, show an unfavourable water balance, and the water balance becomes more favourable north from this area – a reversal of the gradient in Figure 3.8.

Convergence areas include the rain-forest belt although it is shown to be more restricted than in Figure 3.8. Peixoto and Obasi note convergence in the vicinity of headwaters and drainage basins of the large rivers, Niger, Zambezi, Orange, Limpopo and Blue Nile, but the White Nile headwaters lie in the centre of an area of divergence.

Their map also shows convergence over a large tract of the eastern Sahara, the whole of Egypt and a section of the western Sahara. Starr and Peixoto (1958) and Peixoto (1960) consider the western Sahara as a major source of atmospheric vapour on this evidence and discuss the possibility of subterranean supply of water under the desert. At present this is not substantiated.

In general terms the main difference between the water-balance maps obtained through simplified energy-balance theory and vapour-transport theory is the difference in size of the areas showing a positive water balance. The small positive area arising from the energy balance method has already been noted but Figure 3.8 shows a much larger area which includes most of central, eastern and north-eastern Africa including the eastern Sahara desert. Only the south-west sector of West Africa has a positive régime.

Differences between the two maps can be expected. Firstly, Peixoto and Obasi obtained divergence values using actual evapotranspiration whereas potential evapotranspiration has been used in this study. $E \simeq PE$ in equatorial areas and some agreement between the two maps could be expected there but this is not the case. This difference may be due to a comparison of a single year with the mean of several. However, the major differences can probably be attributed to soil moisture deficits and advection effects. It is therefore feasible that a combination of the two methods using data for a comparable year or period of time could be used to evaluate these parameters, their sum effect at least, without resort to ground-based measurements of soil moisture and temperature profiles.

Further advances by the simplified energy-balance approach, along the lines of this study, may be of benefit to the ecologists and to agriculturalists. At the moment, however, the sparsity of stations and the fragmentary nature of their records allow but tentative maps to be drawn. With an increase in the number of observing stations, the accuracy of such maps will improve, and the possibility of aiding the agriculturalist with accurate estimates of irrigation need and, from evapotranspiration data, likely yields, will be enhanced. Concurrent with this there must be continuing research into the individual components of the energy balance, notably evaporation, in the tropical lands of Africa.

APPENDIX 1

List of symbols

E = evaporation
E_R = potential water loss
G = soil heat flux
H = sensible heat flux
L = latent heat of vaporization
L_I = incoming long-wave radiation
L_N = net long-wave radiation
L_O = outgoing long-wave radiation
N = number of possible sunshine hours
P = rainfall
PE = potential evapotranspiration
Ph = photosynthesis
Q = direct component of the solar radiation
Q_A = extra terrestrial radiation
R_N = net radiation
a = linear regression constant
b = linear regression coefficient
n = measured number of sunshine hours
q = diffuse component of solar radiation
α = short-wave reflection coefficient
β = Bowen ratio
ε = 'heating' coefficient

APPENDIX 2

Length of station record

Station	Radiation			Rainfall	
	No. of monthly values		Period	No. of years	Period
	Min.	Max.			
Accra	5	7	1951–7	65	1888–1955
Addis Ababa	1	2	1957–8	37	—
Agadir	0	1	1954	29	1934–55
Kamina (Albertville)	2	4	1954–7	20	1930–53
Algiers	1	1	1961	25	1885–1937
Bambesa	2	2	1959–60	32	1922–53
Bangui	3	5	1957–61	5	1950–5
Beira	3	5	1957–61	39	1913–53
Benin	5	6	1958–63	8	1943–54
Bissau	2	4	1958–61	37	1918–54
Bloemfontein	6	8	1954–61	14	1937–50
Boende	4	6	1956–61		
Brazzaville	4	7	1955–61	18	1935–54
Bulawayo	5	8	1954–61	50	1896–1951
Bunia	2	4	1958–61		
Cape Town	6	8	1954–61	18	1932–50
Casablanca	1	2	1957–8	40	1908–55
Mbandaka (Coquilhatville)	5	7	1955–61		
Dakar	1	2	1953–4	26	1899–1940
Dar-es-Salaam	1	1	1960	49	1893–1954
Douala	0	1	1961	28	1885–1913
Dundo	2	5	1957–61	4	1951–4
Durban	6	8	1954–61	78	1873–1950
El Fasher	3	4	1954–7	17	1918–34
El Giza (Cairo)	2	4	1956–9	42	1904–45
El Oued	2	4	1954–7	26	1913–50
Lubumbashi (Elizabethville)	3	6	1955–60	17	1919–49
Fort Lamy	2	4	1958–61	5	1950–55
Ibadan	0	1	1963–4	14	1940–54
Ifrane	0	1	1958	12	1936–47

C

| Station | Radiation | | | Rainfall | |
| | No. of monthly values | | Period | No. of years | Period |
	Min.	Max.			
Juba	3	5	1957–61	26	1915–47
Kano	2	5	1953–7	39	1921–54
Keetmanshoop	4	6	1956–61	45	1899–1950
Khartoum	3	4	1958–61	46	1900–45
Kimberley	6	8	1954–61	57	1894–1950
Kinshasha	6	8	1954–61	12	1940–53
Lourenço Marques	5	7	1955–61	42	1910–51
Luanda	4	5	1957–61	59	1914–40
Luluabourg	3	7	1955–61	14	1940–53
Lumbo	3	5	1957–61		
Luso	3	4	1957–61	13	1940–53
Malakal	0	2	1960–1	19	—
Malange	3	5	1957–61		
Maun	6	8	1954–61	20	1921–50
Mossamedes	3	8	1954–61	21	1930–53
Nairobi	5	5	1958–62	17	—
Niamey	2	3	1958–60	10	1931–40, 1949–55
Pietersburg	5	8	1954–61	47	1904–50
Port Elizabeth	4	5	1957–61	84	1867–1960
Port Sudan	3	4	1958–61	40	1905–45
Pretoria	5	8	1954–61	12	1938–50
São Tomé	3	5	1958–61	10	1945–54
Kisangani (Stanleyville)	5	8	1954–61	14	1927–41, 1951–53
Swakopmund	1	2	1957–8		
Tamanrasset	14	16	1939–51, 1959–61	15	1925–50
Tananarive	5	8	1954–61	62	1882–1953
Upington	5	6	1956–61	56	1884–1950
Wadi Halfa	0	1	1960–1	39	1902–47
Wad Medani	3	5	1957–61		
Windhoek	6	8	1954–61	60	1891–50
Yangambi	5	6	1956–61		

References

ANGSTRÖM, A. 1956. On the computation of global radiation from records of sunshine. *Ark. Geophys.* **2**, 471–9.

BLACK, J. N. 1956. The distribution of solar radiation over the earth's surface. *Archiv Met. Geophys. Bioklim. B.* **7**, 2, 165–89.

BLACKIE, J. R. 1965. *A Comparison of Methods of Estimating Evaporation in East Africa.* Paper presented at the Third Specialist Meeting on Applied Meteorology, Muguga, Kenya, November 1965.

BLACKWELL, M. J. and TYLDESLEY, J. B. 1965. Measurements of natural evaporation: comparison of gravimetric and aerodynamic methods. *Methodology of Plant Eco-physiology: Proc. of the Montpellier Symposium, UNESCO,* 141–8.

BOWEN, I. S. 1926. The ratio of heat losses by conduction and by evaporation from any water surface. *Phys. Rev.* **27**, 779–87.

BRITISH METEOROLOGICAL OFFICE. 1958. *Tables of Temperature, Relative Humidity and Precipitation for the World, Part IV: Africa, the Atlantic Ocean south of 35°N and the Indian Ocean,* H.M.S.O., London.

BUDYKO, M. I. 1956. *Teplovoi balans zemnoi poverkhnosti,* Leningrad Gidrometeorologicheskoe Izdatelstvo. Translated as *The Heat Balance of the Earth's Surface* by Nina Stepanova, U.S. Weather Bureau, Department of Commerce, 1958.

CARTER, D. B. 1954. Moisture regions of Africa. The Johns Hopkins University Laboratory of Climatology, *Publications in Climatology,* **7**, 4.

DAVIES, J. A. 1965a. Estimation of insolation for West Africa. *Q. Jl R. met. Soc.* **91**, 359–63.

1965b. Evaporation and potential evapotranspiration at Ibadan. *Niger. geogr. J.* **8**, 1, 17–31.

1965c. The use of a Gunn–Bellani distillator to determine net radiative flux in West Africa. *J. appl. Met.* **4**, 4, 547–9.

1966. The assessment of evapotranspiration for Nigeria. *Geogr. Annl* **48**, A, 139–56.

1967. A note on the relationship between net and solar radiation. *Q. Jl R. met. Soc.* **93**, 109–15.

DEACON, E. L., PRIESTLEY, C. H. B. and SWINBANK, W. C. 1958. Evaporation and the water balance. *Arid Zone Research (UNESCO)* **10**, 9–34.

FOGG, G. E. 1966. *The Growth of Plants,* London.

GARNIER, B. J. 1956. A method of computing potential evapotranspiration in West Africa. *Bull. Inst. fr. Afr. noire* **18**, 665–76.

GATES, D. M. 1962. *Energy Exchange in the Biosphere,* New York.

GLOVER, J. and MCCULLOCH, J. S. G. 1958a. The empirical relation between solar radiation and hours of bright sunshine in the high altitude tropics. *Q. Jl R. met. Soc.* **84**, 56–60.

1958b. The empirical relation between solar radiation and hours of sunshine. *Q. Jl R. met. Soc.* **84**, 172–5.

HOUSE, G. J., RIDER, N. E. and TUGWELL, C. P. 1960. A surface energy-balance computer. *Q. Jl R. met. Soc.* **86**, 215–31.

KEAY, R. W. J. 1959a. *An Outline of Nigerian Vegetation,* Government Printer, Lagos.

KEAY, R. W. J. 1959b. *Vegetation Map of Africa*, Oxford.
LABORATORY OF CLIMATOLOGY. 1962. The average climatic water balance of the continents, Part 1. Africa. *Publications in Climatology* **15**, 2.
MCCULLOCH, J. S. G. 1963. *Tables for the Rapid Computation of the Penman Estimate of Evaporation* (cyclostyled), East Africa Agricultural and Forestry Research Organization.
MASSON, H. 1954. La radiation solaire à Dakar. *Bull. mém. Éc. prép. Méd. Pharm. Dakar* **21**, 1.
MONTEITH, J. L. 1965. The photosynthesis and transpiration of crops. *Expl. Agric.* **2**, 1–14.
MONTEITH, J. L. and SZEICZ, G. 1961. The radiation balance of bare soil and vegetation. *Q. Jl R. met. Soc.* **87**, 378, 159–70.
    1962. Radiative temperature in the heat balance of natural surfaces. *Q. Jl R. met. Soc.* **88**, 378, 496–507.
PAGE, J. K. 1961. The estimation of monthly mean values of daily total shortwave radiation on vertical and inclined surfaces from records for latitudes 40°N–40°S. *U.N. Conference on New Sources of Energy, Solar Energy, Wind Power and Geothermal Energy* (mimeographed paper).
PEIXOTO, J. P. 1960. On the global water vapour balance and the hydrological cycle. *Tropical Meteorology in Africa: Proceedings of Symposium jointly sponsored by the World Meteorological Organization and the Munitalp Foundation*, Munitalp Foundation, Nairobi.
PEIXOTO, J. P. and OBASI, G. O. P. 1965. Humidity conditions over Africa during the I.G.Y. *Sci. Rpt.* No. 4, General Circulation Project, Mass. Inst. of Technology.
PENMAN, H. L. 1948. Natural evaporation from open water, bare soil and grass. *Proc. R. Soc. (London) A*, **193**, 210–45.
    1956. Evaporation: an introductory survey. *Neth. J. agric. Sci.* **4**, 8–29.
    1962. Weather and crops. *Q. Jl R. met. Soc.* **88**, 377, 209–19.
    1963. Vegetation and hydrology. *Tech. Comm.* No. **53**, Commonwealth Bureau of Soils, Harpenden.
PEREIRA, H. C. *et al.* 1962. Hydrological effects of changes in land use in some East African catchment areas. *E. Afr. agric. For. J.* **27**, 131.
RICHARDS, P. W. 1952. *The Tropical Rainforest*, Cambridge.
SALVADOR, O. 1964. Contribution à l'étude du rayonnement terrestre au voisinage du sol dans les régions sub-tropicales. *Annls Fac. Sci. Univ. Dakar* **11**, 230.
STARR, V. P. and PEIXOTO, J. P. 1958. On the global balance of water vapour and the hydrology of deserts. *Tellus* **10**, 189–94.
THOMPSON, B. W. 1965. *The Climate of Africa*, London.
THORNTHWAITE, C. W. 1948. An approach towards a rational classification of climate. *Geogrl Rev.* **38**, 55–94.
THORNTHWAITE, C. W. and HARE, F. K. 1965. The loss of water to the air. *Amer. met. Soc. Mono.* **6**, 28, 163–80.

4 Rainfall reliability

S. GREGORY

Introduction

'Moisture is an important factor in all crop-producing areas. It is the all-important factor in the minimal regions, where the average or normal rainfall is generally necessary for successful crop production. In such areas the systems of crop production must be correlated more or less with existing moisture conditions; as a matter of fact, the entire programme of crop production is more or less dominated by the moisture factor' (Klages, 1947). This general statement is of even greater relevance in tropical areas, where high temperatures throughout the year can lead to considerable moisture losses through evaporation and transpiration, while at the same time providing adequate warmth for crop growth all the time. Thus Webster and Wilson (1966) assert that 'rainfall is the most important climatic factor influencing agriculture in the tropics, as it generally has the biggest effect in determining the potential of any area, the crops which it is practicable to grow, the farming systems which can be followed and the nature and sequence of farming operations'. They might also have added that, in association with evaporation and transpiration, it determines whether or not irrigation is both necessary and feasible.

However, the cautionary comment of Worthington (1958), that 'rainfall is the most obvious but by no means the only climatic variable involved in the use of land, and the usual method of its presentation as annual or monthly totals can be very misleading', must also be borne in mind. He further reinforced this point with the statement that 'the deviations from the mean in total rainfall from year to year are often greater [than in temperate latitudes] in the tropics', while Webster and Wilson (1966) stress that 'in many parts of the tropics the rainfall varies greatly from year to year, and to a much greater extent than is commonly experienced in temperate zones'. This importance of the variability of rainfall to tropical agriculture, coupled with an assertion of a frequently abnormal degree of variability, is an oft-repeated theme, either explicitly or implicitly, in most texts concerned with agriculture in low latitudes.

Substantiation of such assertions, and mapped distributions of the degree of variability, are much more difficult to find. Neither of the recent climatological atlases concerned with Africa as a whole (Jackson, 1961; Thompson, 1965) considers this theme, presenting rainfall data only in the

form of averages. The same is true of most of the national atlases for
tropical African territories, save for the *Atlas of the Federation of Rhodesia
and Nyasaland* (Federal Department of Trigonometrical and Topographical Surveys, 1962), while the recent *Oxford Regional Economic Atlas
of Africa* (Ady, 1965) provides no more than incidental indications via its
monthly rainfall diagrams. Almost equally lacking for many parts of tropical
Africa are research publications presenting details of rainfall variability,
although attempts have been made for some areas, such as the FAO/
UNESCO/WHO study on West Africa, recently reported in summary form
by Cochemé (1966), though not published in full at the time of writing.

The only available map depicting the variability of annual rainfall over
Africa as a whole forms part of a world map by Biel (1929) which is used
by Trewartha (1954), and a representation of this appears as Figure 4.1.
This version, based on the mean deviation as a percentage of the average,
is exceedingly generalized, to the point where, except for very broad contrasts (as was the purpose of the original world map), it is more misleading
than helpful. Moreover, it is unfortunate that when this map was recently
used to show such broad-scale differences (Dekker, 1965), a number of
further inaccuracies occurred – areas of very low variability around Lake
Victoria, along coastal Ghana and in part of the Sahara(!) were introduced;
a shading category was omitted in Angola and South West Africa; and a
critical boundary was markedly displaced in Moçambique. Certainly, the
pattern shown in Figure 4.1 should be used neither as an argument in
favour of a close inverse relationship between mean rainfall and variability,
nor as a guide to agricultural possibilities.

Variability studies

The use of the mean deviation, rather than the mathematically more satisfactory standard deviation, is often supported on the grounds that there is
a reasonably constant relationship between the two, such that:

> Mean deviation   = 0·8 standard deviation,
> Standard deviation = 1·25 mean deviation.

The extra labour involved in calculating the standard deviation is perhaps
too much over-stressed, however, in these days of mechanical aids to computation. In either case, it is also usually maintained that the deviation
value varies very closely with its appropriate mean value, Griffiths (1961)
suggesting that over a specified area, the general relationship between mean
and variance (or standard deviation) is such that it is possible to determine
the deviation once the mean is known.

A comparison of average and standard deviation values for annual rainfall over Moçambique (Fig. 4.2A, B) lends some general support to this

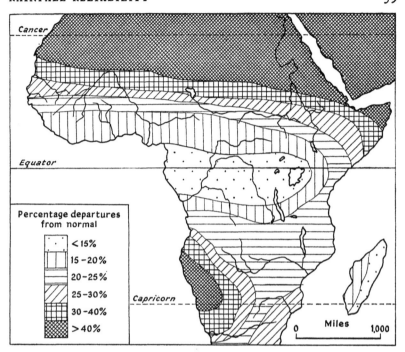

Figure 4.1 *Mean deviation as a percentage of the average* (based on a map by Biel, in Trewartha, 1954).

thesis, but does not fully confirm it. The area of high average falls of over 1,600 mm (63 in) coincides with that of standard deviations of over 400 mm (15·75 in), while conversely the low rainfall areas of the Limpopo and Chengane Valleys in the south display low standard deviation conditions. On the other hand, the dry north-eastern coastlands and the wetter uplands east of Lake Malawi (Nyasa) both have standard deviation values of less than 200 mm (8 in), while the wet uplands east of the Shiré Valley and the dry sections of the Zambezi both fall in the 200–300 mm (8–12 in) range in terms of standard deviation. The more limited degree of relationship between average and standard deviation suggested here, is further confirmed by the fact that although the correlation coefficient between the two sets of values (+ 0·67) is statistically highly significant, the coefficient of determination (correlation coefficient squared) indicates that only some 45 per cent of the variations in standard deviation could in fact be determined from the mean values. Clearly, other factors, perhaps of a locational nature, are also at work.

This can be investigated by mapping not the standard deviation itself but rather its local difference from the basic mean/standard deviation

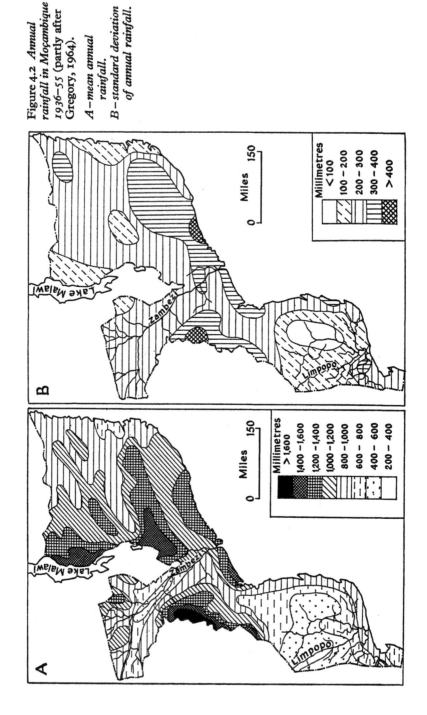

Figure 4.2 Annual rainfall in Moçambique 1936–55 (partly after Gregory, 1964).

A – mean annual rainfall.
B – standard deviation of annual rainfall.

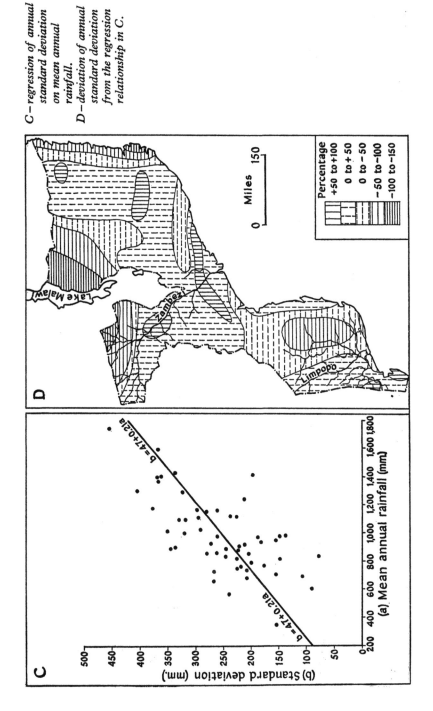

C – regression of annual
standard deviation
on mean annual
rainfall.
D – deviation of annual
standard deviation
from the regression
relationship in C.

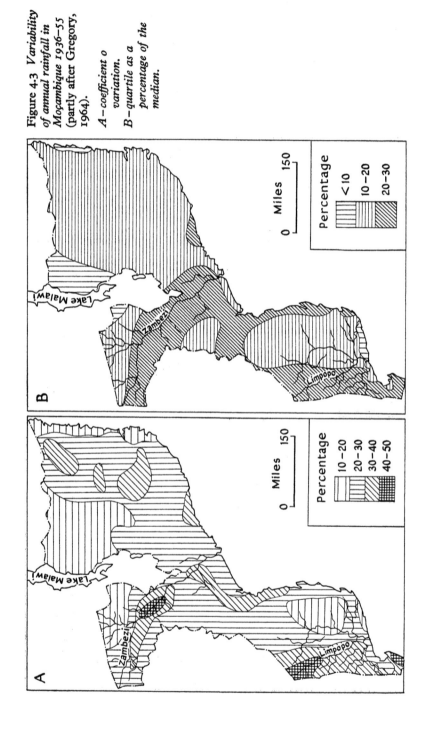

Figure 4.3 *Variability of annual rainfall in Moçambique 1936–55 (partly after Gregory, 1964).*

A – coefficient o
variation.
B – quartile as a
percentage of the
median.

B

Lake Malawi

Zambezi

Limpopo

Miles
0   150

Percentage
<10
10–20
20–30

A

Lake Malawi

Zambezi

Limpopo

Miles
0   150

Percentage
10–20
20–30
30–40
40–50

relationship. The latter can be expressed by the regression equation:

Standard deviation (mm) = 47·0 + 0·21 mean (mm)
or Standard deviation (in)   = 1·85 + 0·21 mean (in),

which is displayed graphically in Figure 4·2C. The resulting map of the
differences from this relationship (Fig. 4·2D) distinguishes between both
the areas of positive and negative residuals, and the areas of low and high
differences. These reinforce the general comments made earlier.

The expression of variability in terms of mean and deviation, rather than
as a percentage relationship between the two, has much to commend it as
a basis for estimating agricultural potential. If, indeed, agriculture is being
carried on in a region 'where the average or normal rainfall is generally
necessary for successful crop production' (Klages, 1947), then variability
in absolute terms of amounts of rainfall is of greater relevance to potential
water need or the possible degree of crop failure than is an expression
solely in percentage terms. This is, in fact, the basis of the probability
estimates which will be discussed later.

The construction of percentage variability maps is essentially a device
by which to represent the spatial pattern of variability, with the influence
of mean value standardized or eliminated. That this effectively happens is
illustrated for Moçambique by the low correlation coefficient between the
average and the coefficient of variation, which at − 0·25 does not reach the
5 per cent level of statistical significance. A comparison of Figure 4.3A,
showing the coefficient of variation (the standard deviation as a percentage
of the average), with Figure 4.2A, showing the mean values, indicates that
percentage values both below 20 per cent and above 40 per cent occur in
the dry areas of the south, while values below 20 per cent apply to both the
drier and the wetter areas of the north.

Despite this, it is still often argued that percentage variability varies
inversely with the mean rainfall. Thus, when speaking of annual rainfall in
Ghana, Walker (1962) stated, 'It will be noted that variability is least over
areas of high rainfall and greatest over areas of low rainfall.' Yet a comparison of the mean rainfall map (Fig. 4.4A) with that of the coefficient of
variation (Fig. 4.4B), shows that variability values of approximately 25 per
cent are related both to the Accra plains with only 30 in (760 mm) of rainfall and to the uplands east of the Volta Lake with more than 60 in
(1,525 mm). Again, the wet areas of the Axim coast in the south-west and
of the Mampong scarp around Mpraeso-Nkawkaw, with mean rainfalls of
80 in and 65 in respectively (2,030 mm and 1,650 mm), return variability
values of approximately 20 per cent, while the drier areas around Tamale,
with mean falls of 40–45 in (1,016–1,143 mm), have a variability of only
15 per cent. Cochemé's comment (1966) that 'the general rule that

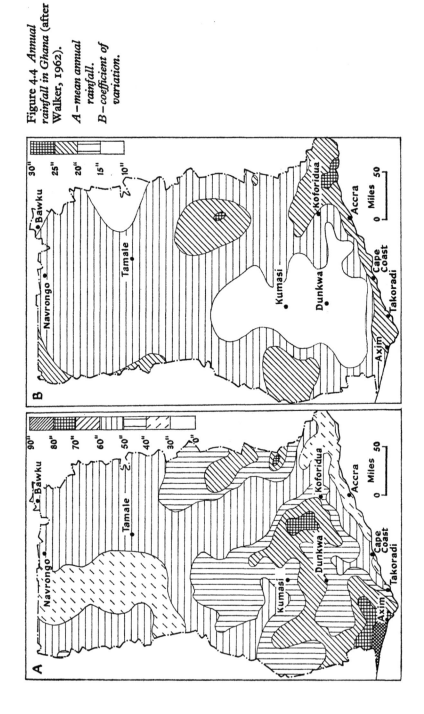

Figure 4.4 *Annual rainfall in Ghana* (after Walker, 1962).

A – *mean annual rainfall.*
B – *coefficient of variation.*

variability increases with decreasing mean annual rainfall applies only very approximately' represents a valid generalization.

Perhaps such statements have been conditioned by the broad-scale differences between semi-arid and very wet areas, represented cartographically in Figure 4.1 and graphically by Conrad (1941), or by the contrasts between dry and wet seasons or months, such as have been depicted by Gregory for Moçambique (1964) and Sierra Leone (1965). This contrast is dominated by very high percentage values for the dry areas or periods, reaching to beyond 100 per cent in some cases. Such large values, however, immediately indicate that the data under review do not fit the normal frequency distribution, for once the standard deviation is greater than 35 per cent it is impossible to find values that are three standard deviations below the mean, while with a coefficient of variation greater than 50 per cent no value can be as much as two standard deviations below the mean. Furthermore, any analysis based on the variance or standard deviation assumes that the data do approximate to a normal frequency distribution – when they do not, then it is not legitimate to employ the standard deviation or the coefficient of variation (nor the mean deviation in absolute or percentage terms) as an expression of variability. The converse of this argument has been employed previously by Kenworthy and Glover (1958) as a diagnostic technique for the definition of a wet period. 'When it was not clear from the mean rainfall régime which months could be generally regarded as wet months, the mean monthly rainfall was compared with its standard deviation, and if the standard deviation was less than some 50 per cent of the mean (in many cases less than 30 per cent) that month was included as part of the rainy season' (Kenworthy and Glover, 1958).

Thus, in dry areas or in dry seasons, rainfall data display marked positive skewness, so that 'predominantly very low values are interspersed with the occasional "wet" dry season' (Gregory, 1964). For much of the time, in fact, rainfall in such areas varies but little, but when exceptions do occur they tend to be considerable ones. One way of representing the degree of variability without incorporating the marked extremes is to use the quartile deviation as a percentage of the median. This possesses limited mathematical value, but it does indicate the percentage range of the central 50 per cent of the occurrences. The resulting pattern for Moçambique is presented in Figure 4.3B, and can be compared with that of the coefficient of variation in Figure 4.3A. The higher variability of both the Zambezi and Limpopo Valleys, and the low to moderate variability of the north and of the Chengane Valley area in the south, broadly reinforce the pattern in Figure 4.3A, though with values of a lower magnitude.

The high degree of variability which the use of the coefficient of variation suggests in the drier areas or periods has led to the assumption, and often the assertion, that tropical rainfall is markedly more variable than that of temperate latitudes. Comments to this effect by Worthington (1958) and by Webster and Wilson (1966) have already been quoted, and similar implications may be drawn from many other publications. While not denying that there are areas of high variability, these are not widespread outside the drier areas of tropical Africa. Thus in Ghana (Fig. 4.4B) 25 per cent is the highest degree of variability, and most of the country returns values of less than 20 per cent – it should be noted that values for Britain range between 10 and 20 per cent too (Gregory, 1955). In Moçambique, annual coefficient of variation values are mainly higher than this, but they only exceed 35 per cent (the critical value suggested above) in very limited, dry areas. In such wet environments as Sierra Leone (Fig. 4.5), where mean annual values range from over 250 in (6,350 mm) to just below 80 in (2,030 mm), variability falls within the limits to be found in Britain. Furthermore, the same is true if one turns to Northern Nigeria between Zaria and Kano (Fig. 4.6). An analysis of the records kept by the British Cotton Growers Association for eight stations for the period 1929–49 shows that, with mean annual falls from below 30 in (650 mm) to just over 40 in (1,015 mm) all stations nevertheless return coefficients of variation of between 13 and 19 per cent.

Before leaving this theme of variability, further cautionary comments are necessary. All climatological studies are, of necessity, based on only a sample period of years. Moreover, the actual period of the sample will vary from one area to another, and from one investigation to another. Direct comparison of any quantitative results obtained, and also the using of such results as a guide to long-term or future possibilities – both of which may be of considerable value in the assessing of agricultural potentialities – is fraught with dangers, unless due allowance is made for the appropriate standard errors. This applies not only to any assessment based on the standard deviations themselves, but even more so to judgements derived from the coefficient of variation (V), into which enter the standard errors of both the average and the standard deviation. Thus, in relation to variability in Moçambique, as shown in Figure 4.3A, Gregory (1964) states that 'with only a 20-year record the standard errors of $V = 40$, 30 and 20 per cent are 6·3, 4·8 and 3·2 per cent respectively. This means that the difference between 20 and 30 per cent barely reaches a level of "probably significant", while the difference between 30 and 40 per cent does not reach this level.' This suggests that care should be taken to ensure that the isopleth interval approximates to the 5 per cent level of significance, so that at least alternate shading categories can be accepted as distinctive areas. In the case of Moçambique the major areal contrasts are therefore between those areas shown in Figure 4.3A in the 'less than 20 per cent' and 'more than 30 per cent' categories.

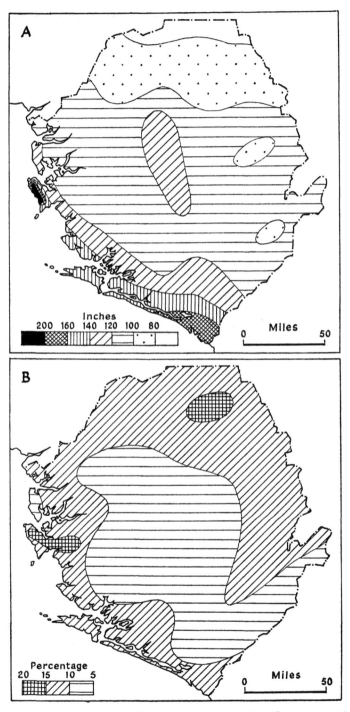

Figure 4.5 *Annual rainfall in Sierra Leone 1941–60* (after Gregory, 1965).
A – mean annual rainfall. B – coefficient of variation.

Similar caution should also be exercised when assessing whether or not any particular location has highly variable rainfall. Harrison Church, for example, uses the annual rainfall values for Bathurst, Gambia, for 1901–18, to illustrate his comment that 'in the interior and towards the northern limit of adequate rainfall, where water is precious, variability is greatest. Average figures should be used only in the knowledge that in any year the

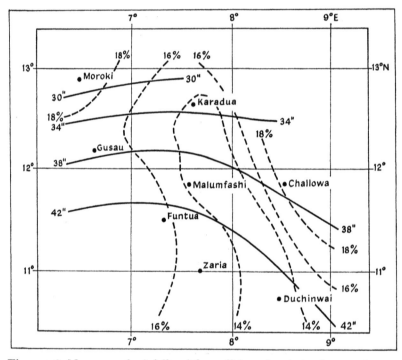

Figure 4.6 *Mean annual rainfall and the coefficient of variation for part of Northern Nigeria, 1929–49* (based on records of the British Cotton Growers' Association, Zaria, Northern Nigeria).

actual rainfall may be about one-half greater or smaller, and that this may happen for several years on end' (Harrison Church, 1963). Apart from the fact that the 50 per cent surplus or deficit in relation to the average applies equally to Britain, the actual variability is not really excessive. With a sample mean of 43·9 in (1,118 mm) and a sample standard deviation of 12·3 in (313 mm), the coefficient of variation becomes 28·0 per cent, somewhat above the 20 per cent upper level in Britain. However, its standard error is 4·67 per cent, and if it were to be compared with the 18–20 per cent range in Britain for the 1901–30 period (Gregory, 1955), the difference could not be shown to be statistically significant. To some extent, this ability to apply such tests provides yet further justification for the use of standard deviation rather than mean or quartile deviations in any study of rainfall variability.

The basic features of this discussion on variability can perhaps best be summarized by comparing monthly, seasonal and annual characteristics for two contrasted locations in West Africa – the Zaria area in Northern Nigeria, using the British Cotton Growers Association records for the 56 years, 1904–59, and the Cavalla area of south-eastern Liberia, for which the Firestone Plantations Company records for the 37 years, 1928–64, are used. As can be seen from Table 4.1, the mean annual values differ as between 45·09 in (1,145 mm) and 103·85 in (2,637 mm), while the seasonal régimes are also markedly dissimilar. At Zaria a concentrated five-month wet season (May–September) with each month averaging more than 4 in (100 mm) is followed by seven months (October–April) each with very little rain. Cavalla, in contrast, averages more than 4 in (100 mm) in all months except January, but the year can be divided into four phases: two wet periods (May–June and September–November) when each month averages approximately 10 in (250 mm) or more, and two intervening drier (though still wet) periods, a longer one from December to April, and a shorter one in July and August. Thus they essentially represent the single maximum 'tropical' and double-maximum 'equatorial' rainfall régimes of West Africa.

Despite these differences, the annual coefficients of variation are virtually the same at just over 16 per cent. There is no sign here either of increased variability with lower mean rainfall or of any excessively high variability at all, and it should be remembered that comparable values occur regionally around Zaria (Fig. 4.6). Cochemé (1966) suggests that such values apply to much of the area south of the 500-mm (20-in) isohyet in West Africa. On a monthly basis, the wet season/dry season contrast at Zaria is as clearly represented in the variability values as in the mean values. The use of the coefficient of variation in this context, as outlined earlier and suggested by Kenworthy and Glover (1958), distinguishes the wet May–September months as being mainly in the 30–35 per cent range, while the dry months vary upwards from 70 per cent. They thus possess skew frequency distributions, especially from November to March, and the percentage variability values are therefore meaningless. Thus in January, with 500 per cent variability returned, 52 of the 56 years had no rain, and in the others, totals never exceeded 0·04 in (1·0 mm) – greater reliability than this can scarcely be expected! Even in March, 32 of the 56 years received no rain, and the average value was exceeded in only 14 years, i.e. no more than 1 year in 4.

In contrast to this, the Cavalla rainfall never reaches these degrees of positive skewness, even in the driest month of January. In fact, it is the

TABLE 4.1 *Monthly and annual average, standard deviation and coefficient of variation values for Zaria (N. Nigeria), 1904–59 and Cavalla (Liberia) 1928–64*

		J	F	M	A	M	J	J	A	S	O	N	D	Year
Average (inches)	Zaria	0·002	0·054	0·215	1·72	4·77	6·19	8·97	12·55	8·88	1·60	0·06	0·027	45·09
	Cavalla	3·29	4·43	6·81	6·68	14·10	16·72	5·49	4·21	12·67	13·98	9·76	5·69	103·85
Standard deviation (inches)	Zaria	0·010	0·144	0·421	1·26	1·74	1·97	2·71	3·73	2·75	1·24	0·22	0·201	7·32
	Cavalla	2·05	2·00	2·93	2·82	6·11	7·60	6·00	3·38	6·17	6·63	3·64	2·50	16·73
Coefficient of variation (%)	Zaria	500	266	196	73·2	36·5	31·8	30·2	29·8	31·0	77·8	367	743	16·23
	Cavalla	62·3	45·2	43·0	42·2	43·3	45·5	109·3	80·3	48·7	47·4	37·3	43·9	16·1

'little dry season' months of July and August which return highest variability values and are also most skew, despite average values of over 4 in (100 mm). The lack of validity of the percentage variability values under these conditions is reflected in the frequency distributions in Table 4.2: out of 37 years, 28 and 27 (in July and August respectively) are below 6 in (150 mm), but occasionally a very wet month indeed can occur, reaching 32·95 in (838 mm) in July 1963. It should also be noted that no month at

TABLE 4.2 *Frequency distribution of July and August rainfall at Cavalla (Liberia), 1928–64*

	Inches				
	0–1·99	2·00–3·99	4·00–5·99	6·00–7·99	8·00–9·99
July	7	12	9	4	1
August	11	10	6	6	3

	Inches				
	10·00–11·99	12·00–13·99	14·00–15·99	16·00–17·99	≥18·00
July	1	0	0	2	1
August	0	0	0	1	0

Cavalla has as low a percentage variability as any of the five wet months at Zaria which, apart from the problems of standard errors, might suggest that wet season months are more reliable and less variable at drier Zaria than at wetter Cavalla.

This contrast is even more marked if the seasonal rainfall is considered as a whole (Table 4.3). There is a statistically significant difference between

TABLE 4.3 *Seasonal average, standard deviation and coefficient of variation values for Zaria (N. Nigeria), 1904–59, and Cavalla (Liberia), 1928–64*

		Average (inches)	Standard deviation (inches)	Coefficient of variation %
Zaria	May–September	41·35	6·70	16·2
	October–April	3·74	1·74	46·4
	March–April	1·93	1·27	65·5
Cavalla	May–June	30·83	7·50	24·3
	July–August	9·69	7·36	76·0
	September–November	36·40	12·47	35·3
	December–April	26·37	5·25	19·9

the variability of the May–September rainfall at Zaria and that of the first main rains (May–June) at Cavalla; also, the skewness of the rains of the 'little dry season' at Cavalla is still apparent, but the long drier season (December–April) is seen to be of relatively low variability. It can also be seen that at Zaria, despite the extreme skewness and apparent high variability of the individual dry season months, the variability and skewness of the dry season rains as a whole are markedly less. The same is true if the critical months of March and April, just before the true rainy season, are considered together (Table 4.3) and compared with their individual characteristics (Table 4.1).

Probability studies

These and many other details of variability are clearly of relevance to any consideration of agricultural potentialities. For any particular crop or economy, however, rather more than this may be desirable. More especially, the likelihood of any required amounts of rainfall occurring can be of critical significance, just as may be the minimum falls that can be expected. The theme of probability and reliability is therefore of considerable relevance, and develops logically from the assessment of deviation and variability already analysed.

The construction and employment of rainfall probability maps as a guide to agricultural potential and policy has in fact been largely pioneered and developed within tropical Africa, and more especially in East Africa. The papers by Manning (1950), Glover and Robinson (1953) and Glover, Robinson and Henderson (1954), and the maps used in the Report of the East African Royal Commission, 1953–5 (Dow, 1955), stimulated considerable interest and activity in this field, at least for some parts of tropical Africa. Excellent summaries of these expanded studies for East Africa are available in McCulloch (1961) and Kenworthy (1964), but in many ways it is surprising that the techniques devised have not been more widely applied elsewhere in the continent.

The concepts underlying these studies are relatively simple. Provided that the frequency distribution of the phenomenon being studied (i.e. rainfall amounts) approximates to the normal curve, and given that an adequately sized sample (i.e. a long enough run of years) is available, then the manipulation of average and standard deviation values permits the estimation of:

1. The percentage probability that some specific rainfall amount will (or will not) be exceeded or be not reached.
2. The amount of rainfall that will (or will not) occur or be exceeded with some specified degree of probability.

Thus, for East Africa, maps showing the percentage probability of a failure to receive an annual rainfall of 20 in (510 mm) and 30 in (760 mm) have been available for more than a decade (Glover *et al.*, 1954; Dow, 1955), while more recently maps showing annual rainfall amounts likely to be reached or exceeded in 90 and 80 per cent of years have been published (East African Meterological Dept., 1961). Comparable maps were also included in the *Atlas of the Federation of Rhodesia and Nyasaland* (Federal Department of Trigonometrical and Topographical Surveys, 1962), while Gregory (1964) presented several for Moçambique. In Figure 4.7B an attempt has been made to combine these three sources, to show for the larger part of eastern tropical Africa the annual rainfall value likely to be reached or exceeded in 80 per cent of years, i.e. the minimum expected rainfall one year in five. To avoid needless complexity, the pattern has been restricted to the 1,000-mm (40-in) and 600-mm (24-in) isopleths, but this allows the broad picture to be seen, a comparison with the average rainfall map (Fig. 4.7A) to be made, and the major areas of reliably high and low annual rainfall to be discerned.

Assessments of such probability characteristics can obviously be of considerable value in the field of agriculture, especially if critical rainfall limits for particular crops are known. For example, in the *Atlas of Uganda* (Department of Lands and Survey, 1962) it is suggested that 'the concept provides a better basis than does the mean annual rainfall for planning and development projects in agriculture, forestry, hydrology, roadmaking and such matters'. This is only true, however, if such assessments are reasonably accurate, and for this the assumption of normality in the frequency distribution, mentioned earlier, is important. For annual rainfall, with mean values in excess of 30 in (760 mm), this is a reasonable assumption, and this also applies very frequently to wet season rainfall too (Kenworthy and Glover, 1958). If mean rainfall values are lower, however, there is a greater possibility of a non-normal skew distribution, as was discussed earlier in terms of variability. This point has been made clearly by Griffiths (1961), speaking of East African conditions: 'There was no apparent geographical grouping of non-normal stations but those with annual means of less than 25 in (635 mm) had a 45 per cent chance of being non-normal and so great caution must be exercised when attempting to apply probability calculations to such stations.'

Non-normality of frequency distribution is even more common if shorter time units, such as the month or week, are investigated. In all such cases it is necessary first to transform the data in order to force an approximation to normality. Both Manning (1956) and Griffiths (1960) support the square root transformation as being the most useful one, Griffiths (1961) arguing that in East Africa this led to virtual normality for 91 per cent of monthly

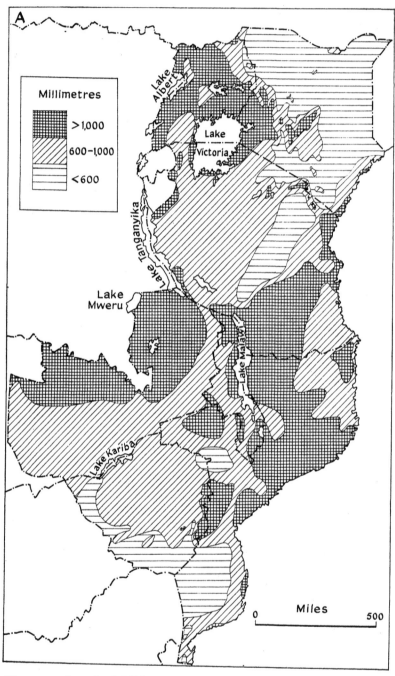

Figure 4.7 *Annual rainfall in eastern tropical Africa.*
A – mean annual rainfall (after Jackson, 1961).

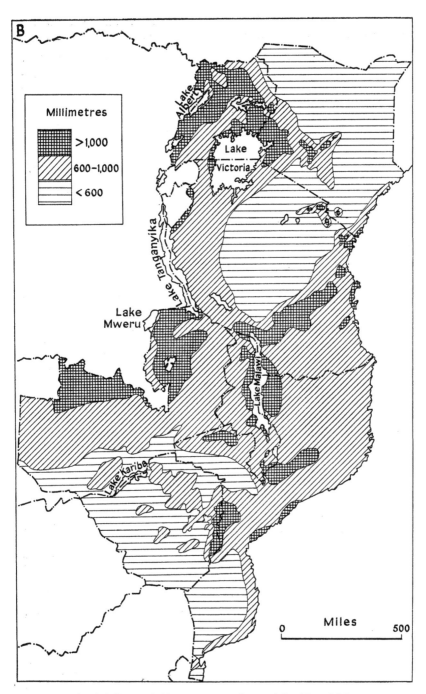

B – *annual rainfall exceeded in 80 per cent of years* (after East African
Meteorological Department, 1961, and Federation of Rhodesia and
Nyasaland, Department of Trigonometrical and Topographical
Surveys, 1962, and Gregory, 1964)

skew data, and that it was preferable to a logarithmic transformation (the other common alternative) so long as the average exceeded o·5 in (12·5 mm). It should be noted, however, that neither of these two common transformations necessarily provides a better estimate in every case of a skew frequency distribution, as can be illustrated from the April rainfall at Zaria (N. Nigeria), the average and standard deviation for which were listed in Table 4.1. The percentage and actual number of occurrences incorporated within plus and minus one and two standard deviations respectively, for the data in its original form and after transformation, are given in Table 4.4.

TABLE 4.4　*April rainfall, Zaria, 1904–59: the proportion of occurrences falling within given limits in untransformed (x) and transformed ($\sqrt{x}$ and log . 10x) data, as compared with the normal frequency distribution*

Form of data	x	$\sqrt{x}$	Log (10x)	Normal frequency distribution
Average	1·72	1·18	1·0837	
Standard Deviation	1·26	0·58	0·4575	
±1 standard deviation	66% (37 items)	71·4% (40 items)	75% (42 items)	68·3% (38 items)
±2 standard deviations	96·4% (54 items)	89% (50 items)	89% (50 items)	95·5% (53 items)

Probability estimates for any particular critical value, or the fiducial limits for any specified probability level, could thus be calculated for Zaria for the year, the wet season (May–September) and the wet season months, feeding the average and standard deviations given in Tables 4.1 and 4.3 into the appropriate formulae.[1] A consideration of seasonal rainfall and its reliability in this area has, in fact, previously been made for neighbouring Samaru (Manning, 1959).

Apart from this need to approximate the data to normality, there is also the question of how accurate the estimates derived from these formulae may be, in the light of the length of records used. The basic discussion of this was provided by Manning (1956), in a fundamental paper on the theme. Manning used the annual rainfall data for Padua, Italy, for the

---

[1] (a) To estimate the probability of a specified fall (x) occurring, when the average ($\bar{x}$) and standard deviation (s) are known, calculate d by the formulae:

$$d = \frac{x - \bar{x}}{s}$$

and refer to a table of the Normal Distribution Function.

(b) To estimate the fiducial limits for a specified probability, obtain the d value appropriate to the probability from the Normal Distribution Function, and incorporate into the formula: $x = d . s . + \bar{x}$　(see Gregory, 1968, pp. 59–67).

227 years 1725–1951, to establish that long-period records approximate to normality. He then took ten successive 20-year periods, for each of which he separately calculated the fiducial limits for $p = 0.5$, i.e. the lower and upper confidence limits, exceeded in each direction 25 per cent of the time, and proceeded to show that this provided an appropriate picture for its own 20-year period. On the basis of this and similar reasoning, he then affirmed that 'for short runs, therefore, when current rainfall is successively incorporated, runs of data of the order of 20 years will provide a precise estimate of expectations *for the period immediately following*' (Manning, 1956, p. 464) [my italics].

However, he later (p. 471) stated that 'the resulting statement of probability would in fact provide an estimate of *long-term crop risk*' [my italics]. This latter is a highly desirable estimate to have available, and is in fact the way in which most probability estimates tend to be used. If this is so, however, the 20-year estimates for the Padua data should be compared with the actual frequencies for the full 227 years, i.e. with the 'long-term' picture. If this had been done, it would have been seen that it is quite possible for a 20-year estimate on occasions to be rather wide of the long-term mark. Thus, for the period 1825–44 at Padua, the upper fiducial limit for $p = 0.5$ was 34·52 in (877 mm), while for the 1725–1951 long-term period it was 38·60 in (980 mm). Moreover, whereas the 1825–44 results suggest that only 25 per cent of the occurrences would exceed 34·52 in (877 mm), in fact over the 227 years of the long-term record, some 46 per cent of the years exceeded this value.

These differences are not peculiar to Padua, nor to the period used; instead, they are the simple result of using a sample period (20 years in this case) to provide an estimate of long-term values (227 years at Padua), without allowing for the attendant standard, or sampling, errors. Moreover, such errors are incorporated into probability estimates from both the sample mean and the sample standard deviation. It is therefore essential that any estimates, either in terms of percentage probability or fiducial limits, should always have their appropriate standard errors quoted along with them.

If fiducial limits are being estimated, the necessary standard error is compounded of the errors of both the mean and standard deviation, such that:

Standard Error (S.E.) of fiducial limit

$$= \sqrt{\left(\begin{array}{c}\text{Variance of} \\ \text{sample mean}\end{array} + \begin{array}{c}\text{Variance of sample} \\ \text{standard derivation}\end{array}\right)} = \sqrt{\left(\frac{s^2}{n} + \frac{s^2}{2n}\right)} = \sqrt{\left(\frac{3s^2}{2n}\right)}.$$

If this were to be applied in the case of the Padua 1825–44 data already discussed above it would be found that the standard error of the estimate of the upper fiducial limit (34·52 in or 877 mm) was 1·56 in (39·6 mm), and thus the long-term value of 38·60 in (980 mm) lies between two and three standard errors above the estimate. In the absence of any knowledge of the long-term value, and with only one sample 20-year period available, it would thus be essential that such a standard error be obtained.

The same argument applies to estimates of probability levels, but in this case the error is in terms of the 'd' value from which the probability is estimated. This means that various other components enter into the formula, along with the variance values above. With cancellations, this formula becomes:

$$\text{S.E. of `}d\text{' value} = \sqrt{\left(\frac{1}{n} + \frac{(x - \bar{x})^2}{2ns^2}.\right)}$$

For example, if one analyses the 1936–55 annual rainfall record for Lourenço Marques, the following values are obtained:

$$\bar{x} = 815 \text{ mm (32·1 in)} \quad s = 225 \text{ mm (8·9 in)} \quad n = 20.$$

From these, it is possible to estimate the probability of obtaining less than 1,000 mm (40 in) in a year, a value which could well be of agricultural significance. This would be

$$`d' = 0·82 = \text{probability 79·4 per cent}$$

and this is the value normally accepted. From the formula above, however, 'd' has a standard error of 0·26, so that

$$`d' + /{-}2 \text{ standard errors} = 0·30{-}1·34 = 61·8{-}90·1 \text{ per cent}$$
$$`d' + /{-}3 \text{ standard errors} = 0·04{-}1·60 = 51·6{-}94·5 \text{ per cent.}$$

Clearly, the possible limits of the true probability are very wide; moreover, the range above the sample estimate is rather smaller than is the range below it, for the error is related to the 'd' value rather than the percentage value.

Quite apart from the importance of realizing the magnitude of these errors when examining conditions in one locality, it is even more important to bear them in mind when comparing different stations or areas. The latter is, of course, an integral part of the construction of any probability map, when isopleth intervals are being chosen, and also enters into all interpretations of such maps. Thus:

> In the choice of isopleth intervals it is again necessary to ensure a statistically significant difference. It was found [for Moçambique] that an interval of 20 per cent probability was the most serviceable for general application, with a 10 per cent interval at the two extremes. Thus, . . . a significant difference can be relied upon between alternate shading

categories, although in some cases smaller differences are in fact significant (Gregory, 1964).

One practical difficulty is that the actual error varies with virtually every station, but in terms of the range of average and standard deviations occurring in tropical Africa over the non-arid areas and during the non-dry periods, and of the length of record that is likely to be available, the most universally satisfactory isopleth interval is of the order – 0 per cent, 10 per cent, 25 per cent, 50 per cent, 75 per cent, 90 per cent, 100 per cent. The wide range of percentage values needed for a statistically significant difference in the middle orders stresses the wisdom of mapping only the upper or lower extremes, as in the case of East Africa (Dow, 1955). These are not only the most likely to be of agricultural significance but are also more likely to yield a rainfall map of statistical significance.

It must be remembered that all these percentage probabilities are theoretically related to an infinitely large population, so that there is no necessary reason for them to apply in detail within any shorter, finite period. An event, such as a deficiency of rainfall, occurring with a 10 per cent probability, should not be expected to occur ten times in a century, and even less likely is it that it will occur once in a decade. Although this is normally understood, there is still a tendency to misinterpret such figures, often unintentionally. Even Manning (1956) says, 'It may be suggested that, at least in peasant agriculture, risk of crop loss from inadequate rainfall more often than once in 10 years would make the venture unduly speculative,' and then draws attention to the map showing the value exceeded 90 per cent of the time. The very use of the term '9 years out of 10' in this context virtually misleads by implication.

However, if the long-term probability of this critical deficiency occurring is 10 per cent, then the probability of such conditions occurring more than once within any particular 10-year period can be assessed from the binomial frequency distribution. The result is that there is approximately a 26 per cent probability that more than one dry year will occur in a particular decade, when the long-term probability is only 10 per cent, while equally there is a 35 per cent probability that there will be no such dry year within a decade. This tends to change the impression created by some probability maps; it must also be realized that, as a result of the standard error of the original probability value, the picture resulting from the binomial analysis may be even more accentuated.

Conclusion

The agricultural implications of these and comparable studies are many and varied. Moreover, they must clearly be related to other climatic conditions, especially the degree of evaporation loss (see Chapter 3) and the characteristics of soil storage and replenishment. These interrelationships must in turn be studied in relation to the moisture requirements of particular crops in specific areas. Thus Manning (1956) suggests that for cotton in East Africa critical values vary from 18 in (460 mm) to 30 in (760 mm) in a season, depending on area; at Namulonge, 23 in (585 mm) are required during the first rains but only 21 in (530 mm) during the second rains, while 25 in (635 mm) over a six-months' period leads to maximum yields. These therefore provide the critical values for which probability assessments may be needed in this context. These relationships have since been developed further (Manning, 1958, 1959), while a summary of significant relationships for the area is provided by Kenworthy (1964).

Elsewhere in tropical Africa the theme of the relationship of rainfall probability studies to agricultural possibilities has received little attention. Yet it may be a perfectly relevant theme even in areas of high rainfall. For example, in Liberia the major climatic contrast between the Cavalla site of the Firestone Plantations Company and its main area at Harbel near Monrovia is the occurrence of a 'little dry season' at Cavalla in July and August, when mean falls of 5·49 in (140 mm) and 4·21 in (107 mm) respectively contrast with those of 18·19 in (462 mm) and 18·11 in (460 mm) at Harbel. The resulting agricultural contrast is the use of the Cavalla site for seed germination and seedling development, for which the break in the rains, i.e. an 'equatorial' régime, is virtually essential. It does not invariably occur in an intensive form, however, for as was seen in Tables 4.1 and 4.3 the frequency distributions for these months, singly or together, are markedly skew. Logarithmic transformation virtually eliminates this, however, so that it is possible to estimate, for example, that there is a 15 per cent probability of more than 10 in (254 mm) in July, or a 22 per cent probability of more than 6 in (152 mm) in August, or again a 26 per cent probability of more than 12·5 in (318 mm) in the two months together. Allowing for two standard errors on either side of the estimate, however, the true value for the latter could lie between 16 and 39 per cent. Alternatively, the July–August rainfall that will be exceeded in no more than 10 per cent of the years can be estimated as 21·07 in (536 mm), or between 18·71 in (475 mm) and 23·43 in (595 mm) allowing for two standard errors. Such analyses, based on seasonal rather than annual conditions, may prove of considerable value.

In many cases, however, analyses of yet shorter periods may prove to be the most important, for 'Detailed local analyses of rainfall conditions are plainly the most useful and to correlate agricultural potentialities with the reliability of annual rainfall over-simplifies the climatic factors concerned' (Kenworthy, 1964). Thus Manning (1956) analyses overlapping three-weekly totals in order to define detailed seasonal characteristics. Such studies may be highly critical, especially in terms of the period of onset of the rains. For example, in such areas as Zaria (Tables 4.1 and 4.2) with a single, relatively short, wet season, the rainfall in the month preceding that in which rainfall is normally adequate for crop growth, i.e. in April, could well repay detailed investigation on a weekly or other short-period basis in terms of specific crops. Such studies in East Africa by Manning (1949, 1950), Glover and Robinson (1953) and Evans (1955) have, in fact, been used to reassess the most suitable planting dates for a variety of crops. In the light of the considerable agricultural potential obviously possessed by these several methods of assessing and analysing rainfall reliability, it is most surprising how little they have been applied outside East Africa. It is to be hoped that, with the greater availability of mechanized aids to computation, this deficiency will be rapidly rectified.

References

ADY, P. H. 1965. *Oxford Regional Economic Atlas of Africa*, Oxford.

BIEL, E. 1929. Die Veränderlichkeit der Jahressumme des Niederschlags auf der Erde. *Geogr. Jber. Öst.* XIV and XV, 151–80.

COCHEMÉ, J. 1966. FAO/UNESCO/WHO agroclimatology survey of a semi-arid area south of the Sahara. *Nature and Resources, Bulletin of the International Hydrological Decade*, II, No. 4, 1–10.

CONRAD, V. 1941. The variability of precipitation. *Mon. Weath. Rev.* **69**, 5–11.

DEKKER, G. 1965. Climate and water resources in Africa. In WOLSTEN-HOLME, G. E. W. and O'CONNOR, M. (eds.), *Man and Africa*, London, 30–56.

DOW, SIR HUGH (Chairman), 1955. *East Africa Royal Commission 1953–5 Report*, H.M.S.O. (Cmd. 9475).

EAST AFRICAN METEOROLOGICAL DEPARTMENT, 1961. *10 and 20 per cent Probability Maps of Annual Rainfall of East Africa*.

EVANS, A. C. 1955. A study of crop production in relation to rainfall reliability. *E. Afr. agric. J.* **20**, 263–7.

FEDERATION OF RHODESIA AND NYASALAND, 1962. *Atlas of the Federation of Rhodesia and Nyasaland*, Department of Trigonometrical and Topographical Surveys, Salisbury.

GLOVER, J. and ROBINSON, P. 1953. A simple method for assessing the reliability of rainfall. *J. agric. Sci.* **43**, 275–80.

GLOVER, J., ROBINSON, P. and HENDERSON, J. P. 1954. Provisional maps of the reliability of annual rainfall in East Africa. *Q. Jl R. met. Soc.* **80**, 602–9.

GREGORY, S. 1955. Some aspects of the variability of annual rainfall over the British Isles for the standard period 1901–30. *Q. Jl R. met. Soc.* **81**, 257–62.

—— 1968. *Statistical Methods and the Geographer*, London.

GREGORY, S. 1964. Annual, seasonal and monthly rainfall over Moçambique. In STEEL, R. W. and PROTHERO, R. M. (eds.), *Geographers and the Tropics: Liverpool Essays*, London, 81–109.

—— 1965. *Rainfall over Sierra Leone*, University of Liverpool, Department of Geography, Research Paper No. 2.

GRIFFITHS, J. F. 1960. Bioclimatology and the meteorological services. *Tropical Meteorology in Africa (Proc. Munitalp/WMO Joint Symposium)*, Nairobi, 282–300.

—— 1961. Some rainfall relationships in East Africa. *Inter-African Conference on Hydrology*, Nairobi, 115–20.

HARRISON CHURCH, R. J. 1963. *West Africa*, 4th edition, London.

JACKSON, S. P. 1961. *Climatological Atlas of Africa*, Pretoria.

KENWORTHY, J. M. 1964. Rainfall and the water resources of East Africa. In STEEL, R. W. and PROTHERO, R. M. (eds.), *Geographers and the Tropics: Liverpool Essays*, London, 111–37.

KENWORTHY, J. M. and GLOVER, J. 1958. The reliability of the main rains in Kenya. *E. Afr. agric. J.* **23**, 267–72.

KLAGES, K. H. W. 1947. *Ecological Crop Geography*, New York.

MANNING, H. L. 1949. Planting date and cotton production in the Buganda Province of Uganda. *Emp. J. exp. Agric.* **17**, 245–58.

—— 1950. Confidence limits of expected monthly rainfall. *J. agric. Sci.* **40**, 169–76.

—— 1956. The statistical assessment of rainfall probability and its application in Uganda agriculture. *Proc. R. Soc. B.* **144**, 460–80.

—— 1958. The relations between soil moisture and yield variance in cotton. *Rep. Conf. Directors, etc., of Overseas Depts. Agric. Col. Off.*, Misc. No. 531, H.M.S.O.

—— 1959. Crop-water relations of cotton at Samaru. *West African Cotton Research Conference, Samaru, Northern Nigeria, Nov. 18–23, 1957, Proceedings*, 181–92.

MCCULLOCH, J. S. G. 1961. The statistical assessment of rainfall. *Inter-African Conference on Hydrology*, Nairobi, 108–14.

THOMPSON, B. W. 1965. *The Climate of Africa*, London.

TREWARTHA, G. T. 1954. *An Introduction to Climate*, New York, 272.

UGANDA. 1962. *Atlas of Uganda*, Department of Lands and Surveys.

WALKER, H. O. 1962. Weather and Climate. In WILLS, J. H. (ed.), *Agriculture and Land Use in Ghana*, London, 7–50.

WEBSTER, C. C. and WILSON, P. N. 1966. *Agriculture in the Tropics*, London.

WORTHINGTON, E. B. 1958. *Science in the Development of Africa*, London.

5 The dry season flow characteristics of West African rivers

D. C. LEDGER

It is now widely recognized that the economic advancement of much of tropical Africa is dependent upon the development of its river systems for irrigation, water supply and other purposes. To a large extent the possibilities for such development are governed by the dry season flow characteristics of the rivers concerned, for these determine both the supply of water that can be made available at the time of greatest need without the construction of expensive reservoir facilities, and the size of reservoir required to increase that supply by any desired amount. Clearly, therefore, an understanding of these characteristics, and of the manner in which they vary from place to place, is fundamental to any appraisal of the water resource situation in different parts of Africa. This essay is offered as a contribution towards such an understanding. It outlines the manner in which the main factors affecting flow régimes in West Africa vary from one part of the area to another, and attempts to summarize what is known about the effects of these variations on the supply of water available for use during the dry season.

The hydrological characteristics of any region are determined by two major groups of factors: climatic and physiographic. The influence of climatic factors depends primarily on the amount and distribution of rainfall, and the effect of wind, temperature and humidity on evaporation. The influence of physiographic factors depends largely on the nature of the geology, the relief, the soil and the vegetation cover. In West Africa both of these sets of factors show marked similarities over broad latitudinal zones.

Climatic factors

The climate of West Africa is controlled by the seasonal migration of the intertropical convergence zone (I.T.C.Z.) separating warm dry air originating over the Sahara from warm moist air moving northwards from the Gulf of Guinea. In January the I.T.C.Z. lies at about latitude 9° north, so that all but the coastal areas are affected by hot, desiccating winds from the north-east. In August the average position of the I.T.C.Z. is about 20° north, and moist south-westerly and westerly winds affect all parts of the region. Walker (1958) has shown that the rainfall characteristics associated with this maritime air follow a distinct zonal pattern. The first zone lies immediately south of the I.T.C.Z. with a width of some 200 miles and is characterized by periodic thunderstorms separated by long spells of hot dry weather. To the south of this zone lies a second one, some 500 miles in width, which is distinguished by the frequent occurrence of well-defined disturbance lines giving short, but intense, periods of rain. This is succeeded to the south by a third zone, from 200 to 300 miles wide, the main feature of which is the occurrence of prolonged periods of low-intensity rainfall, known locally as 'monsoon' rain. Finally, to the south of this lies a fourth zone which is characterized by cool, dry, cloudy weather.

These zonal differences are clearly reflected in the moisture conditions experienced in various parts of West Africa. Places near the coast, for example, lie well south of the I.T.C.Z. for most of the year and experience a long wet season in which much of the rainfall is of the 'monsoon' type, a marked dry spell in July and August and a longer dry period from November until March. Those in the far north, on the other hand, lie under the influence of hot drying winds from the Sahara for most of the year and experience only a short wet period comprising a few intense thunderstorms in July and August.

The variation in rainfall conditions with distance inland is shown in Figure 5.1. The mean annual rainfall decreases steadily with distance inland, falling from an average of about 2,000 mm in the coastal areas of the south to about 50 mm in the semi-desert areas of the north. In general the isohyets follow the parallels, but south of latitude 11° north the pattern is complicated by the effects of relief, which give rise to the higher rainfall experienced in parts of Guinea, Nigeria and Cameroon, and by changes in the orientation of the coastline, which are partially responsible for the relatively low rainfall totals recorded in the coastal areas of Ghana, Togo and Dahomey. The seasonal rainfall pattern shows a gradual transition from the twin maxima of the coastal areas to the short, well-defined, single maximum of the drier northern parts; and it is noticeable that places of similar latitude tend to have similar seasonal patterns regardless of differences in annual rainfall amount.

The decrease in rainfall amount with distance inland is paralleled by an increase in evaporation losses. Work by Garnier (1961), Davies (1966), Roche and Dubreuil (1961) and others suggests that potential losses by evaporation and transpiration range from about 1,200 mm per year in regions of high rainfall to more than 2,500 mm in areas of very low rainfall. The results obtained by these authors also suggest that any month with less than 100 mm of rainfall is likely to be a period of moisture deficiency, and in this essay the term 'dry season' has been taken to refer to that part of the year in which mean monthly rainfall falls below this value. Defined in this

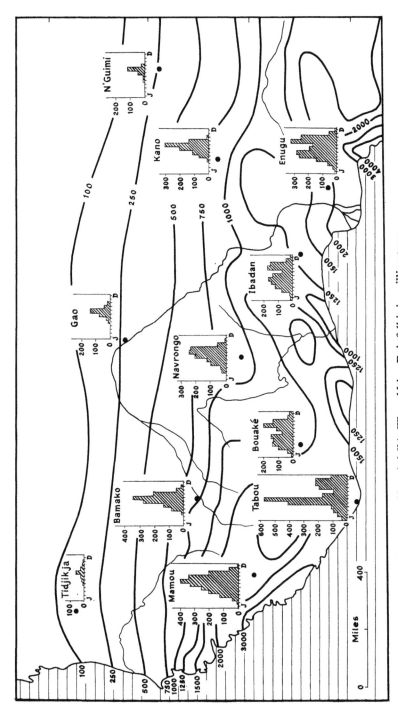

Figure 5.1 *Mean annual and mean monthly rainfall in West Africa. Rainfall is in millimetres.*

way the main dry season period extends from December until March in the
south, and from September until June in the far north. In many southern
areas August is also a moisture-deficient month, but in view of its short
duration this period has not been treated as a dry season for the purposes
of this essay.

Physiographic factors

The tendency for moisture conditions in West Africa to be similar over
broad east–west zones is repeated to some extent in factors such as geology,
relief, soil and vegetation which also have an important influence on river
behaviour.

Geologically, the area falls into two broad regions: a southern one, in
which crystalline Pre-Cambrian rocks dominate the landscape, and a
northern one, underlain largely by sandstones, shales and clays of Cretaceous and later age (Figs. 7.2, 7.3). The boundary between these regions
runs very roughly in an east–west direction at about latitude 12° north. It
should be noted, however, that younger sedimentary rocks also occur in
coastal areas and in a Y-shaped region underlying the lower Niger–Benue
valley; and that schists, quartzites and sandstones of Cambrian and Ordovician age outcrop over extensive areas north of latitude 12° in the Volta
and Upper Niger basins.

The southern region consists essentially of a series of highland blocks
rising to more than 4,000 ft (1,900 m) in the Fouta Djallon, Jos and
Cameroon Highlands, which are separated from each other, the coast and
the sedimentary areas of the north by rolling plains from 600 to 2,500 ft
(180–750 m) in altitude. These highlands form the main hydrographic
divide in West Africa, separating rivers flowing directly to the sea from
those flowing northwards for at least the first parts of their courses. In the
areas of greatest relief these rivers are torrential in character, flowing in
narrow valleys containing numerous falls and rapids. In flatter areas, however, many of the larger rivers – particularly those flowing northwards –
have developed extensive floodplains.

The northern area comprises a series of vast depositional plains separated
from each other by shallow flat watersheds. These plains are believed to
have been caused by a slight downwarping of Cretaceous age which has
since been partially filled by detritus carried by rivers flowing into the area
from the south. Because these plains are extremely flat, many of the rivers
crossing them – for example, the lower Senegal, the middle Niger, the
Bani, the Rima, the Komadugu Yobe and the Logone – are characterized
by vast swampy floodplains and anastomosing channel systems.

Although from a hydrological viewpoint the soils of West Africa are very
complex, showing great contrasts in infiltration capacities within very short

distances, the over-all soil pattern is not unlike that of geology and relief. North of latitude 12° north the soils over large areas have been derived from drift material deposited during the last arid phase. Hence they tend to have a high clay content. Farther south, sedimentary soils are predominant, and since these are frequently deeply weathered they are generally much more permeable than those typical of more northerly areas.

The distribution of vegetation in West Africa is determined primarily by the rainfall pattern, and tends, therefore, to be similar over broad latitudinal zones. In the extreme south, where there is a very short dry season, tropical rain forest is the dominant vegetation type. This is succeeded farther inland by derived and guinea savanna consisting of tall woodland and long grass savanna species with forest galleries along the main watercourses. North of the 1,000-mm isohyet the savanna becomes more open with fewer trees and a shorter grass cover. Beyond the 600-mm isohyet the vegetation ceases to be continuous and takes the form of scattered clumps of grass and thornbush which become more and more scanty as rainfall amount decreases.

Catchment size

In order to appreciate the differences of flow behaviour resulting from variations in climate, geology, relief, soil and vegetation, it is important to realize that such behaviour is also influenced by drainage basin size. This influence stems primarily from the fact that streamflow comprises two component parts: direct run-off, which enters the drainage channels during and immediately after periods of rain, and groundwater discharge, which enters such channels more gradually over a prolonged period of time. In small basins, direct run-off usually passes the outlet point within a few days of the rain from which it has been derived, and therefore plays little part in determining the dry season flow characteristics. In large basins, on the other hand, direct run-off takes much longer to pass the outlet point because of the delaying effect of channel and floodplain storage, and can, therefore, greatly influence the shape of the dry season hydrograph. For this reason it is necessary in any comparative study of streamflow conditions to distinguish between large and small rivers.

Streamflow data

The streamflow data used in this study have been taken largely from the most recent volumes of the *Annuaire Hydrologique* published by the Office de la Recherche Scientifique et Technique Outre-Mer (ORSTOM) (1949–60). Additional information, however, has been obtained from a variety of other sources, the most important being the studies of the Niger undertaken by ORSTOM (1960) and the Netherlands Engineering Consultants (1958), the records of water-levels and discharges published by the

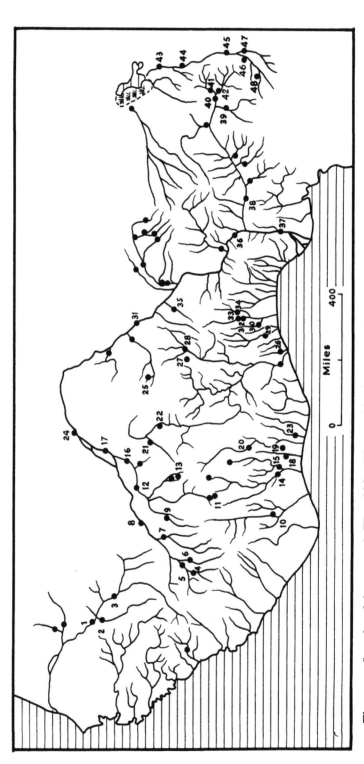

Figure 5.2 *Location of river gauges in West Africa.* The numbered gauges are those for which data is given in Tables 5.1. and 5.2. and in Figures 5.3., 5.5. and 5.6.

Federal Department of Inland Waterways, Nigeria (1955–9) and Rodier's book *Régimes Hydrologiques de l'Afrique Noire à l'Ouest du Congo* (1964). The locations of the stations for which data have been obtained are plotted in Figure 5.2, which indicates that despite large gaps – notably in Sierra Leone, Liberia, Ghana and southern Nigeria – the network is sufficiently

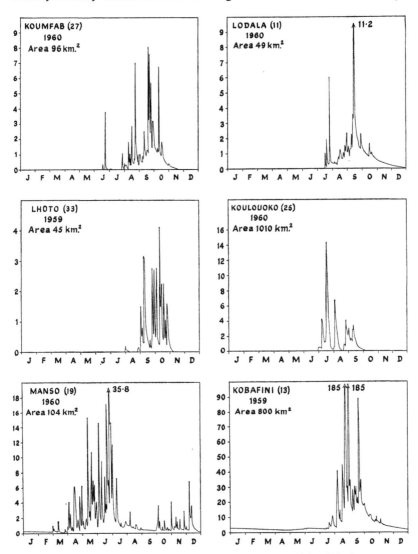

Figure 5.3 *Mean daily discharge of small rivers in West Africa.* Discharges given in cubic metres per second. The location of rivers in Figure 5.2. is shown by the number in brackets after each river.

dense to provide information for rivers of all sizes in most of the topographic and climatic regions occurring within West Africa.

Small rivers

From the time that direct run-off ceases at the end of one wet season to the
time that it begins again in the next one, the flow of small rivers in West
Africa depends upon the amount of water draining to them from groundwater storage. Over most of the area physiographic and climatic factors
combine to keep this amount to a minimum. The rocks underlying most
catchments are impermeable, the soils have low infiltration capacities and
evaporation losses during dry periods are high. Hence the flow of small
rivers falls away very rapidly after periods of rain, and, as is shown by the
hydrographs for the Koumfab, Lhoto, Koulouoko and Lodala, frequently
ceases altogether during the first month of the dry season (Fig. 5.3). The
hydrographs for the Manso and the Kobafini indicate, however, that some
areas at least are capable of sustaining permanent flow throughout the dry
season. These areas occur most commonly in the southern part of West
Africa where the impermeability of the underlying crystalline rock is often
offset by the presence of a deep mantle of weathered material possessing
an appreciable groundwater storage capacity. In regions with a long wet
season, a large water surplus and relatively low evaporation rates this
storage capacity is often sufficient to maintain a dry season flow of the order
of from one to two litres per second per square kilometre. In drier areas,
however, permanent flow occurs only occasionally, the most notable examples being in the vicinities of Bamako and Bandiagara where the drain
age channels are incised into deeply weathered Cambrian and Ordovician
sandstones.

The areas in which groundwater discharge to small rivers usually occurs
throughout the dry season are shown in Figure 5.4. They include Sierra
Leone, most of Guinea, Liberia, southern and western Ivory Coast, south-
western Ghana, the southern half of Nigeria and central and southern
Cameroon. It is noticeable that they do not include the central and eastern
parts of the Ivory Coast, south-eastern Ghana, southern Togo and southern
Dahomey, where, despite a long wet season, the water surplus is generally
insufficient to produce significant groundwater recharge. In these, and
other, parts of West Africa streams are normally dry for most of the dry
season, the period without flow ranging from about two months in the
extreme south to more than nine months in the far north.

Large rivers

The extent to which the dry season hydrographs of large rivers in West
Africa reflect those of the smaller rivers flowing into them depends on the

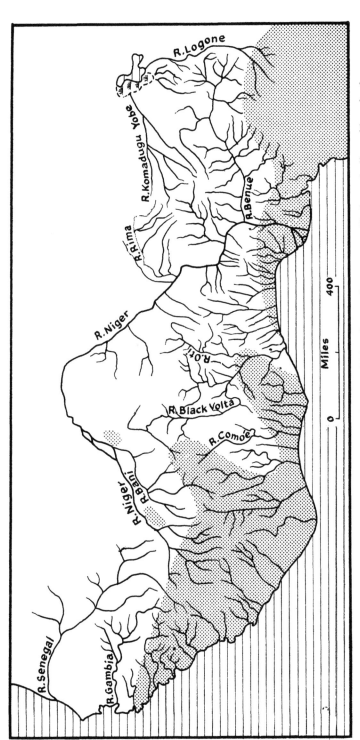

Figure 5.4 *Areas of permanent groundwater discharge in West Africa. Areas of permanent discharge indicated by stipple.*

TABLE 5.1 Flow data for large rivers not possessing extensive floodplains

River	Station	Number of station in Figure 2	Catchment area in km²	First month if dry season	Mean monthly discharge in cubic metres per second during stated month after the start of the dry season												Mean annual flow in cubic metres/ second	Percentage of total flow occurring during dry season
					1	2	3	4	5	6	7	8	9	10	11	12		
Bia	Ayamé	23	9,320	December	35	11·1	9·7	23	47	78	192	151	37	64	152	105	76	15
Sassandra	Guessabo	10	35,400	November	379	163	83	54	54	73	81	148	309	571	1,101	835	322	21
Bandama	Brimbo	14	60,300	November	416	152	52	28	20	38	57	118	232	499	1,095	1,149	322	19
N'zi	Ziénoa	15	33,150	November	141	28	9·6	4·2	3·6	16·0	34	93	126	95	185	322	89	19
Comoé	Anaissué	20	70,200	November	311	103	32	14·8	14·7	24	39	67	196	409	982	924	261	18
Agnéby	Agboville	18	4,660	November	7·6	2·6	0·6	0·3	0·6	3·2	10·7	28	30	3·6	5·4	13·1	9	15
Mono	Tététou	29	20,500	October	277	54	11·0	3·0	1·3	2·8	5·1	9·9	51	164	215	343	95	21
Ouémé	Pont de Savé	32	23,600	October	464	117	11·7	2·6	0·2	0·4	0·3	1·2	32	142	405	691	156	19
Zou	Atcherigbé	30	6,950	October	81	13·8	0·5	0·02	0·03	0·3	1·2	7·7	21·7	54	43	75	25	15
Okpara	Kaboua	34	7,600	October	176	47	6·4	1·3	0·3	0·2	0·1	0·7	9·4	34	81	154	41	16
Falémé	Kidira	2	28,180	October	435	127	44	19·0	11·3	4·7	1·9	18·0	109	642	892	193	28
Senegal	Galougo	3	127,000	October	1,495	575	246	131	80	42	20	15	106	539	2,161	2,585	669	30
Senegal	Bakel	1	232,700	October	1,710	560	230	129	77	46	22	11	122	569	2,351	3,429	774	30
Niandam	Baro	4	12,600	October	621	353	145	79	44	38	31	48	145	353	524	816	261	30
Milo	Kankan	6	9,900	October	485	231	106	55	34	27	27	47	104	273	491	672	216	30
Sankarani	Gouala	7	35,300	October	1,235	535	221	117	68	45	37	47	114	316	815	1,433	417	31
Niger	Kouroussa	5	18,000	October	710	379	181	104	60	31	20	25	104	244	426	840	241	29
Niger	Koulikoro	8	120,000	October	4,564	2,089	872	407	197	102	69	98	368	1,250	3,204	5,292	1,549	31
Baoulé	Bougouni	9	15,700	October	467	182	72	32	19·8	7·7	2·4	1·8	5·3	41	346	543	144	31
Pendjari	Porga	28	22,276	October	336	45	7·8	3·6	1·8	0·9	0·6	3·0	11·4	23	160	419	85	24
Alibori	Kandi-Banikoara	35	8,165	October	93	4·3	1·0	0·4	0·2	0·03	0·04	7·4	8·7	26	125	235	42	20
Mayo Kebbi	Cossi	41	26,000	October	147	93	59	18·6	7·4	2·3	0·6	13·8	54	128	232	327	91	25
Faro	Safai	39	23,500	October	978	258	82	35	20	18·0	26	99	290	578	798	1,224	369	23
Benue	Riao	42	31,000	October	689	90	24	11·5	5·0	2·0	0·4	5·6	40	221	745	1,423	272	24
Benue	Garoua	40	64,000	October	857	173	62	26	12·0	4·8	2·3	177	79	329	1,083	1,909	381	24
Benue	Makurdi	30	305,000	November	3,575	1,055	700	500	300	280	715	1,855	3,860	5,440	10,420	10,550	3,150	24
M'Béré	M'Béré	48	7,100	November	97	56	36	25	20	24	46	89	138	215	303	264	110	28
Pendé	Doba	47	15,600	October	467	148	47	28	13·6	8·4	7·8	12·0	22	101	317	565	145	30
Logone	Moundou	46	34,900	October	1,060	385	162	110	83	61	54	99	156	324	998	1,451	412	30

channel and floodplain storage capacity of the basins concerned. In part, as was explained earlier, this capacity is a function of basin size. In part, however, it is clearly also a function of channel and floodplain morphology; and in this respect West African rivers fall into two distinct categories: those of the northern plains which have vast floodplains, and those of the areas farther south which do not.

The main features of the dry season flow régimes of large rivers in the southern part of West Africa are illustrated by the mean monthly flow data presented in Figure 5.5 and Table 5.1. It is clear that in general terms these rivers behave in a manner similar to that of the smaller rivers flowing into them, for their discharges fall rapidly to a very low level once the wet season has ended. The data for the Bia, Sassandra, Upper Niger (Kouroussa), Niandam, Milo, Sankarani and M'Béré, for example, indicate that base flows exceeding one litre per second per square kilometre can normally be expected from catchments lying entirely within areas of high rainfall and permeable subsoil. Those for the Ouémé, Zou, Okpara, Pendjari, Baoulé, Mayo Kebbi and Upper Benue (Riao), on the other hand, demonstrate that little, if any, base flow can be expected in rivers whose basins lie outside such areas, although it should be noted that where these rivers have wide, sandy beds – for example in Northern Nigeria – there is often a significant sub-surface flow.

But while their over-all flow patterns during the dry season are similar, it is apparent that large rivers take appreciably longer than small ones to recede from high wet season to low dry season flow. The period of such recession varies in length from between two and three months. This means that the period of really low flow begins later and lasts for a shorter time than in small rivers. It also means that a considerable part of the total flow from large basins passes the outlet point during the dry season. As Table 5.1 shows, this proportion varies from about 15 to 20 per cent for rivers flowing southwards towards areas with increasingly short dry seasons, to about 30 per cent for those flowing northwards towards areas with progressively longer dry seasons. Rivers, such as the Benue, which flow in a more or less east–west direction have values lying approximately half-way between these two extremes.

These differences between large and small rivers are slight, however, compared to those arising in the northern part of West Africa where, because of the feeble gradients, many rivers are able to carry only a small proportion of the water flowing into them from headwater areas. As a result, most of this water overflows into the surrounding plains, transforming them into huge, shallow lakes ranging in area from a few hundred square kilometres in smaller basins, such as those of the Black Volta above Kouri and the Rima, to as much as 25,000 sq. km in the inland delta zone of the

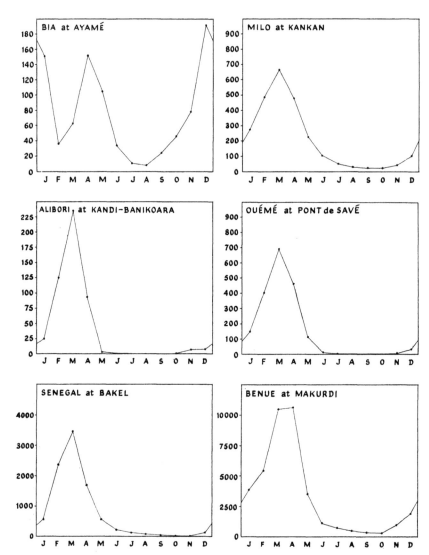

Figure 5.5 *Mean monthly discharge for large West African rivers not possessing extensive floodplains.* Discharges given in cubic metres per second.

Niger basin. Later, when the inflow rate begins to fall, much of the water stored in these plains is released back into the main channels, thereby sustaining flow at points downstream.

This storage process has several important effects on the dry season flow régimes of the rivers affected by it. Firstly, it flattens and elongates the flood wave so that the length of the period of low base flow diminishes with distance downstream. Secondly, it delays the rate at which the flood peak moves downstream so that the early part of the dry season is a period of rising, rather than falling, discharge. Thirdly, it increases the amount of base flow so that the minimum flow in these rivers exceeds the minimum flow into them from headwater areas.

As Figure 5.6 and Table 5.2 indicate, these effects have their greatest impact in the Niger basin with the result that the Niger has an extremely

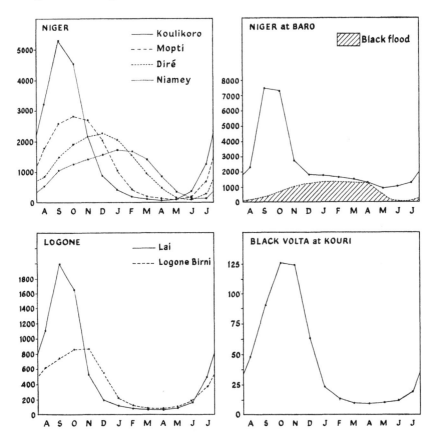

Figure 5.6 *Mean monthly discharge for large West African rivers possessing extensive floodplains.* Discharges given in cubic metres per second.

TABLE 5.2 Flow data for large rivers possessing extensive floodplains

River	Station	Number of station in Figure 2	Catchment area in km²	First month if dry season	Mean monthly discharge in cubic metres per second during stated month after the start of the dry season												Mean annual flow in cubic metres/ second	Percentage of total flow occurring during dry season
					1	2	3	4	5	6	7	8	9	10	11	12		
Black Volta	Kouri	21	20,000	October	125	122	63	23	12·9	9·6	8·9	9·4	11·8	18·3	47	90	45	50
Black Volta	Boromo	22	58,000	October	93	75	64	40	24	13·2	10·1	11·4	19·7	34	79	113	48	51
Bani	Douna	12	101,600	September	2,535	2,546	1,261	433	177	105	69	43	31	45	187	1,250	726	55
Bani	Sofara	16	129,400	September	1,326	1,538	1,405	938	405	177	100	59	38	50	174	828	586	62
Niger	Mopti	17	281,600	September	2,585	2,814	2,687	2,027	1,031	416	190	109	82	170	689	1,774	1,219	65
Niger	Diré	24	330,000	September	1,498	1,899	2,161	2,299	2,057	1,556	979	450	152	87	282	866	1,189	85
Niger	Niamey	31	Unknown	September	1,067	1,254	1,412	1,588	1,736	1,725	1,419	868	347	130	124	516	1,011	90
Niger	Baro	36	Unknown	November	2,710	1,800	1,710	1,600	1,485	1,280	900	1,000	1,280	2,285	7,420	7,300	2,525	35
Niger	Onitsha	37	Unknown	November	6,300	2,860	2,140	1,850	1,715	1,650	1,600	2,280	4,550	8,425	17,720	20,000	7,000	22
Logone	Lai	45	60,320	October	1,647	566	197	120	87	65	63	91	164	487	1,102	1,992	551	42
Logone	Bongor	44	73,700	October	1,750	741	263	145	88	62	55	79	155	443	1,036	1,713	547	49
Logone	Logone Birni	43	76,000	October	868	878	552	217	127	85	69	99	178	383	604	756	403	58

complicated dry season flow régime. The most important changes in the flow pattern occur during the river's passage through the inland delta region, where the storage is sufficient to lengthen the base of the flood wave by two months and to delay the occurrence of peak flow by three months. Thus at Diré – where the wet season ends a month earlier than it does at the upper end of the inland delta zone – the discharge continues to rise during the first four months of the dry season, and the low flow period lasts for only three to four months. As a result of these changes about 85 per cent of the total flow at Diré occurs during the dry season. It should be noted, however, that the improved distribution of flow in time is achieved only at the expense of enormous water losses by evaporation and seepage, the mean flow out of the delta zone being only 50 per cent of that discharged into it by the Upper Niger and the Bani.

After Diré, the river begins to flow in a south-easterly direction towards progressively wetter areas. Between Diré and Niamey this change of course has little effect on the river's régime because no important tributaries join the Niger in this stretch. Thus at Niamey the dry season flow characteristics are similar to those at Diré, the only important differences being that the flood peak occurs approximately one month later, the period of low flow is a few weeks shorter and the mean flow at the end of the dry season is somewhat higher.

From Niamey onwards, however, the river begins to receive tributaries such as the Alibori whose main period of flow is from June until October. As a result, the wet season again becomes the period of highest flow, and the dry season becomes a period of falling discharge, until at Baro the régime has the same over-all pattern as that of the Upper Niger at Koulikoro. But because the flood from the upper basin – known locally as the 'Black Flood' – arrives in Nigeria during the latter part of the dry season, the flow from the end of December onwards is very much higher at Baro than Koulikoro. Between January and March, for example, the flow at Baro remains at a more or less constant level of about 1,700 cu m per second, and although the discharge falls off from April onwards, the recession from the peak of the 'Black Flood' has hardly started before the first floods of the wet season arrive from local tributaries. After Baro, the régime changes very little, except that the minimum discharge continues to increase for the reasons outlined above, until at Onitsha the flow exceeds 1,600 cu m per second throughout the dry season.

But while the effects of floodplain storage are most pronounced along the Niger, they are by no means negligible along other northern rivers possessing large floodplain zones. The Senegal, Bani, Black Volta, Rima, Komadugu Yobe and Logone, for example, all possess floodplains large enough to lengthen the period of high water by from one to two months, to delay

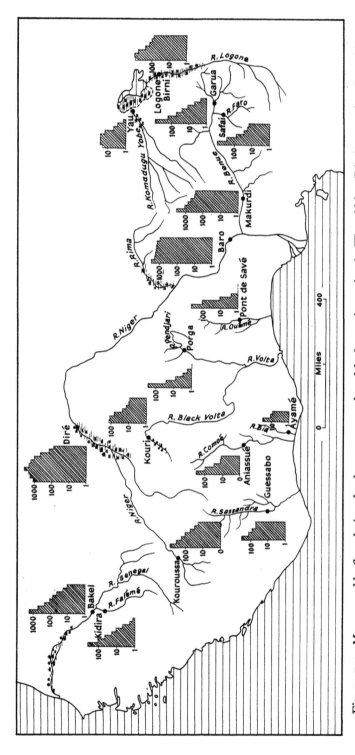

Figure 5·7 *Mean monthly flow during the dry season at selected hydrometric stations in West Africa. Discharges given in cubic metres per second.*

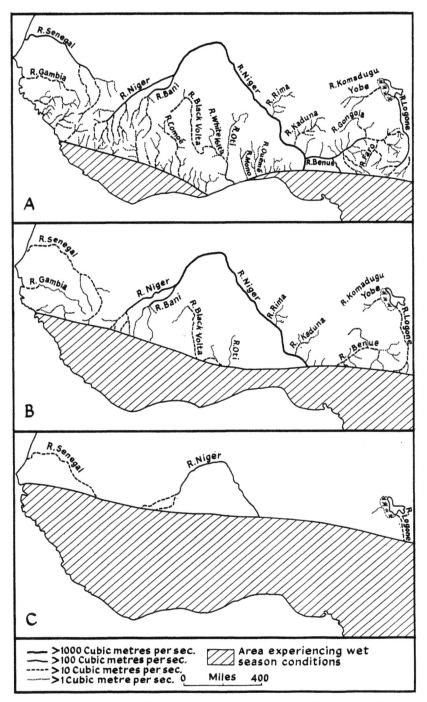

Figure 5.8 *Discharge of West African rivers during the fourth (A), sixth (B) and eighth (C) months of the dry season.* Discharges in cubic metres per second have been estimated by extrapolation of data for all stations shown in Figure 5.2.

the passage of peak flow by a similar amount and to effect a significant increase in the volume of base flow. As a result, in all of the basins for which data are available, over 50 per cent of the total flow at downstream stations occurs during the dry season. As in the case of the Niger, however, these changes in the flow régime are achieved only at the cost of large water losses by evaporation and seepage. Between Lai and Logone Birni, for example, the mean flow of the Logone diminishes from 551 to 401 cu m per second, and while data are not available for other rivers, it is believed that most of them suffer proportionately similar losses during their passage through the floodplain zone.

Dry season water availability

It is apparent from the preceding discussion that most rivers in West Africa contain considerable quantities of water during the first one or two months of the dry season regardless of their régime characteristics. In areas where the dry season is longer than three months, however, the régime differences outlined earlier assume great importance, for, as Figures 5.7 and Tables 5.1 and 5.2 indicate, significant flow beyond the third dry season month occurs only in those rivers whose régimes are substantially affected by groundwater and floodplain storage. The pattern of water availability resulting from these differences is shown in Figure 5.8.

Map A presents the flow situation experienced during the fourth month of the dry season. It shows clearly that even after a relatively short dry season period the mean flow of rivers over large parts of West Africa is already below 1 cu m per second. The network of rivers whose flow exceeds this amount is still relatively dense in southern areas, but in northern areas flows over 1 cu m per second occur only in the Senegal, the Gambia, the Niger, the Bani, the Black Volta, the Rima, the Komadugu Yobe and the Logone. Maps B and C show that the pattern of water availability during the sixth and eighth months of the dry season in those areas experiencing dry seasons of these lengths is essentially the same as it was during the fourth month, although, of course, flows are very much lower in most cases. It is noticeable, however, that the flow of a number of tributary rivers in the Senegal, Volta, Rima, Kaduna, Benue and Komadugu Yobe basins falls below 1 cu m per second between the fourth and sixth dry season months.

The most striking feature of the dry season water availability pattern shown in Figure 5.8 is the immense importance of the Niger compared to other West African rivers. In its upper reaches the river receives a well-sustained base flow from headwaters in the Fouta Djallon Highlands, and by the time it reaches Koulikoro the Niger already possesses a very favourable dry season régime, the mean flows at this station during the fourth, sixth and last months of the dry season being 407, 102 and 69 cu m per second respectively. These flows, however, are small compared to those experienced farther downstream as a result of the river's passage through the Inland Delta zone. At Diré, for example, the mean monthly flow exceeds 1,000 cu m per second for the first six months of the dry season, is still as high as 450 cu m per second during the eighth month, and falls below 100 cu m per second only in the tenth, and last, dry season month. Conditions become still more favourable in the lower reaches, and by Baro the mean flow of even the lowest month exceeds 900 cu m per second.

Although their flows are far smaller than those experienced at points along the middle and lower reaches of the Niger, several other rivers also contain large quantities of water throughout the dry season period. The most prolific of these is the lower Benue, which experiences a mean end of dry season flow in excess of 250 cu m per second over a considerable distance. But the Logone and the lower Bani are also characterized by fairly high end of dry season flows, that of the former being about 55, and that of the latter about 30, cu m per second. It is interesting to note that during the earlier part of the dry season the Bani at Douna exhibits characteristics almost identical to those of the Niger at Koulikoro, the mean flows during the fourth and sixth months being 433 and 105, as compared with 407 and 102, cu m per second.

The only other rivers worth noting are the Senegal, the Black Volta, the lower Kaduna and the lower Komadugu Yobe, all of which experience mean monthly flows of 10, or more, cu m per second during at least the first six months of the dry season in the areas through which they flow.

The rivers experiencing appreciable mean monthly flows after the third month of the dry season, however, are few in number, and the most important point emerging from this study is that most places in West Africa are far removed from significant supplies of surface water for a considerable part of the dry season. Clearly, therefore, the possibilities of providing even modest supplies of river water throughout the dry season depend, over very large areas, upon the feasibility of providing adequate reservoir facilities at an economic cost. That this can be done for hydro-electric schemes has been proved in a number of cases. That it can be done for irrigation and other purposes related to agricultural development is still a matter for considerable doubt.

References

DAVIES, J. A. 1966. The assessment of evapotranspiration for Nigeria. *Geogr. Annl.* **48**, A.3, 139–56.
GARNIER, B. J. 1961. Maps of the water balance of West Africa. *Bull. Inst. fr. Afr. noire* **22**, 3, 709–22.

NETHERLANDS ENGINEERING CONSULTANTS. 1958. *River Studies of the Niger and the Benue and Recommendations for their Improvement*, The Hague.

NIGERIA. 1955–9. *Records of Waterlevels and Discharges*, Federal Dep. Inland Waterways, Lokoja.

OFFICE DE LA RECHERCHE SCIENTIFIQUE ET TECHNIQUE OUTRE-MER. 1949–60. *Annuaire Hydrologique*, Paris.

1960. *Monographie du Niger*, Paris.

ROCHE, M. and DUBREUIL, P. 1961. Resultats obtenus sur les bacs Colorado de l'Afrique de l'Ouest d'Expression Française. *Inter-African Conf. Hydrol. Commission de Co-operation technique en Afrique au sud du Sahara, Publ. 66*, 139–43.

RODIER, J. 1964. *Régimes Hydrologiques de l'Afrique Noire à l'Ouest du Congo*, Paris.

WALKER, H. O. 1958. *The Monsoon in West Africa*, Ghana Met. Dep., Dept. Note 9, Accra.

6 Geomorphology and land classification in tropical Africa

M. F. THOMAS

Any geographical study of landscape must include a rational division of the land surface upon which other distributions have been fashioned. Similarly, plans for regional development should include geographical analyses of those elements of the total landscape complex that are relevant to the directions which development is intended to follow. For any particular project there may be a single direction, such as the development of agriculture, forestry or rangeland, of communications or industry. But over a period of time, and especially within large areas, development must usually be multi-directional, involving new crops, new roads and settlements, and perhaps new industrial activity. While individual projects may be preceded by specialized surveys on a detailed scale not generally practicable for a whole territory or region scheduled for development, the information gained may be of limited value to the understanding of wider areas or of other aspects of the environment. Thus, where comprehensive development is envisaged for large regions (of several thousand square miles), detailed studies are usually preceded by reconnaissance surveys which rely heavily upon the interpretation of medium-scale (1:20,000–1:70,000) aerial photography. It is implicit from what has been stated above that such surveys should be relevant to possible plans for development in various sectors of the economy, in any portion of the territory designated (which might be an entire country).

It can be argued that for large areas of tropical Africa the principal direction of improvement is, or should be, in agriculture (including pastoralism and forest conservation) and that regional surveys should therefore be conceived and organized with this in mind. However, agricultural development frequently depends upon improvements in communications and water supplies, and upon a knowledge of, if not changes in, social organization and traditional agricultural practice.

It can therefore be argued that no single type of reconnaissance survey will produce adequate information upon which to base planning proposals. Land capability assessment may be the most important guide for agricultural development in tropical Africa, but it cannot be used alone for the implementation of complex regional development plans. Prediction of land capability has generally been based, at the reconnaissance level, upon small-scale (1:500,000) mapping of land systems or land types conceived as ecological units of the landscape combining common features of land form, soil and vegetation. Valuable though these surveys may prove to be, it will be argued here that they do not necessarily represent the most appropriate approach to African environmental conditions and development problems. However, the importance of land systems surveys is acknowledged in two different ways: first, in that much of what follows has been stimulated by the concepts and methods used in them, and second, because one product of the mapping system suggested below would in fact be a map of similar kind and scale.

The argument which follows is in fact based upon two premises: first, that surveys of the natural environment provide only a part of the basis upon which over-all policy can be formulated, and second, that they should be of a scale appropriate to development on the ground as well as to planning in the office. While acknowledging the need for reconnaissance (and also detailed) surveys of existing social and land-use patterns in an area, this study is concerned with reconnaissance surveys of the natural environment. More specifically it is the purpose of this discussion to inquire into the place of geomorphology in development planning in tropical African environments, and to consider in some detail the problems, both theoretical and practical, encountered in the production of land-form maps and maps embodying more complex units of the landscape.

It is now widely accepted that some form of land classification is required before soil or other detailed surveys are attempted in an area, if only to indicate those smaller and more manageable regions which would repay more detailed study. The best known and most widely practised of such survey methods are those embodied in the reports of the Division of Land Research and Regional Survey of the Commonwealth Scientific and Industrial Research Organization of Australia (C.S.I.R.O.). These are concerned with conditions in Australia and New Guinea, where parallels with parts of tropical Africa are striking. The Land Resources Division of the Directorate of Overseas Surveys (DOS) is now actively applying similar methods in tropical Africa. The Natural Resource Survey of Malawi undertaken by Young (1965, and Chapter 13) is also similar in purpose.

Working quite independently Beckett and Webster (1965) of the Department of Agriculture at Oxford developed a system of terrain classification in conjunction with the Military Engineering Experimental Establishment (MEXE). Since then the interested groups in the United Kingdom, Australia and South Africa met to form a common terminology of Land Classification (Brink et al., 1966). The principles underlying these schemes embody geomorphological and ecological concepts which are discussed below. However, the reports of the C.S.I.R.O. and the DOS include maps only of small scale (1:250,000–1:500,000) indicating land systems which are not specifically geomorphological units.

Geomorphological mapping *per se* has not been applied to development surveys, and most of the work has been achieved for small areas by individual geomorphologists. Systematic mapping has been attempted in Senegal by J. Tricart and his colleagues from the Centre du Géographie Appliquée, Strasbourg (Tricart, 1961), and in Uganda by J. Doornkamp (1969) in conjunction with the Geological Survey. An important difference between land-system surveys and geomorphological maps is one of scale. The latter are generally plotted on scales of 1:50,000 or 1:100,000 and final production at a scale less than 1:250,000 would not normally be envisaged.

The smaller scale of the land-system surveys, and the complex land units developed for them, have been designed to economize on time and expenditure and to produce a readily understandable document for planning purposes. The C.S.I.R.O. reports have normally embodied chapters on geomorphology, but those produced by the DOS for parts of Africa have not so far attempted this (Bawden and Tuley, 1966).

Before discussing the schemes mentioned above, and the use of geomorphological units for reconnaissance mapping, certain general considerations regarding environmental surveys are pertinent.

Soil mapping and reconnaissance surveys

It can be argued that the soil is the element of the land surface which most directly affects plant growth, and that vegetation communities (with which the soils are intimately interrelated) are also more directly concerned with agricultural development than is the actual form of the land. With this few geomorphologists would disagree, but it does not, however, follow that the soil type is necessarily the most appropriate unit of the land surface that can be used in regional surveys. The detailed mapping of soils proceeds by the process of digging profile pits, followed by field and laboratory analysis. The results obtained reveal a great deal about the soils sampled, but cannot *of themselves* be used to predict the extent of particular soil series or the patterns of different occurrences of the same soil. On the other hand, the reconnaissance mapping of soils from aerial photographs proceeds more from known or assumed relationships of soils with relief, vegetation communities and drainage conditions than from the ability to recognize soil types directly from the photographic prints.

For rather different reasons the ecologist is also handicapped in mapping plant communities from aerial photographs. Vegetation throughout most of tropical Africa has undergone considerable modification as a result of pastoral and agricultural land use, and in areas even with quite low densities of population the mapping of 'natural' vegetation is impossible. The alternative of mapping existing vegetation/land-use associations poses problems that are only now receiving detailed attention (see Chapter 8). Such studies may reveal much about the ecology of agricultural land use, but it is questionable whether they can be classed as studies of the natural environment. For although this environment is continually being altered in significant ways by man's own activities and those of his domestic animals, certain aspects are but little affected over periods of time measurable in tens or hundreds of years. These are principally the macro-climate and the rocks and landforms of the earth's surface. Of these 'conservative' elements of the environment the land form is most obvious on air photographs and most readily understood from their interpretation.

For these reasons, and for others, geomorphology has provided the basis for land-system mapping in Australia and for soil mapping in such countries as Uganda (Chenery, 1960) and Northern Nigeria (see Chapter 7). For not only is land form readily interpreted from aerial photographs, but the pattern of landforms[1] can be used to predict spatial patterns of soils and to explain in part the patterns of vegetation. The relationsip between soils and topography has been too often debated to require much elaboration here. It may be emphasized that the use of the *soil catena* as a primary mapping unit following the work of Milne (1935) has now become widespread throughout Africa. Often the term *association* is used to designate areas within which a particular catena is present, but this term has also been used, as in Western Nigeria (Smyth and Montgomery, 1965), for soils grouped on the basis of a common parent material. As defined by Ellis (1932) and used by Clarke (4th edition, 1957) the *soil association* is a group of 'topographically related soils on one geological parent material', so that it correctly refers to areas within which both parent material and relief have uniform characteristics. The emphasis placed on solid geology by the Soil Survey of Western Nigeria may be contrasted with the geomorphological approach to soil mapping adopted in Uganda and Northern Nigeria. This observation will be taken up again later in the discussion.

Whatever the basis of preliminary mapping from aerial photographs, final maps are most reliably based upon detailed field analyses of smaller, sample areas within each land system or class. In these detailed surveys areal units are generally based upon one of several disciplines, but principally upon geology, pedology, geomorphology or ecology. It is, in fact, at

---

[1] Following Savigear (1965), p. 514 a *landform* is regarded as a feature of the earth's surface with distinctive form characteristics which can be attributed to the operation of particular processes on earth materials of particular lithological and structural character. The term land form refers only to the form of the land.

this level of detail that the fundamental principles of land division must be worked out, and not at the level of reconnaissance mapping from aerial photography.

Such detailed studies result in the designation of small, fairly homogeneous units of the natural environment, and although they can seldom be mapped at scales less than 1:25,000 they do none the less underlie most methods of mapping and classification of the features of the land surface, and it is at this level that fundamental differences in approach between the various field scientists become most sharply defined. A wide variety of such units have been defined for both theoretical and practical purposes.

Although these principles form the logical basis of much work in this field, it must be realized that initial division of an area into land systems or classes may be on the basis of pattern recognition, and this may be regarded as a form of classification which leads to the recognition of 'polythetic' units in the sense of Sokal and Sneath (1963). In such units no single feature may be regarded as diagnostic, but the sharing of a number of common characteristics defines the class. The recognition of such photographic patterns contributes to the concept of the land system as used by the C.S.I.R.O. The significance of such patterns is difficult to define and describe, because of the large number of features, both natural and man-modified, which comprise them. Where population density is significant and land use has modified the natural vegetation, it is doubtful whether such patterns can be accepted as representing in any readily understandable way, the environmental qualities of *land*.

Fundamental units of the natural environment[1]

One of the best known of such areal units is the concept of the *site* introduced by Bourne (1931) as 'an area which throughout its extent has similar local environmental conditions' (p. 16). Clarke (1957) used the concept of the *soil-site* and suggested that 'a unit of land suitable for a single system of utilization may be called a "site" ' (p. 13). He also made the point that 'if all soil forming factors or site characteristics are known, then the soil profile characteristics may be foretold' (p. 13). Such site characteristics include properties of climate, parent material, vegetation, morphology and, significantly for this discussion, age. Moss (1968) has suggested the term *habitat-site* as 'a unit of the land surface with a characteristic slope form which constitutes a distinct habitat for plant growth' (p. 53). All these ideas regarding 'site' are *environmental*: that is to say they concern those properties of a uniform area of land that affect soil development and plant growth.

Such ideas are close to but not identical with *ecological* concepts such as

[1] See Table 6.1 and Appendix A.

the *ecotope* (Troll, 1963) and *tessera* (Jenny, 1958). Of these the *tessera* has been most clearly defined as 'a three dimensional element of the landscape' (p. 5). Thus 'the "thickness" of a tessera is given by the height of the vegetation plus the depth of the soil' (p. 5). It should be added that this is a sampling unit which may be defined areally in an arbitrary manner. Thus while a site is a unit of the *land surface*, an ecotope or tessera is a unit of the *ecosystem*, and is therefore an element of the total landscape with all its complexity and detailed variation, both areally and in profile. It is probably the inherent complexity of such concepts that accounts for the absence of any carefully formulated hierarchy of areal units among the propositions of their authors. Bourne (1931) discussed the *region* but without rigorous definition. The greater the number of variable elements within the environmental unit the more truly unique is each occurrence, and the more difficult it becomes to combine smaller units into larger or to define the limits of variation within individual occurrences. Although polythetic units may be recognized from aerial photographic patterns or from detailed field analysis, it is likely that one element of the landscape pattern will be regarded as a parameter for the delimitation of areal units of the total environmental complex.

The is essentially the approach of the land-system surveys of the C.S.I.R.O. which, while incorporating pattern recognition in their environmental divisions, finally base their mapping upon geomorphological principles. The land systems so defined are regarded as having identity with soil and vegetation regions. The *land system* is a unit of higher order than those considered above, is mapped on scales of 1:250,000 or less, and is defined as 'an area or group of areas, throughout which there is a recurring pattern of topography, soils and vegetation' (Christian and Stewart, 1953, p. 76). The geomorphological basis of this scheme is emphasized by Christian (1957, p. 76): 'A simple land system is a group of closely related topographic units, usually small in number, that have arisen as the products of a common geomorphological phenomenon. The topographic units thus constitute a geographically associated series and are directly and consequentially related to one another.' A *complex* land system is defined as 'a group of related simple land systems' and along with the *compound* land system (which is an arbitrary group for mapping purposes) forms the basis of most of the maps produced by the Division of Land Research and Regional Survey of the C.S.I.R.O.

The recurrent pattern which comprises the land system is made up of individual *land units*, defined by Christian (1957, p. 76) as parts of the land surface 'having a similar genesis and [which] can be described similarly in terms of the major inherent features of consequence to land use – namely topography, soils, vegetation and climate'. It is made clear that the complexity of such units is variable, depending upon the nature of inquiry, and may vary from a whole mountain to a unit comparable with Bourne's *site*. On the other hand, Mabbutt and Stewart (1963, p. 102) define land units simply as 'slope segments of characteristic form, declivity and position'. They are therefore morphological units. In fact it is clear from Mabbutt and Stewart that land-system mapping proceeds from the premise that geomorphology provides the basic survey units and that these are found to correspond closely with soil and vegetation boundaries of significance to land use. However, *land units* are not mapped as such by the survey teams and it may be questioned whether, at this scale, the distributions of the various elements of the environment are coincidental. At the level of land systems it appears from the work of the C.S.I.R.O. that these patterns do vary sympathetically.

The units discussed here are environmental or ecological in concept. Fundamental units of the earth's relief are *morphological*, and although these are considered further below, they should also be included here. Both Wooldridge (1932) and Linton (1951) have referred to *flats* and *slopes* as the fundamental units of relief. These are small, uniform *slope units* (see Chapter 13) bounded by breaks or inflexions of slope. Linton has used the term *site* in this sense and the *morphological unit* of Savigear (1965) is conceived as a fundamental unit of relief.

Two separate issues emerge from this discussion. The first of these concerns the scale of mapping. Fundamental units of the land surface require a mapping scale of at least 1:25,000 which is quite impracticable for reconnaissance purposes; on the other hand, land-system mapping on a scale of 1:500,000 does not permit the use of the maps in the field or the representation of actual features of the land surface on the maps. The second issue concerns the conceptual basis of mapping programmes, and in particular the problem of defining and mapping units of the environment suited to the requirements of reconnaissance surveys.

The question of scale

The argument for small-scale surveys of large areas concerns the availability of staff and of money, and the urgency of the proposed planning. In tropical Africa qualified staff are few, money may be limited and many land-use problems are of great urgency. But it is implicit in land systems surveying that more intensive surveys will follow, as development is planned for the more favourable areas. Such intensive work requires a much larger mapping scale, and may quite possibly involve the use of the same air photography as that analysed for the preliminary surveys. This also presupposes that large areas may be neglected because of poor soil conditions or other environmental hazards. But, although this may be justified in sparsely peopled areas, it cannot be within areas of significant population density (whatever this may be). For many countries in Africa land-use planning is as much a problem of the densely settled areas as it is of the sparsely peopled regions. The situation in Australia, where the opening up of more or less empty areas is contemplated, finds parallels only over limited areas of tropical Africa. Furthermore, the progress of topographic mapping on medium scales – generally 1:50,000 – has been so rapid over much of tropical Africa during the last ten years that it is no longer necessary to formulate reconnaissance surveys on the basis that the territory is largely unmapped. Of course large areas remain to be mapped at scales larger than 1:250,000, but within the next ten years we may expect these areas to have been reduced substantially.

It is therefore arguable that, although small-scale maps should be produced to facilitate over-all planning, more detailed maps presented at scales of from 1:50,000–1:250,000 should form the basic documents of the survey. These would have been mapped initially on the scale of the photography or of the available topographic maps – probably 1:40,000 or 1:50,000. Such maps could show land units or some other basic division of the land surface, and they would provide suitable base maps for later specialist work. Although they would take longer to produce, they could be made with the same resources and manpower as the smaller-scale maps.

Subsequent surveys, however, would have not just a reconnaissance map as a guide for intensive work, but a sure foundation of maps suitable in scale for planning on the ground, and it will be argued subsequently that geomorphic mapping provides the best basis for these.

The choice of mapping unit

This, therefore, impinges on the second issue: the nature of land or land-form units to be mapped on this larger scale. Mapping on scales greater than 1:50,000 would be impracticable for several reasons: air photo cover is usually at scales approximating to 1:40,000, topographic map coverage will not be at a scale greater than 1:50,000 for large areas, and the use of a larger scale would, for these and other reasons, take a prohibitive amount of time. Furthermore, the more detailed the mapping divisions the finer become distinctions between contiguous units with the result that such distinctions could not often be made from the interpretation of aerial photographs. Representation of units on a scale of 1:50,000 restricts recognition to areas of several thousand square yards. Such units can seldom be completely homogeneous. The *sites* of Bourne, for instance, required a mapping scale of at least 1:25,000. Similarly, the individual *flats* and *slopes* regarded by Linton (1951) as the fundamental units of relief require a similar scale. The refined method of morphological mapping devised by Savigear (1965) really requires a mapping scale of at least 1:10,000 and is in any case based upon detailed field measurement. It thus emerges that it is not possible to map from aerial photographs the fundamental morphological or ecological units as recognized in the field by the writers quoted.

Another approach to this problem has been made by Beckett and Webster who, in a report to MEXE in 1965, recommended the *facet* as 'the largest portion of terrain that can be conveniently treated as one block for purposes of *moderately extensive* land use or construction' (p. 8). This is not really a definition but establishes a basic premise for delimitation that fits well into schemes of land development. The term 'moderately extensive' is intended to include the development of B-roads, mixed farming, field drainage and soil conservation, but excludes intensive uses such as A-roads, irrigated farming and urban drainage. Such facets must be mappable at 1:50,000 and contain internal variations of a simple kind. They are defined in physiographic terms, and must be recognizable from air photographs without ground check. These authors found that alternative methods of definition carried a low recognizability rating, but emphasized that physiographic terms must be precise and based on 'their intrinsic properties and hence not on their inferred genesis' (p. 7). To achieve this the facets were recognized according to (a) morphology or configuration; (b) surficial materials; (c) water régime; and (d) direction, kind and degree of lateral variations of (a), (b) and (c) within the unit. In a more recent report (Brink *et al.*, 1966) the *land facet* is designated as 'the smallest unit of description and consists of one or more land elements grouped for practical purposes' (p. 9). It 'is a part of a landscape which is reasonably homogeneous and fairly distinct from surrounding terrain' (p. 9). The *land element* is equivalent to Bourne's site and is 'for practical purposes uniform in lithology, form, soil and vegetation' (p. 9). It emerges that the land facet, like the land unit, is defined primarily in practical terms; it is not completely uniform, neither is it specifically a landform.

Initially, the land facets were grouped into *recurrent landscape patterns* (Beckett and Webster, 1965), but more recently the term *land system* has been adopted to describe a 'recurrent pattern of genetically linked land facets' (Brink *et al.*, 1966, p. 10). It is pertinent to comment that the objectivity of the earlier work of Beckett and Webster (1965) has been lost with the introduction of a genetic element into the recognition and definition of the land system. The distinction between this concept of the land system and that used by C.S.I.R.O. is difficult to make, not least because one is in a sense derived from the other. Two points, however, emerge from this report (Brink, *et al.*, 1966): first, the land system is defined primarily in physiographic terms, and would appear in effect to be a *land-*

form system. Second, such systems are regarded as being more or less arbitrarily defined according to the scale of the survey (1:250,000–1:1,000,000).

The basic mapping units used in the scheme discussed above have in common definitions concerned with homogeneity or degree of internal variation which may be adjusted according to the purpose and scale of the survey. Different occurrences of any one unit, for instance the land unit of C.S.I.R.O., may therefore exhibit a wide range of size and complexity. If, however, an approach is made to the problem of geomorphological division which seeks to recognize *genetic* units in the terrain, then the element of arbitrariness is reduced, while the recognition of the units becomes a matter of subjective interpretation. Conscious attempts to avoid the genetic approach because of its inherent problems, must also result in a loss of significance, because important aspects of land form are not fully investigated.

Lueder (1959) attempted to enumerate *unit landforms* recognizable from aerial photographs. This unit was defined as 'a terrain feature . . . usually of the third order . . . that may be described and recognized in terms of typical features wherever it may occur' (p. 20). However, no attempt to avoid 'inferred genesis' is made, and the landforms are classified according to the dominant process responsible for their formation.

Herein lies an important problem, for without a genetic basis facet or land form mapping lacks the connecting link with other elements in the total environment, or so it can be argued. But statements as to origin and genesis of forms undoubtedly involve greater subjectivity and the use perhaps of less precise or contentious terminology. Mabbutt and Stewart (1963) claim that 'the genesis of the land system can . . . normally be most readily expressed in geomorphic terms. For these reasons, land system mapping is commonly based on geomorphic principles' (p. 100). The genetic content of such mapping refers to lithology in erosional systems and formative process in depositional systems and is held to 'provide an understanding of the land system itself, in that it stresses causal interrelationships of all the elements in the land system complex' (p. 101). Mabbutt, however, lists separately the chronological principles in mapping. These include the relative ages of land surfaces and forms, and the presence or absence of inherited features from previous periods. Such principles are really also genetic and involve a greater degree of subjectivity.

In summary, land-form mapping may either be strictly *morphological* in the manner of Savigear (1965) who emphasizes form or shape to the virtual exclusion of other properties of the land surface, or *physiographic* involving the use of shape, surface materials and drainage as in the case of Beckett and Webster (1965), or *geomorphological* using all these attributes plus statements as to relative ages and the possession of features attributable to different phases in the development of the land surface.

It emerges from this discussion that several decisions have to be taken concerning the mapping of basic landscape units or complexes: firstly, a decision concerning scale; secondly, a decision concerning the basis of the mapping, and thirdly, a method for mapping on the chosen scale. It has already been suggested that scales between 1:250,000 and 1:50,000 have advantages that may outweigh additional time or expense. It has also been asserted that at large scales the correspondence between land form, soil and vegetation units may not be good. By this it is implied that boundary drawing on the basis of complex land units may be impossible and that one element of the environment must be selected for mapping and that the degree of correspondence between this and the other elements of significance to development may in part be predicted from a basic knowledge of the phenomena concerned, and in part discovered later during intensive surveys of a specialized character, *using the base maps supplied by the reconnaissance survey*. If this is accepted, then there are powerful arguments for using geomorphology as the basis for the initial survey. It should be added that, of the conservative elements of the natural environment, the land form is not only the most easily recognized from aerial photographs but it is also the most consistently conspicuous on the ground, except perhaps in areas of thick forest. The geomorphological map is therefore well suited as a base map for subsequent, geological, soil, vegetation or land-use surveys.

These arguments do not justify the use of geomorphology for the division of the land surface into regions for development planning. The further claim which such authors as Mabbutt and Stewart (1963) and Verstappen (1964) have made is that geomorphology provides the key to an understanding of other distributions more directly relevant to land use.

Certain systems of land-form mapping, notably those of Savigear (1965) and Beckett and Webster (1965) attempt to avoid problems of interpretation by the use of purely descriptive techniques. However, if prediction of soil and vegetation patterns is possible from the land-form map this must be due to interaction between these elements of the environment, in terms of formative processes and their effects within a space–time continuum. Form reflects genesis in any case, as Mabbutt and Stewart (1963) point out, but there is a strong case for attempting an assessment of genesis and evolution in terms of relative ages of different parts of the land surface. This is supported by Verstappen (1964) who claims that 'a purely descriptive approach evidently is not satisfactory, as it is the very insight into the genesis of the geomorphological features which clarifies their physical properties and their economic potentialities' (p. 26).

A purely descriptive framework for land-form mapping must either be extremely detailed (viz. Savigear, 1965) or very limited in value, for unless the properties of particular landforms or parts of landforms can be predicted from their genesis, form and position in the landscape, they must be accurately described in detail at every occurrence. Thus, unless some assumption is made concerning the genesis of, say, pediment slopes within homogeneous granular materials, there is no basis for assuming that different occurrences of the same slope form will have other properties such as soil water régime and grain size in common. In observational science there is always the risk of being wrong, particularly when features are studied indirectly from the air photograph, but the danger of this risk must not overshadow the value of being correct! A correct interpretation of the land form, in terms of its genesis and sequence of development, provides the understanding whereby the form units within an area may be related one to another in terms of formative processes. It carries with it the ability to predict the occurrence of certain land units with any given land-form system. This is emphasized in the MEXE report of 1966 and also by Mabbutt and Stewart (1963).

The geomorphic principles of land-system mapping by the C.S.I.R.O. have been summarized by Mabbutt and Stewart (1963):

1. *Morphologic*, based upon external form and upon relief and slope categories in particular.

2. *Genetic*, by which a common source is traced, this usually being lithologic in erosional systems or according to a common process in depositional systems. It is this element that affords a rationale in mapping.

3. *Chronologic*, wherein the relative and possibly absolute ages of land surfaces are recognized and inherited features of former conditions are identified.

4. *Dynamic*, by which is meant the mode and rate of landscape change and the recognition of dominant formative processes.

This group of principles also underlies the work of Verstappen (1964), but this author emphasizes the hierarchy of forms in terms similar to Linton (1951) and specifically mentions *morphometric* properties of landforms. This is contained within the morphologic principles of Mabbutt and Stewart (1963) but implies a greater degree of precision in measurement of form.

Several points need to be emphasized here. First, the principles put forward by Mabbutt and Stewart (1963) concern mapping on a scale of 1:500,000 for land systems and not on the larger scale advocated here. This does not in any way affect the validity of these principles, but it may affect the emphasis placed on one or another of them in the mapping programme.

For instance, the precise measurement of form can only be a feature of larger-scale mapping, as on small scales a degree of generalization is necessary. For instance, average or typical slopes may be quoted, but actual slopes will not be mapped. At a scale of 1:50,000 some morphometric techniques may be applied to such features as drainage composition, slope and local (relative) relief. Secondly, the establishment of an hierarchy of land-form units is very relevant to the problem of producing maps of varying scale and detail for regional planning purposes. The C.S.I.R.O. avoid this problem by mapping only land systems, but mapping at scales of 1:50,000 or 1:100,000 demands that these should be divided into rationally defined smaller land-form units.

An hierarchy of land-form units

In Table 6.1 a comparison is made of the various environmental mapping units discussed above, so that morphological, soil and ecological units of varying scales may be compared. Table 6.2 indicates the writer's suggested hierarchy of morphological or land-form units which are discussed below. Exact comparisons according to the scale or size of units shown in Table 6.1 are not possible for two reasons. Units which are different in concept are difficult to compare in any case, but units which may be defined *ad hoc* for convenience in use are difficult to categorize. The land unit of the C.S.I.R.O. falls into this class and so does the facet of Beckett and Webster (1965).

The mapping units suggested in Table 6.2 are an attempt to recognize genetically significant and therefore real units of terrain. The boundaries of such units may be hard to define, but their existence should be indisputable. Thus the *site* (Linton, 1951), the *morphological* unit (Savigear, 1965) and the *land element* (Brink et al., 1966) are in reality all names for small, uniform *slope units* (see Chapter 13) bounded by inflexions or breaks of slope. In most landscapes these small features cannot be mapped at medium scales. Higher in the hierarchy correspondence between different units is less certain. Possibly this is because, as Linton (1951) has pointed out: 'nature offers us two inescapable morphological unities and two only: at one extreme the indivisible flat or slope, at the other the undivided continent' (p. 215). Thus neither the land unit nor the land facet has been defined rigorously. Furthermore, neither is specifically related to the concept of the *landform*.

Lueder (1959) used the term *unit landform* in this context and this is comparable with the use of the term *landform* by Savigear (1965) and Young (Ch. 13). In practice, land facets and land units are likely to be recognizable landforms. Thus Beckett and Webster (1965) use terms such as 'river floodplain' and 'floodplain terrace' for describing particular facets

TABLE 6.1 A comparison of mapping units for the description of the natural environment [1]

Suitable mapping scales	Morphological and physiographic[a] units	Environmental and ecological units	Soil units
At least 1:10,000	Site (Linton, 1951) Morphological unit (Savigear, 1965) Slope unit (Young, Chapter 13)	Site (Bourne, 1931; Clarke, 1957) Tessera (Jenny, 1958) Ecotope (Troll, 1963) Habitat-site (Moss, 1968) Land element (Brink et al., 1966)	Soil Individual (Soil Survey Staff 1960) Soil series (Soil Survey Staff 1951, 1960) Soil type (Soil Survey Staff, 1951 1960)
1:25,000 to 1:50,000	Facet (Webster, 1963; Beckett and Webster, 1965)	Land unit (Christian and Stewart, 1953; Christian, 1957, 1959; Mabbutt and Stewart, 1963) Land facet (Brink et al., 1966)	Soil series } Soil type } (as above: these units may span two categories) Soil complex } (Soil Survey Staff, 1951)
1:50,000 to 1:100,000	Stow (Linton, 1951) Unit landform (Lueder, 1959) Landform (various definitions: e.g. Howard and Spock, 1940; Savigear, 1965; Young, Chapter 13)	Land unit (as above: this unit may be included in several scale categories according to use)	Soil association (Ellis, 1932) Soil catena (Bourne, 1931) Soil complex
1:100,000	Complex unit landform (Lueder, 1959)	Land unit (as above)	Soil association
1:250,000 to 1:500,000	Tract (Linton, 1951) Recurrent landscape pattern (Beckett and Webster, 1965) Relief unit (Young, Chapter 13)	Region (Bourne, 1931) Land system (Christian and Stewart, 1953; Christian, 1957, 1959) Land system (Brink et al., 1966)	Soil association. Soil-mapping units above the level of the association do not have obvious topographic scale equivalents
1:500,000 to 1:1,000,000	Section (Linton, 1951) Major relief unit (Young, Chapter 13)	Complex land system (Christian, 1957) Land region (Brink et al., 1966)	

[1] A glossary of authors' definitions for the units shown above appears in Appendix A, p. 527.

[a] The term physiographic is used here to include morphology, texture of surface soil and drainage conditions. In certain instances it is difficult to allocate units to their appropriate category because they are polythetic (Sokal and Sneath, 1963). This is particularly so for the units proposed in reports to MEXE (Beckett and Webster, 1965; Brink et al., 1966). However, on the basis of the descriptions in these reports the units have been classified as above. The writer's suggested hierarchy of landform units appears as Table 6.2.

in the Oxford area. But if the term unit landform is adopted the size and complexity of the feature is dictated by the environment and not by the nature of the survey. It is true that the concept of the landform may vary according to the operator, and may be difficult to apply in specific instances, but the whole body of geomorphological theory and observation is then available for interpreting the terrain pattern. Lists of landforms (equivalent to the concept of the unit landform) are available particularly in the literature on geomorphological mapping. Lueder himself provides such a list (1959), and another notable example is found in the discussion by Klimaszewski (1963) of Polish geomorphological mapping. A discussion of the classification of landforms by Howard and Spock (1940) also quotes many early schemes. In fact it is strange that so little reference is made to the literature on geomorphological mapping by the authors of reports on land classification.

This introduces the question of the terminology for the actual units described, for only at the site level can a full descriptive definition be given to the features of terrain. A recognition that 'landforms' as such exist enables mapping to proceed at medium scales suggested in this study. But description of landforms frequently introduces ambiguity, because of uncertain origins and evolution of forms described according to a genetic terminology that has wide variations in descriptive meaning from author to author. The term 'tor' is a classic of this confusion, and 'pediment' is another. Thus a pediment may be defined simply as a concave footslope below any hillslope,[1] or it may be restricted to rock-cut features carrying a veneer of transported debris.[2] It is of considerable importance to know whether a so-called pediment slope is due to hillslope retreat in uniform materials or to the juxtaposition of different materials; whether it carries colluvial material or simply wash debris; whether it is rock-cut or bevelled across a deep-weathering profile. Thus without adequate definition and qualification the term pediment cannot convey all the essential properties of particular pediment slopes. On the other hand, all pediments share certain common features of slope form and relationships with contiguous facets of the land surface. The abandonment of the term therefore does not assist in the solution of this problem, for if we introduce a descriptive terminology then those common properties that always relate to pediments cannot be conveniently summarized. The only rational approach to these difficulties is to attempt two guiding rules in the descriptive scheme: firstly, that all accounts or reports should carry a glossary of defined terms; and secondly, that all landform terminology should include adequate descriptive qualification.

[1] See Pallister (1956, footnote to p. 82).
[2] See King (1948, p. 86; 1962, p. 143).

E

TABLE 6.2 *Terminology and definitions for an heirarchy of land-form units*

(1) *Site* – Sites are fundamental units of relief and are not susceptible to sub-division using morphological criteria (micro-morphological divisions such as might be determined from a close examination of the soil surface are excepted from this definition). The term site derives from Bourne (1931) and, more directly from the physiographic use of the term by Linton (1951). The term *slope unit* (Young, Chapter 13) is equally acceptable, and the concept corresponds closely with the *morphological unit* (Savigear, 1965) and the *land element* (Brink, *et al.*, 1966). Although defined in terms of slope form (rectilinear, convex or concave) and declivity, sites will normally be uniform in lithology (which may commonly be defined in terms of the texture of the surface soil). This is so, because minor changes in lithology give rise to minor changes in slope form that mark a change to another site. Sites thus lie between breaks or inflexions of slope, and commonly occupy areas from less than 1 sq m to perhaps more than 100 sq m. In general they may be mapped only at scales around or greater than 1 : 10,000.

(2) *Facet* – Facets are relief units exhibiting a high degree of homogeneity and which are genetically single features within the landscape. Simple and minor variations in lithology and slope may occur (several sites will normally be combined to form a single facet). Although the term was advanced by Webster (1963) as a physiographic equivalent to Bourne's site, following Wooldridge (1932), the use of the term by Beckett and Webster (1965) is similar to that suggested here. Savigear (1965) uses the same term to describe plane surfaces only. A facet is generally only one part of a unit landform (below). Examples might be crest slopes between adjacent valleys, deposi-tional surfaces of river terraces, simple forms of pediment and so on. They can generally be mapped consistently at a scale of 1 : 25,000, but not always at 1 : 50,000. Facets are probably the smallest units of relief that can be mapped from aerial photographs, using normal methods of interpretation.

(3) *Unit landform* – A unit landform was defined by Lueder (1959, p. 20) as 'a terrain feature . . . usually of the third order . . . that may be described and recognized in terms of typical features wherever it may occur'. *Landforms* have been described in similar terms by Howard and Spock (1940), Savigear (1965) and Young (Chapter 13). The *land unit* used by the C.S.I.R.O. will commonly be of this order but may include facets and landform complexes (below). A unit landform will generally include a number of facets and a great many separate sites, but a few unit landforms may consist of a single facet. Viewed conversely, some facets may be regarded as unit landforms, and pedi-ments may in certain cases be examples of this overlap in definition. The term is used in the same sense as *stow* has been used by Linton (1951). Thus the most obvious and frequently occurring landform is the single-cycle river valley. In the tropical African landscape, duricrusted mesas, tor groups, single bornhardt domes and fixed dunes are examples. This morphological unit is of critical importance in that each unit landform is likely to correspond with a single soil association or catena. The terminology for landform descrip-tion must be precisely defined for each area. Unit landforms may usually be mapped at a scale of 1 : 50,000.

(4) *Landform complex* – The landform complex finds no exact parallel in the literature cited. Its existence is recognized by allusions to complex land units and complex unit landforms, but no definitions for these are offered. It may

be argued that such a morphological unit is one of convenience, but certain landforms occur more often as complexes than as simple units (groups of tors are an example), while others assume complex forms as a result of multi-phase development. Thus a two (or more) cycle-river valley may be classed as a landform complex which contains two unit landforms: the old valley and the present stream valley incised below it. Such forms may have common features throughout a landform system (below) and in certain cases it may be necessary to map landform complexes rather than individual unit landforms at scales of 1 : 50,000–1 : 100,000. However, arbitrary grouping of unit landforms is not valid within this scheme.

(5) *Landform system* – This arises out of the definition of the *tract* (Linton, 1951), the *land system* of the C.S.I.R.O. and the *recurrent landscape pattern* and land system of MEXE (Beckett and Webster, 1965; Brink *et al.*, 1966). The *relief units* of Young (Chapter 13) also fall within this category. Such systems are defined here solely in geomorphic terms; they exhibit a repeated pattern of unit landforms and/or landform complexes. It is therefore implicit that the landform system will include a limited range of facets which comprise the landforms of the system. As the C.S.I.R.O. and other organizations have shown this type of unit is basic to regional surveys for development purposes. Landform systems may correspond with lithological outcrop patterns and/or with geomorphological zones in the landscape. Where a single geomorpho-logical zone extends across several lithological groups, variations in landform pattern may arise that suggest the need for the recognition of *complex landform systems* within which sub-types are *simple landform systems*. This distinction follows similar reasoning concerning the land systems of C.S.I.R.O. Examples are shown in Figures 6.2. and 6.3. Distinctive types of landform system associated with tropical shields are *etchplains*. Etchplains are classified in Table 6.3. Landform systems may be mapped at scales from 1 : 250,000–1 : 500,000.

(6) *Landform region* – Comparable with the *section* of Linton (1951) this is an area within which all the landforms are systematically related through structural or other factors. It corresponds with the *land region* (Brink *et al.*, 1966) and the *major relief unit* of Young (Chapter 13). The landform region will contain two or more related landform systems. The landforms developed as a result of the denudation of a large igneous intrusion, or of a series of uniformly dipping sedimentary strata are possible examples. A group of related etchplains might, within the context of tropical shield lands, be considered as a landform region.

Land-form units of higher order are not considered here as being necessary to mapping on scales of 1 : 500,000 or greater. Some discussion of possible terminology, however, appears in Linton (1951) and Brink *et al.* (1966).

Thus a pediment can be defined as a generally concave replacement slope occurring below the main hillslope and beyond any zone of talus accumula-tion. Individual occurrences might be described variously as 'rock-cut pediment with transported debris', 'deeply weathered pediment with super-ficial layer of transported debris', 'pediment slope carrying thick colluvial deposit over weathered bedrock' and so on. On a map all pediment slopes would be indicated similarly, but might carry an alphabetic code to distinguish different types. Still greater precision can be achieved by giving dimensional information. In a given area pediments are likely to fall within a narrow range of slope and length values which can be quoted.

On maps of 1:50,000–1:100,000 units below the unit landform cannot be represented consistently. It should be realized, however, that some unit landforms will consist of a single facet, while others may comprise several smaller facets. It might be necessary at times to map features that span from a single facet to a landform complex. In the latter case some internal consistency among the facets within the landform complex could be diagrammatically or otherwise indicated in the accompanying report.

In summary it can be said that it is desirable first to establish a scheme of terrain classification that is based on fundamental principles and not only on expediency. It is then necessary to decide at which level of complexity mapping can be achieved. It is clearly desirable that the units mapped should be as homogeneous as scale permits on the one hand, while on the other hand it is important to select as the basic mapping unit one which can be represented throughout the terrain under analysis. It might, in fact, be possible to map facets in one area but only landform complexes in another. In general the aim should be to map unit landforms.

The questions to be asked concerning the methods for mapping such forms are answered in part by the existing literature and examples of geomorphological maps, but unfortunately there is no real agreement among these concerning priorities in mapping. Many geomorphological maps are based upon detailed field and laboratory analysis and include lithological subdivisions within unconsolidated materials that cannot be achieved within the framework of aerial photographic interpretation (viz. Tricart, 1961). It is not an aim of this study to recommend any particular scheme of cartographic conventions for the mapping of unit landforms. It may, however, be stated that a distinction should be made in mapping between morphometric (morphologic) properties, genetic (lithologic) properties and chronologic relationships between landforms. In general it is likely that shape should be represented by linear forms on the map, lithology by stipple or appropriate shading, and age by colour overprinting, but none of these is immutable.

Landform systems in tropical Africa

A more pertinent problem to this inquiry is the basis for the recognition of units of larger size and greater complexity; in particular the basis for the recognition of landform systems. Because these are in certain respects comparable with C.S.I.R.O. land systems and the MEXE recurrent landscape patterns and land systems, precedent exists for their definition. But recognition is often based upon descriptive patterns from print laydowns and forms the initial division of terrain into unlike areas; only later can these units be said to be built up from the combinations of smaller units, to confirm and elaborate earlier predictions. It is in this field that questions of genesis and evolution may become problematical and complex, and yet vital to a proper understanding of the patterns described.

It has already been noted that 'recurrent patterns' of land units or facets have formed the basis for the recognition of land systems, and this is equally the case with the landform system which exhibits a repeated pattern of unit landforms or landform complexes. In a sense such units require no further elaboration and if the geomorphological map indicates the unit landforms, then the delimitation of larger units can be achieved simply by inspection of the changing patterns of landforms. But greater value can be obtained from such higher order divisions if their genesis and relations with neighbouring landform systems are appreciated. This requires that we use more than descriptive and local terminology to characterize the regions. In fact this is achieved in the Land Research Series reports of the C.S.I.R.O., in which the land systems are generally arranged either according to lithology as in the study of the Hunter Valley (Story *et al.*, 1963) or to both lithology and geomorphological history as in the studies of the Alice Springs Area (Perry *et al.*, 1962) and of the West Kimberley Area (Speck *et al.*, 1964). These latter reports have particular relevance to the shield areas of Africa.

In his study of the geomorphology of the Alice Springs Area, Mabbutt (1962) lists the land systems under four main classes:

1. Erosional Weathered Land Surface.
2. Partially Dissected Erosional Weathered Land Surface.
3. Erosional Surfaces formed below the Weathered Land Surface.
4. Depositional Surfaces.

As many as 88 land systems fall within these categories, but they are grouped at a lower level of complexity according to dominant morphological or lithological characteristics. Thus within the Erosional Weathered Land Surface are found the following:

(*a*) mountains of gneiss and granite;
(*b*) hills and plains of granite, gneiss or schist;
(*c*) peneplains on weathered granite, gneiss or schist;
(*d*) sandstone and quartzite ranges;
(*e*) peneplains or undulating terrain on weathered sedimentary rocks.

An important group of surfaces is recognized within the Partially Dissected Erosional Weathered Land Surface, and these exhibit the phenomenon of stripping of regolith from deep weathering profiles. This process of stripping has particular relevance to the morphology of tropical shields. Wright

(Speck *et al.*, 1965) has also classified the land systems of the Tipperary Area within a similar framework. This kind of classification establishes genetic relationships between adjacent land systems and enables a certain degree of prediction concerning soil and other properties of the land surfaces. In most of the C.S.I.R.O. reports a chapter is devoted to theoretical geomorphology and it is clear that in many cases an understanding of this makes possible the systematization of the other data.

Recent work by Doornkamp (1966) and Doornkamp and Temple (1966) in Uganda establishes consistent genetic relationships between terrain types (not specifically designated as landform systems or regions) that have been found to influence soil patterns in a profound and detailed manner (Chenery, 1960; Radwanski, 1960; Ollier and Harrop, 1960). Over a wide area three 'landscapes' are found: the Upland, Inselberg and Lowland landscapes. These are influenced in their distribution by lithology, but are fundamentally the result of tectonic warping, leading to the partial stripping of an ancient deep regolith beneath the Upland Landscape, and the development of a Lowland Landscape across the truncated weathering profiles. The Inselberg Landscapes were exhumed over suitable rocks during this process. The work of Ollier on the soils (1959) and the inselbergs (1960) of this area supports these conclusions.

It is therefore possible to map not only genetically defined landform units on maps of medium scale but also genetically related land or landform systems at smaller scales. In practice, however, land systems have been recognized first from the landform patterns on print laydowns of aerial photography, and they may be defined in empirical and descriptive terms. Two related problems thus emerge from this discussion. First is the possible recognition within the terrains of tropical Africa of genetically related landform systems that once described from a particular area can be recognized on the same criteria elsewhere; second is the basis of descriptive definition and the application of quantitative principles to this.

These considerations are especially relevant to studies of terrains within which slope characteristics do not vary greatly. Thus while Young (1965, and Chapter 13) has accorded to slope a major role in the defining of his natural regions in Malawi, Taylor *et al.* (FAO, 1962) did not find it possible to define *land types*[1] in a part of Western Nigeria on this basis. Such differences in approach, also referred to above, can in part be accounted for in terms of an understanding of the geomorphic processes operating in and the geomorphic development of different parts of the continent. The *catena* concept of Milne (1935) has become basic to the mapping of soils throughout Africa, and this is in effect a statement of the influence of local relief forms upon soil development. However, in certain areas, notably in

[1] Land types are similar in concept to land systems.

Western Nigeria, *soil associations* have been defined principally on lithological criteria (Smyth and Montgomery, 1965). It seems worth while to
inquire into possible reasons for this, and to see whether any general statement concerning the variety of landform systems found within tropical
Africa can be regarded as widely valid.

The importance of deep chemical weathering of rocks under tropical
conditions is well attested in the literature (Ollier, 1959; Barnes, 1961;
Thomas, 1966), and the widespread development of laterite (plinthite,
duricrust or cuirass) on surfaces of low relief, within zones of seasonal
climate is equally well known (Prescott and Pendleton, 1952; Maignen,
1960; Alexander and Cady, 1962, etc.). In large measure the two phenomena occur together and may be associated with ancient surfaces of
planation. Both greatly affect the morphology and soils of an area and
immediately focus attention upon the weathering profiles of the rocks
rather than upon the rocks themselves as parent materials. In so far as the
'peneplain laterites' (du Preez, 1949) are fossil features they may be regarded
as soil parent materials rather than as a part of the soil profile of the present
day. The influence of deep weathering upon soils is often acknowledged in
a general way (d'Hoore, 1964), but only a few authors have attempted to
consider this matter specifically and in detail (Spurr, 1954; Ollier, 1959).
On the other hand, the mapping of soil associations on geomorphic grounds
in Uganda (Chenery, 1960; Ollier and Harrop, 1960; Radwanski, 1960)
and of land systems in Australia has acknowledged this influence. Similarly, certain examples of landform mapping, notably by Mabbutt (1961a,
1965a), and also by Wright (1963), Thomas (1965), Thorp (1967) and
Doornkamp (1966) have also emphasized the importance of deep weathering in the development of African and Australian land surfaces. This emphasis upon deep weathering implies the recognition of a suite of distinctive landforms and soils produced by the dissection of the weathering
profiles, and it has been suggested elsewhere that these may be systematically related and described in specific terms (Thomas, 1965, 1966). The
value of this concept to landform system mapping will be considered
further below.

Another factor leading to an emphasis on geomorphic mapping principles
is the occurrence of drift deposits over wide areas of the northern savannas
in West Africa and also elsewhere on the continent. This topic is dealt with
in detail by Pullan (Chapter 7). On the other hand, where fossil laterites
and deep weathering are less extensive the influence of solid geology becomes of much greater importance.

The concept of a morphological classification of African land surfaces
has attracted the attention of a number of writers (Harpum, 1963; d'Hoore,
1964; Thomas, 1965) and may be considered here in more detail. D'Hoore

recognized zones of *departure, transference* and *accumulation* of erodible rock materials, and was able to distinguish land surfaces:

(i) surfaces too level to permit important movements,
(ii) zones of accumulation,
(iii) zones of transference, and
(iv) the rejuvenated zones of Africa.

(d'Hoore, 1964, p. 43)

Somewhat similar ideas have been embodied in work by Clayton (1958) and Moss (1968). The other approach to African land surfaces is an arrangement according to age in the manner of King (1962) who lists the chronology not only of erosional surfaces but also of depositional surfaces. There is therefore a degree of correspondence between classes (*i*) and (*ii*) of d'Hoore and this distinction made by King. Classes (*iii*) and (*iv*) then become the dissected areas between preserved remnants of old planation surfaces and zones of littoral or interior sedimentary accumulation.

Further inquiry into the character of these land surfaces reveals that certain characteristics depend upon age and on the nature of the tropical African environment. It will also become apparent that these features belong to the comparatively stable areas of the continental shield, but are less obvious, if present at all, in the (largely peripheral) rejuvenated zones. However, for the interior plateaux and basins and the gently tilted peripheral zones of West Africa at least, it may be possible to formulate a classification of surfaces that has wide validity and practical significance.

This notion depends upon the importance of deep weathering and the recognition of certain important and recurrent features of most deeply weathered terrains. Büdel (1957) was the first to draw attention to the existence of 'double surfaces of levelling' within tropical plainlands, where denudation could be regarded as operating not only on the land surface itself but also by means of continued chemical decomposition, upon what has come to be called the *basal surface of weathering* (Ruxton and Berry, 1957) or *weathering front* (Mabbutt, 1961b), which may descend as much as 200 ft (exceptionally over 300 ft) below the land surface between outcrops. The relationship of this basal surface with the land surface forms the basis for the study of deeply weathered terrains both over crystalline and also over sedimentary rocks where similar features may be found (Wright, 1963; de Swardt and Casey, 1963). In crystalline rocks the two surfaces do not always undulate sympathetically across the landscape (Thomas, 1966). However, the basal surface is frequently found as a well-defined, narrow, transitional zone between highly weathered and little altered rock. As a result the highly erodible weathering products commonly become stripped from the underlying rock surface as a result of climatic or tectonic changes in the controls over landform evolution (Thomas, 1965).

A further feature of tropical weathering profiles is the widespread occurrence of fossil laterite deposits on the older and gently undulating surfaces. The best-preserved plains are generally coincident with the survival of these laterite duricrusts, though their precise relationships to planation surfaces remain in doubt (de Swardt, 1964; Trendall, 1962; Thomas, 1965). Notwithstanding problems of interpretation it remains meaningful to refer to many of these deposits in terms of their association with planation surfaces of particular age. D'Hoore (1964) among others recognizes the common occurrence of extensive primary laterite on the African or early Cenozoic surface (King, 1962).

In fact it is probable that the planation surfaces of the African continent are all essentially similar in character, though they undoubtedly vary in the extent and depth of weathering, and in the extent and types of laterite deposit present. Differences will exist that reflect the influence of varying climatic (especially rainfall) régimes on the one hand and the effects of time on the other. Thus the deepest weathering profiles are likely to occur beneath the older planation surfaces, where these are well preserved. Similarly, deeper weathering is likely to be found within the wetter climatic zones if other factors such as age and rock type are constant.

These differences, however, have become obscured by the interaction of the two major factors (rainfall and time) and by the effects of climatic oscillations that have brought both wetter and drier conditions to most areas of tropical Africa during the Pleistocene period. In West Africa, for instance, the older surfaces occur well inland, within sub-humid, and even semi-arid climatic zones. Younger land surfaces, formed in the southern parts of West Africa, have experienced wetter conditions, but at the same time they have suffered accelerated erosion resulting from the southerly tilting of the basement. It is not possible, therefore, to predict the extent or depth of weathering from climatic data unless the age of the land surface and the character of the rock are similar in any examples used for comparative purposes.

Recent tectonic deformation of the African shield has been of an epeirogenic or cymatogenic[1] character and has not, generally, resulted in tectonic relief as such. The East African Rift Valleys are notable exceptions. On the other hand, the arching and uplift of the eastern and southern parts of the continent have resulted in a contrast between 'High' Africa and 'Low' Africa, both in terms of altitude and of the range of landforms

[1] A cymatogen is described by King (1962) as an undulating orogeny producing wave-like formations of large dimensions in the crust.

produced. The much greater altitudinal separation of planation surfaces found along the eastern rim of the continent has led to the formation of a much more diverse relief than is typical of the interior plateaux and basins such as are present in Uganda, or the less elevated plains and plateaux of western tropical Africa. In the discussion which follows, many of the arguments apply mainly to these lower and interior areas but have restricted relevance to regions of severe tectonic deformation.

Etchplains and landform systems

Thus over wide areas of tropical Africa comparatively stable conditions have persisted for a very long time, permitting very deep weathering-profiles to be formed. Interruptions appear to have been in the form of comparatively minor elevations of the crust, and have resulted primarily in the erosional stripping of the regolith produced by weathering, accompanied by a limited degree of incision into the underlying basal rock surface. Wayland (1934) used the term *etchplain* to describe plains that had formed in this manner, and emphasized the essential difference between this process of producing a plain from a plain and that of peneplanation (or pediplanation) of an area of diversified, tectonic relief forms. Because both deeply weathered land surfaces and stripped, rocky terrains are formed by the processes outlined, a certain ambiguity in the application of the term etchplain has arisen. Mabbutt (1961a, 1965a) embodied this concept in his studies of relief development in Western and central Australia, and used the term etchplain to describe 'stripped landsurfaces'. Perhaps, because of a similar interpretation, Thornbury (1958) and Bishop (1966) suggested that the diagnostic features of etchplains are found over very limited areas, and Bishop has advocated the abandonment of the term on these grounds. But if both weathered land surfaces and surfaces exhibiting varying degrees of erosional stripping are included within the concept of the etchplain, then a very wide area of Basement Complex rocks at least falls within the definition. It is probable that many areas of sedimentary rocks can also be classified in similar terms.

A wide range of terrain types or landform systems can therefore be described as varieties of etchplain. A series of such etchplain types has been described elsewhere (Thomas, 1965), each being qualified by a suitable descriptive term. A development of this scheme is described in the following table (6.3).

Although these appear to be the principal types of etchplain and etch-surface, many other modifications can be envisaged. Burial of the *in situ* weathered plain may occur as a result of aeolian, alluvial or lacustrine deposition, while important areas of colluvial material exist elsewhere. To some extent these are features of etchplains in that the low relief encourages

TABLE 6.3 *A classification of etchplains and etchsurfaces*

A. *Lateritized etchplains* are surfaces of low relief underlain by extensive primary laterite deposits. Ideally they should exhibit a few signs of recent steam incision, and should carry few, if any, residual hills above their general level. Most examples, however, exhibit some degree of steam incision below the laterite horizon, usually as a result either of recent warping or of climatic and vegetational changes of Pleistocene age.

B. *Dissected etchplains* are therefore recognized as the first stage in modification of lateritized etchplains, following accelerated steam erosion, resulting in the formation of duricrust breakaways overlooking well-defined valleys between gently sloping tablelands. Exposures of the basal surface of weathering in the form of tors or domes is unusual except in localized areas of shallow weathering, but rock may be exposed extensively in the more important stream channels.

C. *Partially stripped etchplains* follow further dissection and stripping of the deep regolith. At this stage much of the former laterite cover will have been removed; a few mesas of primary laterite may survive, but most laterite deposits will be of a secondary, concretionary type. Rock outcrops in the form of tors and low domes may be common, and occasional larger bornhardts may appear. Over wide areas soil development will be on truncated weathering profiles: from the mottled, pallid or transitional zones containing corestones and unweathered rock fragments.

D. *Dominantly stripped etchplains and etchsurfaces* represent a further development of the stripping process, when all but the deepest basins of weathering and areas of least resistant rock have become stripped to form a complex multi-convex surface of exposed rock and shallow regolith, carrying important areas of lithosolic soils. Where the basal surface is markedly uneven (see Thomas, 1966) this type of terrain merits the term *etchsurface* rather than etchplain. Such stripped surfaces are usually a result of important tectonic warping of a cymatogenic nature, or are restricted to areas of open savanna vegetation where widespread stripping of hillslopes is possible. Localized occurrences of this type of relief coincide with particularly resistant rock types such as migmatitic acid gneiss.

E. *Incised etchsurfaces* occur where the basal surface of weathering has not simply been exposed but has become subject to widespread modification by steam erosion. This will frequently occur in conjunction with dominant stripping and as a result of important base-level changes. Climatic oscillations will seldom be competent to produce important modifications of the basal surface (see Mabbutt, 1961a). In many instances the incised etchsurface is absent or only developed over small areas, and the stripped etchplains abut against true escarpments, formed either as a result of differential erosion or because of tectonic upheaval and critical height failure.

F. *Pedimented etchplains and etchsurfaces* evolve from stripped or incised etchplain types, when stream incision ceases and valley sides undergo slope retreat beneath shallow weathering profiles (Jessen, 1936; Ruxton and Berry, 1961). It is doubtful whether this type of landform system can be distinguished from the stripped and incised types on air photographs.

accumulation of water-borne sediment on lower slopes and in the semi-enclosed basins that are quite common in such terrains. Accumulations of aeolian material and extensive alluviation resulting from warping or other factors, however, produce more fundamental changes in morphology and soil parent material. It is debatable whether the term etchplain can be usefully applied in these cases. However, in Uganda the alluviation of valley floors around Lake Victoria has not obscured the basic features of etchplain morphology and the term *Aggraded Etchplain* might well be applied. Similarly, the aeolian drift in northern Nigeria overlies lateritic horizons that influence soil profile development in important ways.

Where the basic morphology is a result of the twin processes of etching and stripping, but where substantial modifications have arisen from the operation of other processes, an *ad hoc* terminology for the qualification of the term etchplain may be usefully used. There are, of course, important areas where vestiges of ancient weathering profiles are absent and the inheritance of the present landforms from etchplain conditions cannot be construed from available evidence. In these areas, notably within the Rift Valley complex, and in the marginal areas of High Africa discussed above, landform systems must be denoted in different terms.

The introduction of the etchplain concept into this discussion has been made because between the dissected, upwarped margins of the continent and the extensive depositional surfaces of its interior basins lie vast areas of basement complex and related rocks that give rise to terrains that are distinctive yet elusive in their diagnostic characteristics. It is believed that the elaboration of the etchplain concept of Wayland (1934) may assist in the understanding and designation of terrain types (and therefore landform systems) within this province of tropical Africa. It is within these areas, moreover, that the landforms and soil parent materials most clearly reflect the effects of tropical weathering processes, on account of the absence over lengthy periods of major tectonic disturbance. Small-scale shifts in base level or in climate have here given rise to the range of land-surface types described in this section.

The further relevance of this classification lies in its application to the systematic description of landform systems which fall within the etchplain types enumerated above. Such a description will have value only if widely applicable and capable of further elaboration as work proceeds in types of terrain unfamiliar to the writer. Its value lies in the ability to predict from the terminology used, the type of terrain encountered in an area, its dominant soil characteristics and catenary sequences.

In order that this may be achieved, additional information concerning the lithology and the age of the land surface must be given: for example, a dissected etchplain may occur in an area of biotite gneiss where there are remnants of an early Cenozoic planation. The relevance of age in landscape terminology remains controversial. Spurr (1954) attempted to classify the soils of Tanzania (Tanganyika) according to the geological dating of the major planation surfaces, but other writers such as Mulcahy (1961) have pointed out that land surfaces of great absolute age are rare, and that the soils developed on most planation surfaces are also much younger than their supposed geological dating. Nevertheless Spurr (1954) pointed out that 'many of what appear today to be catenas formed under existing conditions include among their members truncated profiles of senile soils developed during past erosion cycles in an extremely different environment from that in which they now occur'. Similarly, where soil parent materials are pre-weathered regoliths, the extent and depth of weathering and the proportion of weatherable minerals are clearly related to the absolute age of the planation surface as Ollier (1959) has demonstrated.

Etchplains and laterite deposits

Although it is necessary to take account of age, etchplains of particular type should exhibit broadly similar features wherever they occur. Some of these have been outlined in Table 6.3, but some further elaboration is possible in terms of characteristic catenary patterns and the occurrence of particular types of laterite.

Although much controversy still surrounds the classification and description of laterite deposits, the work of du Preez (1949) on the laterites of Nigeria remains one of the few attempts to classify these deposits in terms both of their characteristic morphology and of their occurrence in the landscape. The basic division between a *peneplain laterite* (or primary deposit) and a *lower lying or pediplain laterite* (usually but not necessarily a secondary formation) has been recognized by most writers. The work of de Swardt (1964) implicitly accepts these categories, by recognizing an Older and a Younger laterite throughout most of tropical Africa. Although this classification is not entirely acceptable, because much local detail cannot easily be fitted into this framework, it does suggest that a broadly similar sequence of laterite formation, dissection and re-formation has occurred throughout a very wide area.

Thus a dissected or partially stripped etchplain may exhibit a pattern of facets which reflects this type of modification. Low mesa hills will be capped by the Older or peneplain laterite, and bounded by breakaways and pediment slopes leading down to concretionary deposits of Younger or pediplain laterite forming benches above a recent valley containing the contemporary stream. This morphology records a sequence of events leading to the dissection and partial stripping of a deep regolith, and may be the result of both tectonic and climatic changes. But the widespread occurrence of this sequence suggests that it is climatically induced. It is not suggested that this is the only sequence, or that differences do not exist in the morphology of etchplains according to climatic and vegetational zoning, and over rocks of varying composition. It is, however, presented as a framework or model which may be compared with actual occurrences, and if necessary modified as a result of more detailed knowledge.

Etchplains and soil catena types

The character of unit landforms in an area and the arrangements of facets across these will be reflected in the nature of the soil catenas. Thus, while it is relevant following Spurr (1954) and Ollier (1959) to refer to the age of the land surface and hence of the soil parent materials, the detailed morphology, as has long been recognized, greatly influences the pattern of soils. Moreover, the extent and character of stripping also results in varying catenary patterns.

A recent attempt to classify catenas into a few common categories fits well into this analysis. Thus Moss (1968) classifies soil catenas in tropical Africa:

    I *Catenas with rock outcrops*
        1. Associated with extensive areas of relatively flat land with (*a*) or without (*b*) extensive pre-weathering.
        2. In mountainous areas.

   II *Catenas with hard laterite*
        1. With the hard laterite as an upper slope or summit feature with massive laterite (*a*) or with detrital laterite (*b*).
        2. With the hard laterite as a lower slope feature.

  III *Catenas without rock outcrops or hard laterite*
        Found over a wide range of rock types from the very acid to ultra basic and being subdivided accordingly.

  IV *Soil associations on relatively flat land* (i.e. not forming toposequences)
        1. At high levels.
        2. At low levels.

These catenas should accord with the occurrence of particular types of etchplain. Thus catenas with rock outcrops will occur principally within etchplains of varying degrees of stripping – the position of the outcrops often being predicted from the degree of waste removal.

The mapping of unit landforms on the field sheets will in any case be the basis of confirmation for the recognition of etchplain and catenary types.

From air photographs it is generally possible to map the more important outcrops of crystalline rocks, and it is frequently possible to map summit laterites within savanna areas.

Etchplains and lithology
The association of particular etchplain types and catenary sequences with the occurrence of rocks of particular lithology is a further element in this approach. Many acidic igneous and metamorphic rocks produce areas of characteristic inselberg landscape within which deep weathering is confined to narrow zones or small areas between outcrops, and laterites are seldom found as summit features. Such terrains are stripped etchplains, but their geology is important to an understanding of the environment found in these areas. By contrast amphibolitic rocks and phyllites seldom form any outcrops at all but readily become lateritized. Thus many of the most striking lateritized etchplains showing de Swardt's Older and Younger laterite types are in fact found over amphibolitic rocks.

It might therefore be argued that geology rather than hypothetical geomorphological relationships should provide a more reliable key for predicting landform and soil patterns. Two points need to be made here: one of a practical nature, the other theoretical. First, geological mapping covers only a small proportion of the African continent and is proceeding very slowly indeed in certain areas. Furthermore, much of the geology is mapped in very general terms and is often based, where outcrops are few, upon inference from soil and landform patterns. Secondly, the degree to which geology becomes an overriding influence appears to be related to the degree of recent dissection. Old planation surfaces may be deeply weathered and lateritized over a wide variety of rocks; while more recent zones of dissection show the rocks etched out into relief according to their resistance to weathering and erosion. It is within these areas that the contrasts between lithologic types become most apparent. In south-western Nigeria stripped and lateritized etchplains frequently correspond with such lithologic differences. As a result, the distinction between partially and dominantly stripped etchplains within this area is often between rocks of varying resistance to weathering. Where the basal rock surface is not merely stripped of its regolith cover but is strongly affected by surface erosion, the designation of etchplain types becomes inappropriate, for the process of etching is no longer of major importance to an account of the landforms of such an area. The scheme for landform system mapping suggested here is therefore to be regarded as an approach to the systematization of our knowledge concerning a certain range of terrain types that occur widely on the African shield. It is not suggested as a final or comprehensive classification of landform systems. Particularly within the zones of high relief slope categories

become more important and the approach of Young (Chapter 13) particularly relevant.

Quantitative description for landform systems

One remaining problem is the provision of a quantitative basis for the
recognition of terrain types. This is a problem for several reasons. One is
practical, for measurement from air photographs is time-consuming and,
unless adequate corrections are made for tilt and other distorting factors,
liable to error. Others are more theoretical and concern the nature of the
measurements necessary to the recognition of landform systems. Slope
categories such as those used by Young (1965) are clearly important, yet
across a wide area in Western Nigeria the FAO (1962) found only slight
variations of slope. They comment: 'The landscape is remarkably uniform.
As a whole the country is rolling with slopes of 2–6 per cent. The rises in
each land type are much the same height and the valleys the same shape
and depth' (p. 12). Yet within this area there are important differences in
the quality of land. The gently undulating plains are, in certain areas,
broken by abrupt inselberg hills (mostly bornhardts) (Figs. 6.1 and 6.2).
The values for the bounding slopes of these landforms are high, but they
do not constitute a part of the agricultural land. Many other outcrops in
this stripped etchsurface are less abrupt and form rock pavements. One
important distribution in such areas is in fact the disposition of the rock
outcrops and rocky hills. Soil patterns closely follow this distribution and
the better soils are often found immediately around the inselbergs.

Quantitative description for such an area (or any other) can, in this context, perform two main functions. Firstly, it can give precision to the description of the morphologic properties of the landform system. Secondly, diagnostic features of the system can be determined, in order to distinguish it
clearly from its neighbours. In this latter context the proportionate area of
exposed rock or of laterite cuirass might be important to the designation
of etchplain type.

A more empirical approach to the establishment of the most important
form elements of an area has been made, *inter alia*, by Strahler (1954), who
distinguished between dimensional properties of terrain (such as stream
length, length of overland flow and relief) and properties of shape as
expressed by dimensionless ratios (such as surface slope, relief ratio,[1] and
the ruggedness index[2]). A more detailed assessment of the properties of
drainage basins is found in Horton (1945), and a further discussion of mor-

---

[1] The relief ratio Rh is the ratio of the total relief of a basin to its longest dimension parallel to the principal drainage line (Schumm, 1954).

[2] The ruggedness index $H = \dfrac{D \cdot r}{5,280}$, where D is the drainage density and r the
relief of the drainage basin.

phometric properties of terrain appears in Melton (1957). Melton pointed out that the interpretation of these measurements involves an understanding of a large number of variable factors. Not all of these can be accounted for in a reconnaissance survey. Furthermore, most techniques of morphometric analysis relate to drainage basins, but landform systems cannot easily be defined in these terms. Thus drainage basins of the third or fourth order generally transgress lithologic and geomorphic boundaries of significance to the present task. Also, as Lewis (1959) has pointed out, morphometry concerns form elements and not landforms *in toto* as units of the landscape. This means that they are not easily applied to the mapping units suggested in this study.

Perhaps as a result of some of these difficulties, and of the practical problems of scale and representation of data, geomorphic mapping and morphometric analysis have remained in large measure exclusive fields of study. Similarly, geomorphic mapping and land systems surveys have remained largely separate subjects of research. It is not, however, justifiable to reject morphometric techniques on the grounds discussed here, and selective use of certain properties of terrain morphology may greatly strengthen the objectivity and precision of landform-system mapping.

A tentative approach to this problem has been made at a simple level for an area of inselberg landscape south of Iseyin in Western Nigeria (Figs. 6.1, 6.2). Geomorphological mapping of this areas from 1:40,000 aerial photography can be resolved into two major unit landforms: stream valleys and domical rock outcrops. The outcrops represent only that proportion of the total which is visible on the photographic print. Those of small area and negligible relief are generally missed. Thus, although this is designated as a part of a dominantly stripped etchsurface having shallow weathering profiles and numerous rock outcrops, the proportion of exposed rock which can be measured off the map is only 7 per cent. On the other hand, the distribution of these rock domes and pavements on the map is important, in that it gives an accurate, though not very detailed guide to the distribution of shallow, weakly developed soils (especially lithosols) and to the character of the dominant soil catena types.

In order to derive some other properties of this land system from the sample mapped, the stream net and the unit landforms were plotted on to graph paper. On the basis of a 0·1-in grid the incidence and length of outcrops were measured and compared with the total length of grid line sampled. The incidence of stream channels along the grid lines was also counted. From these simple measurements certain properties of the land form were derived. Thus the incidence of streams, expressed in number per mile, is 0·92 and the reciprocal of this figure, 1·08, gives the mean valley width in miles (1·73 km). Since the occurrence of rock outcrops was

measured along the same grid the proportion of transverse valley profiles
having rock outcrops could be estimated at 25·66 per cent. This again neg-
lects small outcrops, and the mean length of outcrop encountered along
these profiles is 0·41 miles (0·66 km). These quantities give some indication

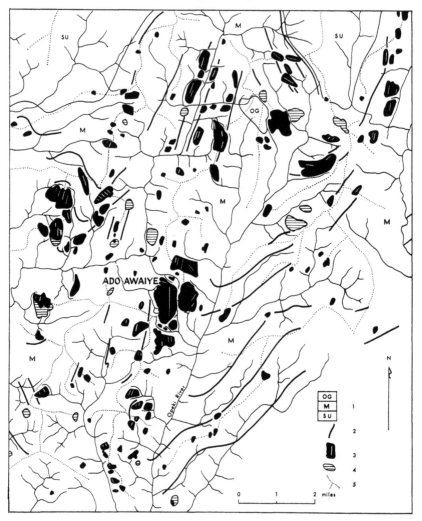

1. Rocks of the Basement Complex: OG, Older Granite; M, Migmatite; SU, undifferentiated schist
and quartz-achist. 2. Structural trend lines in the basement rocks. 3. Inselberge (mainly bornhardts)
with extensive rock exposures. 4. Other abrupt hills with substantial soil cover. 5. Interfluves.

Figure 6.1 *Landform map of the area around Ado Awaiye, Western Nigeria.*
Originally plotted at a scale of approximately 1 : 40,000 from infra-red aerial
photography by Canadian Aero Service Limited.

of the character of this region, though they could not be used for detailed analytical work. The stream incidence in this area was compared with drainage density measured by the conventional method and showed close correspondence, the drainage density being 0·99. Measurement of relief was not attempted as no accurately contoured map was available for this area at the time of writing. To these properties sample measurements of slope taken in the field can be added to give greater precision to the designation of landform systems. The properties of this landform system and a generalised example of a dissected etchplain are given in tabular form in Table 6.4. This tabulation has much in common with those used by the C.S.I.R.O. and the detailed work by Dowling and Williams (1964), and Mabbutt (1965b); and it is clear that additional information on climate, vegetation/land-use patterns, and even soil-texture class and other properties may be added to those shown here.

Land form patterns and geology

This study has been specifically concerned with certain types of terrain but the general question of landform system patterns as they are seen on the aerial photograph and as they are related to geological outcrop patterns is appropriately touched upon here. Haantjens (1965) has classified land-system photo patterns:

1. 'Recurring' patterns with a multidirectional arrangement of land-forms.
2. 'Catenary' patterns where there is a unidirectional arrangement of units in the system.
3. 'Irregular' patterns lacking in apparent spatial organization.

Among the different types of etchplain all three patterns may be represented. Recurrent and catenary patterns will dominate within the less-stripped landform systems, where streams flow over a more or less uniform regolith. But within stripped etchplain types (Fig. 6.1) and in strongly dissected areas irregular patterns may reflect the influence of detailed lithological variations.

Geomorphic zones and lithologic distributions may interact in the landscape to produce actual patterns of landforms and landform systems that are quite complex. There are frequently differences of scale or level at which each influence becomes dominant. Thus the major morphological zones (complex landform systems) of an area are often a result of its geomorphic history, while subdivisions (simple landform systems) may be a result of lithologic factors. At a more detailed scale, the actual morphology of unit landforms, and especially of stream valleys, again reflects not only lithology but also the geomorphic history of the area. In many areas soil survey work has recognized the influence of geology on the one hand, and of detailed morphology on the other, but the broader geomorphic zonation of terrain has been neglected. This may possibly be due to the fact that the survey has been confined to one major geomorphological zone, but it may also reflect a lack of concern for geomorphological history or evolution. The broader influence of geomorphology is always through the genetic and chronologic sequence, and will be missed if this aspect of terrain study is denied its place in the survey work.

An example of terrain patterns organized in this way is given in Figure

TABLE 6.4 *Descriptive notation for landform systems*

A. Dissected (lateritized) etchplain

Dominant rock type: medium-grained biotite gneiss.

Planation surface remnants: possibly early Cenozoic (African) on summits; late Cenozoic or Pleistocene features within larger valleys.

Landform complex: laterite mesas associated with two-phase stream valleys.

Unit landforms: (1) Laterite mesa with pediment.
                (2) Inner valley with active channel.
                (3) Rock domes (bornhardts).

Morphometry: drainage density 0·6; mean valley width 1·66 miles (2·6 km); exposed laterite at least 15 per cent of area, and occurring on 45 per cent of transverse valley profiles; rock outcrops less than 0·1 per cent of area.

Detailed morphology:

Unit landforms	Facets
(1) Laterite mesa	(a) Crest slope of primary vesicular laterite; slopes generally less than 2 per cent cambered at edges; mean area 1,000 square metres
	(b) Breakaway: hardened laterite cliff or steep rubble
	(c) Deeply weathered pediment with veneer of fine laterite rubble and concretionary pisolitic gravel; slopes 2–7 per cent, length 200–400 m
(2) Inner stream valley	(d) Laterite bench of concretionary secondary laterite; slope 2–4 per cent
	(e) Minor laterite breakaway: hardened laterite forming break of slope
	(f) Pediment slope; often deeply weathered, but with occasional outcrops, 100–200 m; slope 2–7 per cent
	(g) Alluvial valley floor
(3) Rock domes (bornhardts)	(h) Convex rocky slopes up to 70 per cent or more; relief 10–50 m

Soil catena types: most soil catenas have massive laterite cuirass as summit feature (over 50 per cent). Many catenas with 2 laterites – facets (a) and (d) – perhaps 25 per cent. Less than 1 per cent of catenas with rock outcrops.

B. Dominantly stripped etchplain (see Fig. 6.1)

Dominant rock type: variably migmatised, acid gneiss.
Planation surface remnants: none important.
Unit landforms: (1) Bornhardt domes.
　　　　　　　　(2) Rock pavements.
　　　　　　　　(3) Tors.
　　　　　　　　(4) Open stream valleys without clear evidence of multi-
　　　　　　　　　　phase incision.
Morphometry: Drainage density 0·99; mean valley width 1·08 miles (1·73 km);
　exposed rock surfaces at least 7 per cent of area, and occurring on 25·66
　per cent of transverse valley profiles.
Detailed morphology:

Unit landforms	Facets
(1) Bornhardt domes	(*a*) Many contiguous convex rock surfaces carrying superficial debris; relief exceeds 10 m; bounding slopes generally over 30 per cent
(2) Rock pavements	(*b*) Simple convex rock surfaces; relief less than 5 m, slope less than 15 per cent
(3) Tors (sometimes as landform complexes)	(*c*) Complexes of contiguous joint blocks
(4) Stream valleys	(*d*) Deeply weathered crest slopes; less than 5 per cent
	(*e*) Valley flank (pediment?); generally shallow weathering with quartz gravel surface layer; slopes 5–10 per cent; length 0·4 miles (350 m). Interrupted by facets (*a*), (*b*), (*c*)
	(*f*) Alluvial valley floor

Soil catena types: Many soil catenas with rock outcrops (over 25 per cent);
outcrops occur in all locations (only 44 per cent on interfluves); some
catenas with multiple rock outcrops. Very few catenas with laterite (none
as massive duricrust on summits) – less than 1 per cent.

6.2, and a diagrammatic generalization of this principle is shown in
Figure 6.3. In this area of Western Nigeria the basement geology displays
a marked north–south trend, so that lithological variations occur as a series
of bands or lenses of different materials arranged in this direction. On the
other hand, the principal watershed runs irregularly from west to east and
dissecting streams flow southwards to the coast at the Gulf of Guinea. Tilting of the area towards the south has led to progressive stripping which has
affected the more southerly areas, and belts of more resistant rock. The
interaction between the lithological patterns and the effects of tectonic and
geomorphic evolution has resulted in an irregular arrangement of landform
systems within which the west–east zonation is only evident from the
inspection of a large area of terrain, but the north–south divisions are in
places quite clear, where stripping has proceeded to the basal surface of

weathering over wide areas. The geomorphological interpretation of this area has been discussed elsewhere (Thomas, 1965), but the main outlines can be stated briefly here.

1. *Dissected (lateritized) etchplains.* Within these areas slopes are gently rolling, 1–3 per cent, valleys open and often marshy and there are large areas of shallow laterite and exposed cuirass. 1. A. This subdivision exhibits gritty sands (FAO, 1962) and there are occasional outcrops of acid gneiss or quartzite. 1. B. This subdivision has sandy loams over red-brown clay (FAO, 1962) and no outcrops were recorded. 2. *Quartzite and quartz schist areas.* These areas are stripped etchsurfaces, but are classified as a distinctive landform system. Slopes are steeper than in adjacent areas with some prominent quartzite ridges. 3. *Partially stripped etchplains.* In these areas slopes range from 2–8 per cent, laterite outcrops are rare and occasional inselberge appear. 3. C. In this area valleys are open, soils are grey-brown coarse sands (FAO, 1962), and outcrops are quite common. 3. D. Although at a similar stage of development this division has few outcrops and soils are sandy loams (FAO, 1962). 3. E. In this area quartzitic rocks occur but they produce few characteristic features. Shallow weathering profiles are common. 4. *Dominantly stripped etchplains.* These areas show many outcrops which create irregular slope profiles, though over regolith the range is similar to adjacent areas (3–7 per cent). Soils are mostly shallow grey-brown coarse sands (FAO, 1962). In 4. F outcrops often take the form of important bornhardt domes, whilst in 4. G they are more commonly small rock pavements. Major inselberg groups are shown in black. The insert refers to the area shown in Figure 6.1.

Figure 6.2 *Etchplains and landform systems in a part of Western Nigeria.*
This map incorporates some information from the Land Use Survey of the Oyo-Shaki Area (FAO, 1962).

The dissected lateritized etchplain in the northern zone is the remnant of an early Cenozoic surface (possibly the 'African') which has been tilted southwards and stripped of much of its deep regolith cover, except around the present watershed area and over rocks particularly susceptible to continued decomposition. This stripping has revealed ridges of quartzitic rock and has exposed wide areas of resistant migmatite (Fig. 6.1) as dominantly stripped etchsurfaces. The tilting and uplift have not been of the order met with in southern Africa and basic rock types have not become stripped of their regolith cover. Amphibolitic rocks commonly preserve old planation

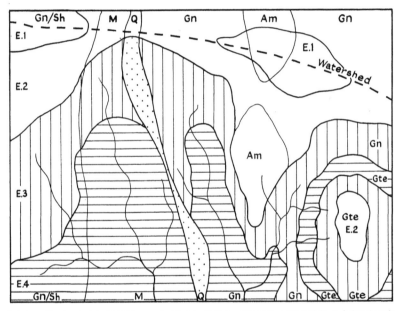

E. 1. Lateritized etchplain. E. 2. Dissected etchplain. E. 3. Partially stripped etchplain. E. 4. Dominantly stripped etchplain. Gn/Sh, Gneiss/schist complex; M, Migmatite; Q, Quartzite; Gn, Banded gneiss; A, Amphibolite; Gte, Granite. The distribution of etchplain types is influenced by the interaction of the two principal factors: the development of the drainage pattern (mainly N–S) and the arrangement of lithological types in the Basement rocks (N–S). The north to south flowing drainage gives rise to east–west zonation, except close to the main drainage lines. Individual etchplain occurrences (simple landform systems) may be described: E. 4., Gn.; E. 4., Gte; E. 2. Am. and so on.

Figure 6.3 *Hypothetical relationships between etchplains and lithology.* Shading categories are as in Figure 6.2.

surfaces by virtue of their well-developed, laterite cuirasses, and where these have been destroyed by erosion, renewed weathering has prevented stripped etchplains from emerging.

Summary and conclusions

A scheme of landform mapping for reconnaissance surveys, and its application to certain types of terrain in tropical Africa, have been discussed with certain aims in view. These may be summarized here. The primary aim is to produce two documents of value to development surveys: firstly, a geomorphological map of a scale that enables it to be used in the field as a base map for other specialized surveys, especially of soils, and secondly, a map of landform systems similar in content to the land system maps of the C.S.I.R.O. The emphasis on geomorphology leads to maps of terrain and not of 'land' in the sense used by Christian (1959). This may seem less useful to any attempt to understand the natural environment as a whole, but if mapping is attempted at a scale of 1:50,000–1:100,000 the plotting of soil boundaries and variations in the vegetation/land-use communities introduces complex problems, particularly at a reconnaissance level. Some of these are conceptual problems, particularly those concerning the vegetation communities (see Chapter 8). Soil and ecological surveys, however, might be carried out as soon as the geomorphological maps are prepared, or it may be possible to incorporate soil and ecological information at the scale of the landform system maps. It is still suggested, however, that land systems as defined by the C.S.I.R.O. may be difficult to define in areas of significant population density. It may be expected that soil patterns, where they have not been substantially affected by man's influence on the vegetation cover, will exhibit a spatial pattern similar to the landform distribution. Herein lies one value of the geomorphological survey; another is the ease with which landform maps can be used in the field, where they can be seen to define visible features of the terrain, probably more clearly than the contoured, topographic map used alone. Necessary geological information can also be conveyed on landform maps, but without non-lithological detail.

Not all the suggestions made in this study are likely to be acceptable to every situation, but certain general conclusions may be stated:

1. In the mapping of terrain an hierarchy of landform units must be formulated, which does not rest upon expediency, but upon the observed organization of landforms. The question of practicability will affect the units which can be mapped consistently throughout a particular area, but no doubt should exist concerning the nature and complexity of these units.

2. The organization of landforms into landform systems reflects both genetic factors such as rock lithology and weathering processes, and also chronologic factors of landform development. These latter may include the duration of weathering penetration, formation and destruction of laterite sheets, phases of erosional stripping and other events in the erosional (or depositional) history of an area. The results of these factors

may be of as much importance to soil development, and ultimately to land-use proposals, as are the simple morphologic properties of individual landforms. A purely descriptive scheme of terrain analysis is therefore unlikely to be adequate for land classification studies.

3. Landform systems are also linked to each other as a result of geomorphological evolution within much wider areas which may be landform regions. Some value may accrue from the ability to *predict* landform and soil properties from an understanding of the interrelationships between individual landform systems. One model for such an understanding within a certain range of terrains developed over basement complex rocks (and probably many other types) in tropical Africa, is the sequence of weathering and erosional stripping which gives rise to related series of etchplains. The theory of this development is discussed in some detail elsewhere (Thomas, 1965). Other kinds of relationship between contiguous landform systems may be determined for different types of terrain.

Finally, it may be observed that, whether or not the specific suggestions made in this study are found to have wide application, there is clearly a need to bring together work being achieved within the fields of theoretical geomorphology, morphometric analysis and the designation of land systems for development purposes. Solution of problems posed by this situation has importance both to the geographical study of environment and to the practical aspects of land classification and regional development in tropical Africa.

Acknowledgement

I am grateful to Dr R. Webster of the University of Oxford for reading the draft of this chapter and for his comments and constructive criticisms.

References

ALEXANDER, L. T. and CADY, J. G. 1962. *Genesis and Hardening of Laterite in Soils*, Tech. Bull. U.S. Dep. Agric. Soil Conserv. Serv. 1218.

BARNES, J. W. (ed.). 1961. *Mineral Resources of Uganda*, Bull. Geol. Serv. Uganda 4.

BAWDEN, M. G. and TULEY, P. 1966. *The Land Resources of Southern Sardauna and Southern Adamawa Provinces, Northern Nigeria*, Land Res. Div., Directorate Overseas Surv., Tolworth.

BECKETT, P. H. T. and WEBSTER, R. 1965. *A Classification System for Terrain*, Rep. Military Engineering Experimental Establishment 872, Christchurch.

BISHOP, W. W. 1966. Stratigraphic geomorphology: a review of some East African landforms. In DURY, G. H. (ed.). *Essays in Geomorphology*, 139–76, London.

BOURNE, R. 1931. *Regional Survey and its Relation to Stocktaking of the Agricultural and Forest Resources of the British Empire*, Oxf. For. Mem. **13.**

BRINK, A. B., MABBUTT, J. A., WEBSTER, R. and BECKETT, P. H. T. 1966. *Report of the Working Group on Land Classification and Data Storage*, Rep. Military Engineering Experimental Establishment **940**, Christchurch.

BÜDEL, J. 1957. Die 'Doppelten Einebnungsflächen' in den feuchten Tropen. *Z. Geomorph*. N.F. **1**, 201–25.

CHENERY, E. M. 1960. *An Introduction to the Soils of the Uganda Protectorate*, Mem. Res. Div. Ser. 1. Soils, **1**, Uganda Protectorate Dep. Agric., Kampala.

CHRISTIAN, C. S. 1957. The concept of land units and land systems. *Proc. 9th Pacific Sci. Conf. 1957* **20**, 74–81.
    1959. The eco-complex and its importance for agricultural assessment. In KEAST, A., CROCKER, R. L. and CHRISTIAN, C. S. (eds.), *Biogeography and Ecology in Australia*, Monographiae Biol. VIII.

CHRISTIAN, C. S. and STEWART, G. A. 1953. *Survey of the Catherine-Darwin Region*, 1946, Land Res. Ser. C.S.I.R.O. Aust. **1**.

CLARKE, G. R. 1957. *The Study of the Soil in the Field*, 4th edition, Oxford.

CLAYTON, W. D. 1958. Erosion surfaces in Kabba Province, Nigeria. *Jl W. Afr. Sci. Ass.* **4**, 141–9.

DE SWARDT, A. M. J. 1964. Lateritisation and landscape development in equatorial Africa. *Z. Geomorph*. N.F. **8**, 313–33.

DE SWARDT, A. M. J. and CASEY, O. P. 1963. *The Coal Resources of Nigeria*, Bull. Geol. Surv. Nigeria **28**.

D'HOORE, J. L. 1964. *Soil Map of Africa Scale 1 : 5,000,000. Explanatory Monograph*, Publ. Com. Co-operation Technique au Sud du Sahara **93**, Lagos.

DOORNKAMP, J. C. 1966. *The Relationships between the Geomorphology and the Tectonic Deformation in a Part of South-West Uganda*, unpublished Ph.D. thesis, Nottingham University.
    1969. *Geomorphological Maps of South-West Uganda at 1 : 250,000*, Geol. Surv. Dep. Uganda, (in press).

DOORNKAMP, J. C. and TEMPLE, P. H. 1966. Surface, drainage and tectonic instability in part of southern Uganda. *Geogrl. J.* **132**, 238–52.

DOWLING, J. W. F. and WILLIAMS, F. H. P. 1964. The use of aerial photographs in materials surveys and classification of landforms. *Civil Eng. Problems Overseas, 1964*, Conf. Inst. Civil Engrs.

DU PREEZ, J. W. 1949. Laterite: a general discussion with a description of Nigerian occurrences. *Bull. Agric. Congo-Belge* **40**, 53–66.

ELLIS, J. H. 1932. A field classification of soils for use in a soil survey. *Scient. Agric.* **338**.

FAO. LAND USE SURVEY TEAM. 1962. *Report on the Land Use Survey of the Oyo–Shaki Area, Western Nigeria* (mimeographed paper), Food and Agric. Org., Ibadan.

HAANTJENS, H. A. 1965. Practical aspects of land system surveys in New Guinea. *J. trop. Geogr.* **21**, 12–20.

HARPUM, J. R. 1963. Evolution of granite scenery in Tanganyika. *Rec. Geol. Surv. Tanganyika* **10**, 39–46.

HORTON, R. E. 1945. Erosional development of streams and their drainage basins; hydrophysical approach to quantitative morphology. *Bull. Geol. Soc. Amer.* **56**, 275–370.

HOWARD, A. D. and SPOCK, L. E. 1940. Classification of landforms. *J. Geomorph.* **3**, 332–45.

JENNY, H. 1958. Role of the plant factor in the pedogenic functions. *Ecology* **39**, 5–16.

JESSEN, O. 1936. *Reisen und Forschungen im Angola*, Berlin.

KING, L. C. 1948. A theory of bornhardts. *Geogrl. J.* **112**, 83–7.

1962. *The Morphology of the Earth*, Edinburgh.

KLIMASZEWSKI, M. 1963. The principles of geomorphological mapping in Poland; and, Landforms list and signs used in the detailed geomorphological maps, Problems of Geomorphological Mapping. *Inst. Geogr. Polish Acad. Sci., Geogr. Stud.* 46, 67–71 and 139–77.

LEWIS, G. M. 1959. Some recent American contributions in the field of landform geography. *Trans. Inst. Br. Geogr.* **26**, 25–36.

LINTON, D. L. 1951. The delimitation of morphological regions. In STAMP, L. D. and WOOLDRIDGE, S. W. *London Essays in Geography*, **11**, 199–217.

LUEDER, D. R. 1959. *Aerial Photographic Interpretation*, New York.

MABBUTT, J. A. 1961a. A stripped landsurface in Western Australia. *Trans. Inst. Br. Geogr.* **29**, 101–14.

1961b. 'Basal Surface' or 'Weathering Front'. *Proc. Geol. Ass.* **72**, 357–9.

1962. Geomorphology of the Alice Springs area. In PERRY, R. A. *et al.*, *1962 Lands of the Alice Springs Area, Northern Territory, 1956–57*, Land Res. Ser. C.S.I.R.O. Aust. **6**, 163–178.

1965a. The weathered landsurface of central Australia. *Z. Geomorph.* N.F. **9**, 82–114.

1965b. *Report on a Visit to Northern Nigeria*, Soil Sci. Lab., Dep. Agric., Univ. Oxford.

MABBUTT, J. A. and STEWART, G. A. 1963. The application of geomorphology in resources surveys in Australia and New Guinea. *Revue géomorph. Dyn.* **14**, 97–109.

MAIGNEN, R. 1960. Soil cuirasses in West Africa. *Sols Afr.* **4**, 5–42.

MELTON, M. A. 1957. *An Analysis of the Relations and Elements of Climate, Surface Properties, and Geomorphology*, Tech. Rep. 11, Off. Naval Res., New York.

MILNE, G. 1935 Some suggested units of classification and mapping, particularly for East African soils. *Soil Res.* **4**, 183–98.

MOSS, R. P. 1968. Soils, slopes and surfaces in tropical Africa. In MOSS, R. P. (ed.). *The Soil Resources of Tropical Africa*, Cambridge.

MULCAHY, M. J. 1961. Soil distribution in relation to landscape development. *Z. Geomorph.* N.F. **5**, 211–25.

OLLIER, C. D. 1959. A two-cycle theory of tropical pedology. *J. Soil Sci.* **10**, 137–48.

1960. The inselbergs of Uganda. *Z. Geomorph.* N.F. **4**, 43–52.

OLLIER, C. D. and HARROP, J. F. 1960. *The Soils of the Eastern Province of Uganda*, Mem. Res. Div. Ser. 1: Soils 2, Uganda Protectorate Dep. Agric., Kampala.

PALLISTER, J. W. 1956. Slope development in Buganda. *Geogrl. J.* **122**, 80–87.

PERRY, R. A. et al. 1962. *General Report on the Lands of the Alice Springs Area, Northern Territory, 1956–57*, Land Res. Ser. C.S.I.R.O. Aust, **6**.

PRESCOTT, J. A. and PENDLETON, R. L. 1952. *Laterites and lateritic soils*, Tech. Comm. Commw. Bur. Soil Sci. 47.

RADWANSKI, S. A. 1960. *The soils and land use of Buganda*, Mem. Res. Div. Ser. 1: Soils 4, Uganda Protectorate Dep. Agric., Kampala.

RUXTON, B. P. and BERRY, L. 1957. Weathering of granite and associated erosional features in Hong Kong. *Bull. geol. Soc. Am.* **68**, 1263–92.

    1961. Weathering profiles and geomorphic position on granite in two tropical regions. *Revue géomorph. Dyn.* **12**, 16–31.

SAVIGEAR, R. A. G. 1965. A technique of morphological mapping. *Ann. Ass. Am. Geogr.* **55**, 513–38.

SCHUMM, S. A. 1954. *Evolution of Drainage Systems and Slopes in Badlands at Perth Amboy, New Jersey*, Off. Naval Res. Proj. NR. 389–042 Tech. Rep. **8**.

SMYTH, A. and MONTGOMERY, R. F. 1965. *Soils and Land Use in Central South-Western Nigeria*, Ibadan.

SOIL SURVEY STAFF, 1951. *Soil Survey Manual*, U.S. Dep. Agric. Handbook 18.

    1960. *Soil Classification. A Comprehensive System*, 7th Approximation, U.S. Dep. Agric.

SOKAL, R. R. and SNEATH, P. H. A. 1963. *The Principles of Numerical Taxonomy*, San Francisco.

SPECK, N. H. et al. 1964. *General Report on Lands of the West Kimberley Area, W.A.*, Land Res. Ser. C.S.I.R.O. Aust. **9**, Melbourne.

    1965. *General Report on Lands of the Tipperary Area, Northern Territory, 1961*, Land Res. Ser. C.S.I.R.O. Aust. **13**, Melbourne.

SPURR, A. M. M. 1954. A basis of classification of soils of areas of composite topography in central Africa, with special reference to the soils of the southern highlands of Tanganyika. *2nd Inter-Afr. Soils Conf., Leopoldville, Doc.* **7**, 175–92.

STORY, R. et al. 1963. *General Report on the Lands of the Hunter Valley*, Land Res. Ser. C.S.I.R.O. Aust. 8, Melbourne.

STRAHLER, A. N. 1954. Quantitative geomorphology of erosional landscapes. *Cr., 19th Int. Geol. Congr. Algiers 1952*, Sec. **13**, 341–54.

THOMAS, M. F. 1965. An approach to some problems of landform analysis in tropical environments. In WHITTOW, J. B. and WOOD, P. D. (eds.). *Essays in Geography for Austin Miller*, Reading, 118–44.

    1966. Some geomorphological implications of deep weathering patterns in crystalline rocks in Nigeria. *Trans. Inst. Br. Geogr.* **40**, 173–93.

THORNBURY, W. D. 1958. *Principles of Geomorphology*, New York.

THORP, M. B. 1967. Joint patterns and landforms in the Jarawa granite massif, Northern Nigeria. In STEEL, R. W. and LAWTON, R. (eds.), *Liverpool Essays in Geography*, London, 65–83.

TRENDALL, A. F. 1962. The formation of apparent peneplains by a process of combined lateritization and surface wash. *Z. Geomorph. N.F.* **6**, 183–97.

TRICART, J. 1961. Notice explicative de la carte géomorphologique du delta du Senegal. *Mem. Bur. Recherche Géol. Min.* **8**.

TROLL, C. 1963. Landscape ecology and land development with special reference to the tropics. *J. trop. Geogr.* **17**, 1–11.

UNSTEAD, J. F. 1933. A system of regional geography. *Geography* **18**, 175–87.

VERSTAPPEN, H. TH. 1964. Geomorphology and the conservation of natural resources. *Publs. int. Train. Cent. aer. Surv.* B–**33**, 24–35.

WAYLAND, E. J. 1934. Peneplains and some other erosional platforms. *Am. Rep. and Bull. (1933), Geol. Survey Dep. Protectorate of Uganda,* Notes 1, 74, 366.

WEBSTER, R. 1963. The use of basic physiographic units in air photo interpretation. *Trans. Symp. Air Photo Interpretation, Inst. Arch. Photogram, Delft,* 14.

WOOLDRIDGE, S. W. 1932. The cycle of erosion and the representation of relief. *Scott. geogr. Mag.* **48**, 30–6.

WRIGHT, R. L. 1963. Deep weathering and erosion surfaces in the Daly River basin, Northern Territory. *J. geol. Soc. Aust.* **10**, 151–63.

1965. Geomorphology of the Tipperary Area. In SPECK, N. H. et al., *General Report on the Lands of the Tipperary Area,* Land Res. ser. C.S.I.R.O. Aust. **13**, Melbourne.

YOUNG, A. 1965. *The Physical Environment of Northern Malawi, with Special Reference to Agriculture,* Dep. Agric., Malawi.

7 The soil resources of West Africa

R. A. PULLAN

Introduction

There is no resource more important to West Africa than soil. Whatever its capabilities may be in terms of supporting plant and animal life (and these are very variable), and however maligned it may have been in this respect (often quite unjustly), the soil has furnished indirectly adequate food for the population. Agricultural exports have dominated the economies of West African countries for many years and will continue to do so. Food taken from lake, river and ocean environments forms only a small part of the food supply for a small percentage of the population of this area and the provision of food by hydroponics, by which means plants are fed artificially and supported mechanically, is unknown.

The use of the soil resources must be undertaken in such a way that they are conserved and not exploited, for exploitation can mean soil destruction through erosion. Soil erosion may increase rapidly in the near future. Population growth is bringing about the breakdown of the traditional farming system of shifting cultivation. This is most marked where population density is already high and the fallow period is being shortened as land becomes scarcer. Yields are declining, more land has to be cultivated to maintain supplies of food, and a greater surface of soil is exposed to possible erosion. The continued practice of the annual burning of fallow land and regrowth woodland is a further factor preventing the optimum use of the fallow period for restoring the nutrient supply of the topsoil in the savanna areas (Nye and Greenland, 1960). However, soil erosion is not yet a serious problem throughout West Africa though it is apparent around the larger villages and towns. Elsewhere it is contained by the practice of shifting cultivation except in areas with steep slopes.

There are many large areas in West Africa where expansion of the cultivated area can be achieved using traditional methods of cultivation without putting the soil at hazard. All that may be required is the provision of roads and permanent rural water supplies. Elsewhere declining yields may be halted and normal yields increased by several methods. Many agronomic practices which bring about increases in yields are not related to soils directly. They include the control of weeds, disease and animal pests at all stages of growth, and the improvement and regulation of water supply to plants. Indeed, decline in the supply of nutrients from the soil is only one and possibly a minor factor in declining crop yields under shifting cultivation and its relative importance is very variable. However, it can be tackled through the addition of artificial fertilizers and the results of the fertilizer trials programme of the FAO Freedom from Hunger Campaign, initiated in 1961, are encouraging (Richardson, 1965). Traditional methods of maintaining soil nutrients by the application of compound sweepings have been restricted to the fields adjacent to the compounds or villages with few exceptions. There are larger areas of permanently cultivated farms about the larger towns in the northern savannas. The best-documented example is Kano, in Nigeria, where town refuse and night-soil composted with compound sweepings are carried to the fields during the dry season. Soil erosion has been slight about Kano for the terrain is flat, the topsoil is sandy and readily absorbs rainfall, and deflation by wind is restricted by the presence of many economic farm trees throughout the fields and by the provision of low hedges. Other methods such as the night herding of nomadic cattle on specific fields and the more recent introduction of mixed farming in areas of the Sudan savanna in Northern Nigeria have not resulted in the practice of fallowing being abandoned.

The great increase in the use of chemical fertilizers that is to be expected in West Africa in the near future will rapidly extend the problem of soil erosion. The tendency will be for fields to be cultivated for longer periods of time and if the use of fertilizers goes hand in hand with improved agronomic practices, as might be expected, the problem of weeds, diseases and pests will be controlled and permanent cultivation will be achieved, and much more land will be available for food production. However, the danger of soil erosion will then be greater unless conservation is also undertaken. Some permanent cultivation is found about Zaria, some seventy miles south of Kano. Here the rainfall is higher and the soils have a finer texture than at Kano. They tend to form an impermeable cap during the dry season and rainfall infiltration is slow. The result is that sheet wash and gully erosion are rapidly destroying large areas of soil and the problem is aggravated by the presence beneath the topsoil of a lateritic ironstone horizon in the soil profile.

Whatever methods are used to increase food production in West Africa they must be compatible with soil conservation. Conservation involving earthworks is very expensive, and if these cannot be constructed the implementation of rational land-use plans should ensure that soils liable to erosion should not be cultivated but used in other ways. The formulation of land-use plans demands that the soil pattern should be mapped and the physical and chemical characteristics of the different soils described. Only if this precedes fertilizer programmes, the opening up of unpopulated areas, the creation of permanent cultivation and the rational planning of cultivation, grazing and forestry, will the soils be conserved.

The proper utilization of the soil resources of any area necessitates the provision of basic data on soils. Firstly, the major soil groups have to be described and identified within a classification, which may have to be modified to accommodate them. Their physical and chemical characteristics must also be determined so that the range and variation within the soil groups is known. Secondly, the soils have to be mapped.

The nature of the basic information on the distribution of soils

Basic soil data in West Africa are derived from surveys which fall into two broad categories; detailed and reconnaissance.

Detailed soil surveys usually cover areas of the order of tens of square miles and are undertaken to provide specific information for research projects on experimental agricultural stations or for agricultural projects in which there will be a high capital outlay as in irrigation schemes. The maps will show soils at series, or even type and phase, level, at scales of 1:10,000 or greater. The area will be sampled by auger on a rectangular grid at intervals ranging from 100 to 300 ft, depending on the complexity of the soil pattern and the time available. A smaller interval may be required on alluvial soils, especially if the soils are believed to be saline to varying degrees. Field measurements of permeability and conductivity may be taken together with replicate samples throughout the profiles of the soil series identified. Topographic surveys will also be made.

Reconnaissance soil surveys cover hundreds of square miles and are frequently carried out systematically in areas bounded by map sheet lines. The aim is to identify all the areally important soil series which can then be described and their properties recorded after analysis. Methods vary considerably and are often influenced by the availability or non-availability of contoured topographic and geological maps and air photographs, by the degree of accessibility, and the time available. In areas of savanna it is usual to determine the areas to be sampled by undertaking a comprehensive interpretation of the air photographs and delimiting those areas of different terrain which might represent soil landscape units. These units should show uniformity of landform association, vegetation and drainage pattern. Changes in land-use pattern are also used as a possible indication of soil changes but must always be regarded tentatively until the relationship is established in the field. A programme of auger sampling is then devised, using existing roads and tracks wherever possible, to include all the soil landscape units. The positions of cut traces or, more probably, suitable footpaths can be chosen along which samples may be taken where no road crosses a unit. Areas which, from experience, are recognized as having

F

predominantly shallow or stony soils, lateritic ironstone outcrops, or complex alluvial soil associations may be avoided in the sampling programme if they are areally unimportant or time is short. Auger sampling at intervals of 500 ft may be undertaken along roads and tracks to prove that the unit, previously defined, exists. More detailed sampling at intervals of 100–300 ft may be undertaken to show the importance of each soil series along catenas or toposequences of soils. Profile pits are dug in the soil series identified at typical locations and the soil described and sampled for analysis.

In areas of rain forest the value of air photographs decreases and regular sampling along all roads and major pathways may have to be undertaken without previous selection.

The soil map scale frequently conforms to the scale of existing published map series. In Nigeria 1:100,000 is used; in Chad 1:200,000 has been chosen. At this scale it is usual to show soil associations rather than soil series. The soil association may be a group of soils which have a catenary or toposequence relationship. On the other hand, it is possible to use associations of soil series which have a common parent material. Where variations in the parent material take place over short distances or where there is no simple relationship of soils to topographic position, soil complexes may be mapped. The complexes and associations are given local topographic names for easy reference. A soil series may be found in one or more soil associations. If parent material is used as the basis for association mapping the soil map may be expected to look like the geological map unless other criteria are used for a further breakdown. The use of soil association based on landscape units is more flexible and can be subdivided in terms of parent material should an important lithological boundary divide a particular landform association. In Northern Nigeria, association mapping on the basis of landforms has replaced that of parent material.

A different approach is held by the pedologists of the Office de la Recherche Scientifique et Technique Outre-Mer, Dakar (ORSTOM). They use the basic classification of Aubert (1963) based on profile characteristics in which the major soil groups are subdivided in terms of significant pedological features, such as colour or the presence of concretions, and then on the basis of parent material. Soil associations are devised in terms of series falling into the same sub-groups. If two or more soil groups are found together a complex is devised by combining the representative colours and symbols of these soils. This system has the advantage of portraying the distribution of the major soil groups and their subdivisions, though the relationship of these within the complexes is often not clear as the soil memoir describes the soils within the classification rather than map areas with associations of soils.

The major soil groups that are found in West Africa have been adequately described in terms of their profiles, and their physical and chemical properties have been determined for some series within the soil groups. The range and variation at series level is not yet known. Fournier (1963) and d'Hoore (1964) have provided comprehensive general surveys of African soils,[1] and a great amount of new material from recent surveys in many countries in West Africa is rapidly filling gaps in the general picture. This framework can be used by agronomists in their studies of fertilizer application to crops in the different agricultural regions.

The distribution of the major soil groups is known only in very general terms. This information is based on detailed and reconnaissance soil surveys. These are few in number and widely scattered throughout West Africa. There are still large areas without information which has been systematically obtained. The compilation of national and continental soil maps demands, therefore, a large measure of extrapolation and, before these small-scale maps are used, the basis of the extrapolation must be understood. This may be implicit in the nature of the classification of soils that is used.

The general pattern of West African soils

Soil distribution over Africa has been mapped at the scale of 1:20 million by Marbut and Shantz (1923), Schokalskaya (1953) and FAO (1962), but the high degree of generalization inherent in the use of this small scale, inadequate basic data, and the attempted application of soil classifications and pedogenetic concepts derived from the study of non-tropical soils, make these maps of little use for any purpose.

More recently the Commission for Technical Co-operation in Africa (C.C.T.A.) has published a *Soil Map of Africa* at the scale of 1:5 million together with a descriptive monograph by the general co-ordinator of the project, d'Hoore (1964). This map uses mapping units based mainly on the French classification applied to West Africa (Aubert, 1963) but incorporating also some elements of the Belgian classification (Sys, 1960) which was devised for Congolese soils as a replacement for the unsatisfactory application of an American system (Kellog and Davol, 1949). The adoption of a predominantly French system in West Africa is logical for the largest areas covered by existing soil surveys were in territories where the pedologists of ORSTOM had worked. In Ghana and Northern Nigeria the classification devised by Charter (1957) was being used, while an independent classification had been devised by Da Costa and Azvedo (1960) for use in the Portuguese territories. These and other applied classifications have been correlated with the French system as far as possible though many difficulties exist.

[1] A brief description of the major soil groups found in Africa is given in Appendix B, pp. 531-536.

Twelve major soil groups are defined and these have been divided into sixty-two soil elements. The elements are individual soil groups divided in terms of parent material, or, where this is impossible owing to the complexity of the lithological pattern, this may be unspecified.

The different soil elements are combined according to their importance within the mapping areas and two hundred and seventy-five different mapping units are used. Ideally a soils map at this scale should derive information from national soil maps and should be concerned with the correlation of soil classifications and the rationalization of mapping units. Very few such maps existed at the time of compilation in West Africa and only two had been published. National soil maps were prepared in draft and submitted for incorporation into the continental soil map. Some of these have been published since (Boulet *et al.*, 1964; Bocquier and Gavaud, 1964; Martin and Segalen, 1965; Tomlinson, 1965). Yet it must be remembered that many of these were prepared with very little basic soils data. The draft map for Northern Nigeria was prepared from survey data covering only 17 per cent of the country. Two tentative revisions have been made since, and new survey data has been incorporated. There has also been a reappraisal of the classification of certain soils and in particular those on sandy parent materials.

Extrapolation of soil boundaries has to be made at two levels. The first, and most difficult, is at the major group level. The difficulty is least when the major soil group is directly related to parent material and to physiography, greatest where it is related to the interplay of changing environmental factors over a period of time going back as far as the early Pleistocene at least and possibly, in the case of those soils with lateritic ironstone horizons within their profiles, much earlier than this. The second level is the breakdown of the soil groups using parent material, one factor which greatly influences the variations of soils within the great soil group. In this case geological maps are of some help, especially if they detail the lithology of the rocks as well as their age.

Hence the boundaries (essentially transitional zones) between the ferrallitic, ferrisolic and ferruginous tropical soils vary in reliability. The designation of areas of lithosolic soils, eutrophic brown earths and young alluvial soils presents little problem.

The generalized soil map for West Africa (Fig. 7.1) is derived from the *Soil Map of Africa* (d'Hoore, 1964) with some revision. It emphasizes the dominance of ferruginous tropical soils in the eastern and central parts of West Africa. Ferrallitic soils dominate the western parts. Though there is a clear relationship in a general sense between the areas of high, evenly distributed rainfall and the ferrallitic soils, climatic parameters for these soils cannot be devised nor is there a precise relationship between either

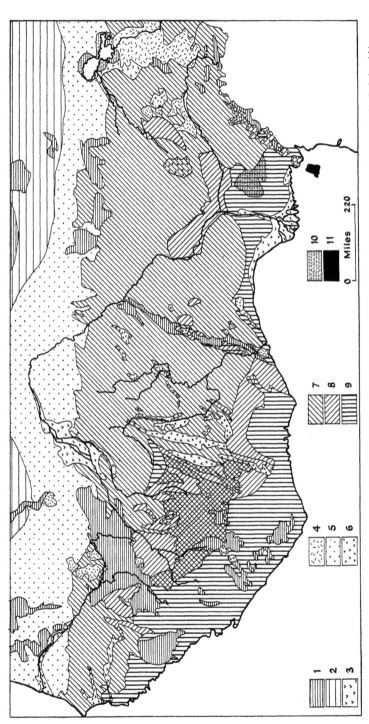

1. Lithosols and raw mineral soils.   2. Desert soils.   3. Predominantly hydromorphic soils developed on recent and late Pleistocene alluvium.   4. Vertisols.   5. Sub-arid brown and reddish brown soils.   6. Eutrophic brown soils.   7. Ferruginous tropical soils.   8. Ferrisols.   9. Ferrallitic soils.   10. Organic hydromorphic soils.   11. Volcanic complex.

Figure 7.1 *Distribution of major soil groups in West Africa* (simplified after d'Hoore, 1964, with alterations in Northern Nigeria).

potential or existing vegetation and the major soil groups. There is an increasing body of evidence available to substantiate the theory of the contraction and expansion of the rain forest, savanna and semi-arid climatic/vegetation zones in West Africa during the Pleistocene. It is also generally accepted, though proof is generally lacking, that many of the landforms of West Africa are older than the Pleistocene and can be dated within the Cainozoic. On the other hand large areas within the Sudan and Sahel zones of West Africa have landforms which originated in the Late Pleistocene and yet have experienced changes in climate of considerable importance. It is not surprising that palaeosols have been identified throughout West Africa and it is exceedingly difficult to assign dominant pedogenetic features to a particular climate.

One group of soils bears a much closer relationship to present climate than the others. These are the brown and reddish-brown semi-arid soils. Work on these soils has been summarized by Bocquier and Maignien (1963) and detailed descriptions given by Maignien (1959b), Boulet *et al.* (1964) and Bocquier and Gavaud (1964). The accumulation of free calcium in the middle and lower horizons of these profiles is directly related to the rainfall and temperature régimes while the low but well-distributed organic matter content is related to the present vegetation of very open tree savanna with *Acacia* spp. and a cover of mainly annual grasses and other herbs whose root systems provide an annual increment of organic matter.

The division of the major soil groups into mapping units is based in part on the types of parent material from which they are derived. There is no doubt that this facilitated extrapolative mapping but the influence of parent material on both the physical and chemical variations within the soil groups can be demonstrated. It may also be the prime pedogenetic factor where the profile is young and less well developed.

Figures 7.2 and 7.3 have been compiled from existing geological maps and geological memoirs. They show the distribution of sedimentary and crystalline rocks and the pattern of these is of great importance in understanding the soils map of West Africa. Igneous and metamorphic rocks dominate the areas of heavier rainfall with the exceptions of the areas of recent alluviation along the rivers and coasts and in particular the Niger delta, the Cretaceous sediments of the Niger and the Benue troughs, and the Palaeozoic Voltaian and Pre-Cambrian Buem sediments in Ghana and Upper Volta. The rocks of the Basement Complex are not well known and large areas have not been mapped and so remain undifferentiated. However, air photographic interpretation allows the differentiation of contrasting lithological areas and these, allied to detailed ground survey, allow the differentiation of the following geological provinces:

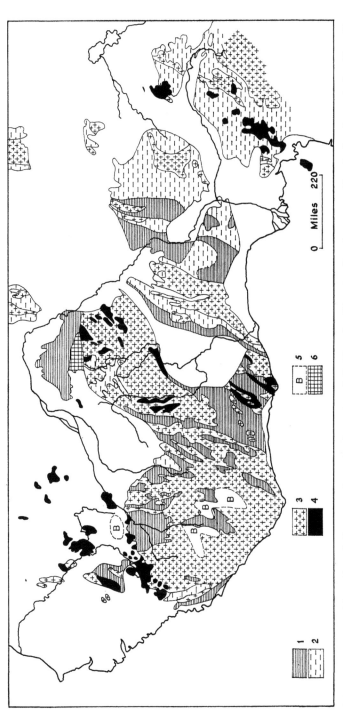

1. Metasediments comprising schists, phyllites, meta-siltstones and quartzites. 2. Undifferentiated Basement Complex composed largely of gneisses and migmatites with Older Granites. 3. Granites and granitised rocks. 4. Metamorphosed lavas, dolerites and basalts. 5. Scattered small basalt and dolerite outcrops. 6. Weakly metamorphosed dolomites.

Figure 7.2 *Distribution of crystalline rocks in West Africa.* The key shows the rock types which predominate.

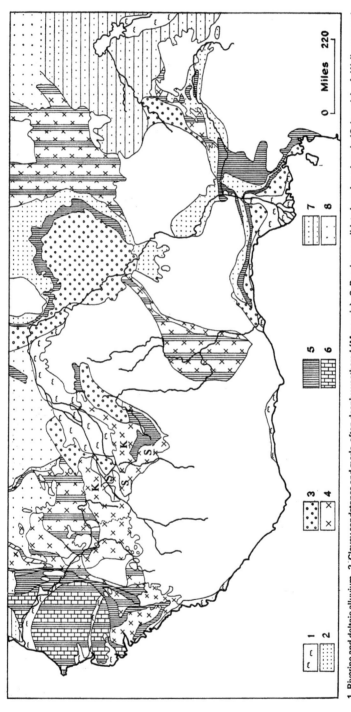

1. Riverine and deltaic alluvium. 2. Clayey sandstones and marls, often deeply weathered (Mesozoic). 3. Poorly consolidated sands and marls (Cainozoic). 4. Well cemented grits and sandstones (early Palaeozoic). 5. Predominantly shales and clay-shales. 6. Calcareous marls and clays of marine origin. 7. Unconsolidated clays, sandy clays and sands, with surface dunes and fine alluvium. 8. Loose, frequently unstabilized sands (Recent).

Figure 7·3 *Distribution of sedimentary rocks in West Africa.*

1. A complex of Pre-Cambrian gneisses, migmatites and coarse-grained metasomatic rocks into which granites of Late Pre-Cambrian and early Palaeozoic age have been intruded as bosses and batholiths. Different degrees of granitization have occurred. Syenites, granodiorites, diorites and quartzdiorites are also found together with quartzschists and quartzites though areally these are of restricted importance. This geological province is important in the eastern half of West Africa. It may be subdivided into small units of restricted occurrence which also contain intrusions of what have been called the Younger Granites, of Jurassic age. These form ring complexes and are invariably associated with volcanic activity and lava fields of Tertiary age, which may be acidic or basic in character. The close assemblage of many granitic intrusions and volcanic activity not only produces a distinctive geological assemblage but has also affected the gemorphological evolution of these areas. Thus the Jos Plateau and the Eastern Highlands of Nigeria belong to this sub-province and both rise to form plateaux above 4,000 ft (1,300 m) with hills rising to 5,000 ft (1,600 m). Figure 7.4 illustrates the geological pattern of part of the Jos Plateau. Rarely do the lava fields form distinctive units like the Biu Plateau in Nigeria.

2. Zones and enclaves of metasediments occur throughout the granite/gneiss/migmatite province. The rocks consist of quartzites, quartzschists, schiste, phyllites, shales and metamorphosed lavas (greenstones) believed to be of Late Pre-Cambrian and Early Palaeozoic age. The strata are highly folded and dip steeply so that lithological changes are frequent. The pattern (Fig. 7.5) is usually one of elongate troughs striking north–south, though transverse pitching domes and basins have been described from the Yelwa area in Northern Nigeria (Pullan and de Leeuw, 1964). Small elongate meta-sedimentary zones form rock pendants in the Ivory Coast, but their distribution is irregular though they conform to the general north–south trend.

3. Large areas in which granites and granitized metamorphic rocks predominate are found in the western half of West Africa. The geology of Sierra Leone and Liberia is dominated by granites though zones of schists are found. In Mali and the Ivory Coast the pattern is more diverse and roof pendants of schists and gneiss are more important. They may underlie the highest terrain as in the Nimba Mountains. Particular note should be made of the hypersthene granite province of the Man Massif.

4. Dolerites, representing horizons within large laccoliths outcrop over a restricted area of Basement Complex in Guinea and over Palaeozoic sediments in Senegal and Mali.

These geological provinces and the variations which are found within them are of great importance in understanding the distribution of the soil groups. In particular they explain the distribution of rocks rich in ferro-magnesian minerals. Thus ferrallitic soils developed over these rocks are rare, for these soils occur mainly in the province dominated by acid granitic

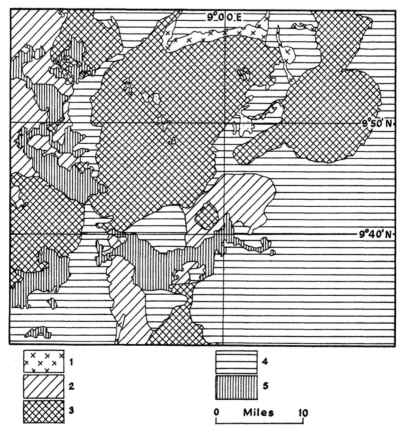

1. Quartz porphyry and granite porphyry. 2. Older Granite. 3. Younger Granite. 4. Undifferentiated gneiss with some schist. 5. Newer Basalt.

Figure 7.4 *Lithology of part of the Younger Granite Province, Jos Plateau, Northern Nigeria* (after Geological Survey of Nigeria).

rocks. However, they are found over metasediments in the Ivory Coast and humic ferrallitic soils are associated with the calc-magnesian granitic province of the Man Massif. Where the metasediments are associated with high mountains with highly dissected topography, such as the Nimba Mountains, lithosolic and rock debris soils are found, and upstanding

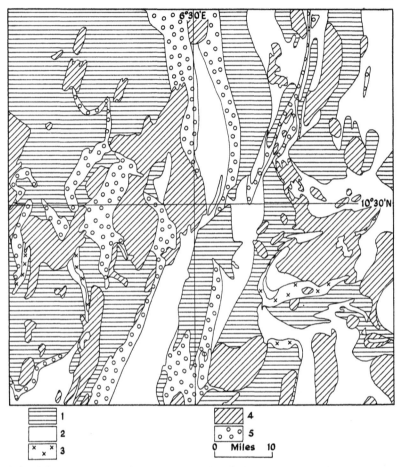

1. Banded migmatite and gneiss with some granite. 2. Schists, phyllites and meta-siltstones.
3. Amphibolite. 4. Older Granites. 5. Quartzites and granulites.

Figure 7.5 *Lithology of part of the Metasedimentary Province, around
Kusheriki, Northern Nigeria* (after Geological Survey of Nigeria).

quartzite ridges in areas of metasediments are usually associated with such
soils.

The presence of ferro-magnesian-rich rocks in areas dominated by acidic
rocks may bring about a change in major soil group. In the Ivory Coast,
ferrisolic soils are associated with schists in areas dominated by ferruginous
tropical soils. Dabin *et al.* (1960) have described eutrophic brown soils on
dolerites and greenstones in areas dominated by ferrallitic and ferrisolic
soils, and a similar relationship is also found in Guinea. Eutrophic brown
soils are regarded as juvenile soils developed on base-rich rocks and

associated with areas of recent river dissection. They are shallow and stony and being constantly eroded, so that they can maintain their character without developing into deeper, excessively leached ferrallites or ferrisols. It is more usual, however, to find the eutrophic brown soils on basalts or greenstones in areas dominated by ferruginous tropical soils. Eutrophic brown soils are associated in catenary sequence with vertisolic soils in areas with 30–40 in (750–1,000 mm) of rainfall annually. They have been described from the Biu Plateau in Nigeria and from north-west Mali, on basalts and dolerites respectively. Similarly, basalts underlie eutrophic brown soils in northern Cameroon (Laplante, 1954a), but in the wetter Mambilla Plateau to the south Hope (1966) has described humic ferrisolic soils over both basalts and gneiss, while Laplante (1954b) has described what appear to be ferrisols on the basalts in the Cameroon Highlands with a rainfall over 80 in (2,000 mm). At a lower elevation, Hope and Clayton (1958) have described ferrisolic soils over both gneiss and basalt at Jema's immediately south-west of the Jos Plateau in an area with a mean annual rainfall of 67 in (1,675 mm). Such examples indicate the importance of the length of time for pedogenesis in determining the type of soil that has evolved and this is a reflection of the geomorphological stability of the landforms.

Table 7.1 has been compiled to show the variations in nutrient status of ferruginous tropical soils developed under similar conditions on different parent materials within an area of metasediments near the Middle Niger in Nigeria.

The distribution and lithology of sedimentary rocks is shown on Figure 7.3. Variations in lithology within the metasediments could not be shown on Figure 7.2 for folding and subsequent erosion have produced changing patterns. By way of contrast the sedimentary sequences vary little, generally speaking, and consist predominantly of sandstones. Thus the Voltaian succession in Ghana is composed of 1,200 ft (370 m) of sandstones and 1,300 ft (400 m) of variable sandstones, shales, tillites and rare limestones (Furon, 1963). Palaeozoic sedimentary rocks cover large areas of the Taoudenni Basin in Mali, Guinea and Upper Volta, and again sandstones predominate. There are 750 ft (230 m) of variable sandstones, including the Sotuba sandstone, followed by 300 ft (90 m) of micaceous shales and sandstones in the Bamako region. In the Bobo–Dioulasso region there are 3,000 ft (920 m) of sandstone, including the Koutiala Sandstone, and less than 300 ft (90 m) of variable shale and limestone (Haughton, 1963). All these areas are relatively unaffected by earth movements and though faulting has occurred in the Fouta Djalon region, in Guinea, the gently inclined or horizontal strata ensure that sandstones dominate as the soil parent material.

The Cretaceous sediments of the Benue trough in Nigeria are more

TABLE 7.1 *Cation exchange capacities and base saturation percentages for soils developed on rocks of different lithology in the metasediment province of the Middle Niger, Northern Nigeria*

	Soil depth (inches)	Granite		Gneiss		Schist		Phyllite		Cretaceous sandstone	
		CEC[1]	Saturation %	CEC	Saturation %	CEC	Saturation %	CEC	Saturation %	CEC	Saturation %
Auna survey (Pullan and De Leeuw, 1964)	0–6	11	36	4	62	7	81	2	100	4	66
	6–12	9	45	3	62	11	69	—	—	3	60
	12–24	10	45	3	62	28	67	15	100	2	54
	24–36	12	49	2	57	28	76	15	100	5	40
	36–48	10	64	7	71	29	77	23	100	7	30
	48–60	10	78	7	100	32	86	28	100	6	66
Bussa survey (Klinkenberg, 1965)	0–6	4	100	4	40	8	67	9	69	1	100
	6–12	3	100	7	38	7	68	6	51	1	100
	12–24	6	48	13	94	10	66	15	38	3	60
	24–36	6	48	15	48	6	76	13	58	5	30
	36–48	6	68	14	55	6	76	13	58	8	26
	48–60	5	72	—	—	—	—	—	—	8	26
Yelwa and Dukku surveys (Pullan and de Leeuw, 1964)	0–6	4	51	—	—	8	69	Not present		5	50
	6–12	4	66	8	40	9	49			6	58
	12–24	4	100	5	63	9	53			5	34
	24–36	11	49	6	81	13	57			6	25
	36–48	11	49	9	52	23	74			13	15
	48–60	11	44	5	100	—	—			13	15

[1] CEC (cation exchange capacity) – measured in m.equiv./100 g.

varied. Marine, and estuarine shales and sandstones alternate with thin limestones. Many of the sandstones are saline. Dissection of well-formed domes and synclines has exposed all variations in lithology in the Gongola Valley but, in the low-lying plains about the lower Benue, sandstones predominate in an area of gently dipping strata. Vertisolic and hydromorphic soils have developed on the heavier textured clays and shales and similarly the clay–shale sediments of Palaeocene age, which underlie the Lama–Hollis depression in Togo and Dahomey, have developed vertisolic soils. Vertisols have also developed on low-lying areas of base-rich crystalline rocks in Ghana (Brammer, 1962) and Togo (Leneuf, 1954).

Tertiary sedimentation has been concentrated in the great continental basins of the north where thick sandy clays form lacustrine deposits of the Continental Terminal. These generally poorly consolidated sediments cover large areas in the Middle Niger, outlying areas of the Upper Niger and the Wawa Bush area, east of Bauchi in Nigeria. Contemporary sediments of the Senegal Basin are marine in origin and consist of marls, limestones and clays. However, much of the area is covered with a thin and patchy but pedologically important cover of dune sands (Maignien, 1961a).

Quaternary sedimentation in the Chad Basin and elsewhere consisted of a predominantly lacustrine clay or sandy clay succession with sandy zones which are now important acquifers. Over 2,000 ft (470 m) is found in the Chad Basin, but once again the surface sediments have been reworked by wind in the Late Pleistocene to form a fossil dune field. The fine dune sands are interspersed between fine-textured recent deposits of the interdune depressions. Extensive recent lacustrine and lagoonal clay plains have been described south of Lake Chad (Pullan, 1964a) and their high montmorillonitic clay content has favoured the development of vertisolic soils.

The presence of large areas of unconsolidated or deeply weathered clayey sandstones is one of the most important characteristics of West African pedology. This fact has also provided the pedologist with some of his most difficult problems in terms of classification. Large areas of sandstones are covered by lateritic ironstone horizons. Where these form a cuirass at the surface, the 'soils' are classified as raw mineral soils. Large areas of sandstone in Guinea and Mali are so covered. Where colluvial and sedentary soils form a mantle, the lateritic ironstone may prevent or retard translocation of clay down the profile and so a well-structured B horizon will form. Under such conditions ferruginous tropical soils may form though they may have a low-base saturation, more typical of ferrisolic soils. Pullan (1964b) has described these soils from the Wukari area south of the Benue River. Where the lateritic ironstone horizon has been dissected and destroyed, the underlying deeply weathered sandstones and the colluvium, produced by the breakdown of the lateritic ironstone (Moss, 1965), present a freely drained impoverished parent material which develops a poorly differentiated profile whatever the climate. Ferrallitic profiles have developed on the deep Cretaceous sands of the Igala Plateau south of the confluence of the Niger and Benue Rivers (Story, 1961). The mean annual rainfall is 55 in (1,375 mm) and the vegetation is derived savanna woodland. Moss has described similar profiles on similar parent material north of Lagos in an area with a mean annual rainfall which varies from 45 to 70 in (1,125–1,750 mm). These soil profiles have been described by Smyth (1963) who has given some analytical detail. However, there is an abrupt transition to ferruginous tropical soils when the Basement Complex is reached, though climatically the areas are very similar. Smyth has described the difficulties of classifying the Basement Complex soils on gneiss for, though they have a high level of base saturation and shallow profiles above weathered rock, they show very little sign of a structural B horizon and the clay fraction has been shown by Nye (1955) to be kaolinitic.

In Nigeria clayey sandstones extend along the Niger Valley as the Nupe Sandstones and after a break at Yelwa they extend as the Illo Sandstones. In Sokoto Province the Rima Sandstone represents a deep sandy deposit with only a small clay fraction. Recent work on the soils of the Nupe Sandstone, reported by Klinkenberg and Higgins (1966), suggests that these soils have ferrisolic profiles. They extend far into the area dominated by ferruginous tropical soils and an analogous situation is also found with the deep sandy soils on Eocene Sandstone west of the Gongola River. Analytical data for Nupe and Illo Sandstone soils given by Klinkenberg (1965) and Pullan and de Leeuw (1964) shows that these deep, largely undifferentiated profiles have very low base saturation percentages below the topsoil. Tomlinson (1961) has described deep undifferentiated profiles on Rima Sandstones which also have very low base saturation percentages but which may show some colour differentiation indicating the formation of a weakly leached upper horizon.

The area of ferrallitic soils in southern Mali and Upper Volta (Fig. 7.1) is not documented but may be similarly explained.

Pleistocene and Recent sediments along the major valleys in the savanna areas show clearly the differentiation in profiles resulting from variable length of development. In the Niger the sandy soils of the high-braided river terraces have weakly differentiated tropical ferruginous soils. Sandy levee soils of the more recent floodplains have mineral hydromorphic profiles while juvenile soils are found by the present river course. Similar associations are found in the more arid areas of the Central Niger Delta in Mali, the lower Senegal River in Senegal and the Yobe–Hadejia River system in Niger and Nigeria. Many soils of all textures within these areas are also saline. Salt accumulation continues slowly today, but may have been important during the past periods of greater aridity. The alluvial deposits themselves may be referred to the pluvial periods of the Pleistocene and there are numerous fossil floodplains associated with watercourses that flowed south through Niger to join the Hadejia and Yobe Rivers and also the Niger River. Fossil saline accumulations make the prediction of the occurrence of salinity hazards in alluvial soils very difficult. They have been recorded much farther south than might be expected in the Sokoto flood-plain where the streams are permanently flowing and the annual rainfall is over 30 in (750 mm) (Pullan, 1961).

The close relationship between the marine–estuarine alluvium and the development of acid sulphate soils (juvenile soils on fluvio-marine deposits) under mangrove swamp in Sierra Leone, Portuguese Guinea, Senegal and Gambia typifies the general theme that the younger the soil profile is, the closer the relationship with the parent material.

A twofold criticism of the choice of parent material as a dominant factor in soil differentiation can be made. Firstly, many West African soils have profiles which are developed in material of mixed parentage. Colluvial and aeolian topsoils may not always be derived from the underlying weathered rock. Similarly, lateritic ironstones frequently contain much iron of abso-lute accumulation which has been released by weathering higher up the local or regional slope. Such complex profiles frequently contain a horizon of quartz gravel over the weathered Basement Complex rocks. This is derived from the quartz veins which are found in the rocks at irregular intervals and it is spread over the whole slope by soil creep under present conditions or by fluvial action during pedimentation over a long period of slope evolution. The segregation of the gravel to form a stone line may be attributed to subsequent burying by colluviation or to termite activity. The gravel frequently separates the colluvial-biological and sedentary weather-ing sections of soil profiles which may be distinguished by a marked textural discontinuity.

Secondly, many West African soils may be developed on material which has been pre-weathered to a considerable depth (Thomas, 1966). The pedological significance of this has been discussed theoretically many times but there are no conclusive studies of the degree of transformation of the weathered material at different depths in deep weathering profiles. It is important to know whether there is progressive alteration of primary minerals up the profile or whether granular disintegration only has taken place.

It may be concluded from this brief analysis that, though there is a zonal pattern to the major soil groups of West Africa with respect to the ferral-litic, ferruginous tropical and the brown and reddish-brown soils, a direct relationship between soil, climate and vegetation is found only in the case of the semi-arid area with brown and reddish-brown soils. Parent material influences the character of the younger soils and may also, in the case of sandy sedimentary rocks, in areas with more than 30 in (750 mm) of rain annually, be a major factor in the creation of ferrisolic soils with an azonal distribution.

The detailed soil pattern

The fundamental soil unit for the purpose of description is the soil series. The scale of map which can be used to show soil series varies with the complexity of the environment as a whole. Many of the soil series have a re-petitive association with each other which is also related to their geomor-phological position. The practical application of soil resource knowledge demands that the distribution of the soils of varying character is known. Over large parts of Africa this pattern can be related directly to landforms and because of this the use of landform–soil associations is particularly useful, not only as an aid to the mapping of soils through air photograph interpretation but also as a shorthand method of describing the soil pattern where this is complex. In particular it is important because it allows soil maps at reconnaissance scales of 1:100,000 or smaller to contain, by impli-cation, much greater detail. These can be illustrated by sample detail sur-veys in very restricted areas which are typical of soil associations mapped at the smaller scale. Two examples of the way this may work can be given by way of illustration:

1. The presence of shallow soils (lithosols) or of lateritic ironstone horizons at or near the surface is an important factor in determining the successful utilization of the soil resources of such an area. The frequency and pattern of occurrence can determine whether the area has to be conserved under woodland, with or without grazing, or whether it is possible to combine both these with restricted agriculture. Knowledge that a specific pattern occurs within a particular soil association as mapped on the reconnaissance soil map allows a rapid appraisal of the possible use and therefore of the value of the soil resources of that area.

2. In certain West African savannas it is possible to use the alluvial soils associated with broad shallow valleys (*fadamas*) for either wet-season rice cultivation or small-scale irrigated vegetable gardens. The former system is best operated if tractor ploughing on contract to farmers is used while in the latter case the communal use of small pumps is desirable. The possible provision of these is dependent not only on there being suitable soils available but also on the pattern and acreage of the alluvium. The small *fadamas* cannot be shown on reconnaissance soil maps but frequently the drainage pattern, which is related to the

geomorphological evolution and the geological history of the area, can be schematized by detailed samples and the possibilities of implementing the various schemes determined on the basis of adequate soil–landform descriptions and soil association mapping.

This approach to soil–landform associations may be illustrated by considering the large area occupied by ferruginous tropical soils in Northern Nigeria. Within this area all the geological provinces described earlier are to be found. Important pedological differences may be related to parent material. There are lithosols, raw mineral soils, lateritic iron-stone crusts, surfaces covered by detrital laterite and alluvium. Some of the underlying rocks have been deeply weathered. Inselbergs may occur frequently, rarely, or be absent, and are associated with distinctive terrains. The area is situated several hundreds of miles from the sea and has escaped major Pleistocene rejuvenation. It has been geologically stable for a very long time and has undergone epeirogenesis rather than orogenesis. Climatic changes have brought about only minor alterations to the extensive undulating plains and smaller plateau surfaces from which rise single or groups of inselbergs and larger hill masses. Near the larger rivers of the Niger and the Benue, zones of increased dissection and lower surfaces can be found, while there are large alluvial infills in the Chad Basin to the north-east. The plains have a landscape associated with the formation of primary and secondary lateritic ironstone horizons which have undergone differing degrees of dissection. In the Chad Basin there has been alluviation and modification of the original fluvial and lacustrine landforms by aeolian action during the Pleistocene to a varying degree. The geomorphological evolution of the area is particularly important in determining the soil patterns throughout this area. Examples are given below of distinctive patterns of soils associated with high-level undissected plains, with and without inselbergs, and dissected plains where the underlying metamorphic rocks give rise to complex relief and soil patterns, or a simple pattern on sedimentary rocks. In contrast to these areas, where older landforms are losing their dominance in the landscape and therefore younger soils are occupying increasingly greater areas, the Chad Basin is a large area of young landforms and soils. However, though these areas have apparent morphological uniformity the soil pattern reflects minor oscillations in climate during the late Pleistocene and this is illustrated by transects.

Some soil patterns from the high plains of Hausaland and its dissected fringe

A relatively undissected surface is well developed about Zaria, in Nigeria, situated near the focal watershed of drainage systems to the Chad Basin,

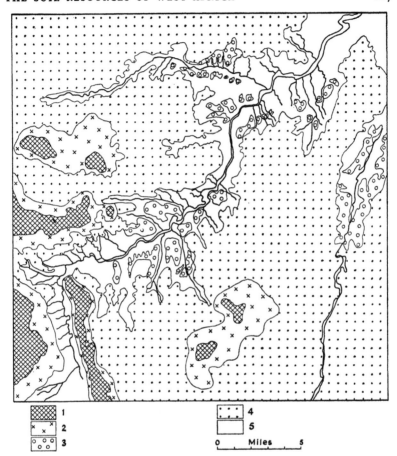

1. Primary, vesicular, lateritic ironstone, forming a surface cuirass. 2. Relatively steep slopes of the upper pediment cut across the mottled zone of the weathering profile and littered at the surface with ironstone blocks and nodules (raw mineral soils). 3. Secondary, lateritic ironstone, partially hardened and exposed at the surface. Poorly cemented, well rounded nodules and concretions. 4. Variable thickness of colluvial mantle, possibly with fine aeolian material added (ferruginous tropical soils). 5. Gullied slopes cut into the deep weathering profile, with rock exposures in the stream channels (raw mineral soils).

Figure 7.6 *Detailed soil pattern typical of the High Plains of Hausaland, Zaria, Northern Nigeria.*

and to the Niger via the Sokoto–Rima which drains initially to the north-west, and the Kaduna system which drains more directly to the south-west. Inselbergs rise rarely from this surface but whalebacks emerging through the colluvial mantle are not uncommon. Figure 7.6 shows the soil pattern associated with this area. Small residual platforms of primary vesicular lateritic ironstone rise almost imperceptibly 10–15 ft (3–5 m) above the general plain level, or large areas of the primary lateritic ironstone are mantled by colluvium of variable depth. Small scarps are formed, and, beneath these, gentle pediments fall to the shallow valleys which may be streamless and are clay-lined. The pediments are covered by secondary nodular lateritic ironstone which may have become cemented, and this in turn is frequently covered by sandy colluvium. The valleys may be gullied and the secondary ironstone protects vertical banks cut into the weathered gneiss. The gullies extend to the basal surface of weathering.

The presence of a rock inselberg may modify this soil pattern. Figure 7.7 shows the detailed soil pattern about a porphyritic granite inselberg as mapped by Higgins (1958) in Kabba Province, south of the Niger River. Primary and secondary lateritic ironstones, with and without a colluvial cover, are present together with valley soils. About the rock is a 'collar' of often deep, coarse granitic gravel. This is relatively rich in weatherable minerals, and in available nutrients. It is well supplied with seepage water from rock joints and by direct runoff from the inselberg. It is freely drained and is favoured for cultivation in many areas. In such areas as the Jos Plateau, where residual granitic hills are numerous, these raw mineral soils are of great importance in the soil pattern.

Residual inselbergs rising from plains may be contrasted with those grouped along the front of a mountain mass. These hills may reflect the variations in Basement Complex lithology closely as is the case along the Nasarawa Plateau south-west of the Jos Plateau, or they may be more fortuitously located along minor watersheds as is the case along the foothills to the west of the Mambilla Plateau and the Cameroon Highlands (Pugh, 1966).

The depth of incision by rejuvenated rivers increases to the south of Zaria. In the Kusheriki area, 80 miles (128 km) nearer the Niger River, both primary and secondary lateritic ironstones have been recorded (Truswell and Cope, 1963). They have been removed from large areas, however, and the underlying pattern of metasediments and granitic intrusions with migmatitic areas has produced a variable topography of aligned ridges and valleys. A further contrast may be obtained from the Yelwa area where soil landscapes developed over the very variable metasediments and intrusive rocks have been mapped and described in detail by Pullan and de Leeuw (1964). This area is situated to the west of Zaria but is adjacent

1. Deep sandy colluvium of the pediment. 2. Deep colluvium of the lower pediment slope – with scattered iron nodules. 3. Secondary lateritic ironstone formed of cemented nodules. 4. Gravelly raw mineral soils and lithosols formed of well-rotted granite. 5. Hydromorphic clayey soils. 6. Hydromorphic clayey soils derived by eluviation from the pediment. 7. Bare granite of inselbergs or small whalebacks. 8. Primary lateritic ironstone cuirass. 9. Thin sandy colluvium over primary lateritic ironstone. 10. Broken primary lateritic ironstone cuirass with exposure of the mottled zone of a sedentary granite weathering profile.

Figure 7.7 *Detailed soil pattern around a granite inselberg, Kabba Province, Northern Nigeria* (after Higgins, 1958).

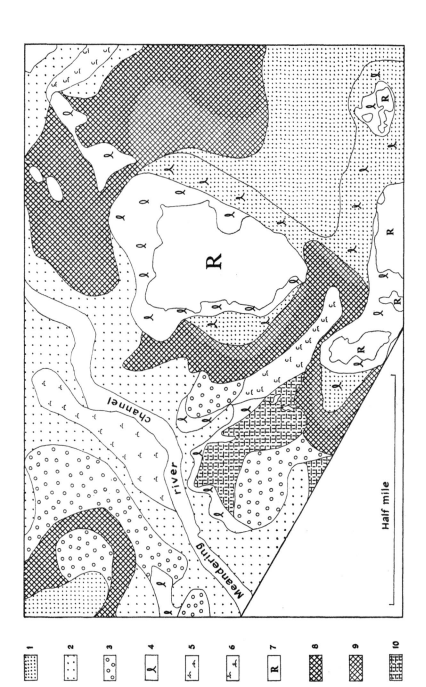

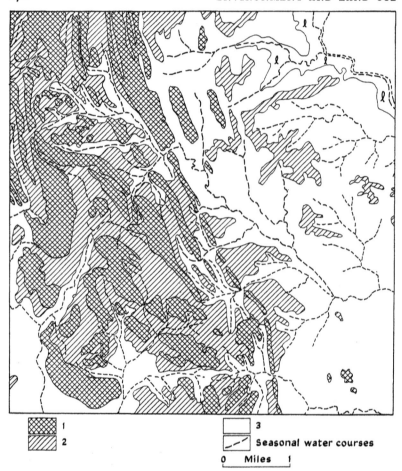

1 █
2 ▨

3 ☐
⌐⌐⌐ Seasonal water courses

0   Miles   1

1. Lithosolic, quartz gravel soils, raw mineral soils and rock outcrops of the steep slopes and ridges.
2. Eroded soils showing truncated, ferruginous tropical profiles of the gentle pediment slopes, together with quartz gravel lithosols. 3. Catenary sequence of ferruginous tropical profiles developed
over schists and characterized by an increasing depth of pedisediment near the stream channels.
Hydromorphic colluvial soils in the valleys.

Figure 7.8 *Detailed soil pattern in an area of metasediments, near Yelwa,
Northern Nigeria.*

to the Niger River. High-level and primary lateritic ironstones are not
found but have been described on the Salka Plateau to the south over Nupe
Sandstone and as smaller residuals on the higher ridges rising to 1,000 ft
(307 m.) at Duku. Figure 7.8 shows the contrast between an area of highly
folded alternating schists and quartzites and highly folded schists. Relative
relief is 200–300 ft (75–90 m). The ridges are covered with small rock out-

crops and mantled by quartz gravel which masks shallow rotted rock in many cases. The valley may have small pediments cut across schists and a soil catena of ferruginous tropical soils has been described. Not all members of the catena are found in the narrow valleys. The lowest catenary member is a colluvial deposit showing hydromorphic influences. Where the lithology is more uniform, low-level pediments have been formed and these are undergoing dissection. Lateritic ironstones have been described from the pediments developed over gneiss but not over the schists, though concretions are commonly found in the profiles. Similarly, the phyllites have been protected by a localized lateritic ironstone which has not developed on granites higher up the same pediment.

Though the high-level primary lateritic ironstone has been destroyed in this area it caps many large mesas developed in the clayey sandstones of the Continental Terminal strata into which the Sokoto River has become incised to a depth of over 100 ft (30 m) near Birnin Kebbi. High ironstone-capped bluffs look over the broad floodplain and form large complex mesas rising 70–100 ft (20–30 m) over plains covered in deep colluvial sands (Pullan, 1961). Secondary lateritic ironstones are not extensive and this may be attributed to the depth of dissection though well-rounded iron concretions are common within the sandy drift. Figure 7.9 illustrates the general pattern of soils in this area. There is a well-established catena in the drift soils. A similar soil landscape has been described by Maignien (1954) in Guinea.

Complexity of soil pattern related to degree of dissection of geologically complex areas can be expected but large areas which have not been dissected can be found in West Africa and the geomorphological pattern may be very simple. Such simple landscapes may represent well-preserved erosion surfaces or they may be areas of sedimentation. However, uniformity of landscape may hide detail of great pedological significance. This may be illustrated by examination of the fossil dune field of Hausaland in Nigeria.

Some soil patterns from the Great Erg of Hausaland

This great erg of Late Pleistocene age can be traced into Niger and Cameroon and is confined to the lacustrine fluvial deposits of the Chad Basin. It is probable that a thin surface veneer of fluvial sands was laid down over the clays and sandy clays of the lake deposits and that these were reworked to form the dunes. Subsequent flooding of the dune field by one or more increases in the size of Lake Chad, together with variable alluviation from rivers entering the dune field from the south and north, during the Late Pleistocene and Recent periods, has covered the interdune depressions with fine-textured deposits (Pullan, 1964a). These in turn have been covered in part by sand washed from the dune slopes. The dunes are

well-defined, longitudinal and aligned from north-east to south-west with
a tendency to turn to east to west. They are confined to the north-west by
the fossilized floodplains of formerly meandering rivers which rose in
Niger and flowed south to join the Hadejia–Yobe River system. Similar
fossil alluvium associated with the former courses of the Hadejia and Yobe
Rivers and the now fossilized Burum Gana are found to the north-east. The

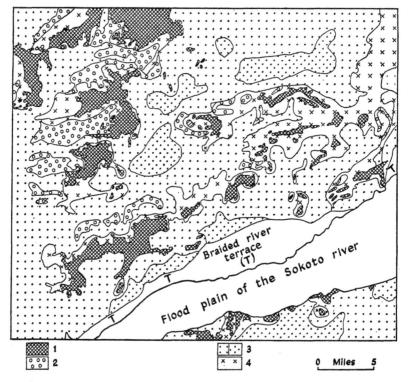

1. Primary lateritic ironstone cuirass (with vesicular structure). 2. Secondary lateritic ironstone
(nodular structure). 3. Sandy drift (shallow/deep). Catena as for northern fossil erg. 4. Lithosolic and
raw mineral soils with rock outcrops.

Figure 7.9 *Detailed soil pattern in an area of deeply dissected lateritic
ironstone cuirass and sandy drift, Birnin Kebbi, Northern Nigeria* (after Pullan,
1961).

dunes are bounded to the south and west by the colluvial and aeolian
drift-covered lateritic ironstone plains developed over Pre-Cambrian
crystalline rocks and Tertiary sandstones. The soil landscapes of the west-
central part of the dune field (Fig. 7.10) have been described by Pullan
(1962a, b).

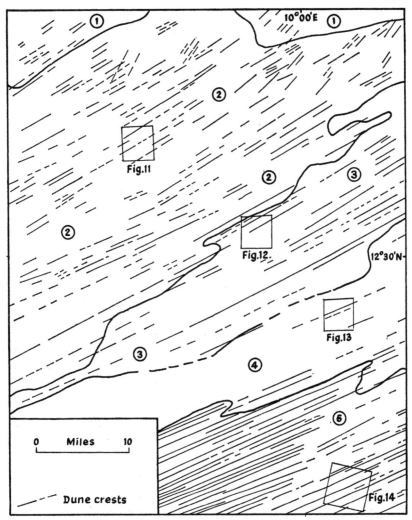

1. Fossil river alluvium. 2. Northern dunes with old alluvium. 3. Old flood-plain of the Hadejia River.
4. Recent flood-plain of the Hadejia River. 5. Southern dunes.

Figure 7.10 *Distribution of the dunes in the west central part of the fossil
erg, Northern Nigeria.* The positions of the detailed soil samples are shown
as small squares. Figure 15 is situated fifteen miles south of Figure 14.

The dune field is divided by the Hadejia River and by its recent and
fossil floodplains. The pattern of the dune frequency varies over the dune
field. The dunes are irregular in width, length and height in the north and
are rarely over 20 ft (6 m) high. The following catena is associated with the
dunes at Gumel (lat. 12.39, long. 09.23).

1. Dune crest and upper slopes

    0–6 in   light yellowish brown, fine sand

    6–82 in   reddish yellow, fine sand.

2. Middle slopes

    0–3 in   grey brown, fine sand

    3–15 in   brown, fine sand

    15–70 in brownish yellow, fine sand

    70–82 in yellow, fine sand, mottled

    82–98 in white, fine sand, mottled.

3. Lower slopes

    0–12 in   very pale brown, fine sand

    12–36 in light yellowish brown, fine sand

    36–60 in very pale brown, fine sand

    60–75 in very pale brown, fine sand, mottled, black

                    manganese(?) concretions

    75–87 in yellow, fine sand, mottled, red iron concretions.

The textural profiles of the interdune soils are very variable and at least five soil series can be identified. The surface horizon is predominantly fine-textured and these soils dominate the soil pattern (Fig. 7.11). The middle soil horizons have alkaline reactions (pH 9) and exchangable calcium and sodium values are high. More rarely sodium values may be low and free calcium carbonate may dominate the middle and lower horizons forming a concretionary horizon known locally as *jijilin*. River water enters these interdune depressions only in the extreme western part of the dune field. Elsewhere they are flooded by rainwater but there is no alluviation today though colluviation from the dune sides is common.

The Hadejia River formerly occupied a much more northerly course through the dune field. The fossil floodplain, from which dunes emerge, stands at a higher elevation than the present floodplain. Nevertheless it is still flooded occasionally by the Hadejia River and some alluviation is taking place. Many of the dunes with lengths up to two miles are well preserved and they have a relative relief of over 25 ft (8 m). The soil pattern is radically different from the interdune depressions of the north for old channels, levees, spill depressions and modified point bar systems can be identified on air photographs, (Fig. 7.12). Profile development has taken place and the important characteristic is that no saline soils have been sampled. However, the values of exchangeable calcium and magnesium are very high in these soils though pH is neutral to slightly acidic. Profiles with accumulations of *jijilin* are common throughout this old floodplain, which is thought to represent alluviation during a pluvial period.

The present floodplain of the Hadejia River is a complex of very recent channels, levees, point bar systems and spill plains near the present river,

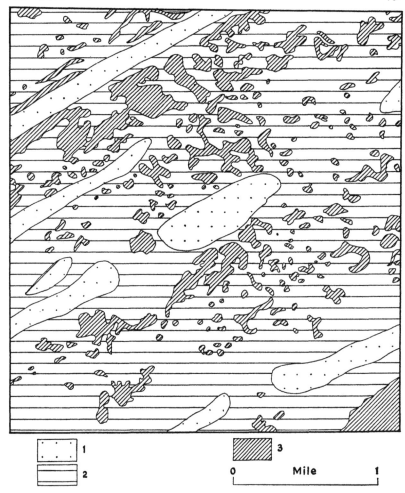

1. Dune soils with a catenary sequence. 2. Sandy top-soils representing colluvial sands re-distributed by flood-water over variably textured hydromorphic soils which may be halomorphic or more rarely calcimorphic. 3. Fine textured top-soils of depressions representing degraded stream channels (often halomorphic).

Figure 7.11 *Detailed soil pattern in the northern dune field.*

together with older series of similar depositional landforms. The pattern is highly complex but is readily identified on air photographs. Figure 7.13 is a section of a much larger area illustrated in Pullan (1962a) and represents a recent, but abandoned, system near the present river now separated from it by the long narrow dune to the north. Profile development is not well advanced in these soils of variable texture in profile. All but the predominantly sandy profiles of the former levees have high exchangeable calcium

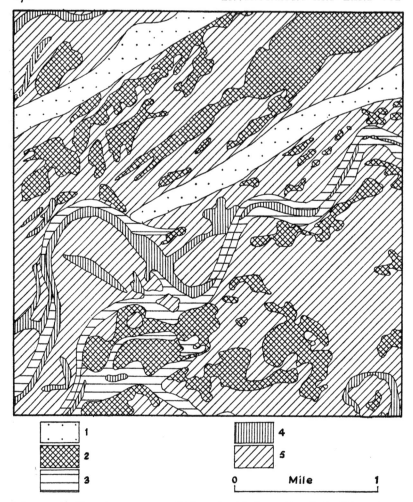

1. Dune sands with a cetenary sequence. 2. Fine textured, hydromorphic soils of spill depressions.
3. Sandy soils of former river levees with weakly developed hydromorphic profiles. 4. Variably textured hydromorphic and calcimorphic soils of old river channels (frequently clayey). 5. Variably textured soil complex of former point bar systems with hydromorphic and calcimorphic profiles.

Figure 7.12 *Detailed soil pattern in the old floodplain of the Hadejia River.*

and magnesium values. Sodium is low throughout most soil horizons
irrespective of the texture, and where it is important in the cation exchange
complex it is less than half the calcium value and usually at a depth
greater than 40 in (1 m). The soils are neutral in the upper part of the profile
becoming moderately alkaline at depth (pH 8·8).

The southern half of the dune field is distinct from the northern half,

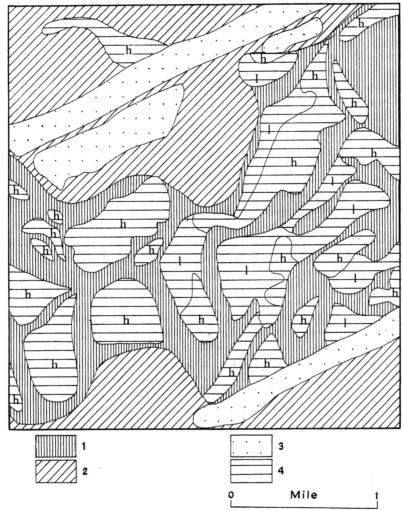

1. Fine textured, low lying, hydromorphic soils associated with old river channels and spill depressions.
2. Variably textured soils with little surface relief, representing spill plains associated with the old meander belt, on which a veneer of sand and clay has been laid. 3. Dune soils. 4. Variably textured soils of high, unflooded (*h*) and low flooded (*l*) point bar systems.

Figure 7.13 *Detailed soil pattern in the new floodplain of the Hadejia River.*

for the dunes are very regularly and closely spaced and can be traced for many miles. They have an average height of 25–30 ft (8–9 m) but have been measured with a relative relief of 35 ft (11 m). Figure 7.14 is taken from the area immediately south of the Hadejia floodplain and represents an area formerly traversed by many diversion channels of the Hadejia River system and its tributaries from the south-west which follow the interdune

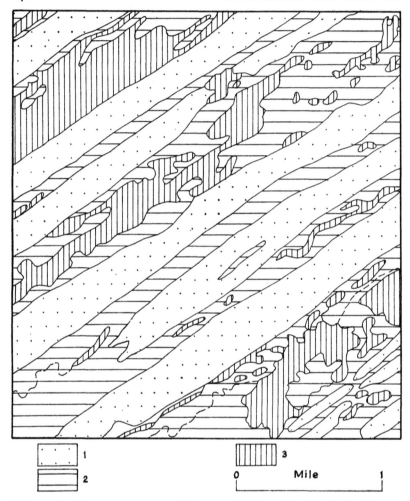

1   2   3

0   Mile   1

1. Dune sands with a catenary sequence. 2. Fine textured hydromorphic soils and colluvial sands.
3. Fine textured hydromorphic soils of lower elevation.

Figure 7.14 *Detailed soil pattern in the southern dunes south of the Hadejia floodplain.*

depressions for long distances. The higher rainfall of this area, the longer period of flow of the rivers together with the flooding of all depressions by river water and the washing of dune sands on to the heavier depression soils and across the interdune depressions appears to have prevented alkaline soils from developing. However, the sampling of these interdune soils has not been exhaustive and saline/alkaline soils may be present. Figure 7.15 represents the southern part of the Southern Dune field as

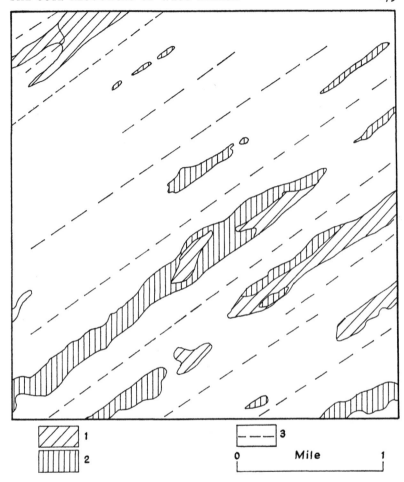

1. Fine textured hydromorphic soils and colluvial sands. 2. Fine textured hydromorphic soils of lower elevation. 3. Dune sands (crest shown) having a catenary sequence.

Figure 7.15 *Detailed soil pattern in the southern dunes near Azare.*

described from the area north of Azare (Pullan, 1962b). Large areas of wash sand cover the interdune depressions and finer-textured soils are less evident. In the extreme southern edge of the dune field lateritic ironstone may be exposed in the longitudinal depressions or may be covered by a thin colluvial sand. It has a coarse conglomeratic structure and is derived from the disintegration of the lateritic ironstone which developed on the Eocene sandstone plain to the south. Low dune forms without marked inter-dune depressions have been identified over this primary lateritic ironstone surface but they rapidly disappear as the superficial material becomes

fine-textured. The profiles developed in these sandy soils are essentially different from those within the dune field having Munsell colours within the
range 2·5 YR in contrast to 7·5 YR within the true dune field. The soil
catena described earlier for the dunes is not found on those dunes over
ironstone and those located within the Hadejia floodplain. In these situations high water-tables are believed to restrict profile development.

    The examples described and illustrated above show the variation which
is found in the soil pattern within the west-central part of the dune field.
Not only does the area covered by dune soils change but the soils associated
with the dunes also change. Further changes can also be described from the
eastern section of the dune field where the relative relief of the dunes is of

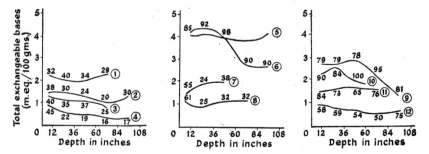

1. and 2. Sandy drift over lateritic ironstone on Eocene sandstone south of Azare (11.40 N., 10.15 E.).
3. and 4. High dunes north of Azare (11.50 N., 10.15 E.). 5. High dune in the recent floodplain of the
Hadejia River (12.20 N., 10.05 E.). 6. High dune in the old flood-plain of the Hadejia River (12.30 N.
10.00 E.). 7. Low irregular dune over ironstone near Allagarno (11.55 N., 11.40 E.). 8. Broad low
dune near Damaturu (11.50 N., 11.55 E.). 9. Low dune at Nguru (12.50 N., 10.25 E.). 10. Corresponding profile in sandy deposits of the old alluvium (13.00 N., 10.20 E.). 11. High dune at Medu
(12.30 N., 9.25 E.). 12. High dune at Babura (12.45 N., 9.00 E.).

Figure 7.16 *Total exchangeable bases (Ca, Mg, Na, K) throughout the profiles
of the dune crest and upper slope profiles for different locations within the west
central part of the fossil dune field, Northern Nigeria.* Figures along the graphs
refer to percentage base saturation.

the order of 5 ft (1·5 m) or less. They are only identified with ease on air
photographs and not in the field. Similarly, the extension of this dune field
to the east and north of the Bama Ridge (Pullan, 1964a) brings the dune
soils into a number of different associations with the montmorillonitic
lacustrine clays which have developed vertisolic profiles (Pullan, in Higgins
et al., 1960). Pias (1960) has also described the soils of this dune field in
Cameroon.

    However, the variations to be found within this dune field are not only
of soil association pattern. Figure 7.16 shows the variation to be found in
total exchangeable bases in the sandy profiles taken from dune crests and
upper slopes, together with the percentage saturation of the exchange
complex at different points within the profiles. The figures are actual

samples and not mean figures derived from replicate sampling on a large scale and as such may not be strictly typical. They represent samples taken from profile pits dug after the consideration of numerous auger samples. In addition two profiles, similar in description to the dune-crest profiles, from old sandy alluvium to the north of the dune field are also included. These soils represent weakly developed ferruginous tropical soils developed in loose sandy parent material and they illustrate clearly the lower values for total exchangeable bases found in the southern (Azare) and western (Babura) parts of the dune fields where mean annual rains of the order of 31–34 in (775–858 mm) compared with the higher base status of these soils at Gumel and Hadejia with 22–25 in (550–625 mm) and about 20 in (500 mm) at Nguru.

Indeed, the low saturation values for the Azare dune soils suggests that they should not be classified as tropical ferruginous soils for the mid-profile saturation and should normally exceed 40 per cent (Table 7.2).

A change in pattern and nutrient status of the dune soils has been shown to be important in the assessment of the soil resources of a single landform/soils unit. However, account must also be taken of the variation to be found in the sandy soils of the upper, middle and lower sections of the dune. The clearly differentiated catena has been described earlier. The catenary sequence of soils described above is identical for the dunes north of Azare. Table 7.2 compares their nutrient status. The Azare profiles reflect the greater leaching of the profiles which may be attributed in part to the slightly greater rainfall, though it will be noticed that the upper and middle slope profiles at Azare have no silt/clay content. Fine sand forms an average of 80 per cent throughout these profiles as opposed to 72 per cent at Gumel. There is also a horizon of loose concretionary ironstones at 40 in (1 m) and rarer concretions are found between 26 and 60 in (65–150 cms) within the lower slope profile at Azare.

The lower CEC[1] values of the upper and middle slope profiles at Azare, the presence of the low CEC values in the middle section of the lower slope profiles at both locations and the large increase associated with the accumulation of calcium and magnesium related to the higher watertable of the adjacent interdune depressions are noteworthy. The higher TEB[2] for the mid-slope profiles compared with the upper slope profiles are apparent but the Gumel example has a proportionately higher increase.

The fossil erg of Northern Nigeria represents the development of a soil pattern related to changes in sedimentation of relatively recent origin together with changes brought about by pedogenesis, in an area where the nature of the parent material has allowed rapid movement of soluble and

[1] Cation exchange capacity.
[2] Total exchange bases.
    G

TABLE 7.2 Analytical data for the catenary dune succession in the fossil erg, Northern Nigeria

		Northern dunes at Gumel						Southern dunes at Azare					
	Depth (inches)	pH	Clay %	CEC¹	P²	TEB³	Satura-tion %	Depth (inches)	pH	Clay %	CEC	P	TEB	Satura-tion %
Upper slope	0–6	7·8	2	3·9	38	1·7	44	0–10	6·0	0	2·0	60	0·9	45
	6–18	6·2	4	3·8	41	1·9	50	10–22	5·6	0	2·5	71	0·5	22
	18–41	5·9	4	3·6	46	2·2	61	22–60	5·2	0	2·4	60	0·5	19
	41–58	5·8	2	2·3	38	1·5	65	60–77	5·7	0	2·0	44	0·3	16
	58–62	5·9	0	3·4	46	2·3	68	77–85	4·4	0	1·9	27	0·3	17
Middle slope	0–3	7·4	2	2·4	43	2·7	100	0–9	6·7	0	1·6	65	1·3	80
	3–15	6·4	2	3·1	41	2·7	87							
	15–39	5·8	4	3·9	41	3·0	77	9–20	6·5	0	2·0	44	1·0	51
	39–48	5·8	2	3·6	38	3·4	94							
	48–70	5·9	8	7·1	35	4·3	61	20–55	5·4	0	2·2	31	0·6	26
	70–82	6·8	2	4·7	20	3·4	72							
	82–98	6·9	0	1·5	10	1·1	73	55–91	5·9	0	1·1	22	0·6	57
Lower slope	0–5	6·8	0	3·7	38	2·1	57	0–7	6·4	2	3·1	49	1·7	54
	5–12	6·1	2	3·9	38	2·2	56	7–16	5·9	4	3·6	38	2·1	58
	12–36	6·0	4	4·0	33	2·5	62	16–26	5·4	2	3·0	44	1·3	43
	36–46	6·3	2	2·0	18	1·1	55	26–60	5·9	2	1·5	33	0·7	47
	46–60	6·8	4	4·9	28	4·0	82	60–63	6·5	0	1·3	11	0·5	35
	60–75	8·8	6	5·7	28	9·1	100	63–76	7·8	8	5·6	17	4·9	83
	75–87	8·5	4	3·7	18	5·4	100	76–89	8·9	16	8·7	49	10·1	100

¹ CEC (cation exchange capacity) given as m.equiv./100 g soil.
² P (phosphorus) given as p.p.m.
³ TEB (total exchangeable bases) given as m.equiv./100 g soil.

non-soluble soil components. The failure of similarly textured soils to develop a marked catenary sequence with a climate drier by 10 in (250 mm) per year farther east, near Lake Chad, suggests that pedogenesis might have been more important during a wetter climate than that of today, and indeed Bocquier and Gavaud (1964) have described many examples of ferruginous tropical soils from the drier areas immediately to the north in Niger Republic which can be regarded as palaeosols.

Changes in the pattern are rapid, of major significance, and related to the evolutionary history of the area which at first sight may appear to be relatively simple. In contrast the area described below illustrates the large measure of uniformity which can be found within the soils of some areas within West Africa.

Some soil patterns on sandstone in the Middle Benue Valley

Recent topographic mapping from air photographs has demonstrated that an undulating plain, some 50 miles (80 km) in width over-all and penetrating deeply into the Eastern Highlands along the major valleys, occupies the Lower and Middle Benue Valley. It has developed over crystalline igneous and metamorphic rocks and is a continuation of the plain described by Pullan (1964b) which crosses the Cretaceous sandstones west of the Donga River. The surface, which is not deeply dissected by the major rivers or their tributaries, falls in height from a variable 650–800 ft (200–46 m) at the mountain front, where a very sharp topographic discontinuity exists, and around outlying inselbergs, to 450–500 ft (138–54 m) near the Benue River. Here some dissection is taking place along both minor and major tributaries of the Benue River.

The surface is underlain everywhere by a primary lateritic ironstone horizon which is essentially vesicular in structure and frequently soft. This horizon is found on either side of the sandstone/crystalline boundary, the presence of loose, large well-rounded quartz pebbles being the only significant indication of the geological boundary, which is highly irregular. The sandstones are incoherent apart from minor outcrops and the crystalline rocks are frequently well weathered so that field identification is difficult beneath the ironstone. This lateritic ironstone is also noteworthy in that it mirrors the undulating surface relief closely. It is rarely exposed, but in areas of recent stream rejuvenation it disintegrates to form a zone of loose nodular gravel at the surface. However, along the rejuvenating streams tributary to major rivers of the Benue system scattered scarps with outcrops may be seen. The ironstone may have hardened and be partly covered by a shallow surface soil.

Recent mapping of the soils in part of this area (Pullan, 1964b) has demonstrated the uniformity of the soil pattern over large areas of the

Cretaceous sandstones. In a survey area of 2,200 square miles, approximately 1,700 square miles are covered by a simple soil association in which the upland soil series is found on interfluve, upper, middle and lower middle slopes, and associated catenary soil members are found only on the lower slopes and valley floors. Variations exist throughout the upland area only in the depth of the lateritic ironstone horizon and texture of the individual soil horizons.

A typical upland profile is as follows:

0–7 in	grey brown, sand
7–15 in	pale brown, loamy sand
15–18 in	light brown, sandy clay loam
18–37 in	reddish yellow (7·5 YR), sandy clay loam, mottled
37 in	weakly cemented vesicular lateritic ironstone.

The lateritic ironstone may be found exceptionally below 80 in (2,000 mm). Horizon boundaries are gradual and smooth. The analytical data for these soils (Table 7.3) illustrate that these soils have ferruginous tropical characteristics. The absence of good structure in the clayey horizons, the gradually changing profile which may be deep, and the low base saturation percentages of some profiles indicate that in this area (present mean annual rainfall, 46 in (1,150 mm)) some profiles have ferrisolic tendencies. Both ferrisolic and ferrallitic soils are developed on very similar parent materials but with different topographic locations immediately to the west and north-west of this area.

The uniformity of soils over this area and for great distances to the east and south (described from the Basement Complex by Hildebrand, 1966) can be related to the uniform topography, lack of river rejuvenation and the presence of the lateritic ironstone. The valley and lower slope soils associated with these upland soils are more varied. The lower slope soils are deep light-coloured sands over vesicular lateritic ironstone.

A typical profile is as follows:

0–9 in	light grey, sand
9–15 in	pinkish grey, loamy sand
15–22 in	pink, sandy loam
22–40 in	light brown, sandy loam, mottled, with few iron nodules
40–55 in	light brown, sandy clay loam, mottled, with few iron nodules
55 in	vesicular lateritic ironstone, hard.

Table 7.3 gives the analytical data for this profile, which represents a soil leached of some clay and bases in which a zone of accumulation is not apparent above the ironstone.

The valley soils vary in texture. Those with deep sandy profiles are believed to represent accumulations of sand derived from upland soils and

TABLE 7.3 *Analytical data for soils of the Wukari area, Northern Nigeria*

Position	Depth (inches)	Clay %	Silt %	Fine sand %	Coarse sand %	pH	P[1]	TEB[2]	CEC[2]	Saturation %
Interfluve upper and middle slopes	0–7	2	6	52	40	6.1	44	2.2	3.8	58
	7–15	4	10	42	44	5.8	33	0.9	2.4	38
	15–18	22	8	32	38	5.3	38	4.0	6.8	59
	18–37	26	10	28	36	5.1	49	4.8	7.3	66
	37–54	LATERITIC IRONSTONE				5.1	55	4.3	10.0	43
Lower slope	0–9	0	0	57	43	5.5	38	1.3	1.9	68
	9–15	6	10	45	39	5.4	17	1.1	2.4	46
	15–22	10	8	44	38	4.7	22	0.9	1.5	60
	22–40	14	8	39	39	4.5	27	1.7	3.1	55
	40–55	20	6	23	51	4.3	33	2.0	5.5	36
	55–60	LATERITIC IRONSTONE								
Valley	0–4	4	8	38	50	6.5	44	2.8	3.9	72
	4–10	2	4	42	52	5.9	17	0.6	1.5	40
	10–17	4	6	35	55	5.5	17	0.5	1.5	33
	17–27	4	4	35	57	5.6	11	0.6	1.9	32
	27–62	0	4	36	60	6.0	6	0.5	0.2	100
	62–64	LATERITIC IRONSTONE								
Valley	0–4	4	8	72	16	5.6	44	3.2	3.3	97
	4–10	6	14	62	18	5.3	22	2.1	2.6	81
	10–19	10	20	53	17	5.4	27	1.7	2.3	74
	19–25	24	10	52	14	4.7	49	2.5	6.5	38
	25–34	32	8	47	13	4.3	27	2.3	10.0	23
	34–45	38	6	44	12	4.2	33	2.6	11.0	24

[1] P (phosphorus) given as p.p.m.
[2] TEB and CEC given as m.equiv./100 g soil.

transported by sheet wash after cultivation, which in this area consists of large mounds built by hoe for yam (*Discorea* spp.) cultivation. A predominantly clayey profile represents accumulation of clay derived from eluviation by subsurface water and by surface wash. Analytical data are given in Table 7.3 for sandy and clayey valley soils which are light-coloured and faintly or distinctly mottled. They have an irregular distribution of loose iron concretions through the profile. The degree of base saturation is very variable, but those soils which have been cultivated have low values and may show signs of leaching also. Humic topsoils develop if the grasses are not burnt annually, as is usually the case.

The soils of this plain have been eroded by small tributaries of the Benue River in a narrow zone near the river. The lateritic ironstone has disintegrated to form horizons of loose nodules either at the surface or buried by sandy colluvium. There is no secondary cemented lateritic ironstone. The underlying rock may be weathered to a shallow depth and the weathering profile may be exposed at the surface. The sandstones and shales, which are saline, have been seen in most pits. These lithosolic or raw mineral soils have little profile development and the sodium salts have not been leached from them. Thus many of these soils have a high sodium content at depths varying from 17 to 55 in (43–138 cm) on colluviated sites. The valley soils are frequently clayey, have a high exchangeable sodium content, and show textural variations in the profile representing changing sedimentation.

A third and areally unimportant contrast has been described about a tributary of the Donga River which has exposed the lateritic ironstone as a scarp along the edge of its floodplain, and the tributary has likewise exposed the primary lateritic ironstone, giving rise to an area in which small ironstone scarps, ironstone gravel-covered pediments and thin colluvial sands over ironstone form a mosaic. This pattern is a microcosm of what the plain would be if soil erosion became a dominant process through improper cultivation.

Conclusion

The sound formulation of plans for the development of agriculture, within the context of comprehensive land-use plans, demands that not only should the range and variation of the soils be known but also that the distribution and pattern of occurrence should be mapped. Soil surveys provide these basic data and it is from medium-scale reconnaissance maps that regional and national soil maps can be compiled. The adequate representation of soils in maps can be achieved only after a suitable soil classification has been devised, but the classification devised for description is not necessarily the best for mapping.

The distribution of the major upland soils can be seen on the *Soil Map of Africa* at 1:5 million. Further generalization (Fig. 7.1) shows that a zonal pattern of the major soils is found in West Africa. These are the ferrallitic, ferrisolic, ferruginous tropical, brown and reddish-brown soils. The distribution of these soils cannot be related directly to present climate and vegetation with the exception of the brown and reddish-brown soils of the semi-arid area. There are also many anomalies in the general soil pattern. Some of these can be related to difficulties in assigning some soils within a soil group of the existing classification when analytical data are inadequate. This is particularly true where the soils have developed on incoherent sandy parent materials such as are commonly found within the Cretaceous, Eocene and Plio-Pleistocene sediments. Recent surveys in the Middle Belt of Northern Nigeria have revealed large areas of soils which are essentially transitional in character and therefore difficult to fit into the existing classification. Other major groups such as the halomorphic, eutrophic brown and vertisolic soils may be related to certain parent materials, while the hydromorphic, weakly developed, rock debris and lateritic ironstone soils may be related to areas with distinctive geomorphological development.

There are few detailed and reconnaissance soil surveys for large areas of West Africa. Extrapolation of boundaries from data that exist has to be undertaken but is difficult. Where parent material is a dominant pedogenic factor this may be used. The boundaries between sedimentary and crystalline rocks are adequately mapped. This is not true of the Basement Complex, however, where many major lithological changes can take place rapidly in areas where metasediments, crystalline metamorphic, acidic and basic igneous rocks occur. The lithological data that are available from the Basement Complex of West Africa are inadequate for the extrapolative mapping of soils with the exception of a few small areas. Yet the Soils Map of Africa uses parent material as a second-order division within the classification of soils. As a result large areas remain undifferentiated. The recognition of four broadly defined geological provinces, namely (a) gneisses and migmatites with granitic bosses and batholiths; (b) metasediments with crystalline metamorphics; (c) massive granite bathlioths with granitized zones; (d) dolerite laccoliths and recent lava fields and volcanic areas, can be used where only general information is available.

It is realized that deep weathering and colluviation may lessen the importance of parent material as a pedogenic factor, but detailed information is lacking and there is no adequate assessment of the part played by these factors on a regional scale.

Two different approaches are possible in the mapping of soils at the reconnaissance level. The main soil groups and their subdivisions may be mapped directly on the basis of a pre-existing classification in which pedogenic characteristics and parent material are most important criteria. Such mapping cannot show the detailed pattern of soils and their inter-relationships. Alternatively, the soils may be mapped as associations which are related directly to landform associations. This method uses the relationship, which has been known for over thirty years, between soils and landforms or soils and facets of a landform. The grouping of soils on such a rational basis allows the area of occurrence of soils to be mapped as well as the pattern. This may be shown directly on large-scale maps and indirectly on medium-scale maps where schematic diagrams and very large-scale extract maps can be given in the accompanying report. This method is also useful in that it allows the rational choice of initial mapping scales and the maximum use of extrapolative mapping, for the delimitation of the associations can be made using air photographic interpretation of land-forms. In areas where access is bad, the stereotyped methods used in mapping soils within existing map sheets are wasteful of time, manpower and materials, while the use of soil–landform associations allows maximum extrapolation between neighbouring map sheets and is the most rational approach to reconnaissance soil survey.

An assessment of the soil resources of an area cannot be made unless the patterns of associations of soils are known. Too little thought has been given to the problem of the representation of pattern. It has either been ignored or inadequately shown. The establishment of the relationships between soils and landforms allows a rational portrayal of soil pattern in terms of soil landscapes. It can be used at all mapping scales without introducing cartographic complexity and this is a decided advantage for fine cartographic presentation can suggest authenticity which does not exist.

Acknowledgement

The author wishes to thank the Director of the Institute for Agricultural Research, Ahmadu Bello University, Northern Nigeria, for permission to include unpublished data, from work undertaken by the author while a member of the Institute.

References

AUBERT, G. 1963. La classification pédologique française. *Cah. off. recher. sci. tech. d'outre-mer*, *Pédologie* 3, 1–7.

BOCQUIER, M. and MAIGNIEN, R. 1963. The brown sub-arid tropical soils of West Africa. *Sols afr.* 8, 371–82.

BOCQUIER, M. and GAVAUD, M. 1964. *Rapport general, édition provisoire, étude pédologique du Niger oriental* (with Carte pédologique de reconnaissance du Niger oriental, 1 : 500,000), Off recher. sci. tech. d'outre-mer, Dakar.

BOULET, M., BOCQUIER, M. and GAVAUD, M. 1964. *Rapport general;* *étude pédologique du Niger central* (with Carte pédologique de reconnaissance du Niger central, 1 : 500,000), Off recher. sci. tech. d'outre-mer, Dakar.

BRAMMER, H. 1962. Soils. In WILLS, J. B. (ed.), *Agriculture and Land Use in Ghana,* London.

CHARTER, C. F. 1957. *Suggestions for a Classification of Tropical Soils,* Misc. Paper No. 4, Ghana Department of Soil and Land Use Survey.

DABIN, B., LENEUF, N. and RIOU, G. 1960. *Carte pédologique de la Côte d'Ivoire au 1 : 2M,* Off. recher. sci. tech. d'outre-mer, Dakar.

DA COSTA, J. V. B. and AZEVEDO, A. L. 1960. Generalized soil map of Angola. *Proc. 7th Int. Congr. Soil Sci., Madison* 4, 56–62.

D'HOORE, J. L. 1954. *L'accumulation des sesquioxides libres dans les sols tropicaux,* Inst. Nat. Étud. agron. Congo-Belge, Sér. Sci., No. 62, Bruxelles.

—— 1964. *Soil Map of Africa, Scale 1 : 5M,* Pub. 93, Commission de Coopération technique en Afrique au sud du Sahara, Lagos.

FAO. 1962. *Africa Survey,* App. 2, Soil Resources of Africa, Rome.

FAUCK, R. 1963. The sub-group of leached ferruginous tropical soils with concretions. *Sols afr.* 8, 407–29.

FOURNIER, F. 1963. The soils of Africa. In *A Review of the Natural Resources of the African Continent,* UNESCO, Rome.

FURON, R. 1963. *The Geology of Africa,* London.

HAUGHTON, S. 1963. *Stratigraphic History South of the Sahara,* London.

HIGGINS, G. M. 1958. *A Preliminary Report on the Detailed Land, Soil and Contour Survey of the Riverain School of Agriculture, Kabba,* Bull. Soil Survey Sect. 5, Ministry of Agriculture, Northern Nigeria.

HIGGINS, G. M., RAMSAY, D. M., PULLAN, R. A. and DE LEEUW, P. N. 1960. *Report on the Reconnaissance Soil Surveys in North-East Bornu,* Bull. Soil Survey Sect. 14, Ministry of Agriculture, Northern Nigeria.

HILDEBRAND, F. 1966. *Report on the Soil Survey of the United Hills Area, Sardauna Province, Nigeria,* Bull. Soil Survey Sect. 13, Institute for Agricultural Research, Ahmadu Bello University, Northern Nigeria.

HOPE, W. A. 1966. *A Report on Localities on the Mambilla Plateau with Particular Reference to the Cultivation of Coffea arabica,* Bull. Soil Survey, Sect. 32, Institute for Agricultural Research, Ahmadu Bello University Northern Nigeria.

HOPE, W. A. and CLAYTON, W. D. 1958. *Report on the Reconnaissance Soil and Vegetation Survey of Part of the Jema's Resettlement Scheme,* Bull. Soil Survey Sect. 7, Ministry of Agriculture, Northern Nigeria.

KELLOG, C. E. and DAVOL, F. D. 1949. *An Exploratory Study of Soil Groups in the Belgian Congo,* Inst. Nat. Étud. agron. Congo-Belge, Sér. Sci., No. 46, Bruxelles.

KLINKENBERG, K. 1965. *Report on the Reconnaissance Soil Survey of Part of Borgu Division, Ilorin Province,* Bull. Soil Survey Sect. 28, Institute for Agricultural Research, Ahmadu Bello University, Northern Nigeria.

KLINKENBERG, K. and HIGGINS, G. M. 1966. *An Outline of Northern Nigerian Soils* (unpublished report), Institute for Agricultural Research, Ahmadu Bello University, Northern Nigeria.

LAPLANTE, A. 1954a. Les sols foncés tropicaux d'origine basaltique au Cameroun. *Proc. 5th Int. Congr. Soil Sci., Leopoldville* **4**, 144–8.

1954b. Les sols latéritiques formés sur les basaltes anciens au Cameroun. *Proc. 5th Int. Congr. Soil Sci., Leopoldville* **4**, 140–3.

LENEUF, N. 1954. Les terres noires du Togo. *Proc. 5th Int. Congr. Soil Sci., Leopoldville* **4**, 131–6.

LENEUF, N. and RIOU, G. 1963. Red and yellow soils of the Ivory Coast. *Sols afr.* **8**, 451–62.

MAIGNIEN, R. 1954. Formations de cuirasses de plateaux, Région de Labe (Guinée). *Proc. 5th Int. Congr. Soil Sci., Leopoldville* **4**, 13–18.

1959a. Soil cuirasses in tropical West Africa. *Sols afr.* **4**, 4–42.

1959b. *Les Sols subarides au Sénégal.* Off. recher. sci. tech. d'outre-mer, Dakar.

1961a. The transition from ferruginous tropical soils to ferrallitic soils in the south west of Senegal. *Sols afr.* **6**, 173–228.

1961b. Sur les sols d'argiles noires tropicales d'Afrique occidentale. *Bull. fr. Étude sol*, numéro spécial, 131–44.

1963. Eutrophic brown soils. *Sols afr.* **8**, 491–6.

MARBUT, C. F. and SHANTZ, H. C. 1923. *Vegetation and Soils of Africa*, American Geographical Society, New York.

MARTIN, D. and SEGALEN, P. 1965. *Carte pédologique du Cameroun Oriental, 1 : 1 million* (2 sheets with memoirs), Off. recher. sci. tech. d'outre-mer, Dakar.

MOSS, R. P. 1965. Slope development and soil morphology in a part of south-west Nigeria. *J. Soil Sci.* **16**, 192–209.

NYE, P. H. 1955. Soil forming processes in the tropics. *J. Soil Sci.* **6**, 51–62.

NYE, P. H. and GREENLAND, J. D. 1960. *The Soil under Shifting Cultivation*, Tech. Comm. No. 51, Commonw. Agric. Bur., Harpenden.

PIAS, J. 1960. *Les Sols du Moyen et Bas Logone, du Bas Chari, des régions riveraines du Lac Tchad et du Bahr Ghazal*, Comm. Sci., Logone, Off. recher. sci. tech. d'outre-mer, Paris.

PUGH, J. C. 1966. The landforms of low latitudes. In DURY, G. H. (ed.), *Essays in Geomorphology*, London, 121–38.

PULLAN, R.A. 1961. *Report on the Reconnaissance and Detailed Soil Surveys of the Birnin Kebbi Area, Sokoto*, Bull. Soil Survey Sect. 17, Ministry of Agriculture, Northern Nigeria.

1962a. *Report on the Reconnaissance Soil Survey of the Nguru–Hadejia–Gumel Area*, Bull. Soil Survey Sect. 18, Ministry of Agriculture, Northern Nigeria.

1962b. *Report on the Reconnaissance Soil Survey of the Azare (Bauchi) Area*, Bull. Soil Survey Sect. 19, Ministry of Agriculture, Northern Nigeria.

1964a. The recent geomorphological evolution of the south central part of the Chad Basin. *Jl W. Afr. Sci. Ass.* **9**, 115–39.

1964b. *A Reconnaissance Soil Survey of Sheet 253 (Wukari) and the Southern Part of Sheet 233 (Ibi), Benue Province* (unpublished draft), Bull. Soil Survey Sect. 29, Institute for Agricultural Research, Ahmadu Bello University, Northern Nigeria.

PULLAN, R. A. and DE LEEUW, P. N. 1964. *The Land Capability Survey Prepared for the Niger Dams Resettlement Authority*, Bull. Soil Survey Sect. 27, Institute for Agricultural Research, Ahmadu Bello University, Northern Nigeria.

RICHARDSON, H. L. 1965. The use of fertilisers. *Paper read at the Symposium on 'The Soil Resources of Tropical Africa', 1965,* African Studies of the United Kingdom.

SCHOKALSKAYA, S. J. 1953. *Die Boden Afrika,* Berlin.

SMYTH, A. J. 1963. Red and yellow soils of Western Nigeria. *Sols afr.* 8, 463–7.

STORY, A. L. 1961. *Report on the Selection of a Site for a Proposed Agricultural Experimental Farm in Igala Division, Kabba Province,* Bull. Soil Survey Sect. 13, Ministry of Agriculture, Northern Nigeria.

SYS, C. 1960. *La Cartographie des sols au Congo et au Ruanda-Urundi, ses principes et ses méthodes,* Inst. Nat. Étud. agron. Congo-Belge, Sér. Sci., No. 63, Brussels.

THOMAS, M. F. 1966. Some geomorphological implications of deep weathering patterns in crystalline rocks in Nigeria. *Trans. Inst. Br. Geogr.* 40, 173–93.

TOMLINSON, P. R. 1961. *Reconnaissance Report on Sheet 51 (Gummi),* Bull. Soil Survey Sect. 12, Ministry of Agriculture, Northern Nigeria.

1965. *Soils of Northern Nigeria with map at approximately 1 : 5 million,* Misc. Pap. Samaru Agric. Res. Station 11.

TRUSWELL, J. F. and COPE, R. N. 1963. *The Geology of Part of Niger and Zaria Provinces,* Bull. Geol. Survey 29, Kaduna, Nigeria.

8 The ecological background to land-use studies in tropical Africa, with special reference to the West

R. P. MOSS

Introduction

The study of rural land use in geography has hitherto been largely characterized by a morphological approach to the phenomena involved, in which the individual facets have been viewed as elements of 'landscape' possessing readily observable properties which have agricultural significance. Classification is achieved on the basis of these intrinsic properties, and sophistication involves merely increasingly detailed subdivision of the basic categories.

In tropical Africa much work has followed this pattern, in part at least in consequence of the increasing availability of near-vertical aerial photographs, which do, however laboriously, lend themselves immediately to detailed morphological interpretation (Stamp, 1956; Brunt, 1958). Nevertheless, real questions of practicability in the tropical African context inevitably arise (Moss, 1960; Chapter 14). More important, however, are the questions which arise when the meaning and significance of the approach to the study of agriculture are considered.

Though the method may have some limited reconnaissance value in certain contexts, it is open to two fundamental objections when it is considered from the point of view of its agricultural significance. In the first place rural land use on individual plots of land is but one expression of agricultural activity. Study of land-use patterns as such, however detailed the categories distinguished, can never reveal the factors which have produced them. Secondly, the significance of individual morphological units can be defined only tentatively without detailed appreciation of the agricultural systems involved. Similar morphological units may have different functions in different systems, and morphologically different units may have similar functions in the system as a whole (Moss 1960; Chapter 14). The important point is that fields or plots are not the natural units by which 'land' is 'used'. They are but one fragmentary expression of agricultural systems, and, indeed, in Britain they are frequently the fossilized remnants of superseded systems.

The land-use pattern is an expression of the agricultural system, representing the interaction between ecological and economic factors in which the vital link is the decision-maker, namely the farmer. Any understanding of tropical land use must thus be obtained by concentrating attention on the study of the factors which affect, and the phenomena which are immediately affected by, the actual decisions of the farmer. Hence an examination of the ecological background to land-use studies must consist in an examination of those ecological facts and relationships which are of demonstrable importance to agriculture. The focus of attention must be the ecological implications of agriculture and forestry, not the description of particular ecological units and distributions in and of themselves.

Using such a focus it is possible to decide the relevance of different kinds of relationship, and to emphasize certain aspects which are of paramount importance. From the point of view of the ecological implications of agriculture, for example, the specious distinction so frequently made between *land use* and *vegetation* is eliminated. Both consist of plants, both are subject to ecological laws, and in most areas of the earth both are influenced to a greater or lesser degree by the human animal and his behaviour. A valid ecological distinction may be made between planted and self-propagated plants, but such a distinction makes the earlier categories no easier to define, especially in tropical Africa. In fact, the only easily applicable distinction which may be made is that *land use* is characterized by more or less regular field patterns, whereas vegetation is not. This also is of little real use in the ecological context.

It is not easy, in the context of tropical African conditions, to separate the plant component of the ecosystem from the soil component. Nor is it a simple matter, in many areas, to distinguish it from the animal component. Furthermore, since many African agricultural systems have striking ecological implications (Ross, R., 1954; Clayton, 1958; Moss and Morgan, 1967), it is often useful and informative to treat man and his agricultural behaviour as an integral part of the ecological complex (Trapnell and Clothier, 1937; Trapnell, 1943, *inter alia*).

Thus the approach adopted in this chapter involves the consideration of the ecological character of the principal plant-soil systems found in tropical Africa, in terms of water, microclimate, nutrients, organic matter and energy. Their variability within the areas they occupy is also considered, and the implications of the relationships which exist between them are also examined where they are of agricultural importance. The role of non-human animals, expressed in grazing relationships, exploitation as game, and the attainment of pest status is also briefly dealt with, and finally, the broad features and ecological relationships of some major diseases are also described.

Character of the plant–soil systems

Perhaps the most striking contrast to be seen in the landscape of Africa is the often sudden, and usually rapid, change from forest-type vegetation and agriculture, with its dense canopy of woody plants, to the savanna-type plant groups, with their more open tree pattern and intervening herbs and grasses. This change in physiognomy is an indicator of a fundamental change in ecological conditions, and is usually accompanied by a no less striking floristic change (Tables 8.1, 8.2, 8.3). It is thus convenient to use the contrast across this boundary as a reference situation from which the variability of the ecological background in tropical Africa as a whole may be described in a comparative manner.

It is important to emphasize at the outset that no attempt is made to reconstruct any ideal climax vegetation, not even to think in terms of a 'savanna ecosystem' or a 'forest ecosystem', though the notion of the ecosystem is implicit in much of the discussion. Ecological considerations of this kind are peripheral to the main purpose, which is simply to describe and evaluate the kinds of relationships which exist at present in real places, and thus to bring out the ecological background to the use of biological resources in tropical Africa as a whole.

Forest plant–soil systems

The forest plant–soil system involves an almost closed cycling of nutrients, associated with a characteristic pattern of water use. Typically it has complex microclimatic relationships which are very distinct from the changes in the surrounding macroclimate, and its high rate of energy use is expressed in a very high rate of litter production, rapid mineralization rates, and a rapid attainment of equilibrium with respect to organic matter relationships. These characteristics are largely related to the considerable bulk, the woody character, and the well-defined structural characteristics which it typically displays.

Water balance and microclimate

The work done on the fundamental features of the water balance of forest plant–soil systems is unfortunately rather scattered. If, however, it is assumed that these features are largely related to the mass and structure of the vegetation, its root development, and the water properties of the soil body, the detailed experimental work on catchments done in East Africa (Pereira *et al.*, 1962a) provides results which may be applied informatively to many of the areas of moderate rainfall (1,000–1,600 mm) in tropical Africa.

TABLE 8.1 *Thicket and secondary forest, and savanna regrowth, near Aiyegbede, Abeokuta Province*

Lists compiled from six samples in thicket and secondary forest, and two samples in savanna regrowth.

	THICKET AND SECONDARY FOREST	SAVANNA REGROWTH
	Over heavy soil with mottled clay at depth	Over sandy soil with hardening pan layer at 50–60 cm
TREES AND SHRUBS		
10–20 m	*Elaeis guineensis*	
3–10 m	*Acacia polyacantha*	
	Albizia zygia	
	Blighia sapida	
	Carica papaya	
	Celtis zenkeri	
	Cola nitida	
	Erythrina senegalensis	
	Holarrhena floribunda	
	Morinda lucida	
	Rauvolfia vomitoria	
Less than 3 m	*Acacia polyacantha*	*Acacia polyacantha*
	Antiaris africana	*Annona sengalensis*
	Carica papaya	*Daniellia oliveri*
	Ceiba pentandra	*Elaeis guineensis*
	Cola millenii	*Ficus capensis*
	Dichapetalum guineese	*Hymenocardia acida*
	Diospyros mespiliformis	*Lonchocarpus cyanescens*
	Diospyros monbuttensis	*Parinari curatellifolia*
	Dombeya buettneri	*Stereospermum kunthianum*
	Ficus exasperata	
	Holarrhena floribunda	
	Lecaniodiscus cupanioides	
	Malacantha alnifolia	
	Monodora tenuifolia	
	Morinda lucida	
	Parkia clappertoniana	
	Ricinodendron heudelotii	
	Theobroma cacao	
	Trema guineensis	
	Vernonia sp.	
	Anchomanes sp.	*Nauclea latifolia*
	Deinbollia sp.	*Tephrosia flexuosa*
	Ehretia cymosa	*Vernonia tenoreana*
	Eupatorium odoratum	*Waltheria indica*

TABLE 8.1: (cont.)	THICKET SITE (cont.)	SAVANNA SITE (cont.)
	Manihot esculenta	
	Melanthera scandens	
	Musa sp.	
	Securinega virosa	
	Solanum verbascifolium	
	Tephrosia bracteolata	
	Triumfetta cordifolia	

CLIMBERS, etc.

	Acacia sp	
	Cardiospermum grandiflorum	
	Combretum zenkeri	
	Dioclea sp.	
	Dioscorea dumetorum	
	Dioscorea preussi	
	Elaeodendron afzelii	
	Gongronema latifolium	
	Hewittia sublobata	
	Motandra guineensis	
	Mucuna pruriens	
	Paullinia pinnata	
	Secamone afzelii	
	Tragia benthami	
	Triclisia subcordata	

HERBS, etc.

	Asystasia insularis	*Aspilia africana*
	Asystasia sp.	*Anana comosa*
	Cissampelos owariensis	*Andropogon gayanus*
	Euphorbia heterophylla	*Biophytum petersianum*
	Neostachyanthus occidentalis	*Lyrsocarpus coccineus*
	Phaulopsis falcisepala	*Cassia mimosoides*
	Psychotria sp.	*Commelina benghalensis*
	Sebastiana chamaelea	*Fimbristylis exilis*
		Imperata cylindrica
		Indigofera hirsuta
		Pennisetum polystachyon
		Tridax procumbens

The basic facts of the water balance, using data obtained from Kericho, Kenya, and Mbeya, Tanzania, are illustrated in Figure 8.1. The forests concerned show important structural and floristic affinities with West African secondary, semi-deciduous forest (Kerfoot, 1962a, b; Keay, 1953, 1959). The principal characteristics of importance are the very high proportion of the total atmospheric precipitation which is transpired by and evaporated from the vegetation, the very low rate of percolation to groundwater, and the very small amount of run-off and stormflow.

TABLE 8.2 *Cocoa plantation and savanna cultivation, near Aiyegbede, Abeokuta Province*

Lists compiled from one sample and transect data in the cocoa plantation, and from four lists and transect data for the savanna cultivation.

	COCOA PLANTATION Over heavy soil with mottled clay at depth	SAVANNA CULTIVATION Over sandy soil with pan at 40–60 cm
TREES AND SHRUBS:		
3–10 m		
Planted	*Theobroma cacao*	
Self-sown	*Antiaris africana* *Spondias mombin*	*Acacia polyacantha*
Less than 3 m		
Planted	*Theobroma cacao* *Citrus aurantium*	*Anacardium occidentale*
Self-sown	*Deinbollia* sp. *Jatropha gossypium* *Securinega virosa*	*Albizia zygia* *Bridelia ferruginea* *Ficus capensis* *Ficus exasperata* *Gardenia ternifolia* *Stereospermum kunthianum* *Eriosema psoraleoides* *Securinega virosa* *Waltheria indica*
HERBS, CLIMBERS, etc.		
Planted	*Xanthosoma sagittifolium*	*Ananas comosa* *Dioscorea* spp. *Manihot esculenta* *Sorghum guineense* *Zea mays*
Self-sown	*Ageratum coryzoides* *Asystasia insularis* *Boerhavia diffusa* *Brachiaria deflexa* *Commelina benghalensis* *Digitaria adscandens* *Euphorbia heterophylla* *Euphorbia hirta* *Tridax procumbens* *Vernonia cinerea*	*Ageratum conyzoides* *Albuca* sp. *Anchomanes* sp. *Andropogon gayanus* *Aspilia africana* *Bidens pilosa* *Biophytum petersianum* *Borreria scabra* *Celosia trigyna* *Commelina benghalensis*

TABLE 8.2
(cont.)

HERBS etc.	COCOA PLANTATION (cont.)	SAVANNA CULTIVATION (cont.)
		Corchorus aestuans
		Corchorus olitorius
		Crepis sp.
		Cyperus aristatus
		Digitaria adscandens
		Eragrostis ciliaris
		Euphorbia heterophylla
		Euphorbia hirta
		Fleurya aestuans
		Imperata cylindrica
		Indigofera hirsuta
		Mariscus umbellatus
		Mucuna pruriens
		Oldenlandia corymbosa
		Pandiaka involucrata
		Phyllanthus amarus
		Physalis micrantha
		Pouzolzia guineensis
		Stylochiton lancifolium
		Synedrella nodiflora
		Talinum triangulare
		Tephrosia bracteolata
		Tridax procumbens
		Vernonia cinerea
		Waltheria indica

TABLE 8.3 *Species lists for adjacent areas of forest and open savanna between Onigbongbo and Idi-Emi, Abeokuta Province*

List compounded from four sample points in the savanna, and two in the forest. The forest, with a completely closed canopy, and few herbs, is floristically similar to Transition Woodland. It is, nevertheless, physiognomically similar to secondary forest.

	FOREST SITE On deep, moist colluvial soil	SAVANNA SITE On shallow sandy soil over hardpan at 10–60 cm
TREES AND SHRUBS 10–20 m	*Afzelia africana*	
	Albizia zygia	
	Anogeissus leiocarpus	
	Ceiba pentandra	
	Celtis brownii	
	Fagara leprieurii	
	Markhamia tomentosa	

TABLE 8.3 (cont.)	FOREST SITE (cont.)	SAVANNA SITE (cont.)
3–10 m		*Afrormosia laxiflora*
		Combretum molle
		Cussonia barteri
		Lophira lanceolata
less than 3 m	*Albizia zygia*	*Afrormosia laxiflora*
	Anogeissus leiocarpus	*Annona senegalensis*
	Antidesma laciniatum	*Cussonia barteri*
	Blighia sapida	*Daniellia oliveri*
	Ceiba pentandra	*Gardenia ternifolia*
	Dichapetalum guineense	*Lannea acida*
	Erythroxylum emarginatum	*Lophira lanceolata*
	Fagara leprieurii	*Oncoba spinosa*
	Fagara zanthoxyloides	*Parinari curatellifolia*
	Holarrhena floribunda	*Syzygium guineense*
	Malacantha alnifolia	*Terminalia laxiflora*
	Markhamia tomentosa	
	Phyllanthus discoideus	
	Rauvolfia vomitoria	
	Sterculia tragacantha	
	Anchomanes sp.	*Aframomum kayserianum*
	Hoslundia opposita	*Cochlospermum* sp.
	Jatropha gossypifolia	*Hoslundia opposita*
	Olax gambecola	*Nauclea latifolia*
	Rytigynia nigerica	*Olax gambecola*
	Sida corymbosa	*Rytigynia nigerica*
	Synedrella nodiflora	
	Triumfetta cordifolia	
CLIMBERS, etc.	*Cissus petiolata*	*Cyanotis longifolia*
	Dioscorea petiolata	*Cyperus schweinfurthianus*
	Mariscus sp.	*Vigna pubigera*
	Uvaria chamae	
HERBS, etc.	*Psychotria vogeliana*	*Aeschynomene lateritia*
	Sida corymbosa	*Alysicarpus* sp.
		Aspilia helianthoides
		Borreria scabra
		Calopogonium sp.
		Cassia mimosoides
		Hibiscus asper
		Indigofera sp. (*spicata?*)
		Loudetia simplex
		Pandiaka involucrata
		Polygala baikiei
		Schyzachium brevifolium
		Tephrosia barbigera
		Vigna multinervis

Comparison between the soil moisture régime of the catchments under forest and those under other types of land use is striking. At Kericho regular sampling over a three-year period showed that there was available water present in the soil at all levels throughout the year, and that the soil was permanently wet below six feet. Even at Mbeya, where the annual precipitation was concentrated into a six- to seven-month rainy season,

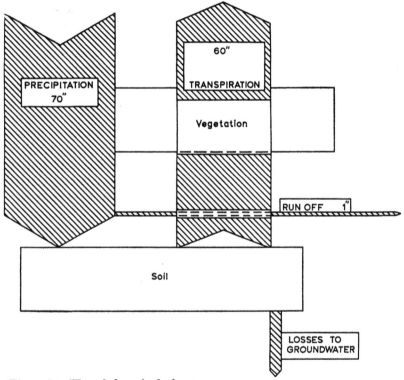

Figure 8.1 *Water balance in the forest.*

wilting point was never reached in the solum at any level in the three years of observation. Outside the forested catchment, in a similar location with comparable soil characteristics and almost identical atmospheric water régime, run-off was much greater, especially as stormflow, and the upper layers of the soil dried out rapidly at the onset of the dry season (McCulloch, 1962a, b; Dagg and Pratt, 1962; Pereira, *et al.* 1962a, b).

The effect of the vegetation is thus clear, and is probably basically similar throughout the areas of moderate rainfall in Africa. Furthermore, this contrast is parallel and related to the microclimatic effects existing beneath forests in Africa. The atmosphere under the forest canopy is protected from the effects of changes outside, and consequently changes in temperature, relative humidity and other variables are much less in magnitude, and lag behind, those outside the canopy. This is related to the decreased air movement effected by the forest (Haddow and Corbett, 1961; Hopkins, 1965a, *inter alia*).

Furthermore, consequent upon the reduced evaporation from the soil surface induced by the high relative humidity of the forest atmosphere and the protection afforded by the litter layer, and the immense quantities of water transpired by the forest, the groundwater level is low, and the layers above it moist, with a low moisture gradient between the free water surface and the soil surface. This is true even in periods of atmospheric water stress. Penman (1963) deals very fully with the general problems of the relationships between vegetation and hydrology, using data from a number or tropical studies, such as those of Bernard (1945, 1954, 1956). In particular he discusses the vexed and important question of the amount of atmospheric precipitation falling over forest areas, especially in Africa, which is derived from evapotranspiration from the vegetation and land surface itself, in relation to the amount which is derived from outside the forest area itself. The problem is to decide how far the hydrological cycle over forest is completely closed, including the levels of the atmosphere outside the forest canopy. Clearly the answer to this question has important bearing on the forest/savanna problem, to be discussed later, and it would seem fair to conclude that the high values for transpiration over forests make a substantial contribution to the maintenance of forest in areas of marginal rainfall.

The importance of root development to the operation of the water cycle is considerable. This is discussed in general by Penman (1963, pp. 74–6), and the effect of the extent of root development upon resistance to drought is clearly shown. Thus soil factors which influence the extent and character of root development are of basic importance to the stability of the system, especially in areas with a significant dry season. The sensitivity to soil factors such as compaction, sudden texture change and localized enrichment of nutrient sources has been amply shown in the case of many tree crops in tropical forests (Baeyens, 1938, 1949; McKelvie, 1962). It is likely that forest trees display similar sensitivity, though their root systems are probably always more extensive than those of tree crops (McKelvie, 1962; Kerfoot, 1962c), often extending to depths of 30–40 ft (9–12 m) below the soil surface.

Thus the forest plant–soil system is strongly buffered against changes in atmospheric conditions, especially with relation to variability in water supply, air movement and changing humidity. The effect of seasonality is thereby much reduced and to a considerable degree the system tends to maintain the conditions it requires for its persistence, both in space and time.

Nutrients

Water movement is the vehicle by means of which nutrients are cycled in the plant–soil system. Hence it is not surprising that under forest there is an almost closed nutrient cycle, with little loss to groundwater, and negligible losses in surface and near-surface lateral movements. Furthermore, the cycling is not markedly seasonal, as it is in savanna (Nye and Greenland, 1960; Vine, 1968). The basic features of the nutrient cycles under forest are illustrated in Figure 8.2, and the amounts of nutrients stored in the various parts of the system in Figure 8.3. Both these figures have been constructed from data provided by Nye and Greenland (1960), and the following points require emphasis:

1. The role of rainwash from the leaves is significant in conveying nutrients to the soil.
2. The important contribution of the subsoil to the amounts of nutrients taken up by the plants must be pointed out. This is related at least in part to the extensive root development of the forest vegetation, which determines the volume of soil from which the plants derive their necessary resources. If through the inhibition of root development a smaller volume of soil is exploited, then competition between the individual plants is increased, and the vigour of the plant component of the plant–soil system diminishes. Hence in such situations the system is more vulnerable to change induced either by climatic fluctuations or by the indirect effects of human activity.
3. The accumulation of nutrients in the first five years after cultivation is rapid, during the normal succession through the fallow period. This is related to the proportions of foliage to woody growth at various stages of development of the fallow, since accumulation is much more rapid in the leaves than in the woody parts. Nevertheless Figure 8.3 reveals clearly that significant amounts of nutrients are held in the wood, and it is important to remember that these are made available to crop plants when the trash is burned prior to cultivation.
4. Finally, the rapid rate of mineralization of organic matter is significant, amounting on average to an annual turnover of 170 per cent of the weight of litter at the soil surface at any one time. This high turnover indicates the high rate of energy use typical of tropical forests, and Bray and Gorham (1964) have shown that rates of litter production in tropical forests are from 60 to over 100 per cent greater than the rates

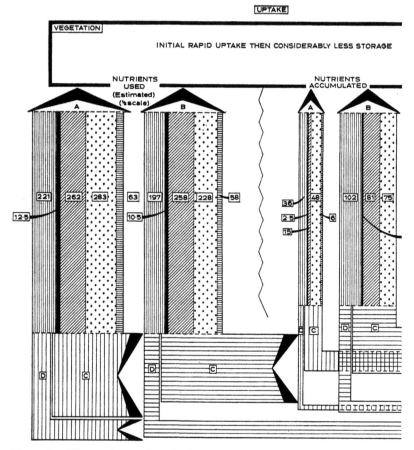

Figure 8.2 *The nutrient cycle under forest in West Africa* (after Nye and Greenland, 1960) in lbs/acre.

characteristic of their cool temperate counterparts, in the absence of limitation due to water or nutrient deficiencies.

Organic matter relationships

The last observation immediately raises the question of the characteristics of organic matter relationships under tropical forests. These are illustrated in Figure 8.4, and the following points of interest may be made. Leaching and erosion losses are low and thus, like the nutrient cycles, the organic matter cycle is almost closed. Furthermore, the resilience of the organic matter cycle is important. Disturbances affecting the rate of addition of litter to the soil, and hence also the rate of organic matter production and

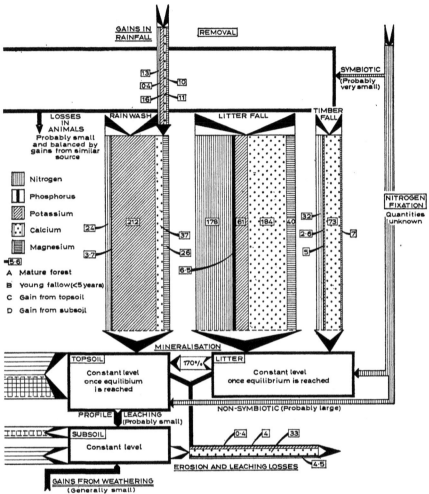

the level of organic matter in the soil, are quickly remedied once normal litter additions are resumed. It is also likely that the organic matter content of the soil may remain high, even if such disturbances are prolonged (Laudelout, 1961; Vine, 1965). Finally, a considerable contribution to the level of organic matter in the soil is made by root slough and exudate. This is very important in relation to cultivation, since these roots usually remain in the soil, often alive. Thus these releases of organic matter continue, and afford a significant source of nutrients during the cultivation period. About 5,000 lb of roots slough, exudate and dead root material are added to every acre under mature forest each year (Nye and Greenland, 1960).

Thus under a forest cover the plant–soil system is an extremely complex,

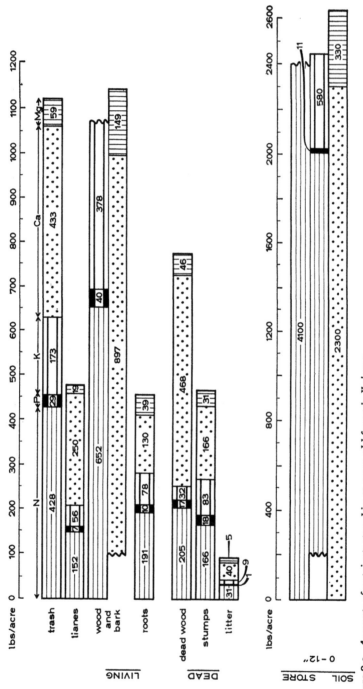

Figure 8.3 *Amounts of nutrients stored in 40-year-old forest in lbs/acre.*

resilient and stable ecological entity, strongly buffered against change induced by environmental effects, notably the seasonal and diurnal climatic changes, and possibly by secular changes as well. Furthermore, even though the forest towards the drier margins may be classified on floristic grounds as 'semi-evergreen' (Richards, 1957, p. 340), this hardly affects the issue at all, at least in West Africa, since the proportion of deciduous trees is small, both in species and number. It is certainly too small to give a seasonal aspect to the forest comparable with the tropical deciduous forests of India and Burma. Hence, this change in floristic composition does not drastically

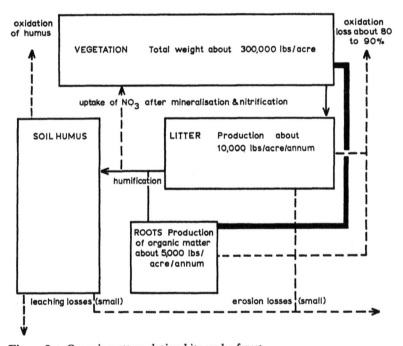

Figure 8.4 *Organic matter relationships under forest.*

affect the effectiveness of nutrient accumulation, the efficiency of their use, or the limited range of the seasonal fluctuations in the activity of the cycles.

Faunal relationships of forest

In developed forest the fauna is rich, and strongly layered as a result of the layering of food niches and microclimate induced by the structure of the vegetation. This layering is displayed by mammals (Rahm, 1961), by birds (Cansdale, 1960), by reptiles (Villiers, 1950; Cansdale, 1961) and by insects (Birket-Smith, 1960, *inter alia*). Furthermore; owing to complicated zoogeographical factors, and also in part to the distinctiveness of the

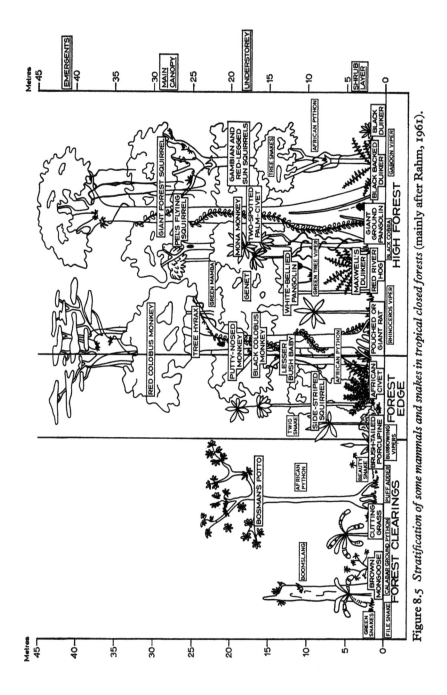

Figure 8.5 *Stratification of some mammals and snakes in tropical closed forests (mainly after Rahm, 1961).*

African Civet – *Civettictus civetta*
Black-backed Duiker – *Cephalophus dorsalis*
Black Colobus Monkey – *Colobus polykomos*
Black Duiker – *Cephalophus niger*
Bosman's Potto – *Perodicticus potto*
Brown Mongoose – *Crossarchus obscurus*
Brush-tailed Porcupine – *Atherurus africanus*
Cutting-grass or Agouti – *Thryonomys swinderianus*
Demidoff's Galago or Lesser Bush Baby – *Galago demidovii*
Gambian Sun Squirrel – *Heliosciurus gambianus*
Genet – *Genetta tigrina*
Giant Forest Squirrel – *Protoxerus strangeri*
Giant Ground Pangolin – *Smutsia gigantea*
Maxwell's Duiker – *Philantomba maxwellii*
Mona Monkey – *Cercopithecus mona*
Pel's Flying Squirrel – *Anomalurus peli*
Pouched or Giant Rat – *Cricetomys gambianus*
Putty-nosed Monkey – *Cercopithecus ascanius* (*nictitans*)
Red Colobus Monkey – *Colobus badius*
Red-legged Sun Squirrel – *Heliosciurus rufobrachium*
Red River Hog – *Potamochoerus porcus*

Side-striped Squirrel – *Funisciurus leucostigma*
Tree Hyrax – *Dendrohyrax dorsalis*
Two-spotted Palm Civet – *Nandina binotata*
White-bellied Pangolin – *Phataginus tricuspis*
African Python – *Python sebae*
Beauty Snake – *Psammophus sibilans*
Black Cobra – *Naja melanoleuca*
Boomslang – *Calamelops unicolor*
Burrowing Vipers – *Atractaspis* spp.
Calabar Ground Python – *Calabaria reinhardtii*
File Snake – *Mehelya poensis*
Gaboon Viper – *Bitis gabonica*
Green Mamba – *Dendroaspis viridis*
Green Tree Snakes – *Gastropyxis smaragdina Hapsidophys lineata*
Green Tree Viper – *Atheris chloraechis*
Puff Adder – *Bitis arietans*
Rhinoceros Viper – *Bitis nasicornis*
Tree snake – *Boiga blandingii*
Twig Snake – *Thelotornis kirtlandii*

habitat, the lowland forest mammalian and avian faunas are more or less distinct from both the savanna and the montane faunas (Dekeyser, 1955; Moreau, 1966; Booth, 1960). Figure 8.5 illustrates this layering with respect to mammals and reptiles (*Ophidia* only). A point of major interest, which poses an interesting evolutionary problem (Moreau, 1966), is the development of a distinct bird, and to a lesser extent mammal, fauna in secondary forest and farm areas.

Human activity in the forest has three main effects on the fauna. First, the mammalian and reptilian faunas are severely depleted by the exploitation of some species, such as the duiker and the larger rodents (notably the agouti and the pouched rat) for food, and the destruction of others, especially the snakes, because of their danger to humans and domestic animals. Secondly, some animals, such as the Agama lizard (Harris, 1957) and the geckos, are favoured by the provision of an increase of feeding or shelter niches or by a reduction of predators. And finally, feeding niches for many insects are reduced in number, while man himself provides an increase of host or reservoir organisms for various parasites causing disease. The former often results in the concentration of feeding on crop plants (as in the case of the mirids), and a consequent pest problem; the latter produces fundamental changes in disease pattern, as has happened in Nigeria with reference to schistosomiasis as a result of increasing movements of population occasioned by improved transport facilities (Collard, 1964).

Hence, in considering the forest in relation to land use and agriculture, the faunal relationships have a significant claim for study and evaluation.

Variability within the closed forest zone

The closed forest zone of tropical Africa, though limited in latitudinal extent, nevertheless includes wide variation in rainfall amounts and distribution. So far the discussion has centred on the zone of moderate rainfall, but substantial areas in south-east Nigeria, Cameroon, Gabon and the Congo have an annual precipitation total in excess of 1,600 mm (Fig. 8.6), coupled with the absence of dry season. In these areas the forest is undoubtedly floristically distinct (Richards, 1957), and possible also structurally different (Keay, 1953; Richards, 1957). It is also clear that the moisture characteristics are likely to be different, and especially that leaching losses are likely to be greater, a situation which is emphasized by the fact that the soils of this zone are generally classed as excessively leached (Vine, 1949, 1951, 1954b, 1956; D'Hoore, 1964).

Within the forest zone also there is evidence that soil physical factors, especially moisture relationships, either in conjunction with, or independently of, chemical characteristics, affect the floristic composition of the vegetation (Richards, 1957). Though the most striking examples have been reported from Trinidad (Beard, 1946), examples have been given from Nigeria (Sykes, 1930), and the occurrence of well-defined freshwater swamp communities is amply demonstrated (Keay, 1953, 1959; Chipp, 1927).

Although the mangrove vegetation (*Rhizophora* spp., with *Avicennia nitida* and *Laguncularia racemosa* in West Africa only) is not related to that of the closed forest zone, in extensive areas of the coastal belt, especially in West Africa, it is a significant ecological unit characterized by its adaptation to saline conditions. It is frequently zoned in relation to frequency of flooding and salinity (Keay, 1953; Rosevear, 1947), and in some parts has been extensively developed for rice growing, but with respect to agriculture and land use in relation to Africa as a whole, it presents but a very small part of the total ecological situation.

The principle variability in the forest belt is, however, the result of human exploitation of the biological resources of the communities, and it is to this aspect that the final section must be devoted.

Effects of cultivation on closed forests

The two principal husbandry practices involved in the agricultural use of tropical closed forests are the shifting or rotational cultivation of food crops, and the development of peasant and commercial plantations of permanent tree crops such as cocoa (*Theobroma cacao*), oil palm (*Elaeis guineensis*), rubber (*Hevea* spp.) and coffee (*Coffea* spp.). Both, in their simplest forms, are closely adapted to the ecological background in which they have developed.

Cultivation of food crops breaks up the cycles of water and nutrients, and destroys the structure upon which microclimatic distinctiveness depends. Nevertheless, under normal practices, the disturbance is of short duration and limited extent, being restricted to small discrete patches, and lasting only for three or four years. It is likely that the difficulty of clearing woody vegetation is in part responsible for this (Moss and Morgan, 1968). Furthermore, most woody roots are left in the ground, facilitating a supply of nutrients to the soil and thence to the crop by sloughing, exudation and decomposition, and also ensuring rapid regeneration of woody growth after the cultivation cycle. The widespread practice of maintaining a dense cover of adventitious plants also tends to restrict the more extreme effects of the disturbance of the microclimate. It is also important to point out that many of these adventitious plants are not strictly 'weeds', but provide products of technological, medicinal or dietary use, such as dyes, drugs or flavourings (Moss and Morgan, 1968). Lastly, the continued presence of living roots in the soil, together with the effects of the extensive rooting volume of the adjacent forest, act as a buffer against major change in

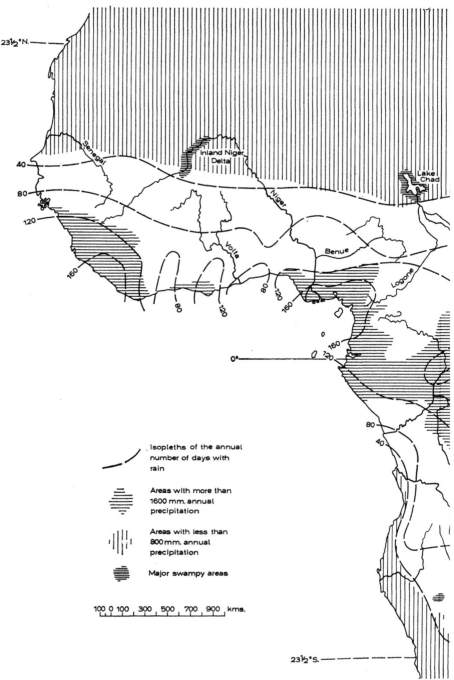

Figure 8.6 *Variability in atmospheric water supply in inter-tropical Africa.*

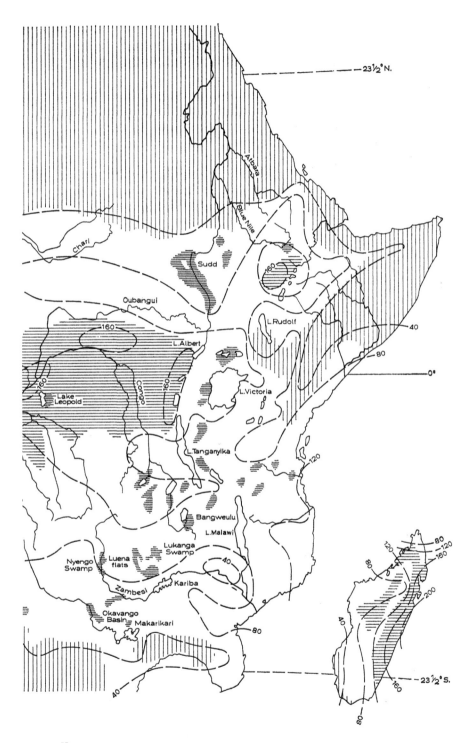

H

groundwater levels and soil moisture characteristics at depth. The size of the cultivation patches is thus critical in relation to the disturbance they bring to the balance of the closed forest as a whole.

The succession following cultivation is well documented (Ross, R., 1954; Clayton, 1958). It is important to notice that woody regrowth establishes itself rapidly under most conditions, that the nutrient and water cycles are quickly re-established, and that fallows of five to ten years are normally adequate to restore the status of the soil to a level sufficient to sustain a new cycle of cultivation (Nye and Greenland, 1960; Vine, 1968). In certain circumstances, however, the vigour of the fallow may decrease, grasses invade, and a change to grass-with-trees occur. This situation is dealt with in more detail later.

Tree crops disturb the balance in a different way. Most are cultivated relatives of forest trees, though not all indigenous to Africa, and even when they are grown in pure stands (which is often not the case) they tend to maintain the cycles and microclimatic characteristics so typical of closed forest, though in a simpler and attenuated form. Nevertheless nutrient levels do decline, and it is possible that the generally shallower rooting system of these trees may have a significant effect on water relationships (Hartley, 1968). The nutrient effect may be serious, especially under conditions of efficient and high-level production such as may be achieved on some commercial plantations. As Hartley (1968, p. 156) succinctly states: '. . . rotational food crop and fallow agriculture . . . is essentially one of alternate losses and rebuilding, while permanent crop agriculture is characterized by greater soil stability and by longer-term losses'.

It is possible that the tendency, especially on peasant plantations, to grow permanent crops of this kind in association with forest trees as deliberate shade, or simply to leave large forest trees standing because of the problem of destroying them, contributes to the apparent ecological stability of this form of land use. It is significant to notice, however, that the planting of a ground cover crop beneath the trees, which might be viewed as a step towards more nearly simulating the conditions under closed forest, has proved valuable only under some crops, notably oil palm and rubber (Hartley, 1968). In relation to the pest problem the situation is however, clearer. Pest attacks on food crops undoubtedly increase towards the end of the cultivation phase, and the new fallow quickly begins to restore balance. Under tree crops insect pest problems are continually present, but often can be considerably mitigated by growing the crop under heavy shade (Williams, 1953), as in the case of the cocoa mirids.

It is important to point out, however, that chemical control of such insects is now possible, and the consequent elimination of shade trees, coupled with the use of fertilizers, has produced remarkable increases in yield (Table 8.4) (Cunningham and Lamb, 1959; Cunningham, et al., 1961), at least in relation to cocoa (but see also, for rubber, Haines, 1949). The ecology and control of other pests and of insect-borne diseases, such

TABLE 8.4   Cocoa yields at Tafo, Ghana, under various shade and fertilizer treatments (after Hartley, 1968)

SHADE	I	O	I	O
INSECTICIDE	O	I	O	I
FERTILIZER	O	O	I	I
YIELD (lb/dry cocoa/acre/year)	857	2,263	1,059	3,090

Notes:
   Data from factorial experiments at Tafo, Ghana.
   1 indicates presence of factor, o indicates absence.
   The fertilizer used was a compound of nitrogen, phosphate, potash and magnesium. Magnesium is included in addition to the main nutrients since tree crops in tropical forests frequently show symptoms of magnesium deficiency, and this can affect responses to the other nutrients.

as the swollen-shoot viruses, is equally significant and complex (Johnson, 1962b; Dale, 1962; Wharton, 1962; Entwistle, 1962). In some cases also pests and diseases form interacting complexes which have considerable significance, as in the case of mirid attacks, fungus infection of mirid lesions, and attack by swollen-shoot viruses borne by the mealy-bug *Pseudococcus* (=*Planococciodes*) *njalensis*, which is itself 'farmed' in tented colonies by certain species of ant (especially *Pheidole megacephala*). Other mealy-bugs, such as *Planococcus citri* and *Ferrisiana virgata*, may also be important to the transmission of the virus (Johnson, 1962b; Tinsley, 1964).

   Consideration of pest and disease problems of other perennial and annual crops certainly reveals similar complex patterns of ecological relationships. These may frequently be traced back to the reduction of the complexity of the forest ecosystem which is the inevitable consequence of cultivation, or to the introduction of new microclimatic or feeding relationships.

   Before passing on to consider relationships in the savanna areas, some mention must be made of forestry as a form of land use. Exploitation of the closed forest zone for timber is achieved principally by the creation and management of forest reserves, by means of which the natural ecosystem is controlled and used (Ross, J. K., 1957; Foggie, 1962; Streets, 1962). The principal timbers are therefore the indigenous hardwoods, that is the *Khaya* mahoganies, the *Entandrophragma* mahoganies, obeche (*Triplochiton scleroxylon*), odum (*Chlorophora excelsa*), makore or bako (*Mimusops heckelii*), *Afrormosia elata*, and the so-called luxury timbers such as the African Walnut (*Lovoa trichilioides*), and kusia or opepe (*Nauclea diderrichii*). Attempts to grow pure stands have generally not been successful, and this is probably related to the complex germination and regeneration patterns characteristic of tropical forests (Mildbraed, 1930a, b; Aubreville, 1938; Richards, 1939, 1957). These patterns also imply that silviculture in forest ecosystems managed for timber must be directed towards the encouragement of the regeneration of these hardwood species, which generally characterize only the later stages of the long succession following felling or tree fall. Attempts to introduce exotic species, such as teak (*Tectona grandis*), have rarely been successful from the point of view of timber production, but this with other species forms a dominant and increasing source of poles, firewood and small timber. This class of product is normally produced from fixed plantations of teak, cassia (*C. siamea*) and gmelina (*G. arborea*), which are coppiced regularly (Ross, J. K., 1957; Foggie, 1962; Foggie and Piasecki, 1962). It has been suggested that the difficulty of growing many exotic species to timber size is related to soil physical characters (Moss, 1957), a fact which may be analogous to the possible influence of these characters on the floristic composition of mature rain forest, which was mentioned earlier (Sykes, 1930).

Apart from the obvious importance of timber as an export commodity, and as an item of international trade, the demand for firewood, poles and small timber in the rapidly growing towns, especially in West Africa, has in some areas resulted in a considerable increase in the use of land for coppice plantations. In Abeokuta and Oyo Provinces in Western Nigeria, for example, this demand, coupled with the requirements for flue-curing of local tobacco, has produced a proliferation of teak plantings, especially around islands of forest in the main belt of savanna (Moss and Morgan, 1968). The ecological significance of this is that teak is not greatly susceptible to accidental fires, and when it has developed a canopy it rapidly kills out herbs and grasses.

Savanna plant–soil systems

The savanna areas of inter-tropical Africa cover at least 50 per cent of the total land surface. This is considerably greater than that of the closed forest areas, but the extended discussion already devoted to the closed forest makes it easier to emphasize by contrast the broad similarities of the savanna, before dealing with its considerable internal variability. In this context the term savanna connotes simply the group of tropical vegetation types which consist physiognomically of a more or less continuous layer of grasses and herbs, with a tree layer of variable density. As the trees become denser, these savanna types merge into savanna woodland, which in some cases may be devoid of grasses, and structurally strongly resembles closed secondary forest (Moss and Morgan, 1968). With decreasing tree density the savanna types merge into open grasslands or steppes.

Contrasts between forest and savanna

The most striking and obvious difference between closed forest and savanna is in the sheer bulk of the vegetation. Nye and Greenland (1960) quote figures for oven-dry weight (excluding roots) of 300,000 lb per acre for forest, compared with only 60,000 lb for savanna. Furthermore, more than 75 per cent of this latter figure is woody material. If the bulk of the roots is also taken into account the contrast is even more striking.

The difference in bulk is associated with a contrast in structure. Except in savanna woodland of the denser types, there is no closed canopy of woody plants, and the shrub layer is not continuous either. The grass–herb layer varies considerably in floristic composition, and also in height and continuity, and, finally, these vegetation types are characterized by an almost complete absence of climbers.

These differences in structure and bulk imply, in the first place, important differences in nutrient and water properties. These are accentuated by the seasonal aspect of many of the savanna plants. At Mbeya, for example, in comparison with the situation under forest already described, sampled plots under cultivation and regenerating bush rapidly dried out to a depth of 3½ ft as soon as the rains ceased, and remained at wilting point or very near it for the remaining four months of the dry season (McCulloch, 1962b; Pereira, et al. 1962a, b). This effect, due principally to the vegetation difference in relation to the climatic régime, might be expected to be accentuated by the inherent seasonal character of many savanna plants. Noirfalise (1956) has provided an excellent account of the microclimate of derived savanna, and Lawson and Jeník (1967) have recently demonstrated very detailed effects on vegetation patterns which result from microclimatic differences.

This seasonality of the savanna with respect to water characteristics has important effects on nutrient cycles. These are themselves seasonal, compared with the relative continuity of those under forest, and water deficiency at the beginning and end of the rains may have a marked inhibiting effect on nutrient uptake by the vegetation (Hannon, 1949). These effects may be reinforced by the occurrence of soils which show a rapid response to fluctuations in the atmospheric water supply (Moss and Morgan, 1968).

The contrast between the nutrient storage of savanna and forest is revealed by a comparison of Figure 8.7 with Figure 8.3. Though grass and herbs form such an important part of the vegetation, the role of the trees is the most significant. Not only do they store the bulk of the water and nutrients, and presumably also pump up nutrient ions from the subsoil in a similar way to forest trees, but they also contain a much higher proportion of calcium than the grasses and herbs. Apart from the obvious importance of this fact in cation exchange reactions, limitations in calcium supply seriously affect the phosphate balance of the soil (Laudelout *et al.* 1954;

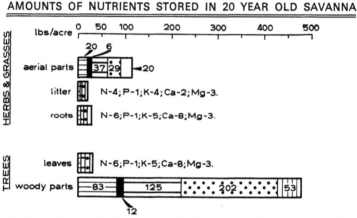

AMOUNTS OF NUTRIENTS STORED IN 20 YEAR OLD SAVANNA

Scale and symbols the same as in figure 3, for the plant component.

Figure 8.7 *Amounts of nutrient stored in 20-year-old savanna* (lbs/acre).

Nye, 1958a, b; Nye and Hutton, 1957). Furthermore, the inhibiting effect of certain grasses, notably *Andropogon* spp., on nitrate formation is also important and may be related to microbiological factors (Vine, 1968).

Thus tree density in the savanna is an important variable affecting fundamentally the behaviour of the plant–soil system. Within the broad category of 'savanna' there may be a variation with respect to water and nutrients from conditions approximating to steppe grassland to those more nearly analogous to closed forest, as in the case of some savanna woodlands (Table 8.5).

Variability in the savannas

The variability of the savanna areas of Africa may be related to three broad categories of factors. In the first place, and especially in West Africa and Uganda, there is undoubtedly a change, expressed especially in floristic terms, with climate; this is revealed in the latitudinal vegetation zones of West Africa (Keay, 1953, 1959) and in relation to the local climatic effects of relief in Uganda (Langdale-Brown, 1960). Secondly, there is frequently a topographic zonation, related to the change in soil

TABLE 8.5 *Dense transition woodland, near Ipapo, Oyo Province*
List compiled from two stands. Though floristically related to Transition
Woodland, the structure of the vegetation strongly resembles forest, with a
closed canopy, climbers, and a poorly developed herb layer.

TREES: dense canopy at about 10 m, with some species, notably *Acacia polyacantha, Afrormosia laxiflora, Anogeissus leiocarpus Daniellia oliveri, Lannea
acida, Lonchocarpus laxiflorus, Parkia clappertoniana, Phyllanthus discoideus,
Terminalia glaucescens* and *Vitex doniana* occasionally emergent above this.

TREES AND SHRUBS

Acacia polyacantha
Afrormosia laxiflora
Albizia glaberrima
Albizia zygia
Annona senegalensis
Anogeissus leiocarpus
Bridelia ferruginea
Butryospermum paradoxum
Crossopteryx febrifuga
Daniellia oliveri
Dicrostachys cinerea
Ficus capensis
Gardenia ternifolia
Hymenocardia acida
Lannea acida
Lonchocarpus laxiflorus
Nauclea latifolia
Parkia clappertoniana
Phyllanthus discoideus
Pseudarthria confertiflora
Securidaca longipedunculata
Securinega virosa
Terminalia glaucescens
Vernonia colorata
Vernonia tenoreana
Vitex doniana

CLIMBERS AND HERBS, etc.

Ampelocissus leonensis
Aspilia africana
Borreria scabra
Cassia mimosoides
Cissus petiolata
Desmodium velutinum
Clematis hirsuta
Mucuna pruriens
Sida corymbosa
Sporobolus pyramidalis
Stylosanthes mucronata
Lantana rhodesiensis
Phyllanthus muellerianus

physical character coresponding to slope variation (Morison *et al.*, 1948; Duvigneaud, 1951), or to the occurrence of soils better supplied with mineral nutrients (Cole, 1963). Finally, there are the patterns related to the influence of animals, including man. These include, besides cultivation and burning, grazing, trampling, browsing and the favouring of certain plants at the expense of others, either voluntarily or involuntarily. These latter effects may be locally important in terms of dispersal, increased germination rates by passage through an animal's gut, and the creation of exceptionally favourable germination sites by the dropping of faeces pads by grazing animals (Turner, 1967; Vesey-Fitzgerald, 1963; Ramsay and de Leeuw, 1964, *inter alia*).

Thus there is very wide variability in the ecological character of the savanna areas of Africa. As Phillips (1959) has pointed out, they are associated with much more significant internal differences than the forest zones, and this intensifies the inherent problems of rational development. It may be that ignorance of the detail of the heterogeneity of tropical closed forests, owing largely to the complexity of their floristic composition and the difficulty of its rationalization, may be in part responsible for this, but it is certainly not by any means entirely so. Furthermore, as has been argued already, the basic similarity of the structure of tropical closed forests imposes a functional unity which tends to mask and override other differences.

Man and animals in the savanna

A fundamental distinction between the human use of the forest and the savanna is to be found in the keeping of livestock. Savanna areas until recent times, especially in East Africa, were characterized by large herds of ungulates and other grazers and browsers, with their associated predators – which is but one aspect of the significant faunal distinction between forest and savanna (Dekeyser, 1955; Rosevear, 1953; Cansdale, 1960; Moreau, 1966, *inter alia*). In most areas these herds have been to a great degree replaced by herds of wandering cattle, usually of the Zebu type, with their herdsmen. Hence the predator food niches are now principally occupied by man.

This change has frequently resulted in fundamental changes in the vegetation, such as the development of dense thickets of *Acacia hockii*, *Hoslundia opposita* and *Solanum* spp. in Uganda (Turner, 1967). This results from the reduced fire intensity induced by heavy grazing, the greater selectivity of grazing and browsing as a result of the food preferences of the cattle, and the dispersal of plants to new habitats by both the cattle and the herdsmen. Overgrazing itself leads to its own peculiar problems, and a further effect of the replacement of multispecific occupation of herbivore niches by a more or less monospecific situation of animals in close association with man has been the spread of diseases such as trypanosomiasis, with its complex three- (agent–vector–host) or four- (agent–vector–reservoir–host) factor ecological complex (May, 1961).

These factors, among others, have led some authorities to favour game farming as an alternative to ranching and herding as an acceptable land-use system in many savanna areas (FAO–UNESCO, 1963). It is argued that, owing to the large number of cohabiting species (often as many as twenty kinds each with its own food preferences), such an ecological system supports higher values of standing crop biomass (the live weight of the animals being farmed) per square mile than does ranching. Furthermore, the habitat is not degraded since this total load is distributed over the whole of the plant component, and not concentrated on the graminaceous element as in the case of cattle. There is also evidence which suggests that the nutritional efficiency of wild ungulates is greater, and that populations of different species of fast-growing and early maturing animals of this kind represent a rapid rate of turnover which efficiently exploits this greater nutritional efficiency. Finally, many species are tolerant of a lack of free water, which is an additional advantage in the drier areas and during the dry parts of the year. Thus on such ecological grounds it may be argued that game-farming may be the most efficient use of land and agricultural resources in many drier savanna areas, particularly in East Africa. It may, however, be that the problems of cropping, marketing and processing, and above all of the social and traditional background into which such developments must be fitted, may prove insuperable obstacles, except in a few areas.

Cultivation in the savanna areas is traditionally shifting or rotational, with a graminaceous or herbaceous fallow period. The form and efficiency of the fallow is very variable, from coarse low-value grasses such as *Imperata cylindrica*, through better grasses, notably *Pennisetum purpureum*, to well-developed herb–shrub fallows with *Tridax procumbens*, *Tephrosia* spp. and *Indigofera* spp. (Taylor *et al.*, 1963; Moss and Morgan, 1968). Compared with woody fallows, grass–herb fallows are an inefficient way of restoring the nutrient status and improving water balance. From the point of view of the cultivator, however, the great adantage of savanna over forest is the ease of clearance. In some situations the low productivity of savanna cultivation per acre may be compensated for by the fact that much larger areas can be brought under cultivation, and this may be an important factor in crop choice (Moss and Morgan, 1968).

Cultivation in the savanna also introduces the complexities of fire clearance, with the associated possibilities of vegetation change (Langlands, 1967). At this stage it is important to notice that indiscriminate burning of the savanna and the wide, uncontrolled spread of fires are probably not characteristic of cultivating communities, though they may be in hunting activities. Indeed, there is evidence which suggests that destructive late burning immediately prior to the rains is often strictly controlled in its effects and extent, by the pulling of grass and herbs to create a fire-break, where fire is likely to be seriously damaging. The much less destructive early burn is less strictly controlled (Moss and Morgan, 1968). Even the suggestion that burning is non-selective, affecting all savanna communities equally, and indiscriminate in its effects, probably needs considerable qualification in many areas of tropical Africa.

Permanent crops are not a common feature of savanna areas. Even cotton, which is a perennial, is not commonly cultivated for long periods, in view of the increase in pest attacks produced by long cultivation (Irvine, 1950). The one major exception to this rule are the fibres of *Agave* spp., especially sisal (*A. sisalana*), which with their productive life of seven or eight years and their xerophytic character form an ideal plantation crop in the drier savannas of East Africa (Cobley, 1963).

Pest attacks on crops, especially the cereals, are probably not directly related to the change in ecological conditions induced by cultivation though the replacement of a complex system by an inherently simpler one undoubtedly favours the spread of insect pests and fungal diseases such as the rusts. The rise of certain birds, such as the passerine *Quelea quelea*, to pest status over large areas of Africa is probably related to the provision of more easily accessible and concentrated source of food rather than to the depletion of alternatives (Russell, 1968; Moreau, 1966; Morel and Bour-lière, 1955, 1956).

The effects of increasing aridity

Away from the Equator the availability of atmospheric water decreases steadily, both in actual amount, and in increasing length and intensity of the dry period. These deficiencies are also accompanied by an increasing unreliability of occurrence (Fig. 8.6). In the areas of moderate rainfall to which most discussion has so far been devoted the problems created by this tendency are not paramount, though even here an improvement in water supply to crops can produce remarkable results (Russell, 1968). Towards the more arid margins, however, these problems override all others. The trees in the savanna become smaller, though not necessarily less frequent, and the proportion of thorny species may tend to increase (Keay, 1959). It is important to notice, however, that the absence of trees and the emergence of steppe grassland are not generally associated with increasing atmospheric aridity but with the influences of edaphic or anthropogenic factors (Duvigneaud, 1951).

In these areas water conservation in cropped land is fundamental to successful agriculture. Many parts are used simply for extensive grazing, but extension and improvement of arable agriculture is dependent upon the use of effective husbandry methods, which have been described in detail by Russell (1968). First, soil management, involving reduction of run-off, by development of tilth and protection of the soil surface by a mulch must be practised. It is necessary to remember, however, that a living mulch competes with the crop for available water, and that dead organic material is often used in the tropics for thatching or livestock feed. Thus efforts must generally be concentrated on reducing run-off velocity by contour-ridging or tie-ridging. The unreliability of the rainfall, however, creates real problems in that, though the ridges are effective in heavy rainstorms, in years of above-average amounts problems of poor aeration and waterlogging may occur. Alternation of tied and non-tied furrows has been proved successful when carefully executed (Lawes, 1961). Second, crop management and plant breeding can also make a major contribution. This involves tailoring the crop to fit the conditions, when these are understood in terms of reliability and amounts of rainfall, and the physiological demands of the crop (Manning, 1949; Forsgate et al., 1965). Another fundamental problem is that of sowing at the right time (Akehurst and Sreedharan, 1965). Competition from weeds needs to be eliminated, and the spacing of the crop must be the optimum to use the soil moisture most effectively. Finally, most crop plants display sufficient genetic variety to make it possible to develop varieties which are adapted to particular sets of ecological conditions. Two lines of development are possible: in areas of short but fairly reliable rains short-season varieties need to be developed; whereas in areas of unreliable rainfall the character to be selected in breeding must be true drought-resistance (Russell, 1968). The real question in these situations is, however, whether it is economic to undertake the improvements and practices necessary for development.

Thus, in these more arid areas, the selection of the most appropriate husbandry system and the ecologically most suitable crop varieties is vital for obtaining acceptable yields. But understanding of the ecological background is fundamental to this selection, and thus to the provision of a successful agriculture.

Relationships between forest and savanna

Full discussion of the problem of the ecological status of savanna is beyond the scope of this paper, and is not directly relevant to its focus of interest, but in so far as there are thought to be significant relationships between cultivation pressure, fire clearance, and the development of savanna from closed forest, it is necessary to review certain of its main aspects.

In particular local situations the relative distributions of forest and savanna have been shown to be related to a variety of factors. Some have argued for a dominance of fire (Aubreville, 1949; Keay, 1953, 1959; Hopkins, 1965b), others for a combination of fire and cultivation pressure (Clayton, 1958, 1961), and yet others for the overriding significance of edaphic factors, reflected in a correspondence of vegetation types with slope form and landforms (Cole, 1963). Furthermore, Tinsley (1964) has described an ecological complex involving cocoa in Nigeria in which mirid attack, fungus infection and development of the swollen-shoot virus combine to induce a change from cocoa plantations to grass-with-trees. In some areas also there is a conscious effort on the part of the indigenous cultivator to preserve woody vegetation, by concentrating on the cultivation of permanent tree crops in forest outliers (which incidentally also have the advantage of being at a distance, with an intervening barrier of savanna, from present areas of infection by mirids and swollen-shoot), and by strictly controlling burning practice in the contiguous savanna (Moss and Morgan, 1968). In some areas nutrient factors may also be important, and in any case in most situations these factors may be expected to have an influence on the vigour of the regeneration of woody fallow.

In general, therefore, it is useful to conceive the boundary between forest and savanna in any particular location as expressing the balance of a whole complex of interacting factors. These are illustrated in Figure 8.8. The possibility of a change from forest to savanna conditions occurring in any area can be assessed only by the evaluation of these factors in relation to one another at that location, and by parallel evaluation of their variability in space. For example, detailed study of forest outliers in south-west Nigeria has suggested that under existing conditions of cultivation and burning the present situation is stable. Furthermore, it has been suggested that, in this location, thresholds of maximum cultivation pressure and maximum burning intensity exist at which radical changes may take place (Moss and Morgan, 1968).

Despite the apparent complexity of the problem, it is possible to argue, in West Africa at least, for the existence of three major zones (Fig. 8.9.). In the regions of high and well-distributed rainfall there is a zone in which moisture supply to the plant–soil system is always ample, and in all edaphic and anthropogenic conditions any change from closed forest to grass-with-trees is virtually impossible. In the areas of unreliable and low rainfall with a prolonged dry season, however favourable the soil, cultivation and burning conditions, the moisture supply is always insufficient to support any kind of closed forest. Between the two there is a very extensive and important zone in which the water balance of the plant–soil system itself is the critical factor, and depends largely upon edaphic factors, in relation to cultivation pressure and fire intensity.

This middle zone is of paramount importance to agriculture (Stamp, 1956; Vine, 1954a). Its moderate, well-distributed and fairly reliable rainfall, together with the absence of excessive leaching through the soil, and frequent occurrence of heavy, base-rich weathering products, make it potentially a zone of variable and successful agricultural development (Vine, 1954a; Moss, 1963). Nevertheless it is a zone of soil heterogeneity, in relation to texture, induration, pan development and groundwater levels. It is also, over a considerable proportion of its area, more densely populated than much of the rest of the continent. Thus cultivation pressure

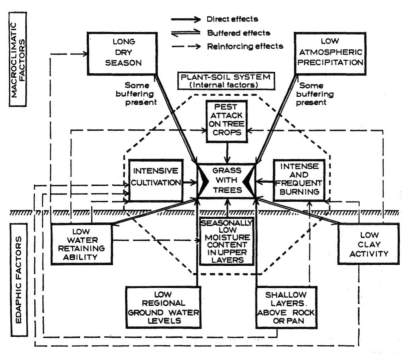

Figure 8.8 *The ecological complex involved in a change from closed forest to grass-with-trees.*

and burning frequency and intensity vary considerably from to place and interact with soil physical character in their effect on the vegetation (Moss, 1961, 1963; Morgan and Moss, 1965). Thus in this zone the ecological pattern is extremely complex, and the evaluation of the possible effects of particular husbandry practices is not at all easy. Most important, it is almost impossible to generalize on a continent-wide scale about the ecological background in this zone, owing to its considerable variability. Furthermore, it is important to notice that in this zone some of the more significant and extensive areas of the transference of active elements in the soil are to be found (D'Hoore, 1964, 1968; Moss, Chapter 14).

Figure 8.10 shows the present distribution of the closed forest system, the principal areas of grass-with-trees, and the present extent of the zone of forest–savanna mosaic. Montane areas are also shown. This shows the main contemporary area of tension between the two plant–soil systems,

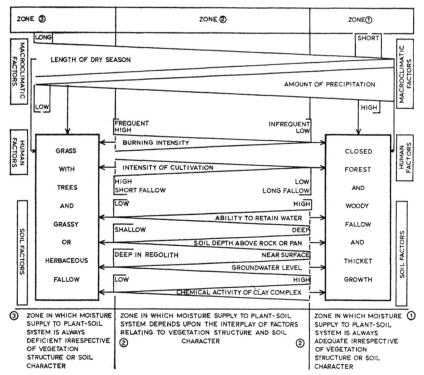

Figure 8.9 *The factors involved in locating the boundary between forest and savanna.*

and should be compared with the zones of moisture régime depicted on Figure 8.6.

Montane regions

In order to complete this outline of the ecological backgound to land use studies in tropical Africa, some mention must be made of the montane regions which form an important element in the geographical ecology of the continent. These regions are complex enough to require a major section to themselves, but in view of their somewhat limited area, especially in West Africa, this is not justified in the context of the present essay. Nevertheless, certain fairly obvious characteristics may be mentioned in order to contrast them with the more extensive regions already described.

The principal factors effecting distinctiveness are the coincidence of climatic amelioration with the occurrence of soils which are well-structured, high in nutrients in the mineral clay, and rich in organic matter. This change in soil character is as much a consequence of the occurrence of base-rich rocks as of the decreased rate of chemical reactions and the slower leaching occasioned by lower temperatures (Vine, 1949, 1954a; D'Hoore, 1964, 1968). Furthermore, the steepness of the slopes and the contrasts of relief imply that changes of local climate and of soil character take place over short distances and within very limited areas.

Thus, owing to lower temperatures and greater activity in the mineral clay, nutrient and organic matter cycles are less vigorous, whereas the water cycle remains largely dependent upon the character of the vegetation. This shows a range of variability from montane grasslands to forest, as well as including highly specialized montane Afro-Alpine plant groups. Furthermore, the considerable local variation in edaphic and climatic factors imposes a complexity of pattern which is impossible to characterize, except on large-scale maps.

These factors produce considerable diversity of pattern in the montane regions, and thus provide the basis for a wide variety of uses. Furthermore, the advantages afforded by the climate and soil have undoubtedly encouraged European settlement in some areas, such as the Kenya Highlands, and the consequent development of a highly sophisticated agricultural technology. On the other hand, in some mountain regions, notably Ethiopia, the fact of sheer inaccessibility has until recently facilitated a resistance to change which has seriously hindered the development of the agricultural potential. It may be argued that it is not only the basic facts of pleasanter climate and more tractable soils that make these areas attractive agriculturally but also the mere fact of variability within small areas. Such variability, where the possiblity of large land-holdings is real, favours the development of a multi-enterprise system in which several profitable lines of exploitation of biological resources may be employed, which would prove a sounder economic proposition in the relatively unstable trade climate of tropical areas, where profit and loss depend largely on the vagaries of world prices of raw vegetable materials for subsequent processing. It is noteworthy that this effect is cushioned in many African states by the institution of the Marketing Board.

Another point of importance is that in these montane regions superior and more profitable varieties of certain crops may be grown. The best example is coffee (*Coffea* spp.), which can be grown over much of tropical Africa, but the best-flavoured (*arabica*) varieties of which can

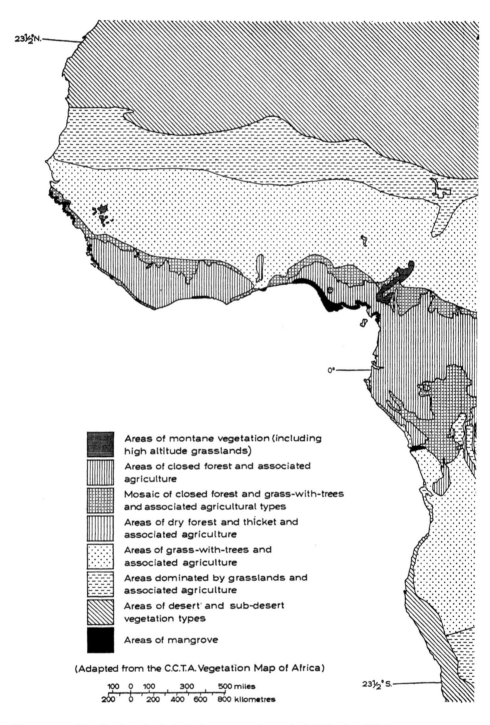

23½'N.—

0°—

23½'S.—

Areas of montane vegetation (including high altitude grasslands)

Areas of closed forest and associated agriculture

Mosaic of closed forest and grass-with-trees and associated agricultural types

Areas of dry forest and thicket and associated agriculture

Areas of grass-with-trees and associated agriculture

Areas dominated by grasslands and associated agriculture

Areas of desert' and sub-desert vegetation types

Areas of mangrove

(Adapted from the C.C.T.A. Vegetation Map of Africa)

100 0 100   300   500 miles
200 0   200 400 600 800 kilometres

Figure 8.10 *Distribution of principal plant systems in tropical Africa* (modified from the Vegetation Map of Africa by Aubreville *et al*, 1959).

23½°N.

0°

23½°S.

be grown only at altitudes in excess of 2,000 ft (610 m) (Irvine, 1950).

These mountain regions thus form a distinct and variable unit, unique in flora, in vegetation variation, and discrete in fauna, especially birds (Moreau, 1966). They thus afford, in their median altitudes at least, an environment capable of varied and profitable exploitation by man.

One further consideration is, however, important. These montane regions are frequently source regions for major rivers whose waters are the basis of agriculture in other parts of the continent. This is especially true of the Nile system in the north-east of the continent, and of the Niger in the west. The possible effects of clearing large areas of montane forest and other self-propagating vegetation for agriculture thus need to be evaluated in a wider hydrological context as well as in the more immediate ecological situation.

The importance of disease patterns

Any evaluation of the ecological background in Africa must necessarily include some mention of the ecology and importance of the many organisms which produce pathological symptoms in human beings. These, such as *Trypanosoma* spp., often have extremely complex relationships with other animals, plants and particular habitat conditions, and may have important chains of relationship with certain forms of land use.

Some organisms, such as the yaws spirochaete (*Treponema pertenue*), are favoured by such conditions as high relative humidity and moist soil conditions, and may also be carried mechanically by particular insects, such as the eye-gnat (*Hippolates pallipes*), which have their own ecological relationships (May, 1958). Other organisms, such as those which cause schistosomiasis (*Schistosoma haematobium* and *S. mansoni*), form three-factor complexes in which the organism passes an essential part of its life-cycle in another animal, in this case freshwater snails usually of the genus *Bulinus* (Malek, 1961). The ecology of these snails is complex, but they are likely to find favourable the freshwater habitats created by man when he initiates small irrigation schemes and similar projects. This could radically change the pattern of the occurrence of the disease in much of West Africa, even though there is some evidence that snail colonies of this genus do not establish themselves very rapidly is some man-made lakes (Thomas, 1966).

Yet other organisms, of which the most studied and important example is the group causing human and animal trypanosomiasis, sometimes form four-factor disease complexes in which there is not only an agent, a vector and a host, but a reservoir of infection also. These also have significant ecological implications, not only in their direct relationships with particular vegetation types and ecological situations but also in the indirect effects produced on human and animal distributions, and consequently on vegetation also (Morris, 1963, 1965). Such complexes, however, often have weak links, by means of which the disease may be controlled or even eliminated (Nash, 1948; Buxton, 1948; Robertson, 1963).

The important point to be emphasized is simply that many, if not most, serious tropical diseases, including malaria, yellow fever, onchocerciasis (river blindness), leishmaniasis, as well as those already mentioned, have strong ecological relationships which may link with, or impinge upon, those relationships involved in agricultural land use. They therefore cannot be neglected in any evaluation of the ecological background to land use in any particular area. Furthermore, they introduce a complicating element which must be strong argument for integrated planning of development in the economic, ecological and medical fields. The significance of disease is shown in Table 8.6, where some common diseases and their hosts and vectors (both mechanical and biological) are listed.

Conclusion

This summary of some of the more important ecological aspects and implications of land use in tropical Africa emphasises four significant general points.

First, land use as a simple morphological concept reveals little or nothing about the basic ecological relationships of agriculture in tropical Africa. Yet it is these which are the essence of the links between man and the land he farms, and they must be evaluated if present relationships are to be understood and future relationships rationally planned.

Second, in planning on an ecological basis it is not sufficient to examine soil and vegetation as separate entities. They operate as two separate but strongly dependent open systems. To neglect the soil is to eliminate the part of the plant–soil system which will remain when the vegetation has been removed for cultivation; to neglect the vegetation is to make it impossible to evaluate, or even recognize, those soil properties which influence and are influenced by it.

Third, the implications of agriculture and land use go beyond mere plants and soils. Pest problems and the effects and patterns of disease, plant, animal and human, are all involved and need to be assessed in any full appreciation of the ecological background to land use study in tropical Africa.

Finally, despite the great variability that is present, the general pattern of that variability can be brought out and studied in terms of the pattern of variation of distinct but related plant-soil systems. This variability may then be mapped on a continental scale, using the vegetation map as a basis. On a local scale also an approach through the plant–soil system is equally

TABLE 8.6 *Some insect disease vectors and carriers in tropical Africa* (after Ene, 1963)

DISEASE	PATHOGEN	CARRIER/VECTOR	HOSTS
Anthrax	Bacterium	*Chrysops* spp. (deer-fly)	Man and cattle
Ascariasis	Nematode	House-fly	Man and domestic animals
Cholera	Bacterium	House-fly	Man
Dengue Fever	Virus	*Aedes* sp. (mosquito)	Man
Dysentery (bacillary)	Bacterium	House-fly	Man
Dysentery (amoebic)	Protozoa	House-fly, cockroach	Man
Elephantiasis	Nematode	*Aedes* sp., *Anopheles* sp., *Culex* sp. (mosquitoes)	Man
Encephalitis	Virus	Various mosquitoes	Man and domestic animals
Filariasis	Nematode	*Aedes* sp., *Anopheles* sp., *Culex* sp. (mosquitoes)	Man
Leishmaniasis	Protozoa	*Phlebotomus* sp. (sand-fly)	Man, dogs, rodents
Loiasis (Loa loa)	Nematode	*Chrysops* spp. (deer-fly)	Man
Malaria	Protozoa	*Anopheles* sp. (mosquito)	Man, apes
Onchocerciasis (river blindness)	Bacterium	*Simulium* spp. (black-fly)	Man
Spirochaetosis	Bacterium	*Pediculus* sp. (louse)	Man
Trachoma	Virus	House-fly	Man
Trypanosomiasis	Protozoa	*Glossina* spp. (tsetse fly)	Man, cattle, game
Tuberculosis	Bacterium	House-fly	All animals
Tularemia	Bacterium	Blood-sucking Diptera (esp. *Chrysops* spp.)	Man, rodents
Typhoid	Bacterium	House-fly	Man
Yaws	Bacterium	*Hippelates* sp. (eye-gnat)	Man
Yellow fever	Virus	*Aedes* sp. (mosquito)	Man

Note: The ecology of many of these vectors and carriers may be changed by various land-use practices. Thus the possible consequences of such practices in terms of disease patterns must be considered in any evaluation for development in agriculture.

fruitful, a point which is elaborated in a case study elsewhere in this volume (Chapter 14).

Thus, despite the complexities, the value of the ecological approach is demonstrated. Detailed evaluation of the geographical ecology of Africa is possible and necessary to any understanding of land use, indeed it is the *sine qua non* of any valid assessment of the resource base for agricultural development in economic terms. It is therefore more validly viewed as a basis for, rather than as a background to, land-use study in tropical Africa.

References

AKEHURST, B. C. and SREEDHARAN, A. 1965. Time of planting – a brief review of experimental work in Tanganyika. *E. Afr. Agric. For. J.* **30**, 189.

AUBREVILLE, A. 1938. La forêt coloniale: les forêts de l'Afrique occidentale française. *Annls Acad. Sci. colon., Paris* **9**, 1–245.

1949. *Climats, forêts, et désertification de l'Afrique tropicale*, Paris.

AUBREVILLE et al. 1959. *Vegetation Map of Africa, South of the Tropic of Cancer*, UNESCO.

BAEYENS, J. 1938. *Les Sols de l'Afrique Centrale, spécialement du Congo Belge*, Brussels.

1949. The bases of classification of tropical soils in relation to their agricultural value. *Tech. Commun. Commonw. Bur. Soil Sci.* **46**, 99–102.

BEARD, J. S. 1946. *The Natural Vegetation of Trinidad*, Ox. For. Mem. No. 20.

BERNARD, E. A. 1945. *Le Climat écologique de la Cuvette Centrale Congolaise*, Inst. natn. Étud. Agric. Congo-Belge, Brussels.

1954. Sur diverses conséquences de la méthode du bilan d'énergie pour l'évapotranspiration. *C.R. Ass. Inst. Hydrol. Sci. Rome* **3**, 161–7.

1956. *Le Déterminisme de l'évaporation dans la nature*, Inst. natn. Étud. Agric. Congo-Belge, Brussels.

BIRKET-SMITH, J. 1960. Results from the Danish Expedition to the French Cameroons (1949–50). XXVII: Lepidoptera, Part 1. *Bull. Inst. fr. Afr. noire* **22**, 521–54; Lepidoptera, Part 2. *ibid.* **22**, 124–983; Lepidoptera, Part 3. *ibid.* **22**, 1259–84.

BOOTH, A. H. 1960. *Small Mammals of West Africa*, London.

BRAY, J. R. and GORHAM, E. 1964. Litter production in the forests of the world. *Adv. Ecol. Res.* **2**, 101–57.

BRUNT, M. 1958. *The Land Use Maps of the Gambia* (mimeographed paper), Directorate of Overseas Surveys, London.

BUXTON, P. A. 1948. *Trypanosomiasis in East Africa*, H.M.S.O., London.

CANSDALE, G. S. 1960. *Animals of West Africa*, London.

1961. *West African Snakes*, London.

CHIPP, T. F. 1927. *The Gold Coast Forest. A Study in Synecology*, Ox. For. Mem. No. 7.

CLAYTON, W. D. 1958. Secondary vegetation and the transition to savanna near Ibadan, Nigeria. *J. Ecol.* **46**, 217–38.

1961. Derived savanna in Kabba Province, Nigeria, *J. Ecol.* **49**, 595–604.

COBLEY, L. J. 1963. *An Introduction to the Botany of Tropical Crops*, London.

COLE, M. M. 1963. Vegetation and geomorphology in Northern Rhodesia: an aspect of the distribution of the savanna of Central Africa. *Geogrl J.* **129**, 467–96.

COLLARD, P. J. 1964. Personal communication.

CUNNINGHAM, R. K. and LAMB, J. 1959. A cocoa shade and manurial experiment at W.A.C.R.I., Ghana. I: First year. *J. Hort. Sci.* **34**, 14–22.

CUNNINGHAM, R. K., SMITH, R. W. and HURD, R. G. 1961. A cocoa shade and manurial experiment at the W.A.C.R.I., Ghana. II: Second and third years. *J. Hort. Sci.* **36**, 116–25.

DAGG, M. and PRATT, M. A. C. 1962. The Sambret and Lagan Experimental Catchments: the relation of stormflow to incident rainfall. *E. Afr. Agric. For. J.* **27**, 31–5.

DALE, W. T. 1962. Diseases and pests of cocoa: virus diseases. In WILLS, J. B. (ed.), *Agriculture and Land Use in Ghana*, London, Chapter 19, 286–315.

DEKEYSER, P. L. 1955. *Les Mammifères de l'Afrique noire française*, Initiations Africains I, Dakar.

D'HOORE, J. L. 1964. *Soil Map of Africa (Scale 1 to 5,000,000). Explanatory Monograph*, Lagos.

    1968. The classification of tropical soils. In MOSS, R. P. (ed.), *The Soil Resources of Tropical Africa*, Cambridge, Chapter 1, 7–28.

DUVIGNEAUD, R. 1951. Les savanes du Bas-Congo; essai de phytosociologie topographique. *Lejeunia, Revue Bot., Liège*, **10**, 1–192.

ENE, J. C. 1963. *Insects and Man in West Africa*, Ibadan.

ENTWISTLE, P. F. 1962. Diseases and pests of cocoa: minor insect pests. In WILLS, J. B. (ed.), *Agticulture and Land Use in Ghana*, London, Chapter 19E, 342–7.

FAO-UNESCO. 1963. *Conservation of Nature and Natural Resources in Modern African States*, Morges, Switzerland.

FOGGIE, A. 1962. The role of forestry in the agricultural economy. In WILLS, J. B. (ed.), *Agriculture and Land Use in Ghana*, London, Chapter 16, 229–35.

FOGGIE, A. and PIASECKI, B. 1962. Timber, fuel and minor forest produce. In WILLS, J. B. (ed.), *Agriculture and Land Use in Ghana*, London, Chapter 17, 236–51.

FORSGATE, J. A., HOSEGOOD, P. H. and MCCULLOCH, J. S. G. 1965. Design and installation of semi-enclosed lysimeters. *Agric Met.* **2**, 43.

HADDOW, A. J. and CORBETT, P. S. 1961. Entomological studies from a high tower in the Mpanga Forest, Uganda. II: Observations on certain environmental factors at different levels. *Trans. R. ent. Soc. Lon.* **113**, 257–69.

HAINES, W. B. 1949. Cultivation and manuring of rubber trees. *Tech. Commun. Commonw. Bur. Soil Sci.* **46**, 217–19.

HANNON, N. 1949. Ph.D. thesis, University of Sydney, quoted in NORRIS, D. O. 1956, Legumes and the *Rhizobium* symbiosis. *Emp. J. exp. Agric.* **24**, 247–70.

HARRIS, V. A. 1957. *Some Aspects of the Behaviour of the West African Lizard, Agama agama agama L.*, unpublished Ph.D. thesis, University of Ibadan, Nigeria.

HARTLEY, C. W. S. 1968. The soil relations and fertilizer requirements of some permanent crops in West and Central Africa. In MOSS, R. P. (ed.), *The Soil Resources of Tropical Africa*, Cambridge, Chapter 8 (in press).

HOPKINS, B. 1965a. Vegetation of the Olokemeji Forest Reserve. II: The climate with special reference to its seasonal changes. *J. Ecol.* 53, 109–24; III: The microclimates with special reference to their seasonal changes, *ibid.* 53, 125–38.

1965b. Observations of savanna burning in the Olokemeji Forest Reserve, Nigeria. *J. appl. Ecol.* 2, 367–82.

IRVINE, F. R. 1950. *A Text-Book of West African Agriculture*, London.

JOHNSON, C. G. 1962a. Capsids: a review of current knowledge. In WILLS, J. B. (ed.), *Agriculture and Land Use in Ghana*, London, Chapter 19B, 316–30.

1962b. The ecological approach to cocoa disease and health. In WILLS, J. B. (ed.), *Agriculture and Land Use in Ghana*, London, Chapter 19F, 348–52.

KEAY, R. W. J. 1953. *An Outline of Nigerian Vegetation*, Lagos.

1959. *Explanatory Notes on the Vegetation Map of Africa, South of the Tropic of Cancer*, London.

KERFOOT, O. 1962a. The vegetation of the Sambret and Lagan Experimental Catchments. *E. Afr. Agric. For. J.* 27, 23.

1962b. The vegetation of the Mbeya Peak Experimental Catchments. *E. Afr. Agric. For. J.* 27, 110.

1962c. Root systems of forest trees, shade trees, and tea bushes. *E. Afr. Agric. For. J.* 27, 24.

LANGDALE-BROWN, I. 1960. *The Vegetation of Uganda (excluding Karamoja)*, Uganda Department of Agriculture Memoirs, Series 2. No. 6.

LANGLANDS, B. W. 1967. Burning in Eastern Africa. *E. Afr. Geogrl Rev.* 5, 21–37.

LAUDELOUT, H. 1961. *Dynamics of Tropical Soils in Relation to their Fallowing Techniques*, FAO, Rome.

LAUDELOUT, H., GERMAIN, R. and KESLER, W. 1954. Preliminary results on the chemical dynamics of grass fallows and of pastures at Yangambi. *Trans. 5th Int. Congr. Soil Sci.* 2, 312–21.

LAWES, D. A. 1961. Rainfall conservation and yield of cotton in Northern Nigeria. *Emp. J. exp. Agric.* 29, 307.

LAWSON, G. W. and JENIK, J. 1967. Observations on microclimate and vegetation interrelationships on the Accra plains. (Ghana.) *J. Ecol.* 55, 773–86.

MCCULLOCH, J. S. G. 1962a. The Sambret and Lagan Experimental Catchments: measurements of rainfall and evaporation. *E. Afr. Agric. For. J.* 27, 27–30.

1962b. Effects of peasant cultivation practices in steep streamsource valleys; measurement of rainfall and evaporation. *E. Afr. Agric. For. J.* 27, 115–17.

MCKELVIE, A. D. 1962. Cocoa: vegetative propagation. In WILLS, J. B. (ed.), *Agriculture and Land Use in Ghana*, London, Chapter 18E, 263–5.

MALEK, E. A. 1961. The ecology of schistosomiasis. In MAY, J. M. (ed.), *Studies in Disease Ecology*, New York, Chapter 10, 261–330.

MANNING, H. L. 1949. Planting date and cotton production in the Buganda Province of Uganda. *Emp. J. exp. Agric.* **17**, 245.

MAY, J. M. 1958. The ecology of yaws. In *The Ecology of Human Disease*, Stud. Med. Geogr. No. 1, Am. Geogrl Soc., New York, Chapter 16, 216–30.

    1961. The ecology of African trypanosomiasis. In MAY, J. M. (ed.), *Studies in Disease Ecology*, New York, Chapter 9, 231–60.

MILDBRAED, J. 1930a. Zusammensetzung der Bestande und Verjungung im tropischen Regenwald. *Ber. dt. bot. Ges.* **48**, Generalversammlungs-Heft, 50–7.

    1930b. Sample plot surveys in the Cameroons rain-forest. *Emp. For. J.* **9**, 242–66.

MOREAU, R. E. 1966. *The Bird Faunas of Africa and its Islands*, New York and London.

MOREL, G. and BOURLIÈRE, F. 1955. Recherches écologiques sur *Quelea qu. quelea* L. de la vallée du Sénégal. Part I. *Bull. Inst. fr. Afr. noire* **17A**, 617–63.

    1956. Recherches écologiques sur *Quelea qu. quelea* L. de la vallée du Sénégal. Part II. *Alauda* **24**, 97–122.

MORGAN, W. B. and MOSS, R. P. 1965. Savanna and forest in Western Nigeria. *Africa* **35**, 286–94.

MORISON, C. G. T., HOYLE, A. C. and HOPE-SIMPSON, J. F. 1948. Tropical soil–vegetation catenas and mosaics. *J. Ecol.* **36**, 1–84.

MORRIS, K. R. S. 1963. The movement of sleeping sickness across Central Africa. *J. trop. Med. Hyg.* **66**, 59–76.

    1965. The ecology of sleeping-sickness. *Discovery, Lond.* **26**, 22–29.

MOSS, R. P. 1957. Some notes on the soils of the Western Region of Nigeria with special reference to plantations of exotic trees. *Br. Commonw. For. Conf. Pap.*, Government Printer, Ibadan, 19–29.

    1960. Land use mapping in tropical Africa. *Niger. Geogr. J.* **3**, 8–17.

    1961. *A Soil Geography of a Part of South-western Nigeria*, unpublished Ph.D. thesis, University of London.

    1963. Soils, slopes and land use in a part of south-western Nigeria: some implications for the planning of agricultural development in inter-tropical Africa. *Trans. Inst. Br. Geogr.* **32**, 143–68.

MOSS, R. P. and MORGAN, W. B. 1967. The concept of the community: some implications in geographical research. *Trans. Inst. Br. Geogr.* **41**, 21–32.

    1968. Soils, plants and farmers in West Africa. In GARLICK, J. P., (ed.), *Human Biology in the Tropics: Proceedings of a Joint Symposium of the Society for the Study of Human Biology and the Tropical Group of the British Ecological Society*, (in press).

NASH, T. A. M. 1948. *Tsetse Flies in British West Africa*, H.M.S.O., London.

NOIRFALISE, A. 1956. *Exploration du Parc National de la Garamba*. Fasc. 6: *Le Milieu Climatique*, Brussels.

NYE, P. H. 1958a. The relative importance of fallows and soils in storing plant nutrients in Ghana. *Jl W. Afr. Sci. Ass.* **4**, 31–49.

    1958b. The mineral composition of some shrubs and trees in Ghana. *Jl W. Afr. Sci. Ass.* **4**, 91–98.

NYE, P. H. and HUTTON, R. G. 1957. Some preliminary analyses of fallows and cover crops at the West African Institute for Oil Palm Research, Benin. *Jl. W. Afr. Inst. Oil Palm Res.* 2, 237–43.

NYE, P. H. and GREENLAND, D. J. 1960. *The Soil under Shifting Cultivation*, Tech. Commun. Commonw. Bur. Soil Sci. 51, Farnham Royal, Bucks.

PENMAN, H. L. 1963. *Vegetation and Hydrology*, Tech. Commun. Commonw. Bur. Soil Sci. 53, Farnham Royal, Bucks.

PEREIRA, H. C. (ed. and compiler), 1962. Hydrological effects of changes in land use in some East African catchment areas, E. Afr. Agric. For. J. 27, Nairobi.

PEREIRA, H. C., DAGG, M. and HOSEGOOD, P. H. 1962a. The Sambret and Lagan Experimental Catchments: the water balance of both treated and control valleys. *E. Afr. Agric. For. J.* 27, 36–41.

1962b. Effects of peasant cultivation practices in steep valleys: the water balance of the cultivated and control catchments. *E. Afr. Agric. For. J.* 27, 118–22.

PHILLIPS, J. 1959. *Agriculture and Ecology in Africa*, London.

RAHM, U. 1961. Esquisses mammalogiques de basse Côte d'Ivoire. *Bull. Inst. fr. Afr. noire* 23A, 1229–65.

RAMSAY, D. MCC. and DE LEEUW, P. N., 1964–5: An analysis of a Nigerian savanna. I: The survey area and the vegetation developed over Bima sandstone. *J. Ecol.* 52, 233–54; II: An alternative method of analysis and its application to the Gombe sandstone vegetation. *Ibid.* 52, 457–66; III: The vegetation of the Middle Gongola region by soil parent materials. *Ibid.* 53, 643–60; IV: Ordination of vegetation developed on different parent materials. *Ibid.* 53, 661–78.

RICHARDS, P. W. 1939. Ecological studies in the rain forest of Southern Nigeria. I: The structure and floristic composition of the primary forest. *J. Ecol.* 27, 1–61.

1957. *The Tropical Rain Forest*, London.

ROBERTSON, A. G. 1963. Tsetse control in Uganda. *E. Afr. Geogr. Rev.* 1, 21–32.

ROSEVEAR, D. R. 1947. Mangrove swamps. *Fm. Forest* 8.

1953. *Checklist and Atlas of Nigerian Mammals*, Lagos.

ROSS, J. K. 1957. Exotic trees in the Western Region of Nigeria. *Br. Commonw. For. Conf. Pap.*, Government Printer, Ibadan, 1–18.

ROSS, R. 1954. Ecological studies on the rain forest of Southern Nigeria. III: Secondary succession in the Shasha Forest Reserve. *J. Ecol.* 42, 259–82.

RUSSELL, E. W. 1968. Some agricultural problems of semi-arid areas. In MOSS, R. P. (ed.), *The Soil Resources of Tropical Africa*, London, Chapter 6, 121–35.

STAMP, L. D. 1956. The World Land Use Survey in relation to intertropical Africa. *Rep. Int. Geogr. Un. Symp. Nat. Resour., Food, and Population in Inter-tropical Africa, Makerere, 1955*, 70–71.

STREETS, R. J. 1962. *Exotic Forest Trees in the British Commonwealth*, Oxford and London.

SYKES, R. A. 1930. Some notes on the Benin Forests of Southern Nigeria. *Emp. For. J.* 9, 101–6.

TAYLOR, B. W., BAKER, R. M., LEEFERS, C. L. and DE ROSAYRO, R. A. 1963. *Report on the Land-Use Survey of the Oyo–Shaki Area, Western Nigeria*, FAO, Ibadan.

THOMAS, J. D. 1966. Some preliminary observations on the fauna and flora of a small man-made lake in the West African savanna. *Bull. Inst. fr. Afr. noire* 28A, 542–62.

TINSLEY, T. W. 1964. The ecological approach to pest and disease problems of cocoa in West Africa. *Trop. Sci.* 6, 38–46.

TRAPNELL, C. G. 1943. *The Soils, Vegetation and Agriculture of North-Eastern Rhodesia*, Lusaka.

TRAPNELL, C. G. and CLOTHIER, J. N. 1937. *The Soils, Vegetation and Agricultural Systems of North-Western Rhodesia*, Lusaka.

TURNER, BRENDA J. 1967. Ecological problems of cattle ranching in Bunyoro. *E. Afr. Geogr. Rev.* 5, 9–20.

VESEY-FITZGERALD, D. F. 1963. Central African grasslands. *J. Ecol.* 51, 243–74.

VILLIERS, A. 1950. *Les Serpents de l'Ouest Africain*, Dakar.

VINE, H. 1949. Nigerian soils in relation to parent material. *Tech. Commun. Commonw. Bur. Soil Sci.* 46, 22–29.

1951. *Notes on the Main Types of Nigerian Soils*, Spec. Bull. Agric. Dept. Nigeria No. 5.

1954a. Latosols of Nigeria and some related soils. *Proc. 2nd inter-Afr. Soils Conf. (Leopoldville)* 1, 295–308.

1954b. Is the lack of fertility of tropical African soils exaggerated? *Proc. 2nd inter-Afr. Soils Conf. (Leopoldville)* 1, 389–412.

1956. Studies of soil profiles at the W.A.I.F.O.R. Main Station and at some other sites for oil-palm experiments. *Jl W. Afr. Inst. Oil-Palm Res.* 4, 8–59.

1968: Developments in the study of soils and shifting agriculture in tropical Africa. In MOSS, R. P. (ed.), *The Soil Resources of Tropical Africa*, Cambridge and London, Chapter 5, 89–119.

WHARTON, A. L. 1962. Diseases and pests of cocoa: black pod and minor diseases. In WILLS, J. B. (ed.), *Agriculture and Land Use in Ghana*, London, Chapter 19D, 333–41.

WILLIAMS, G. 1953–4: Field observations on the cacao mirids, *Sahlbergella singularis* Hagl. and *Distantiella theobroma* (Dist.) in the Gold Coast. Part I: Mirid damage. *Bull. ent. Res.* 44, 101–19; Part II: Geographical and habitat distribution. *Ibid.* 44, 427–37; Part III: Population fluctuations. *Ibid.* 45, 723–44.

Studies of the social environment

9 Peasant agriculture in tropical Africa

W. B. MORGAN

The greater part of Africa's raw material wealth is derived from agriculture, and, despite a measure of European settlement, particularly in East and South Central Africa, that agriculture is overwhelmingly the creation of peasant farmers. Few subjects have invited so much comment and so great a variety of views as African peasant cultivation. It has been criticized as primitive, wasteful of resources and labour, and even destructive of vegetation and soils. Equally it has been defended as a remarkable adaptation to the conditions of African environment, and as a low-cost form of production essential at a given stage of economic development. In many tropical African countries a considerable effort has been made to improve the productivity of peasant agriculture, or, by introducing conservation techniques, to lessen the demands which it makes upon the environment. Attempts have been made even to eliminate the peasantry altogether, and to replace them by commercial small-holders, or by co-operative farmers who share in the operations of large units. The post-war period has seen an over-all reduction in the area devoted to plantation agriculture in tropical Africa. In many cases plantations have been divided into small-holdings. It has, however, also seen the creation of enormous state farms and the introduction of agricultural machinery, which, despite the high costs involved, is increasing its contribution to African productivity.

It is astonishing that so little attempt has been made to examine, describe and analyse the many forms which peasant agriculture has assumed in tropical Africa. So many arguments, criticisms and recommendations for change have been made with only very little evidence, or with a detailed examination of only a few samples of the whole. Much of our information has been provided not by agricultural research workers, but by anthropologists, geographers and economists, many of whom have produced their accounts as a secondary interest. Much of the material is extremely uneven in its nature, lacks quantification or is coloured by a subjective approach. Johnson's criticism of accounts of African agriculture in Southern Rhodesia at the turn of the century, that they 'vary from the patient understanding of the anthropologists to the condescension of the early Europeans' (Johnson, 1963–4), is unfortunately still true of many parts of tropical Africa at the present day.

Of the few attempts made before the Second World War to survey agriculture in Africa on the continental scale, that of H. L. Shantz is the most out-standing (Shantz, 1940, 1941, 1942, 1943). Because of the war it received very little attention, although in retrospect it is perhaps more informative of the state of knowledge of the time than of the state of African agriculture. There were no regions based on the study of peasant agricultural practice. Indeed, the brief account of hoe culture was marred by such remarkable errors as the claim that in such culture the rectangular patch or long strip was absent. Lack of data from several countries made some of the crop distribution maps of limited value. Even so in its attempt to summarize the evidence and to generalize about the African environment, this study represented a considerable advance. More recently we have had, in Allan's *The African Husbandman*, (1965) a detailed account of agriculture in some parts of Central and East Africa, together with a case study from northern Ghana, some studies of pastoralism, and some less detailed comparative material from elsewhere, including commentary on recent developments. Allan focuses attention on a particular aspect of the study of peasant agriculture – the relationship between the number of cultivators, their techniques and the quality of their environment. His approach is essentially ecological, although he is concerned with some aspects of agricultural economics. For Allan the basis of the shifting cultivation, which is still commonly regarded as the chief form of land use in tropical Africa, is the peasant's 'fund of ecological knowledge'. 'His indicator of initial fertility is the *climax* vegetation and his index of returning fertility is the succession of vegetational phases that follows cultivation' (Allan, 1965). One is immedi-ately reminded in this emphasis on climax of the ideas of the landscape geographers, of the conversion of the Natural Landscape (*Naturland-schaft* or even *Urlandschaft*) into a Cultural Landscape, ideas which have perhaps been more fruitful for the ecologists than for the human geograph-ers, for the Natural Landscape has proved elusive, a hypothetical reconstruc-tion from the results of, in most cases, unknown successions of cultural change (Sauer, 1925; Penck, 1927). Of other studies of continental dimensions, by agriculturalists, geographers, economists, ecologists or anthropologists, it may be said that they provide, with regard to peasant agriculture, little of descriptive or analytical value, excepting perhaps in the writings of some of the anthropologists who do list certain details (Baumann and Westermann, 1957). We are much better served by studies of smaller regions, several of which are of considerable value: for example, Buchanan and Pugh's *Land and People in Nigeria* (1955), Barbour's *The Republic of the Sudan* (1961) or Tothill's *Agriculture in Uganda* and *Agriculture in the Sudan* (1940, 1948). Of accounts of the agriculture of individual communities there are numerous examples, but perhaps one could single out for especial mention de Schlippe's study of the Azande

(1956), Béguin's studies in the Congo (1958, 1960), Audrey Richards' descriptions of agriculture in Zambia (1958, 1961), Margaret Haswell's accounts of the economics of agriculture in a Gambia village (1953, 1963), the report by Galletti, Baldwin and Dina (1965) on cocoa-farming families, and Allan's studies in Northern Rhodesia (1945, 1947).

The peasant cultivator

Discussion of tropical African agriculture all too frequently begins with its mode of land use or with the subsistence nature of African agricultural economics, instead of with the peasant cultivator himself. He is, after all, the creator of any system that exists, his are the decisions which effect change, even if prompted by Government or other pressures, and for him agriculture is only a part of his life, even if it is the vital food- and income-producing part. A peasant cultivator is essentially one who depends on his own labour and that of his family. He may, on occasion, hire labour to assist his efforts, but such hired labour is of only minor importance in his economy, and he in no way depends upon it. This is not to deny that there are many African cultivators who do depend on hired labour. By so doing they have ceased to be peasants and have become commercial farmers. The peasant cultivator may receive assistance from other members of his community. 'Group farming' or cultivation by 'age-sets' is a not uncommon feature in tropical Africa for certain phases of agricultural activity, particularly tillage or clearance of thick bush, and especially in cases where speed is important. He may in return add his labour to groups working on his neighbour's holdings or may offer gifts of food or refreshment. The peasant cultivator belongs to one of several hundred communities of enormous social, political and economic variety. His agricultural practice is conditioned by his relationship to the community, just as it is by the ecological limitations of his environment, and by the materials and techniques at his disposal. There is, therefore, in African agriculture, no single model of a peasant cultivator. This does not mean that we cannot generalize about peasant agriculture in tropical Africa, but it does mean that we should understand statements referring to agricultural practice over large areas of Africa purely as generalizations, and not as truths concerning any individual situation.

Productivity

The peasant cultivator is a small-holder, in most cases planting an area of some 2–8 acres, frequently divided into small and sometimes scattered plots. It has been assumed by some observers that holdings are small because of the limitations of technique, particularly the limitations imposed by the clearance of woodland, and by the simple instruments of hand tillage.

Improvement of either or both should make possible much larger holdings, and therefore greatly increase productivity and food supply or incomes. In consequence, attempts have been made in many areas to introduce animal-drawn ploughs, to eliminate woodland clearance by systems of permanent cultivation or even to replace hand labour by machinery. However, a closer examination of peasant agriculture than the mere measurement of holding sizes reveals that few peasants cultivate as much land as even their limited techniques would allow. They cultivate enough to satisfy their wants, and produce what surplus they estimate they can satisfactorily sell, exchange or give to friends or dependents. An important limitation is their own estimate of the value of the result for the labour expended. They are frequently very conscious of the fact that only small returns may come from arduous toil, and, like many other peoples, they may put a high value on leisure and on the need to find time for other activities. That peasants cease to do work when the marginal additions to productivity become very small has been discussed by Clark and Haswell (1964) and analysed statistically for food production among Yoruba cultivators by Galletti, Baldwin and Dina (1956). It may not, therefore, be the case that the size and output of a holding are limited by the fact that hard labour is unable to take in a greater area, as is sometimes assumed. Nor may the frequent lack of animal draught power nor of machines necessarily be the major limiting factor in preventing expansion. A vital limitation is clearly demand, or rather the small size of the markets, generally low incomes and associated low standards of living. Arguments in favour of raising agricultural productivity as such often ignore the lack of an adequate market for the goods produced. Arguments in favour of increasing food production tend sometimes to assume either that African peasants have similar nutritional ideas to our own, or that an apparent shortage is the product of an inefficient agriculture and not of a host of other factors such as poor communications and marketing facilities, problems of taste, habit, storage, seasonal moisture supplies and their variability, or of destruction by pests. Food shortages, even famines, are not uncommon in Africa, particularly in the ecologically marginal areas. The phenomenon of the 'hunger season' is widespread, but is more likely to be removed by improved communications and storage facilities, and by a change in the habit among some communities of consuming very large quantities of food immediately after harvest (Johnston, 1958) rather than by trying to increase agricultural productivity. Again 'kwashiorkor', the protein deficiency disease of very young children, is not to be attacked only by seeking to encourage the production of protein-rich foodstuffs, but by endeavouring to promote changes in child care and feeding habits (Nicol, 1956, 1959b; Welbourne, 1963). The evidence of malnutrition must be treated cautiously as evidence of food shortage. Thus, in Gambia

McGregor concluded that malnutrition was more often secondary to disease than primary (McGregor *et al.*, 1961), while in Nigeria Nicol noted the problem of intestinal and other parasites (Nicol, 1959a). Finally, is it necessarily true that people living on poor diets by the standards of the nutrition research workers put the improvement of those diets before other requirements, such as a bicycle, better tools, clothing, education for their children or medical care? Evidence suggests that in some localities at least they do not. Peasant cultivators, given suitable markets, have proved well able, in certain parts of Africa, to produce surpluses of food or of other agricultural raw materials. The increasing demand for foodstuffs both from export crop growers and from the rapidly expanding urban markets, especially in West Africa, has been met largely by peasant production. Only in Senegal and to a lesser extent in Ghana have food imports in West Africa, for example, assumed any considerable dimension (Morgan, 1963). The growers of the major export crop-producing areas, particularly those producing coffee and cocoa, plant some food crops, but satisfy a large part of their demand by purchases from commercial yam, cassava, maize and rice growers in neighbouring areas. In the Ogoja Province of Eastern Nigeria, for example, the quantities of crops listed in the 1950–1 Sample Census of Agriculture (Nigeria, 1952) gave, even by the most conservative estimate, allowing for all manner of wastage in handling, storage, transport and cooking together with a proportion for seed, a production of nearly 4,700 calories per person per day, or nearly double the probable requirement. This figure did not allow for other crops not listed, such as bananas, nor for animal products, nor for foods obtained by gathering or hunting. Possibly the original estimate of quantities was too large, but Ogoja is a very important food producer for the neighbouring provinces, particularly of yams.

Subsistence agriculture

The African peasant is frequently described as a *subsistence cultivator*, but the term varies greatly in its meaning, from a cultivator whose family consumes all his holding produces, to a cultivator who supplies directly only a part of his family's need and sells the rest of his produce in the market. The subsistence or near-subsistence economy has been claimed as one in which surpluses cannot be marketed. People make gifts of their surpluses, and frequently depend on the system of obligations developed in the large or extended family system to provide insurance against adverse conditions (Bauer and Yamey, 1957). Stages of economic change have been recognized. For example, Van Bath distinguishes:

1. *Subsistence farming*, in which each household produces all the food it consumes.

I

2. *Direct agricultural consumption*, in which most people produce their own food, and also supply it to the non-agricultural population as barter.

3. *Indirect agricultural consumption*, in which the whole non-agricultural population and at least part of the agricultural population satisfy their needs through a market in which farm goods are sold, mainly from districts with agricultural surpluses.

The last stage can be divided into an earlier phase in which over half the working males are agriculturalists, and a later phase in which less than half depend on agriculture (Van Bath, 1963). All three of these stages are represented in the peasant economies of tropical Africa, although the third stage as yet appears only in its earlier phase. Stage one is uncommon and confined mainly to the remoter areas of very low population density in Central Africa. Stage two is the most common, if marketing is substituted for barter. The buying and selling of all manner of goods, including agricultural produce, has been long established in tropical Africa. Admittedly in many areas, particularly in the Congo Basin, Zambia and Tanzania, the existence of only a few small highly scattered nuclei of population, with densities greater than 50 persons per square mile, seriously limited such exchange, or even reduced it to the level of barter (Bohannan and Dalton, 1962; Hodder, 1965a, b). Yet observers have frequently been surprised by the readiness with which the subsistence cultivator may respond to a commercial incentive. Thus Allan noted the surprise and consternation of the Northern Rhodesian (Zambian) Maize Control Board in the 1930s when, after fixing the grain price on the internal market and agreeing to purchase all maize offered for sale, it received nearly five times the estimated quantity (Allan, 1965). Allan supposed the cultivators were selling what he called 'the normal surplus of subsistence cultivation'; that is, the subsistence cultivator normally produces more than his family requires. This would be a strange notion for some of the critics of African peasant cultivation who point to frequent food shortages. It is, however, possible that the cultivators concerned were ready to take advantage of an improvement in the market. Stage three is less common, and is developed mainly in those areas most affected by the industrial and mining developments associated with European investment, particularly in Rhodesia and northern and central Zambia, and in the hinterlands of the great ports. In a large part of West Africa, however, stage three had been reached before the period of European penetration, even before European-made goods had a major effect on the economy.

Peasant agriculture and markets

The markets for agricultural produce in Africa consist of the modern large, usually urban, counterparts of the markets found in Europe; the special markets for certain products, usually export and limited to licensed produce buyers, or even buyers for a state-controlled organization; and finally the traditional markets, many of which have developed to serve new demands and in response to new means of transport.

The growth of urban markets for foodstuffs, fibres, drugs and even leaf wrappings has had a profound effect on peasant agriculture in many areas, in particular around the great capital cities and in those areas where service centres have grown rapidly in response to modern economic changes, and based in part on existing urban traditions. Nowhere have the effects been more profound or widespread than in West Africa, and particularly in Nigeria, Ghana and the Ivory Coast. The Kano close-settled zone, with an innermost zone of population density of the order of 500 persons per square mile, is as much a response to the value of access to the city, of the growth of its markets and of its supply of night-soil, as it is to any locally favourable soil and moisture conditions (Mortimore and Wilson, 1965). In Ghana the demands of Accra and Kumasi for foodstuffs reach out to the northernmost limits of the country, and even beyond into the Volta Republic. In these countries the staple foodstuffs have in many districts a commercial value, which makes their sale an important alternative to home consumption. In such a situation choice of crop, even where subsistence takes the larger part of the produce, may be profoundly influenced by the state of the market.

The development of the export production of vegetable oils and oil seeds, fibres and beverages encouraged the creation of special networks of buyers and of transport for a flow with which the traditional markets were unable to cope. These new networks have had as important an effect on the location of production as ecological factors. Indeed, the somewhat marginal location ecologically of early, and even some later, cocoa planting reflected the development of the market system, and of the location of government seedling distribution centres. The creation of producers' co-operative societies, of organizations for produce inspection and quality control, and of marketing boards has had important influences on the growers, bringing many of them into much closer contact with modern commercialism, and even effecting so profound a change in agricultural practice that many can no longer be classified as peasant cultivators.

Local markets where agricultural produce of the surrounding area may be exchanged at regular intervals, together with imported goods, are widespread in tropical Africa. Evidence that they are traditional features of African economic life is abundant in West, but is fragmentary in East and Central Africa. Hodder suggests that traditional markets may not arise so much out of the demands of local exchange as from the stimulus of long-distance trading contracts. Two necessary conditions were, firstly a sufficiently high density of population, and secondly, a political structure able to maintain peace (Hodder, 1965a). Given these conditions, traditional markets were most likely to be developed in those areas where long-distance trading contacts could be established. Such areas were subject to influences encouraging levels of development above that of mere subsistence. They were in consequence areas where some changes in agricultural practice in response to commercial influences had taken place prior to the development of European colonialism. Such areas included much of West Africa, particularly the Sudanic portions, and the area of the southern or Guinean states ('Eastern Atlantic Zone' of Baumann and Westermann), the Sudanic portions of Equatorial Africa, Ethiopia, the Nile Sudan, Somalia, the coastlands of East Africa, and a number of comparatively isolated areas with evidence of early trading, including the Kikuyu lands, Buganda and a number of locations in the Congo Basin. In parts of Central Africa and for much of Southern and South-eastern Africa traditional markets appear to have been absent. Local markets exist today, but are of very recent origin. This argument, if it is valid, may be of some importance in understanding the distribution of different kinds of agricultural practice in Africa. It is one to which further reference will be made.

Shifting cultivation

Shifting cultivation or shifting field agriculture is the characteristic agricultural technique of much of Central and East Africa, and of some small portions of West Africa (see Fig. 9.1). It 'is characterized by a rotation of fields rather than of crops, by short periods of cropping (one to three years) alternating with long fallow periods (up to twenty years or more, but often as short as six to eight years), by clearing by means of slash and burn, and by use of the hoe or digging stick, the plough only rarely being employed' (Watters, 1960). The length of resting period under self-sown plants of any particular piece of land is governed by the amount of land available, and the degree of restoration of fertility as indicated by the plants it supports. Important factors, therefore, are population density and distribution, the size and type of the market for produce (although shifting cultivation is normally associated mainly with the subsistence economies) and the soil and moisture conditions for plant growth. While, however, the existence of sufficiently high densities of population may preclude shifting cultivation, the existence of low densities provides only a necessary and not a determining condition. Other techniques may be and occasionally are employed. Thus while we may agree with Watters and others that in shifting cultivation 'man's margin of freedom' from the coercive circumstances of the environment is less great (Pelzer, 1958; Watters, 1960), yet in several cases the choice of shifting technique has been demonstrated to be a response to economic incentives rather than to ecological controls. The advantages of short fallow or even permanent agriculture have frequently been over-stressed by those who regard long fallow cultivation as wasteful of land, of vegetation and of soil resources. Shifting cultivation has sufficient advantages to attract at least some cultivators away from permanent systems, using conservation techniques. Thus some hill cultivators in Togo, Dahomey, Northern Nigeria and Cameroon have abandoned terraced hill lands with soils of high fertility for lowlands where extensive methods must be employed. Others have developed shifting cultivation in the plains in addition to permanent cultivation in the hills (White, 1941; Dresch, 1952; Enjalbert, 1956; Netting, 1965). The advantages include generally higher production per man, although lower per unit area, and long rest periods alternating with periods of heavy labour demand, for which in some cases group farming methods were used (Clark and Haswell, 1964; Boserup, 1965). Allan has noted the preference by the Lamba of Zambia for soils of apparent lesser fertility wherever land is ample, but capable of giving satisfactory yields with less labour (Allan, 1965). De Schlippe notes the avoidance by the Zande of more fertile areas because the elephant grass they support is too hard to clear with their small hoes (De Schlippe, 1956).

In shifting cultivation 'fallows' occur not because of a deliberate intention to restore soil fertility, but because land is abandoned when yields become too poor to be worth the effort of cultivation. A return to an old site takes place when the secondary vegetation indicates the likelihood of satisfactory nutrient levels after clearance and burning. In certain cases it is not so much the resting of soil and the restoration of its nutrient status that is important, as the provision of large quantities of vegetable material which can be burnt. This is especially the case where forest or dense woodland is cleared. Elsewhere, however, grassland is preferred even though the material available for burning is less, but in such cases additional techniques may be employed to provide the nutrients required. Frequently grassland as such is the product of intensive local clearance and cultivation, and especially of that intensity of cultivation which high densities of population demand. Burning has been castigated as wasteful of vegetable resources, destructive of high forest, which it replaces by 'degenerate' savanna flora, and destructive of humus and of the better quality grasses for animal feeding. It has also been claimed as the best means of providing nutrients at the right time in wooded or forested areas, and as a device for reducing potential acidity in soils (Jeffreys, 1951; Bartlett, 1955, 1957, 1961; Guilloteau, 1956; Trapnell, 1959; Watters, 1960; Ramsay and Innes, 1963). Clearly burning is essential if nutrients are to be obtained quickly from woody materials or from fibrous grasses, and if extensive methods of cultivation are economically justified. The effects vary greatly according to the date of burning, early burning favouring early recovery of vegetation, and according to whether or not burning is repeated during the dry season. The topping rather than felling of trees practised in some systems also favours early recovery.

In shifting cultivation fields may be well defined at the time of planting and during the growing season, but the system passes on to new land too quickly for anything more than a temporary boundary definition. The return to an old site means the creation of fresh fields, which are normally quite unrelated to former fields. From time to time settlement sites may be shifted, and several observers have suggested that in this feature there is an important distinction from other forms of agriculture. However, it is possible for shifting to take place for long periods before any move of settlement other than a shift within the general area of the site, for social or health reasons, is involved.

Population densities must be low, generally less than 25 per square mile, and in most cases very much less. Settlement, wherever shifting cultivation is the sole or dominant technique, consists normally of dispersed hamlets and compounds, for the existence of villages of any considerable size means that more intensive techniques become worth while on the nearer land. In such cases shifting cultivation exists only in part, and normally as a perimeter-ward practice. In consequence, areas within which shifting cultivation is the dominant technique have few large settlement nuclei, and possess few markets and only very limited trading activities. As the main means of producing staples, chiefly sorghum, *Pennisetum* millets and maize, shifting cultivation is widespread throughout Central and East Africa, but occurs in West Africa only in isolated cases (chiefly south-eastern Ivory Coast with rice as staple, and the Cross River Basin of Eastern Nigeria and Cameroon with the staples, cocoyams and bananas or plantains). Elsewhere it occurs as a peripheral technique in communities where staple crops are produced chiefly by more intensive methods. The variety of crops grown by shifting cultivators, roots, grains and bananas, should be noted. There is no particular association with any type of crop. Both the 'newer' crops, such as cassava and maize, are well represented, as are 'older' crops such as yams and sorghum. In this respect shifting cultivation is no more primitive than any other form of cultivation, and may even, as in certain cases in Northern Nigeria, be locally a recent development in part to produce commercial crops (Netting, 1965).

Rotational bush fallow

The term 'rotational bush fallow' was used by Faulkner and Mackie to denote a form of cultivation in Nigeria, especially in the south-west, in which 'the time in fallow exceeds the time that the land is cultivated', and which takes place in areas of 'moderate density of population, say 100–200 to the square mile'. Where, as in the south-east, land was in cultivation for more than half the time, the system was described as 'broken down already' (Faulkner and Mackie, 1933). A similar notion is expressed by Allan's 'recurrent cultivation' or 'land rotation cultivation', 'signifying a more stable relationship between land and population than that associated with "shifting" cultivation', and with a land-use factor between 4 and 10, most frequently between 4 and 8 (Allan, 1965).[1] In this form of cultivation the aim is to create fallows, or rather a regular system of fallows which are never permitted to revert to woodland or forest. In certain cases, notably in Eastern Nigeria, the fallows are regularly planted with small shrubs which provide good yam poles and confer a right of future clearance and cultivation. Fields are normally rectangular and may have permanent or nearly permanent boundaries. The return to a site of former cultivation means a return to former fields, excepting cases where some change in the field pattern is required due, for example, to a demand for an increase in the cultivated area. In rotational bush fallow systems, fixed settlement is normal, although some movement of houses or compounds within the general area of a site may take place and use-right or even property-right in a given patch of agricultural land is possible. Such rights to a given plot or field may be held by a family or individual within a larger group. With fixed settlements a network of roads and a system of inter-regional trade in addition to local exchange are possible. Commercialism is generally much more flourishing therefore, although the point made above that shifting cultivation may also take on a commercial character in certain circumstances needs to be borne in mind. Quite high population densities may be supported; cases in Southern Nigeria occur with densities as high as 600 persons per square mile. Above that figure, the fallows generally are so short as to be irregular and infrequent in occurrence, so that permanent cropping becomes the dominant practice.

Rotational bush fallow methods are characteristic of what may broadly be termed the Sudanic zone stretching from Senegal in the west to Ethiopia in the east, and most of West Africa. They are uncommon in Central and East Africa, where the few cases shown on Figure 9.1 are either on or near the coast, or in situations of some isolation and crowding – Uganda provides the largest single example. There is a clear contrast between mainly rotational bush fallow in West Africa and mainly shifting cultivation in Central and East Africa. To explain this contrast ecological differences will

[1] Allan's 'land-use factor' is the total number of 'garden areas' required by the system (Allan, 1965, 30), or:

$$\frac{\text{The area in cultivation} + \text{the area in fallow}}{\text{The area in cultivation}}$$

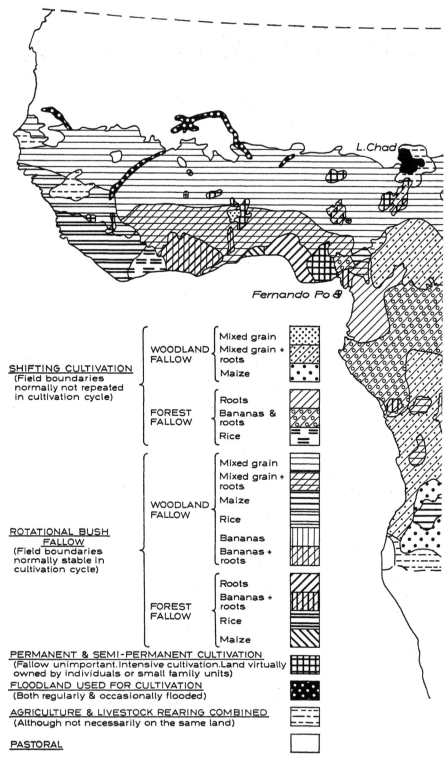

L.Chad

Fernando Po ⊕

SHIFTING CULTIVATION
(Field boundaries
normally not repeated
in cultivation cycle)

WOODLAND FALLOW
- Mixed grain
- Mixed grain + roots
- Maize

FOREST FALLOW
- Roots
- Bananas & roots
- Rice

ROTATIONAL BUSH FALLOW
(Field boundaries
normally stable in
cultivation cycle)

WOODLAND FALLOW
- Mixed grain
- Mixed grain + roots
- Maize
- Rice
- Bananas
- Bananas + roots

FOREST FALLOW
- Roots
- Bananas + roots
- Rice
- Maize

PERMANENT & SEMI-PERMANENT CULTIVATION
(Fallow unimportant. Intensive cultivation. Land virtually
owned by individuals or small family units)

FLOODLAND USED FOR CULTIVATION
(Both regularly & occasionally flooded)

AGRICULTURE & LIVESTOCK REARING COMBINED
(Although not necessarily on the same land)

PASTORAL

"Mixed grain" = sorghum, eleusine, pennisetum & maize grown in various successions.

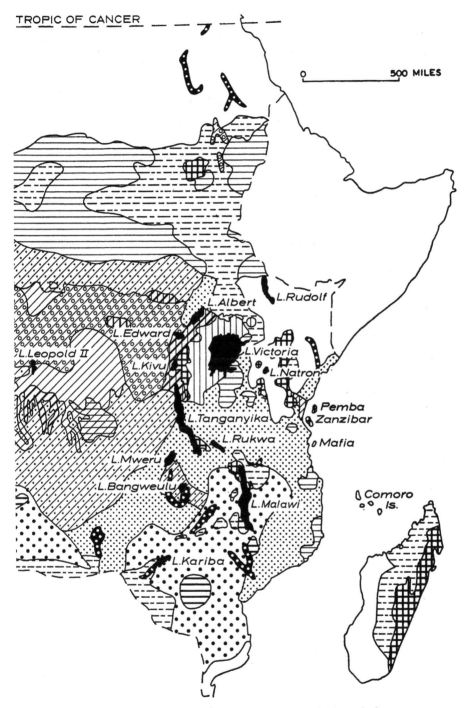

500 MILES

L.Albert
L.Rudolf
L.Edward
L.Victoria
L.Leopold II
L.Kivu
L.Natron
L.Tanganyika
Pemba
Zanzibar
L.Rukwa
Mafia
L.Mweru
L.Bangweulu
Comoro
Is.
L.Malawi
L.Kariba

Figure 9.1 *Peasant agriculture in tropical Africa (staple food crops). A generalized regional division from descriptive data.* (Chief sources: Allan, 1965; Baumann and Westermann, 1957; Forde, D., 1950 onwards; Johnston, 1958; Murdoch G. P., 1959, and 1960. Some distributions shown here differ from those in the sources quoted, due to difficulties of reconciling detail.

not provide any answer. General differences in population distribution and density suggest a fair correlation, but there are cases of rotational bush fallow practice in West Africa where population densities are as low as in East African areas where shifting cultivation is the rule. Perhaps a better correlation may be obtained with the distribution of larger settlement nuclei, of political organization and of markets as described above. The larger settlements frequently have problems of distance which affect the patterns of surrounding land use so that more intensive methods of agriculture such as rotational bush fallow frequently occur near such settlements even though over-all population densities may be low. If West Africa and the Sudanic zone may claim more long-distance contacts and a better-developed commerce than either Central Africa or most of East Africa, then one might expect important differences in settlement patterns and in agricultural practice.

Rotational bush fallow has a rather different relation to vegetation from that of shifting cultivation. With the latter several fallow stages may be observed at any one time, but generally the longer fallows are simply stages in a return to a well-established woodland or forest, within which the signs of earlier cultivation may be difficult to see. Such woodland or forest includes many maturely developed species which have established themselves since the last cultivation, in addition to those trees which were left standing amidst the fields. In the case of rotational bush fallow, maturely developed woodland or forest normally occurs on uncultivated land, usually on land too steep or with soils too poor for cultivation, or on boundary land between rival communities. The limit of the fallows is clearly defined, and these fallows are never permitted to remain uncultivated long enough to support mature trees, other than those which were not felled when clearance took place. All stages of fallow, therefore, including the oldest, differ quite considerably in character from any surrounding woodland or forest. In the savanna woodlands the difference is most marked, as the fallows under rotational bush fallow systems are normally dominated by grasses. The differences in character of forest and savanna woodland fallows are of great importance in agricultural practice, for they result in different nutrient supplies from clearance and burning, and different effort and timing in cultivation. These differences have been recognized in the distinction made between woodland and forest fallows in the classification scheme of the map of peasant agriculture in tropical Africa (Fig. 9.1). The related distributions of shifting and rotational methods give important differences of character between the fallows and the associated vegetation, particularly as between West Africa on the one hand and Central and East Africa on the other. In the savannas, for example, a variety of woody fallow stages is widespread in Central and East Africa, but much less common in the West where grasses rather than woody plants are dominant in many areas. Similarly, the character of the 'high forest' may be much affected by length of fallow, so that in certain areas tall trees are totally absent due to long-continued cultivation with only short fallows, and pure stands of the umbrella tree (*Musanga cecropioides*) or of wild figs (particularly *Ficus exasperata*) occur.

Problems of maintaining fertility under rotational bush fallow, with its more intensive use of land, are, in most cases, more acute than under shifting cultivation. Some cultivators apply animal manures, but normally the inability to integrate pasture and arable land into one system of land use means that manure must be collected wherever stocks are grazed, and carried, sometimes for considerable distances. The extra effort involved rules out such practice, excepting where fallows are very short or non-existent. More frequent use is made of such techniques as crop mixture, rotation and succession. These make possible a greater intensity of production on a plot, thereby reducing the demand on the fallow area available for a given level of production. These techniques are also used in shifting cultivation to make most use of the land so laboriously cleared, but generally they are more elaborately developed by the rotational bush fallow cultivator, whose land resources are so much more limited. In the crop mixture a high density of plants, normally with varying demands on soil nutrients, is obtained. This reduces the labour of weeding, and also takes advantage of the different soil depths and drainage provided by the mounds and ridges, which are the characteristic forms of tillage. The crop mixture may give poor yields for each plant, but can achieve a high productivity per acre. Frequently the members of a mixture are not planted together, but in succession, thus spreading the labour of planting, and enabling especially favoured crops to become well established before other members of the mixture compete for nutrients. For example, cassava, even in areas where it is dominant in terms of quantity, may not be the preferred crop from the point of view of taste or of care applied in cultivation. In any case, it is generally the most tolerant crop grown. Where planted in a mixture with yams, it is frequently interplanted only after the yams have become well established. Where the rainy season is sufficiently long, that is nine months or more, two grain crops may follow one another, the second sometimes being planted after the harvest of the first. For example, two maize plantings are often made in this way, especially in southern Dahomey and Western Nigeria. Rotations of sole crops or of mixtures are quite common, both in regular or nearly regular forms, and in what de Schlippe has called 'pseudorotations' or improvised plantings of crop mixtures, which follow one another quite irregularly (de Schlippe, 1956). Often rotations appear to last for longer periods in areas with a marked dry season than in areas where the dry season is short. In the former, however, the dry season provides an annual rest period, sometimes accompanied by the grazing of livestock on the stubble and crop remains. In the latter the dominant crops are often roots or bananas which may take over nine months to mature, and which, in the case of cassava and bananas, may stay in the ground for a few years. In these wetter areas therefore there is little or no rest period. In addition, there are, except for goats and poultry, fewer livestock to graze and supply manure.

Permanent and semi-permanent cultivation

In this form of cultivation are included practices in which there is either no fallow at all, or in which the fallow proportion is very small, and the occurrence of fallow is frequently irregular. In some cases extremely fertile soils, often weathered from basic volcanic rocks, enable permanent cultivation to thrive with very little attention to intensive methods of production. Normally, however, intensive methods are essential, involve almost individual care for each plant, and employ manures, including night-soil and compost, and the burying of household refuse and crop remains. Woodland, even situated at some distance, may be lopped and the branches burned on the cultivation site. There appear to be three broad classes of permanent cultivation, each with some distinctive characteristics:

1. On compound and kitchen-garden land. This is the most widespread, and normally provides a means of growing a few special crops near to a settlement, so that they may be obtained immediately, easily guarded, or may take advantage of the locally higher fertility. Such cultivation is always supplementary to main crop production by other means.

2. On confined sites, on uplands or islands where the area available is strictly limited. Numerous examples occur throughout tropical Africa, but especially in East and West Africa, where hill refuges have provided easily defended settlement sites for peoples threatened with attack by more powerful communities. Often steep hillsides have to be cultivated, and the result has been the building either of regular terraces or of 'pseudo-terraces', that is irregular forms behind mere lines of stones, resulting more from clearance than any deliberate attempt at wall construction (Floyd, 1964). The most outstanding case of confined island site is that of Ukara in Lake Victoria, where the population density reached 572 per square mile even by 1932, where livestock are housed and fed by hand, and manure has to be carried to the fields, and where a legume is dug in as green manure. An unusual feature is the growing of elephant grass as a fodder crop in pits near the lake shore. Yet for all this effort only a bare subsistence can be attained, and some

members of the community have emigrated to the mainland, while others go on fishing expeditions (Thornton and Rounce, 1945).

3. On not especially confined, yet overcrowded sites, often of quite considerable size, as for example:

(a) Kikuyuland, especially in the Kikuyu Grass and Star Grass zones, where careful rotations are employed, and sheep and goats are stall fed.

(b) Nyanza Province, Kenya, where population densities range from 100 to over 1,000 persons per square mile.

(c) Kigezi, Ruanda and Burundi, where population densities are generally over 200, and locally rise to over 700.

(d) The rice-growing eastern uplands and slopes of Malagasy (Madagascar), especially the alluvial basins of the highlands.

(e) The Kano 'close-settled zone' of Northern Nigeria.

(f) The Ibo and Ibibio lands of Eastern Nigeria.

(g) The Mossi uplands of Upper Volta.

It is in these areas that the greatest problems of overcrowding, of small and fragmented holdings, of low incomes, of large-scale emigration, of overcultivation and of soil erosion occur. The problems have proved especially acute in Kikuyuland where an agricultural revolution to create more efficient holdings is in progress (Clayton, 1964; Jones, 1965) and in Ibo and Ibibioland where an economy based on root crops and the oil palm is in places barely holding its own (Grove, 1951; Morgan, 1955) although the exploitation of considerable mineral oil resources has recently given much-needed assistance.

Floodland cultivation

Irrigation and drainage works are rare in tropical Africa even today. Before 1900 they were virtually unknown. The most important example of the traditional development of surface water control for cultivation is probably that of the polder production of rice in the south-western estuarine lowlands of West Africa, chiefly in Guinea and Portuguese Guinea (not shown on Fig. 9.1). There peoples such as the Balante and Baga have long practised the clearance of mangrove swamps, the creation of enclosed rice-fields and their controlled drainage in order to reduce the salinity and sulphur content of the soils, yet retaining enough off-season salinity to reduce weeds and therefore the great labour of weeding. There are, it is true, considerable irrigation schemes in which the cultivation is done by peasants on small-holdings, such as the Gezira Scheme of the Sudan Republic with nearly a million acres, the Managil Extension, with 800,000 acres, the Office du Niger Scheme of Mali with some 120,000 acres, the Limpopo Scheme in

Moçambique with 40,000 acres, and a number of much smaller units such as Mwea-Tabere in Kenya, the Sabi Valley in Rhodesia and the many irrigation enterprises of the Upper Niger Valley. Even if one adds to this list all the irrigation schemes in which paid labour is used, such as Richard-Toll in Senegal, or the Triangle Estate in Rhodesia, the total area and productivity are minute compared with the area and output of the rainlands (Harrison Church, 1963). Probably more important in total output, although statistically impossible to assess, is the productivity of the floodlands where planting occasionally precedes, but normally follows declining flood levels, and where rich grasslands provide abundant dry season grazing for migrant herds. Floodland cultivation of this kind, normally concentrating on sorghum, maize and sweet potatoes, grown in tiny patches as supplements to the main food supply from rainlands, is widespread wherever suitable alluvial lowlands exist. Occasionally, notably on the southern fringes of the Sahara, wells or pits are dug, and used for the watering of special crops, such as onions for sale in the towns. Rice has become a more common floodland crop in recent years, due mainly to encouragement by government agricultural departments, but exotic rice strains frequently meet severe competition from weeds, suffer badly from pests and are often unsuited to the irregular régimes of African rivers. Floodland cultivation becomes of major importance, even dominant in the agricultural economy, in the areas shown on Figure 9.1, especially in the Senegal and Niger valleys, the 'dallols' or broad floodplains of Niger, the basin of Lake Chad, the basin of the upper Nile and its tributaries, and the floodlands of Central Africa, especially the Barotse Plain, the fringes of Lake Bangweulu, and the middle and lower Zambezi Valleys. The cultivation systems of the Barotse Plain are perhaps among the most highly developed in Africa. There the Kalahari sands act as a vast sponge, allowing different forms of land use at different levels. The more peaty soils are drained by channels and canals. Cattle are kept and their manure used to maintain fertility. Manuring is often carefully controlled by the penning of cattle at night, and the regular movement of the pens until the whole area is covered (Gluckman, 1941; Allan, 1965).

In places of locally high density of population, floodland cultivation may be valued for its high productivity per acre. More generally its value lies in the spreading of work and of harvests. This is particularly true in the areas with only a short rainy season. There the problems include the concentration of planting into one very short period. The harvest labour is likewise concentrated and brief. In consequence, the limit to productivity may not be the area available but the amount that can be done in the season. Lack of work for a large part of the year, except for gathering and hunting, has encouraged that emigration in search of labour elsewhere, which is so characteristic of the short rainy season lands. The possession of floodlands, with their different timings for planting and harvesting, is therefore a most valuable means of extending the working period and increasing productivity. Sometimes a whole series of plantings may be made one after another down a gentle floodplain slope as water levels recede. The period of fresh food is thus extended, and dependence on crop storage reduced. This additional resource may also be of value in cases where the rains are late, or, if early, succeeded by a dry spell before the main rains arrive. It is true that the flood levels may be irregular, as the rains in such areas are highly variable. Yet the irregularities may, for the cultivator, be of a different order, so that a reasonable floodland crop may be obtained in years of poor return from the rainlands.

Pastoralism and mixed farming

The keeping of livestock in herds, particularly of cattle, is generally an activity quite separate from agriculture proper. It depends not on the use of fields as such but on the open range, and throughout most of tropical Africa, has traditionally been transhumant. As such it has not been a part of peasant agriculture proper, although it must have some discussion here for it has important relationships with agriculture, which will become much stronger if attempts to develop mixed farming and settle pastoralists succeed.

The true pastoralists are nomads of the African steppelands, regions of low and irregular rainfall, but supporting for a brief period some of the best quality pastures. The migrating herds change direction according to season. Old coarse grasses are burnt off in advance to encourage a flush of new growth. Certain fixed points must be included in the migration, water-holes, streams, salt licks, the annual meeting-place of the social group, markets for the exchange of produce and the purchase of cloth, ornaments, household utensils and weapons, and usually a fixed base. The fixed base has members of the family resident there for all or most of the year. It may be the place where taxes are collected, and it may possess arable land tilled by members of the family or by the descendants of former slaves. The pure nomad, completely independent of agriculture, hardly exists. Most have some small dependence on cultivation, an important reserve occupation to fall back upon when stock have been heavily reduced by disease or by poor pasture. Even the pure nomad must obtain some corn by exchanging meat produce in the markets of the cultivators. Thus Masai trade with Kikuyu and Fulani with Hausa. For a few weeks the cattle of some Fulani groups are pastured on the stubble of the Hausa fields after harvest. They are encouraged to do so by cultivators who appreciate the value of the manure thus obtained.

Among some peoples crop raising and livestock rearing are of approximately equal importance, or at least while some of their members attach more importance to one, the remaining members attach more importance to the other. For example, Nilotic groups such as the Shilluk, who today are mainly agricultural, once laid more stress on pastoralism, as did the Acholi and the Lango, and Nilo-Hamitic groups such as the Nandi, the Suk, who practise irrigation and pastoralism, the Teso with their plough agriculture, and the Kipsigis with their enclosed fields. In West Africa the Serer both own cattle and cultivate, and have even been claimed to possess a true mixed farming system, on the grounds that their cattle graze regularly on the croplands after harvest, and on the fallows (Pelissier, 1953). However, wherever the herd is of any size, it still needs a considerable area of open pasture and supplementary feeding from the leaves of the 'kad' (*Acacia albida*, the leaves of which appear during the dry season, as do those of some other woodland species). To the south of Lake Chad the Shuwa pastoralists migrate between wet and dry season pastures, but depend greatly on the crops of their permanent settlements.

The failure to integrate livestock and cropping is in part a question of feeding and pasture location, in part of seasonal rhythms and clash of interests, in part of disease distributions, and in part of generally low productivity. In a situation of low productivity a peasantry can hardly reckon the effort of hand tillage worth while in order to feed animals. Fodder crops are therefore rare. Livestock must forage for themselves. If small they look for scraps around the compound, if large they will be taken in herds to self-sown pastures, or, when such pastures are poor, some pollarding of trees will be done to bring leaves to them. Hand-feeding of livestock in stalls is not unknown, but is confined either to the towns or to special cases of very intensive agriculture, as among the Chagga or some small and rather isolated communities in the Mandara Mountains. The best pastures with the most nutritious annual grasses are in the areas with only a brief rainy season. Where the rains last nine months perennial grasses dominate, many of them having a habit of very rapid growth, so that the time when they are most succulent is very short. Tall grass is frequently too coarse to digest. Moreover, the pastures of the wetter areas are easily invaded by shrubs, and must be very carefully grazed otherwise they soon cease to support anything other than plants of only low nutritious quality. In the areas of short rainy season, but where the rains are sufficient for agriculture, the combination of livestock and cropping is difficult, unless there is plenty of land, since both livestock and crops must compete for the same area. Normally the livestock must be removed elsewhere during the rains, and only brought back after harvest in order to ensure that they do not damage the crops. The same problems occur in many floodlands suitable for dry season cultivation, but which, uncultivated, will support rich pasture grasses. The 'bourgou' pastures of Mali and Niger have been affected by the clash of Fulani pastoral interests with those of cultivators such as the Songhay. The chief disease to limit livestock distribution is undoubtedly trypanosomiasis, or sleeping sickness, carried by tsetse flies, and of frequent occurrence in the wetter areas, especially where woodland is abundant. The moist woodlands with low population densities (say less than 50 persons per square mile) are the most seriously affected. These are perhaps the locations which otherwise would appear the most attractive for experiments in mixed farming. Where population densities are high, land suitable for pasture may be required for cultivation. An important problem is the frequent occurrence of tsetse-fly concentrations near surface water needed for drinking. Epidemics of rinderpest, a virus disease of cattle, which spread into West and East Africa from the Nile Valley and Ethiopia at the end of the nineteenth century, have seriously reduced many herds, in several cases forcing herdsmen to give up pastoralism altogether, and practise sedentary agriculture.

Most cultivators keep a few small livestock, particularly goats, chickens and muscovy ducks. Such animals provide an occasional feast, or, among some peoples, a sacrifice on ceremonial occasions. The keeping of pigs and of improved breeds of chicken is increasing. Some schemes for improved egg production, notably in Western Nigeria, have been extremely successful, and hold out promise of greater quantities of protein-rich foodstuffs for peoples often over-dependent on starch foods.

Staple food crops

A thorough discussion of crops is impossible in the scope of this essay, but some word is essential on their distribution and on their role in the agricultural systems. The map of peasant agriculture (Fig. 9.1) illustrates the major zones where a particular crop or group of crops provides the staple food and which, in terms of non-commercial or merely local exchange cultivation, is therefore the dominant crop in any rotation system. The map provides only a very gross generalization. It ignores fibres and other non-food crops, it ignores the considerable areas where crops such as cocoa, coffee, cotton and groundnuts are produced for export, it ignores the great plantations and the areas of mainly European farming, and it gives no idea of relative density, omitting considerable uncultivated or only lightly cultivated areas, or rather showing them as part of the agricultural zones. New schemes for agricultural improvement are similarly omitted. The evidence both for lengths of fallow and for the dominance of a particular crop comes almost entirely from descriptive material, some of which is of doubtful value and difficult to assess. The map therefore provides only a very approximate indication of crop dominance distribution.

The most widespread crop group is 'mixed grain', chiefly *Sorghum* and *Pennisetum* millet, grown mainly by rotational bush fallow in West Africa, and *Eleusine*, *Sorghum* and *Pennisetum* millet grown in various combinations mainly by shifting cultivation in East and Central Africa. These crops are frequently grown mixed, often planted in succession, and even where apparently planted 'sole' may be mixtures of different varieties of the same plant. Spiked or awned varieties of *Pennisetum* millet are gaining in popularity, as fewer children are available nowadays to scare away birds, owing to the demands of education. Certain *Sorghum* varieties are preferred for brewing beer. In some locations in Central Africa *Eleusine* is a minor crop, grown as a hardy poor-yielding famine reserve. Elsewhere, however, especially in East Africa, it is a much less tolerant high-yielding grain, frequently grown as the chief staple. Maize is the staple of southern Central Africa, chiefly of Angola and Rhodesia, grown mostly on the cooler uplands. In West Africa maize dominance is restricted to southern Dahomey, Western Nigeria and the Bamenda Highlands. The crop appears, however, to be more tolerant, especially of humid conditions, than most other grains, and is the most widespread secondary crop in Africa, being grown throughout both the rain forest and savanna areas. Rice is confined to eastern and western extremes. In the east it is the staple of Malagasy (Madagascar), where Asian rice (*Oryza sativa*) was introduced by Malay peoples. It did not spread into the rest of Africa, however, until the era of European invasion. In West Africa, African or 'red-skin' rice (*Oryza glaberrima*) is dominant in Guinea and Portuguese Guinea, and in parts of Liberia, Sierra Leone and the western Ivory Coast. It occurs in upland, swamp and floating forms, and in a large number of varieties, indicating a long-developed rice culture, apparently separate from that of Asia (Portères, 1950). Its boundary with the mixed grain and roots zone and the bananas and roots zone is thought by Miège to be mainly cultural (Miège, 1954). The zone of bananas and roots, particularly the cocoyam, *Colocasia* and *Xanthosoma*, is restricted entirely to the rain forest. These crops occur in a broad, if discontinuous belt across Central and West Africa. The banana, usually in the form of the plantain, the collective term for the varieties which are normally cooked, is the staple of Uganda where its distribution has been ascribed to the spread of Ganda political dominance and the desire locally to copy Ganda ways (McMaster, 1962). In many locations in the rain-forest zone, however, the banana is a secondary or even very minor crop, and the staples are roots, particularly yams, the traditional staple of southern Nigeria, and cassava or manioc, often grown in a mixture with the yams, and of increasing importance as an urban foodstuff (Jones, W. O. 1959). The yam produces a vine which is often supported by trees left standing on the cultivated land or by poles. It is therefore well suited to planting in the rain forest or woody fallow environment. However, it is cultivated well into the savannas, often in association with grain crops. Where few trees or poles are available other supports may be used, such as cassava stems or the stalks of guinea corn, left standing after harvest and woven into a trellis. Cassava has the advantages of considerable tolerance of varying soil conditions and of a wide range of moisture supplies. During dry spells it can remain in the ground without rotting, and recommence growth with the next rains. Propagation is from stem cuttings, so that none of the root need be kept for seed. Yields are high, normally 3–5 tons per acre as a sole crop, and can rise much higher with careful methods, and the use of fertilizers. Disadvantages include its very low protein and vitamin content, and a high concentration of hydrocyanic acid in the 'bitter' varieties, which need considerable treatment before cooking.

Commercial crops

The chief commercial crops grown by peasant cultivators in Africa are perennials supplying oil and oilseeds, especially the oil-palm, and beverages, especially cocoa and coffee. Among the annuals the most important are groundnuts and cotton (normally grown as an annual). The most important producers are countries which have had little or no European settlement, where peasant agriculture has become the mainstay of the export economy and has received greatest encouragement from the Government. Commercial agriculture has been highly developed in countries such as Kenya and Rhodesia, and in the Democratic Republic of the Congo, with its considerable plantations. In these countries peasant agriculture has tended to provide only a small contribution to either the export or the urban markets, that is it has been much more confined to the subsistence sector of the economy. Only in recent years have major efforts been made in these countries to improve peasant productivity, and to encourage the planting of crops for market.

The oil-palm is essentially a plant of the fallows of the rain forest and forest–savanna mosiac or 'derived savanna' zones. Part of Africa's export of palm oil, kernels and kernel oil comes from self-sown palms, and is essentially gathered produce, and a large part, particularly from the Democratic Republic of the Congo, comes from plantations. The biggest exporter and chief area of peasant production is southern Nigeria.

Almost all of Africa's cocoa export comes from West Africa, where in 1964–5 the four countries, Ghana, Nigeria, Ivory Coast and Cameroon, produced nearly three-quarters of the world total. Nearly all of this came from small-holdings, although many of these were commercial farms depending on hired labour rather than peasant holdings proper. Although in its early growth cocoa is associated with shade plants which are also food crops, such as cocoyams and bananas, and although eventually the land may return to food production for the local market, generally cocoa production is quite distinct from the normal system of peasant production including food staples in a given area. It creates a system of its own. This is in marked contrast to the oil-palm, which is grown mainly by people who are also cultivating traditional staples.

Coffee, chiefly of the coarser *robusta* varieties suited to hot and humid environments, is an important peasant crop in the Ivory Coast and Uganda, Africa's two leading producers. In Angola, the third largest coffee exporter of Africa, production comes mostly from European-owned holdings, although African peasant growers are increasing their share of output. In Kenya, Cameroon and a number of other locations, peasant cultivators are planting coffee, including the better quality *arabica* varieties.

The chief groundnut exporting areas are Senegal and Northern Nigeria, where the crop normally takes a regular place in rotation with other annuals. In Senegal the groundnut tends to be dominant, occupying two-thirds or more of the area of many holdings. In Nigeria, despite its commercial importance, it frequently plays a secondary role in the rotations to sorghum and millet. The groundnut is widespread as a minor commercial and local food crop. Attempts to produce it by mechanical means on big plantation or 'peasant partnership' schemes, notably in Tanzania, Senegal and the Nigerian 'middle belt', have proved very costly, and have generally failed.

The most important exporter of cotton in tropical Africa is the Sudan, closely followed by Uganda and Tanzania. Over half of Sudan's cotton by weight, and much more by value, comes from a single irrigation scheme, the Gezira (including the new Managil extension) situated between the Blue and White Niles. Here high-quality long-staple cotton is produced by a partnership between tenant farmers and the Sudan Gezira Board. The tenants provide labour, especially for weeding and harvesting, and the Board is responsible for most of the tillage and spraying, the irrigation channels, and ginning and marketing. In Uganda shorter-staple cotton, although of a high quality, is the chief commercial crop of the drier portions of the country, and especially of the Eastern Province. Cotton is less profitable for the labour expended than coffee, but can be grown in a much greater and rather less densely populated area suitable for agricultural expansion, and is not limited, as is coffee, by international agreement. Although commercial cotton for export is generally restricted in the several countries concerned to certain areas ecologically well favoured and served by adequate rail or road transport and local ginning mills, the crop is in fact widespread, usually playing a secondary role in rotations, and often commanding a good sale in internal markets supplying a local hand-spinning and weaving industry. This is especially so in West Africa where, not infrequently, the local market price for cotton is higher than that offered by the licensed buyer for export.

Africa has a large number of other commercial crops, of minor importance if grown by peasants, but of major importance on large commercial farm or plantation units. Rubber is something of an exception, as Nigeria is now chief exporter in Africa and a large part of its production comes from small-holdings. Small-holdings are also beginning to increase their share of Liberian production. Sugar-cane is a minor peasant crop, but an important crop on plantations. Other peasant commercial crops of lesser importance in tropical Africa include ginger, benniseed, castor, fruit (especially oranges, limes and pineapple), coconut, tobacco (usually for the home market) and many of the food crops, already described, grown for the urban markets, especially rice, cassava and yams.

Minor crops

Concentration on the staple food and commercial crops should not obscure the fact that in several countries a large portion of agricultural production is supplied by the total of the enormous variety of minor crops. Many of these play important roles in the mixtures, successions and rotations, and supply vital proteins or vitamins in the diets. Often minor crops will be found on compound land, where they are readily available in the small quantities required for culinary purposes. This particularly applies in the cases of spices, especially peppers, the leaf plants used to flavour soups or for wrapping, and several of the fruits. Pumpkins, melons, calabashes, okra, brinjal or aubergine, dye plants such as indigo and henna, fibres, including cotton and rama (*Hibiscus cannabinus*) and the many beans and oilseeds, are, however, used in greater quantities and normally appear in the mixtures of the main crop land. In certain areas a few of these minor or secondary crops become dominants, especially where they are commercially highly valued, just as in other cases, the staples and commercial crops, already described, become minor crops of only limited local use. An important minor grain in West Africa – in a few areas a major one – is fonio or acha (*Digitaria exilis, D. iburua*) which exhibits tolerance of a wide range of soil conditions, if giving generally low yields, and is a not infrequent end of rotation crop. Some of the minor crops were once much more important and have been largely replaced by introduced equivalents. This is largely the case with the earth pea (*Voandzeia geocarpa*) which has been replaced in most areas by the groundnut.

Peasant agriculture and the relation of population to land

It is commonly assumed that in peasant countries, particularly in those
whose economies have a considerable subsistence sector, there is a close
relation between population density and the productivity of land. At its
crudest it may be expressed as 'a close connexion between density of
population on the one hand and climate and vegetation on the other . . .'
(Naval Intelligence Division, 1943). In more refined forms it may appear
as a concept of 'critical population density', of 'population–land balance'
or of population pressure (Allan, 1965). While there is undoubtedly a
relationship between population and resources, it is of the most complex
kind. It involves a vast number of factors, many of them non-agricultural,
and it involves understanding a dynamic economic and ecological system
which can, and frequently does, generate changes within itself to meet its
own growth problems. Thus the forecast of a critical population density,
beyond which the existence of a given land-use system is threatened, may
have little meaning in terms of declining living standards or impoverishment
of soil resources, if with increasing population a number of new techniques
are introduced long before the critical density is reached. Moreover, the use
of empirical data creates its own problems, for actual population densities
are normally not critical densities, and may yield only uncertain information concerning them. Thus Buchanan concludes with regard to Nigeria's
population; 'It is, finally, a population whose distribution pattern is
"immature"; the close adjustments of densities to environmental conditions, which is typical of long-settled areas, is here lacking, and the process
of land occupation and settlement is incomplete over much of the Territory' (Buchanan and Pugh, 1955). In view of the frequency of change one
may even question whether the close adjustment of densities to environmental conditions is in fact typical of long-settled areas. Boserup suggested:
'We have found that it is unrealistic to regard agricultural cultivation systems as adaptations to different natural conditions, and that cultivation
systems can be more plausibly explained as the result of differences
in population density . . .' (Boserup, 1965). In a situation of increasing
population density one may therefore envisage a series of changes in the
cultivation systems which may, even in long-settled areas, so affect the
relationships between population densities and natural conditions, as
to make the question of density adjustment purely one of theoretical
interest.

Attempts to calculate the carrying capacity of land therefore provide
models of interest strictly in relation to a given agricultural system.
Practical application of such work can only be envisaged when the analysis
becomes much more complex, and comprehends, in addition, the dynamics

of change. Methods range from simple calculations, such as those of Stamp's estimate of 'maximum' population densities in Nigeria (Stamp, 1938, 1960) or notions such as Urvoy's 'coefficient of density' (Urvoy, 1942) which he related to communities as such, but which could be related to regions defined by agricultural systems, to the more sophisticated analyses such as those achieved by Allan. The Belgian geographer, Béguin, offers models of 'rural space', whereby the production and population 'potentials' of a given area may be calculated. Formulae have been devised to satisfy variable environmental conditions and variable agricultural techniques (Béguin, 1964). The use of crude tonnages in these calculations is unsatisfactory. Better results would be obtained by using starch equivalents or calorific values, especially Stamp's 'Standard Nutrition Unit' which he devised as a means of measuring land resources (Stamp, 1958). Other techniques for the study of the population–land relationship concern evidence of population pressure, of soil erosion, of malnutrition, and of adaptation to population change, especially in the form of shorter fallows, of change in character of fallows, of changes in holding and field sizes, of the use of more intensive or more extensive methods, and of changes in crops and cropping methods (Grove, 1951; Morgan, 1955).

Changing peasant agriculture in tropical Africa

For the whole period of which we have knowledge of African peasant agriculture we have evidence of adaptation and change. In part, changes have arisen not through conscious interference with agriculture or attempts at improvement but indirectly through the introduction of new means of exchange, methods of transport, the slave trade, the encouragement of migration and the introduction of new crop plants. Attempts to 'improve' agriculture began mainly with the effort to encourage the production of certain export crops and raise levels of productivity. Excepting the development of plantations, with which we are not here concerned, the earliest efforts appear to have been made in Senegal at the beginning of the nineteenth century with the creation of a botanic garden and the fostering of peasant groundnut cultivation. Botanic gardens became in several colonial territories the nuclei for Agricultural Departments. These encouraged the introduction of new plants, and supplied growers with both seeds and seedlings, and the necessary knowledge for cultivation. Some of them early encouraged conservation measures, although in the early years of this century the urge to conserve resources and guard against destruction of vegetation and loss of soil came mostly from Forestry Departments. Expert crop inspection, grading, price differentials and the formation of producer societies followed, mostly between the two world wars. Other forms of modification were attempts to control planting dates, enforce the

planting of some crops, and prevent the planting of others, the introduction of ridging, terracing and other erosion control measures, of mixed
farming, green manures and fertilizers, sprays, cutting-out and uprooting
to control diseases and pests, and finally, the planning of elaborate schemes
in which complete and new systems of agriculture were introduced. In the
former Belgian Congo the attempt was made to rationalize or develop
existing agricultural practice in order to produce more and yet maintain
the fertility of the soils. This, the 'paysannat' system, was developed first
of all at the Gandajika experimental station in 1936, and carried farther by
the INEAC organization with its corridor system of clearance, encouraging
fast regrowth and providing adequate sunlight. The corridor methods were
thought mainly suited to rain-forest conditions. In the savannas compact
block lay-outs or 'fermettes' were tried, usually with mechanical aid, and
chiefly to encourage the production of cotton. In the Sudan the Zande
scheme in the south was designed mainly to create more compact and more
intensively farmed holdings with strip-cropping and planned rotations.
In Zambia the Citemene Control Scheme was intended to reduce the
destruction of vegetal resources and create again a more intensive system
(de Coene, 1956; Dumont, 1957; Allan, 1965). Generally, Agricultural
Departments have preferred intensive to extensive methods, and have been
alarmed by the burning of grass and woodland, and the threat of deteriorat-
ing soil resources. In Northern Nigeria and in Dahomey attempts have been
made to create rational agricultural settlements in lands newly occupied by
peoples moving away from overcrowded hill sites, and practising shifting
methods in their new environment. At Shendam in Northern Nigeria the
new scheme has on the whole been successful. Rather less successful have
been the elaborate schemes of partnership between peasant cultivators and
a planning organization providing seed, mechanical equipment, plans for
rotations, fertilizers, insecticides, and even housing, roads and water
supplies. The greatest exception has been the extremely successful Gezira
scheme, but in the floodlands of the Inland Niger Delta of Mali the Office
du Niger scheme has cost far more than the value of its production. So has
the scheme of the Compagnie Générale des Oléagineux Tropicaux
(C.G.O.T.) for the mechanized production of groundnuts in Casamance,
while many other similar schemes have been abandoned (Baldwin, 1957).
Mechanization has proved far too costly in many tropical locations,
despite its advantages where labour is short or where speed in tillage or
clearance is essential. Even in many cases where the use of mechanical
equipment has persisted, as in the rice swamps of Sierra Leone, only
financial subsidies have made it possible (Morgan, H. E. G., 1965). Yet the
use of machinery continues to increase in the hope of raising productivity
per man. Attempts have also been made to create an agricultural élite, a

new class of farmers who will provide an example for others of the advantages of the new techniques. Hence the 'Master Farmers' in Rhodesia, and the 'Farm Settlements' of Western Nigeria (Yudelman, 1964; Takes, 1964; Seager, 1965). Perhaps the most comprehensive and thorough of all the attempts at agricultural change is the effort to revolutionize the Kikuyu economy in Kenya by eliminating fragmentation and fostering agricultural intensification in individual farm units. Group Farming schemes involving co-operative land use had already been tried and failed. What has been attempted in Kikuyuland is the complete agricultural rehabilitation of a people, introducing high-value cash crops on soils which have already been proven capable of withstanding high levels of production intensity, consolidating all holdings, providing plans for farm work and organization to suit varying ecological and economic situations, and introducing new livestock breeds and methods of grazing management (Clayton, 1964). Again, there is an agricultural élite of advanced farmers, whom most Kikuyu can never hope to emulate because of lack of sufficient land and adequate training staff. Even this scheme seems doubtful of success. Allan remarked that he could see 'little evidence of any significant improvement of the general standard of land-use following consolidation'; '. . . the rise of the farmers poses imponderable social and economic problems, and the future is dark with uncertainty. It will be difficult, and may well prove impossible, to prevent re-fragmentation of the great mass of small-holdings and the development of further subdivision, open or concealed' (Allan, 1965). In the development of means for agricultural improvement in Africa we are only at the beginning of dealing with the problems which must be solved.

References

ALLAN, W. 1945. African land usage. *Rhodes–Livingstone J.* 3, 13–20.
    1947. *Studies in African Land Usage in Northern Rhodesia*, Rhodes-Livingstone Pap. 15.
    1965. *The African Husbandman*, Edinburgh.
BALDWIN, K. 1957. *The Niger Agricultural Project, an Experiment in African Development*, Oxford.
BARBOUR, K. M. 1961. *The Republic of the Sudan*, London, 128 et seq.
BARTLETT, H. H. 1955, 1957, 1961. *Fire in Relation to Primitive Agriculture and Grazing in the Tropics*, Annotated Bibliography, 3 vols, Ann Arbor.
BAUER, P. T. and YAMEY, B. S. 1957. *The Economics of Underdeveloped Countries*, Cambridge, 64–7.
BAUMANN, H. and WESTERMANN, D. 1957. *Les Peuples et les civilisations de l'Afrique*, Paris.
BÉGUIN, H. 1958. *Géographie humaine de la région de Bengamisa*, Pub. Inst. Nat. Étud. agron. Congo-Belge.

BÉGUIN, H. 1960. *La Mise en valeur du Sud-Est de Kasai*, Pub. Inst. Nat. Étud. agron. Congo-Belge.

    1964. *Modèles géographiques pour l'espace rural africain*, Académie Royale des Sciences d'Outre-Mer, Brussels.

BOHANNAN, P. and DALTON, G. (eds.), 1962: *Markets in Africa*, Evanston.

BOSERUP, E. 1965. *The Conditions of Agricultural Growth*, London, 28–31, 117.

BUCHANAN, K. M. and PUGH, J. C. 1955. *Land and People in Nigeria*, London.

CLARK, C. and HASWELL, M. R. 1964. *The Economics of Subsistence Agriculture*, London, 82–90.

CLAYTON, E. S. 1964. *Agrarian Development in Peasant Economies: Some Lessons from Kenya*, Oxford.

COENE, R. DE, 1956. Agricultural settlement schemes in the Belgian Congo. *Trop. Agric.* **33**, 1–12.

DRESCH, J. 1952. Paysans montagnards du Dahomey et du Cameroun. *Bull. Assoc. Géogr. fr.* **2–9**, 222–3.

DUMONT, R. 1957. *Types of Rural Economy*, London, 36–50.

ENJALBERT, H. 1956. Paysans noirs: les Kabré de Nord-Togo. *Cah. d'outre-mer* **34**, 137–80.

FAULKNER, O. T. and MACKIE, J. R. 1933. *West African Agriculture*, Cambridge, 44.

FLOYD, B. 1964. Terrace agriculture in Eastern Nigeria, *Niger. Geogr. J.* **7**, 91–108.

FORDE, D. 1950 onwards. *Ethnographic Survey of Africa*.

GALLETTI, R., BALDWIN, K. and DINA, I. O. 1956. *Nigerian Cocoa Farmers*, Oxford.

GLUCKMAN, M. 1941. *The Economy of the Central Barotse Plain*, Rhodes–Livingstone Pap. 7.

GROVE, A. T. 1951. Soil erosion and population problems in South-East Nigeria. *Geogrl J.* **117**, 291–306.

GUILLOTEAU, J. 1956. The problem of bush fires and burns in land development and soil conservation in Africa south of the Sahara. *Sols afr.* **4**, 64–102.

HARRISON CHURCH, R. J. 1963. Observations on large-scale irrigation development in Africa. *Agric. Econ. Bull. Africa* **4**, ECA–FAO Joint Agriculture Division, Addis Ababa.

HASWELL, M. R. 1953. *Economics of Agriculture in a Savannah Village*, Colonial Research Study No. 8, London.

    1963. *The Changing Pattern of Economic Activity in a Gambian Village*, Department of Technical Co-operation, Overseas Research Publication No. 2, London.

HODDER, B. W. 1965a. Some comments on the origins of traditional markets in Africa south of the Sahara. *Trans. Inst. Br. Geogr.* **36**, 97–105.

    1965b. Distribution of markets in Yorubaland. *Scott. Geogr. Mag.* **81**, 48–58.

JEFFREYS, M. D. W. 1951. Feux de brousse. *Bull. Inst. fr. Afr. noire* **13**, 682–710.

JOHNSON, R. W. M. 1963–4. African agricultural development in Southern

Rhodesia 1945–1960. *Food Research Institute Studies* **4**, 165–223, Stanford.

JOHNSTON, B. F. 1958. *The Staple Food Economies of Western Tropical Africa*, Stanford, 206–10.

JONES, N. S. C. 1965. The decolonisation of the White Highlands of Kenya. *Geogrl. J.* **131**, 186–201.

JONES, W. O. 1959. *Manioc in Africa*, Stanford.

LUNAN, M. and BREWIN, D. 1956. The agriculture of Ukara Island. *Emp. Cott. Grow. Rev.* **33**.

MCGREGOR, I. A., BILLEWICZ, W. Z. and THOMSON, A. M. 1961. Growth and mortality in children in an African village. *Br. med. J.*, 1661–6, quoted in HASWELL, M. R., 1963, op. cit., 21.

MCMASTER, D. N. 1962. *A Subsistence Crop Geography of Uganda*, Bude, 28.

MIÈGE, J. 1954. Les cultures vivrières en Afrique Occidentale. *Cah. d'outre-mer* **7**, 25–40.

MORGAN, H. E. G. 1965. The mechanical cultivation of rice in the grasslands of Sierra Leone. *Sols afr.* **10**, 117–21.

MORGAN, W. B. 1955. Farming practice, settlement pattern and population density in South-Eastern Nigeria. *Geogrl J.* **121**, 320–33.
1963. Food imports of West Africa. *Econ. Geogr.* **39**, 351–62.

MORTIMORE, M. J. and WILSON, J. 1965. *Land and People in the Kano Close-Settled Zone*, Department of Geography, Ahmadu Bello University, Zaria, Occasional Paper No. 1.

MURDOCK, G. P. 1959. *Africa, Its Peoples and Their Culture History.*
1960. Staple subsistence crops of Africa. *Geogr. Rev.* **50**, 523–40.

NAVAL INTELLIGENCE DIVISION, 1943: *French West Africa*, I, 238.

NETTING, R. M. 1965. Household organisation and intensive agriculture: The Kofyar case. *Africa* **35**, 422–9.

NICOL, B. M. 1956. The nutrition of Nigerian children. *Br. J. Nutr.* **10**, 181–97, 275–85.
1959a. The calorie requirements of Nigerian peasant farmers. *Br.J. Nutr.* **13**, 293–306.
1959b. The protein requirements of Nigerian peasant farmers. *Br. J. Nutr.* **13**, 307–20.

NIGERIA, 1952. *Report on the Sample Census of Agriculture 1950–1*, Lagos.

PELISSIER, P. 1953. Les paysans sérères: Essai sur la formation d'un terroir du Sénégal. *Cah. d'outre-mer* **6**, 105–27.

PELZER, K. J. 1958. Land utilization in the humid tropics; agriculture. *Proc. Ninth Pacific Congress 1957*, **20**, 124–43, Bangkok.

PENCK, A. 1927. Geography among the earth sciences. *Proc. Am. phil. Soc.* **66**, 621–44.

PORTÈRES, R. 1950. Vieilles agricultures de l'Afrique Intertropicale. *Agron. trop.* **7**, 25–50.

RAMSAY, J. M. and INNES, R. R. 1963. Some quantitative observations on the effects of fire on the Guinea Savanna vegetation of Northern Ghana over a period of eleven years. *Sols afr.* **8**, 41–85.

RICHARDS, A. I. 1958. A changing pattern of agriculture in East Africa: the Bemba of Northern Rhodesia. *Geogrl J.* **124**, 302–14.
1961. *Land, Labour and Diet in Northern Rhodesia*, London.

SAUER, C. O. 1925. The morphology of landscape. *University of California Publications in Geography* **2** (2), 19–53.

SCHLIPPE, P. DE, 1956. *Shifting Cultivation in Africa: the Zande System of Agriculture*, London, 207–10.

SEAGER, A. 1965. Engineering aspects of the Western Nigeria farm settlement scheme. *World Crops* **17**, 20–27.

SHANTZ, H. L. 1940, 1941, 1942, 1943. Agricultural regions of Africa. *Econ. Geogr.* **16**, 1–47, 122–61, 341–89; **17**, 217–49, 353–79; **18**, 229–46, 343–62; **19**, 77–109, 217–69.

STAMP, L. D. 1938. Land utilization and soil erosion in Nigeria. *Geogr. Rev.* **28**, 32–45.

1958. The measurement of land resources. *Geogr. Rev.* **48**, 1–15.

1960. *Applied Geography*, London, 115–19.

TAKES, CH. A. P. 1964. Problems of rural development in Southern Nigeria. *Tijdschr. K. ned. aardrijksk. Genoot.* **81**, 438–52.

THORNTON, D. and ROUNCE, N. V. 1945. *Ukara Island and the Agricultural Practices of the Wakara*, Department of Agriculture, Tanganyika (revised edition), Nairobi.

TOTHILL, J. D. (ed.). 1940. *Agriculture in Uganda*, London.

1948. *Agriculture in the Sudan*, London.

TRAPNELL, C. G. 1959. Ecological results of woodland burning experiments in Northern Rhodesia. *J. Ecol.* **47**, 129–68.

URVOY, Y. 1942. Petit atlas ethno-démographique du Soudan entre Sénégal et Chad. *Mém. Inst. fr. Afr. noire*, 13–14.

VAN BATH, B. H. S. 1963. 1963. *The Agrarian History of Western Europe A.D. 500–1850*, London, 23–25.

WATTERS, R. F. 1960. The nature of shifting cultivation. *Pacif. View.* **1**, 59–99.

WELBOURN, H. 1963. *Nutrition in Tropical Countries*, London.

WHITE, S. 1941. The agricultural economy of the hill pagans of Dikwa Emirate, Cameroons. *Emp. J. exp. Agric.* **9**, 65–72.

YUDELMAN, M. 1964. *Africans on the Land*, Cambridge, Mass., 140–3.

10 Population density and agricultural systems in West Africa

M. B. GLEAVE *and* H. P. WHITE

The problem of the relationship between population density and the carry-ing capacity of the land in different parts of the tropical world is one that has attracted considerable attention in recent decades. Much of this effort has been devoted to discovering a method of determining the critical density above which further increases in population lead to deterioration of the environment. W. Allan, working with specific reference to East and Central Africa, has recently made a notable contribution to the solution of this problem (Allan, 1965). But this is only one aspect of the relationship between man and the land, and in this chapter a rather different viewpoint will be adopted. Here the aim is not so much to explore the effects of increasing population on the environment but rather to explore its effects upon man's methods of using the environment through the medium of his agricultural systems.

The problem under discussion is of more than academic interest to the peoples of West Africa for some of the highest densities of population in the continent are found here. Indeed, four out of ten inhabitants of inter-tropical Africa live in West Africa and it appears that population is growing more quickly than elsewhere in the continent (FAO, 1962a). Further, the problem is of concern because of the agricultural basis of the economies of the West African countries. A varying but always high proportion of the gross national product of the West African nations is derived directly and indirectly from agriculture. And finally, it is of both interest and importance because of the acute state of imbalance between land resources and popula-tion distribution that is characteristic of all West African countries.

The rate of population growth and the imbalance of population and resources merit further treatment here as they are of fundamental impor-tance to the problem. Unfortunately, it is difficult to be precise on either because the reliability of the available statistical material is questionable. However, the Food and Agriculture Organization of the United Nations has recently suggested that the population of West Africa is growing over-all at a rate of 2·5 per cent per year and that, with a few exceptions, this is the rate of growth for each country. The exceptions are Gambia where the rate is said to be 3 per cent, Liberia and Portuguese Guinea with a rate of 2 per cent, while no rate is specified for Mauritania (FAO, 1962a). Since these estimates were formulated, further material has become available which suggests that they may be too conservative. Thus Hunter asserts that the population of Ghana expanded over the inter-censal period 1948–60 at an annual rate of 4·2 per cent (Hunter, 1965). Comparison of the totals given by the Nigerian censuses of 1952–3 and 1963 gives a growth rate of 5·5 per cent per year. But critics of the 1952–3 census claim it to have been an undercount, while those of the 1963 census feel that the total of 55·6 millions is a considerable over-estimate. A mid-year estimate for the country in 1962 suggested a total of 45·4 millions with a growth rate of 2·8 per cent per year (Okonjo, 1966). Whichever rate is the more accurate they are both high. And while there may be reservations on the actual percentages quoted for West African population growth it seems indisputable that population is growing rapidly and 'unless action is taken to accelerate economic progress, this increase will mean added unemployment and underemployment and consequent malnutrition, misery and social unrest' (FAO, 1962a).

This high rate of growth is made worse because of the uneven distribution. Buchanan, writing of Nigeria, has described the population distribution as 'immature' and continues: 'The close adjustment of densities to environmental conditions, which is typical of long-settled areas, is here lacking, and the process of land occupation and settlement is incomplete over much of the Territory' (Buchanan and Pugh, 1955).

The major feature of the distribution of population in West Africa is the tendency towards concentration into discontinuous 'islands' separated by areas of much lower densities. The largest and most important 'islands', such as the cocoa belts of Ghana and of Western Nigeria, Iboland in Eastern Nigeria and the south-eastern part of the Ivory Coast are to be found in the Forest. But here also are the sparsely populated areas of south-western Ivory Coast, of south-western Ghana and the Cross River Basin of Nigeria. There are also important 'islands' in the Sudan zone, notably the Mossi area of Upper Volta and the Sokoto and Kano regions of Northern Nigeria. In the Middle Belt, which Harrison Church (1961) equates with the Guinea Savanna woodlands, such 'islands' are fewer and less important (Gleave and White, 1967). They are confined to those around Korhogo and Bouaké in Ivory Coast, to the Yoruba, Nupe and Tiv lands of Nigeria and to some mountainous areas such as the Central Atacora and Jos Plateau, which acted as refuges in troubled times.

There are, therefore, marked population gradients in West Africa. In Dahomey 48 per cent of the population live in the *Cercles* covering the coastlands and the Terre de Barre, only about 15 per cent of the area of the country (White, H. P., 1966).

The congested areas of the Udi Plateau in Eastern Nigeria, where densities in some districts are as high as 800 per square mile, are cheek by jowl with the Cross River plains where densities fall as low as 50 per square mile in the Cross River District (Udo, 1965a). And it is in the areas of densest population that the greatest growth in numbers is likely to occur. Under these circumstances it seems appropriate to consider in what ways the agricultural systems in West Africa adjust under changing conditions of population density. In view of the complexity of West African agricultural systems we have confined our attention to the area east of the Bandama River which marks the traditional boundary between rice- and yam-based systems (Miège, 1954).

Agricultural practices and systems

In West Africa we may distinguish between three agricultural practices:

1. *Shifting Agriculture* is based on the maintenance of fertility by long periods of bush fallow. It is essentially characterized by the rights over the fallows not being maintained by the individual or family group previously cropping the area, but reverting to the clan or other large group. The land is thus redistributed at intervals. In addition settlements, either single compounds or villages, are moved at intervals (Morgan, 1957).

2. *Rotational Bush Fallowing* has the same basis but in this case the rights in the fallows are continuously maintained by the individual or family (or their descendants) first clearing the land. Only exceptionally is there a redistribution of land or a move of compound or village.

3. *Semi-Permanent and Permanent Cultivation* is distinguished by the fallow periods either being shorter than the intervening cultivation periods or being non-existent.

There is evidence to suggest a connection between population density and agricultural practice, and that the practice becomes modified as density changes. Shifting cultivation is found only where population densities are low, usually less than 25 persons per square mile. Generally rotational bush fallowing is practised where population densities are higher. But increasing population density (or the spread of cash cropping) eventually reduces the length of the fallow periods until the practice must be classified as semi-permanent or permanent. Obviously no absolute rule can be established. Even though population densities are very low, the Konkomba of Ghana maintain individual rights over the fallows for limited periods. Bush fallowing is practised in parts of south-eastern Nigeria where densities exceed 600 per square mile and are exceptionally high by West African standards. However, this suggested direct relationship between population density and agricultural practice may be of fundamental importance for development planning.

Each of these three forms of agricultural practice may also be an agricultural system, but some systems are in fact hybrids of two or more practices. Agricultural systems in West Africa fall into two broad categories. In the first place are those based on traditional methods, though much modified through the influence of cash cropping, market forces and agricultural Extension Services. Those of the second are alien systems, originally introduced with the aid of foreign capital. They include plantations, farm settlement schemes and the state farms of Ghana (Coppock, 1966; Fréchou, 1955). In this chapter, however, we are only considering the first group.

The most common and widespread of these hybrid systems is that between the permanent cultivation of kitchen gardens or of 'compound farms' immediately around the settlements with the shifting cultivation or rotational bush fallowing of the main cultivation areas. Udo (1965a) describes a typical hybrid system in the Calabar plains of Nigeria where compound gardens are cropped continuously with applications of household manure for vegetables and yams, while other staples are raised on the main holding cropped once in seven to ten years. On the Jos Plateau the household farm is divided by euphorbia hedges into tiny fields, permanently cultivated with the aid of household waste. The whole settlement, together with the household land, is surrounded by a thick hedge. Outside are the bush farms cropped for a few years and left fallow for long periods (Grove, 1952).

Similar systems of permanently cultivated 'compound farms' and occasionally cropped 'bush farms' are to be found throughout northern Ghana, whether in the sparsely peopled areas of Western Dagomba (Aikenhead, 1944) or South Mamprusi (Smith, 1945) or the over-populated areas of North Mamprusi (Smith, 1941; Lynn, 1937; Hilton, 1959). It is also to be found in northern Togo (Enjalbert, 1956).

A development of this system is to be found in the forest, where commercial tree crops, a permanent form of agriculture, have been grafted on to normal bush fallowing. The significance of such an addition varies even within small areas possessing some cultural unity. At Oko, a little to the south of Agulu in the oil-palm belt of Iboland, the palm occupies the permanently cropped garden land with under-cropping of staples and vegetables. Household waste and dead leaves are used as manure (Grove, 1951a). Farther south, in Aba Division and Diobu District, the palms occupy not only the garden land but are also found in occasionally under-cropped groves, in plantations and scattered through the areas under bush fallowing (Morgan, 1955a). Similarly, in Lower Dahomey the palms are to be found in groves and also throughout the land used for annual cropping under the bush-fallowing method (White, H. P., 1965).

Another variant of the system is to be found in the Bamenda Highlands of West Cameroon, where the cash crop is *arabica* coffee. This crop was first introduced on the compound land, here used for tree crops such as plantains, avocado pears, mangoes, kola and eucalyptus wood. Subsequent expansion has taken place in the farmland, where the coffee is intercropped with the food crops. It is grown on ridged slopes and fertility is maintained by the hoeing in of grass cut from the fallows (Gleave and Thomas, 1968). But in the Avatime State of Trans-Volta (Ghana) coffee is grown in pure stands on relatively small plots (White, H. P., 1956a).

Cocoa, once the nurse crops have been removed, and kola are normally grown in pure stands. But throughout the cocoa belts of Ghana and Nigera plantations, small and large, are found in conjunction with areas of food cropping by rotational bush fallowing (Hill, 1957; Galleti *et al*, 1956; Hunter, 1963). Frequently, as in Akwapim, population increase has forced down the fallow periods to little longer than the cropping periods. Where cocoa has disappeared through disease, the food crops become the major cash crop (White, H. P., 1958; Hunter, 1961).

There are also the combinations of permanently cultivated hill lands, sometimes terraced, and cultivation in the adjacent plains by rotational bush fallowing. Such a system is being worked by some of the fourteen tribes that live or formerly lived in the Kauru Hills, of Zaria Province Northern Nigeria. Others have given it up on moving to the plains. Intensive agricultural methods are practised now only by those tribes still living in the hills. The Kitimi, for instance, have carefully terraced farmlands on the steep slopes of the interior valleys in which they live. But they also have farms in the plains which they cultivate by bush fallowing (Gleave, 1965).

Similarly, the Kofyar people cultivate food crops in their traditional lands, often terraced, in the hills on the southern margins of the Jos Plateau, using intensive techniques, but they also cultivate a belt of plains land between three and four miles wide at the foot of the escarpment, using bush-fallowing techniques to produce cash crops, chiefly yams (Netting, 1965). This is a common pattern where densely settled upland and sparsely settled plain are in juxtapostion (White, S., 1944; Froelich, 1949, 1952; Enjalbert, 1956, *inter alia*). Such a dual system may persist for some time (in the case of Rumaiya it lasted for about twenty years), or it may be a temporary phase in the abandonment of the hills for the plains. This pattern is also repeated in Northern Ghana and Togo (Hilton, 1959; Enjalbert, 1956).

Besides the numerous rotational bush-fallowing permanent cultivation hybrid systems, we can also distinguish hybrid systems involving different forms of permanent cultivation, though these are rare. The best example is to be found in the Kano 'close settled' zone. Here the irrigated bottom-

K

lands, the *fadamas*, are cultivated in conjunction with permanent cropping of the 'uplands' of the interfluves (Mortimore and Wilson, 1965).

Agriculture in areas of low population density

Agriculture in areas of low population density tends to be more extensive, because there is abundant land available. It thus tends to have more of a shifting element, for fallow rights revert to the group and the land is reallocated for cultivation at the end of the fallow period where there is no pressure on land. In the eastern part of Dagomba District in Ghana's Northern Region, where population density is nowhere above 49 per square mile and over wide areas is below 25, land is owned by the community but the members of the community can farm freely in its territory. Land is allocated by the chief or his appointee for farming each year. Once assigned to an individual the land remains his for the period of cultivation, usually three or four years, but reverts to the community with the onset of the fallow period (Ghana, 1959; Pogucki, 1950). It may be argued that such a tenurial practice is a function of cultural inheritance rather than of population density. This, while being possible, seems unlikely in view of the widespread prevalence of the practice among many different tribes in West Africa.

Other features of the agricultural systems in areas of low population density reflect the lack of pressure on land resources. Only a very small proportion of the land is under cultivation at any one time. In the eastern part of Dagomba District in Ghana approximately 5,000 square miles was surveyed of which less than 1 per cent was under cultivation in any one year. This was much less than the proportion, about 4 per cent of the total, taken up by non-agricultural uses. Thus a very high proportion of the land is not being used at any one time other than for various stages of grass production. But, in spite of cattle-keeping being common among the farmers in the area, it was estimated that 85 per cent of the grassland is not being used at all. Indeed, few farmers use their livestock for making a living except that bulls are occasionally sold to pay taxes. The gross under-utilization of the resources of the area is reflected in the relatively low level of the interference by man in the vegetation (Ghana, 1959).

Everywhere land for cultivation is broken from the fallow and is used to grow crops until yields begin to fall. There is thus considerable variation in the length of time for which land is cropped. It is shortest on naturally poor soils and in the more distant, less accessible portions of the village lands, while the more fertile soils and those immediately around the village are cultivated for a longer period. Thus in the savanna areas of Yorubaland, plots on the poorest soils are cropped for only one year and are then abandoned, but most of the land is cultivated for at least two years before it is allowed to revert to fallow (FAO, 1962b). In Dagomba, the cultivation period is commonly three or four years, but in the Cross River district of Eastern Nigeria, with equally low population density and with high forest vegetation, the land is only cultivated for one year. Where the natural fertility of the soil is maintained by the fallow period, fallows are long in areas of low population density (Nye and Greenland, 1960). The length of the fallow period does, however, vary and it is shortest in the most accessible areas. In the sparsely populated portions of the Yoruba grass-lands, the Dagomba area, and the Cross River district the fallow period is rarely shorter than eight years, and it may be as long as thirty or even fifty years. In the more densely settled areas it will fall below the minimum suggested (Vine, 1954).

The size of holdings also varies in spite of the fact that there is abundant land available. Generally, only sufficient land is cultivated to satisfy the food requirements and other needs of the family. Nevertheless, the norm shows some variation in size as between the forest zone and the savanna zones. A subsistence area of half an acre per head of population is usual in the forest regions of equatorial and sub-equatorial climate (Allan, 1965). In the savanna regions this is probably increased to three-quarters of an acre, possibly a little more. But owing to the paucity of statistics these figures should only be considered as tentative. For a family of five, two and a half acres are required in the forest but almost four acres in the savanna. Because of cash cropping, whether of food crops or export crops, normal holding sizes are rather larger than this but the question of cash cropping will be considered later in the chapter.

Rather more data is available on holding sizes, although the accuracy of it is often open to serious doubt. The broad validity of the suggestion being made is, however, indicated in Table 10.1, though, as the Bulletins point out, there are numerous inaccuracies in the data.

Holdings usually consist of small 'farms', or plots, scattered through the village territory at distances up to five miles from the settlement. There is a marked tendency, however, for farms to be located as near the village as possible or near footpaths. In south-western Ghana it has been shown that, following road building, fallows shorten near the roads and increase away from them (Ahn, 1958). The cultivated farms are usually intermixed with recent fallows and with regenerating bush. But the pattern of cultivated land in the village territory is incipient rather than well-developed because of the small proportion of land under cultivation at any one time. It is, therefore, best described as a mosaic of cultivated patches interspersed with fallows and uncultivated land which reduces the erosion hazard.

Such a pattern is found in the Gonja villages of Western Gonja on Ghana's Northern Region. But interspersed with them are Lobi villages with a different pattern of land use (Manshard, 1961). Here the family farming unit clears a large area close to the compounds for its food-crop farm which will supply its food needs. Farther away from the settlement are the 'bush' farms on which individuals grow crops for sale. The com-

TABLE 10.1 *Nigeria: percentage of farms by size and by province*

Farm size (acres)	Forest		Guinea savanna		Sudan savanna	
	WET Benin	DRY Ibadan	Niger	Benue	Kano	Sokoto
Under 0·25	2	3	—	2	2	2
0·25–5·0	5	8	5	6	4	10
0·50–1·0	11	16	6	12	14	12
1·0–2·5	42	42	21	27	27	26
2·5–5·0	24	18	25	28	26	29
5·0–10·0	11	11	22	20	19	17
Over 10·0	5	2	21	5	8	4
Sampling error ± per cent	15?	15?	23	17	22	17

After Agricultural Sample Survey Bulletin 3, Northern Nigeria 1957–8 and Bulletin 4, Western Nigeria, 1958–9

munal cultivation of one large block of land is found elsewhere in West Africa. It is found, for instance, around Wum in the Bamenda Highlands of West Cameroon where the growing of food crops is largely the responsibi-lity of the women. The land is cultivated only for one year and the one large cultivated block is moved around the village lands from year to year.

Finally, crop rotation is widely practised where the land is cultivated for more than one season. The character of the rotation varies from crop belt to crop belt. Space precludes a consideration of crop distributions in West Africa which is covered adequately elsewhere (Johnston, 1958; Harrison Church, 1961; Irvine, 1953; Buchanan, 1952, *inter alia*).

Changes in tenurial practice

With increasing population density the practice of reallocation of the fallow land at the conclusion of the fallow period gives way to the practice of the cultivator's maintaining his rights in the land while it lies fallow. The culti-vator thus breaks the fallow for a second period of cultivation and, if the fallow/cultivation cycle is short, for subsequent periods also. In such situa-tions the land is still communally owned but the rights in it are vested in individuals or families. Eventually, pressure on land becomes so great that land acquires a market value, individual ownership becomes accepted and the buying and selling of land becomes possible and even common (Biebuyck, 1963; Labouret, 1941).

This sequence can be illustrated by reference to Hausaland. Under the traditional customary law land is held by the village community and allocated by the Headmen, but occupiers are guaranteed cultivation rights over their holdings. Any improvements made can thus be passed on from one generation to the next. But the interests of the community are safeguarded by the lapsing of such rights should the land be left fallow for too long. This is still the custom in areas of low to moderate densities below 150 per square mile, where pressure on land resources is minimal (Prothero, 1957a; Baldwin, 1963).

In the Kano close-settled zone, where, near to Kano City, population densities rise to over 500 persons per square mile, the tenurial system is on paper similar. In practice, however, in densely settled areas gift, purchase, pledging, loan and lease may all be practised, and indeed, it seems that there is an active land market operating. In three village areas near Kano 'land tenure is on a family basis but more individualized than in the south' (Mortimore and Wilson, 1965). Here gift of land is made often by a father to his grown-up son either by dividing the family holding or by purchasing a new holding from outside the family. In fact, the normal method of increasing the size of the holding to support a growing family is by purchase. Purchasing of extra land may offset the effects of partible inheritance, but it also creates a class of rural landless. Land was being bought and sold forty years ago and the last twenty years has seen considerable rises in the price which land fetches (Mortimore and Wilson, 1965).

A further instance of the effects of increasing population density on tenurial practices within a single tribal area may be cited. Similar variety to that found in Hausaland is to be found within the area occupied by the Nupe of central Nigeria. Their lands are cut into two parts by the Kaduna River. To the west is Trans-Kaduna and here the traditional land-tenure system with land vested in the tribe or village still pertains. The area had an over-all population density at the time of Nadel's study in the mid-1930s of 17·6 per square mile. But in Cis-Kaduna, that is districts east of the Kaduna River, a complex system of ownership and land transfer had evolved which in extreme cases amounted to outright purchase for money. In this part of Nupeland, which was conquered by the Fulani and contained Bida, the capital, over-all population density in the mid-1930s was 55 per square mile (Nadel, 1942). While Fulani influence may be invoked to account for the differences in tenurial practices in the two portions of Nupeland it seems more reasonable to accept Nadel's suggestion that the transformation of traditional practices in Cis-Kaduna was brought about by increasing pressure on land resources. Indeed, even in

Trans-Kaduna disputes occur about the very valuable land along the river and stream courses and in the more densely settled areas these disputes are particularly bitter.

In southern Ghana the system of land tenure became modified at an early date in response to increasing population pressure. Throughout the area individual rights were maintained over the fallows. At the same time, in the forest, even if technically there was no sale of land, cocoa trees once planted could be used as pledges for loans, and even bought and sold outright (Pogucki, 1952, 1954, 1965). But the most interesting development has been the spread, through pressure of population, of the Krobo and Akwapim peoples of south-east Ghana throughout the forest. This has been achieved through outright purchase of land. The emigrants, formed into 'companies', bought large blocks of land and distributed them among the company members, in proportion to their share of the capital. Account is not taken as to the depth of the holdings but only the frontage, each participant receiving so many 'ropes' and clearing back, thus producing a strip pattern of land holding (Hunter, 1963). The formation of these *huzas* began in the extreme south-east of the forest (Field, 1943), but later the system, with modifications spread through the southern forest, though not into Ashanti (Hill, 1963).

Changes in land use and land-use patterns

With increasing population density, more land is taken into cultivation. Conversely the proportion of bush, which is little cultivated, and which may be used only once in fifteen or twenty years, diminishes.

Thus at Soba in Zaria Province, Northern Nigeria, where population density in 1952 was 103 per square mile about 6 per cent of the village area was under cultivation (Prothero, 1957a, b). The land-use pattern at Soba is characteristic of large parts of the savanna land of West Africa. The dominant feature of it is the arrangement of uses in a number of concentric rings centred on the village. The first, immediately outside the village, is the zone of permanent cultivation; the second is the ring of land either under cultivation or in fallow; the third and final ring is one of thick bush in which there are four small hamlets each with its own ring on a smaller scale.

Such a pattern of concentric zoning is also found in Central Ghana, but the uses are here arranged in a somewhat different order. The first ring, immediately around the settlement, contains the permanently cropped garden plots, the second ring is a grazed grassland area formerly cultivated but now exhausted; the third zone is that of cultivated land interspersed with fallow; and finally, the fourth ring is a zone of unfarmed bush (Wills, 1962). It has been suggested that such a concentrically zoned arrangement of uses is characteristic of areas in Hausaland with population densities between 50 and 150 to the square mile (Grove, 1961).

Where densities are below about 50 per square mile this pattern may be only partially developed. But where densities get higher than about 150 per square mile the concentric zoning of uses breaks down, as each comes under pressure from a neighbour. Thus in the case of Central Ghana the ring of cultivated land and fallow, or as it may be described, the 'land rotation area', comes under pressure from the expansion of permanently cultivated land, and within it fallow periods become shorter. But, in turn the ring of unfarmed vegetation separating settlements is absorbed into the land rotation area and may finally disappear. Such changes represent intensification of the use of land and are also found where the land-use pattern is not concentrically zoned. Thus on the more southerly cuesta of the *Terre de Barre* of Togo and Dahomey population densities have so increased that there is now no land not under crops or very short fallows (White, H. P., 1965).

The changes in land-use pattern outlined are reflected in the proportion of land devoted to different uses. Data is again scarce and fragmentary but two surveys have been carried out in the densely populated parts of Hausaland, enabling a comparison with conditions in Soba (Luning, 1961; Mortimore and Wilson, 1965).

Land-use statistics for three villages in Katsina Province suggest that cropland and fallow have spread over the whole cultivable area (Luning, 1961). Although there was a considerable proportion of grazing and woodland, varying from 29 to 41 per cent of the total area, this does not represent a reserve of cultivable land. It is scrubland usually developed on shallow soils too poor to sustain cultivation. In one village area, however, some of the most suitable areas are being farmed for a short period and then left in fallow for a long period, but in another there has been no change in the land-use pattern between 1952 and 1960. Further, the proportion of the land under fallow is very low but related, at least in part, to population density.

A similar, though perhaps more extreme state of affairs, has been revealed in a survey of the whole of three village areas and parts of two others located to the north of Kano City and about five miles from it (Mortimore and Wilson, 1965). The survey area forms part of Ungogo District which has a population density of 609 per square mile, but in the village areas surveyed the density is 915 per square mile. Here cultivated land occupies 83·3 per cent of the area, bush and grazing land only 7·5 per cent. The much lower percentage of bush and grazing land in the survey area near Kano compared with that of Katsina Province reflects the differing physical conditions in the two areas. Both have soil developed on extensive deposits of wind-borne sand and upon weathered rocks. But in the Katsina survey area there are long strips of laterized ironstone about a mile or two wide and placed about five miles apart. Overlying these are only thin soils which cannot hold cultivation. In the Kano survey area, by contrast, laterite is nowhere close enough to the surface to preclude cultivation. The small uncultivable area is found on steep slopes or results from sheet-wash, gullying or seasonal waterlogging. Thus in the two densely populated survey areas of the Hausa north of Nigeria there is a much closer adjustment to the minutiae of physical conditions than in the moderately settled areas of which Soba is typical.

Changes in agricultural technique

Increasing closeness of settlement is accompanied by changes other than those in tenurial practice, land use and land-use patterns. Fallow periods become shorter and the relationship between their length and population density has been indicated by W. B. Morgan for sample areas near Aba and Port Harcourt. In the less densely settled areas, as in Etche (density 187), the fallow periods are often of nine to fifteen years. In Ndoki and the Ogwe district of Asa (density 306) seven year fallows are usual but in Nbosi (density 436) the most common fallow period is five years (Morgan, 1955a).

With the decline of fallow periods, and of fallow acreage, other means have to be developed by the cultivators of wresting a living from the soil. There are two alternatives open. The first is to practice traditional agricultural techniques, countering the ever-declining fertility of the soil by adjusting standards downwards. Thus less demanding crops, more tolerant of poor soil conditions, replace more favoured crops in the rotations. In the eastern portion of Egbaland, an area of the southern root-crop economy, better-yielding and more palatable varieties of 'male' yam are replaced by inferior 'female' ones until soil fertility declines to such an extent that even these cannot be cultivated. Cassava, a less favoured food, is then introduced into the rotation as a replacement for 'female' yams (Mabogunje and Gleave, 1964).

Farther north, in the area of the northern grain economy, guinea corn and millet are the main food crops where fallows are long enough to maintain the fertility of the soil. But where population pressure has reduced the fallow period to dangerous levels, the poverty-tolerant cereals *acha* and *tamba* replace the guinea corn. An extreme instance of such a change-over has occurred in Gyel village area of the Jos Plateau in Northern Nigeria where in 1953 population pressure was aggravated by the loss of 14 per cent of the area to tin mining. Arable land accounted for 60 per cent of the total area and 40 per cent of this was devoted to *acha*, 28 per cent was fallow and only 16 per cent was devoted to millet (Davies, 1946). There is also some evidence to support the view that in order to offset declining yields as the fertility of the soil diminishes farms tend to be increased in size so as to maintain the necessary output of food (Allan, 1965). If this is the case, and further investigation is needed to verify it, the process of soil depletion is hastened unless intensification of methods occurs.

The second alternative is to adopt intensive techniques, notably composting and the use of manure. Most West African peoples are familiar with such techniques, for they are used on a small scale on the compound land. It is only as population pressure increases that it becomes necessary to apply them to a higher proportion of the cultivated land. This also can be illustrated by reference to the densely populated Ibo and Ibibio areas of Eastern Nigeria. At Oko, where the population density is about 530 per square mile, the compound land occupies 1,400 acres of the village area, while the outside farmlands occupy about 1,600 acres. But where population density exceeds 1,000 per square mile, as in parts of Orlu, Okigwi and Awka Divisions, the compound land occupies almost the whole of the village area (Grove, 1951a). Further, the compound land at Oko is expanding at the expense of the outer farmlands. Oil-palms are planted in the open fields after manuring of the ground, and new compounds built when the palms are established (Grove, 1951b). This process has recently been described with reference to the Ozubulu village area, south-west of Awka (Udo, 1965b). The villages of the Ozubulu village group were, in 1930, separated by farmlands and other unoccupied lands such as *juju* groves. At the present time the settlements have coalesced and the former farmlands have been almost entirely replaced by intensively cropped compound land. Furthermore, the traditional method of rotational bush fallowing has been replaced by permanent cultivation in which manure, household and vegetable waste are used to maintain soil fertility.

The use of manure is also characteristic of the densely populated parts of Northern Nigeria. In the villages of Katsina Province discussed earlier, dung produced by cattle, horses, donkeys, goats and sheep is applied to cropland. The number of animals kept by each farming family increases as the families get bigger. Goats and sheep graze the farms in the close season, returning home each night. During the farming season they are tethered in the compound and the manure is collected in heaps. Household refuse, ashes and the droppings of other animals are added to the heap. Those farmers with large holdings usually need to supplement their own supplies, so that there is an active market for manure in the villages. In the villages near Kano, manure produced in this way on the holdings is supplemented by supplies brought from the city by the donkey-load. The application varies from holding to holding, but on a small sample averaged 1·6 tons per acre. Other examples of the use of manure are the experiments, not very successful in terms of the number of farmers converted to the system, to encourage mixed farming in northern Ghana (Lynn, 1946) and Northern Nigeria (Coppock, 1966).

Advanced agricultural techniques often associated with terracing of steep slopes are also found in the upland areas of West Africa, to which tribes withdrew in formerly disturbed times. The detailed features of these systems vary from tribe to tribe and upland area to upland area and a brief summary of them in Togo, Dahomey and Nigeria has recently been provided elsewhere (Gleave, 1966). The most advanced of them is probably that of the pagans of Dikwa Emirate who withdrew to the Mandara Mountains. The system is based on mixed farming, use of manure on carefully terraced slopes, crop rotation and the husbanding of useful trees (White, S., 1944. More accessible accounts based on White's study can be found in Buchanan and Pugh, 1955, and Allan, 1965). These advanced techniques enabled population densities of at least 130 per square mile to be maintained in the unpromising environment of the Mandara Mountains in pre-British days. But in many areas of dense population, the forest of Yorubaland, of Ghana, of the Ivory Coast and in the *Terre de Barre*, there is no manuring.

Intensification by the introduction of new crops into areas that are unsuited to cropping traditional crops is also a feature of West African agriculture. In many areas of both forest and savanna the possibility of increasing crop acreage and food supplies by the growing of rice in swamplands has been developed. Thus swamp rice is now grown in the forest areas of Ivory Coast east of the Bandama River, in south-western Ghana and in southern Nigeria where root crops and not rice are traditionally the staple crops. In Eastern Nigeria, for instance, rice output trebled between 1950 and 1959, the main areas for swamp rice production being in the Cross River Basin from Afikpo to Ogoja and in the Anambra Valley in Onitsha Province (Coppock, 1966). Rice output has also expanded recently in Western Nigeria but here swamp rice is more restricted. New land has been opened up for its cultivation in the valley of the Owuru River in Abeokuta and Ijebu Provinces. In the savanna lands of northern Ghana and Northern Nigeria swamp rice cultivation is being expanded in the *fadama* lands as in the Sokoto–Rima Valley as well as in large-scale schemes such as the Shemenkar Valley scheme of Plateau Province of Northern Nigeria.

Farm size

The question of farm size and pattern can be conveniently considered here. It has already been pointed out that in areas of low population density there is considerable variation in size of holding in spite of land being freely available. This is equally true of areas of moderate and dense population (Prothero, 1957b; Luning, 1961; Mortimore and Wilson, 1965). But whereas the size of holdings is a matter of choice where there is no pressure on land resources, in congested areas it is a matter of wealth. For it appears that the large holdings in congested areas are 'owned' by those able to purchase land to offset the division of holdings resulting from inheritance. The large 'owners' may be full-time farmers or they may be primarily interested in other pursuits, working their holdings either by hiring labour, or by the *abusa* or share-cropping system of Ghana (Hill, 1957).

The major difference between the sparsely settled and the congested areas is in the structure of the holdings. In the former case the holding may consist of a few scattered farms in the village lands whereas in the latter case it may consist of many small farms widely scattered in the village area. The size of plots resulting from subdivision has recently been analysed for a small survey area in the Bagango Valley near Bamenda in West Cameroon (Table 10.2). The small plot-size and the scattering of the plots that make up the holdings can be illustrated by reference to conditions in Kimbaw also in the Bamenda Highlands of West Cameroon. Here the average

TABLE 10.2 *Plot size in Bagango Valley, Bamenda Province*

Size (acres)	Total area (acres)	No.	Area %	No. %
Less than 0·2	6·9	48	6·8	24·5
0·2–0·4	13·4	47	13·3	24·0
0·4–0·6	25·2	52	25·1	26·5
0·6–0·8	14·3	21	14·2	10·7
Over 0·8	41·0	28	40·6	14·3
Total	100·8	196	100·0	100·0

After Gleave and Thomas, 1968, Table 2.

acreage cultivated by each woman is 1·34 acres and the average number of plots is eight (Kaberry, 1952). These figures, however, mask considerable variations. At one extreme was a woman farming 1·04 acres with only four plots all within twenty minutes walk of her hut. At the other extreme was a woman who cultivated 1·8 acres in fourteen different plots. Nine of these lay within ten minutes walk of the compound; two were within thirty minutes; two were an hour away; and one very large tract was ninety minutes walk over two steep hills.

Finally, the increasing permanence of agriculture in areas of dense population is reflected in the permanence of the plots, and increasingly these are now being hedged or fenced. The euphorbia hedges of the pagan areas of the Jos Plateau are of long standing in this respect, formerly having, in part at least, played a defensive role. The rectangular fields of the densely populated Hausa areas of Northern Nigeria are often bounded by thatching grass, by other useful plants, notably the economically useful trees, or by euphorbia hedges. Mud walls or guinea-corn stalks are also used to fence in cassava left in the ground through the dry season.

In southern Nigeria there is widespread evidence of strip fields (Morgan, 1955b) and in the densely settled parts of the Eastern Region these are increasingly being fenced, whether in farm or compound land, to protect crops from livestock (Morgan, 1955a). Although fencing of garden land is common in the forest, that of farmland is confined to the Eastern Region. But whether the land is held in the strips of the *huzas* or in blocks it is more and more coming to be fixed in outline, landmarks being provided by trees and bushes often specifically planted for the purpose.

The hedging and fencing of fields and farms, while being undertaken for practical reasons, is creating in West Africa a number of areas having distinctive rural landscapes in a sub-continental area which is largely 'open'.

The expansion of cash cropping

This chapter is concerned with the consequences of population growth upon agricultural systems. Obviously there are many other factors, cultural borrowings, soil conditions and rainfall régime, with important consequences. One of these, the spread of cash cropping, must be considered at this point. Firstly, this is because its consequences are so similar to those of population growth that the two are inseparable as causative factors. But secondly, there is a degree of mutual reaction between the two factors. Economic development, which over large areas of West Africa must be based on cash cropping, creates conditions, such as availability of medical services, favourable to population growth. Conversely, population growth leads to a search for occupations alternative to agriculture, which in turn diverts crops from the subsistence to the exchange economy.

It is possible to distinguish three stages in the modification of agricultural systems resulting from the introduction of cash cropping, the first two also, of course, being brought about by population growth. The first is the fixation of shifting cultivation. In the case of tree cropping this is obvious enough, but annual crops grown for cash also lead to fixation. They must be grown near to evacuation routes which are themselves fixed, while the spread of cash crops, like population growth, gives land a scarcity value. The only real exception is the growing of yams for cash sale in parts of the Middle Belt. In the area between Ejura and Kete Krachi in Ghana this is still practised under shifting methods.

The second stage is the intensification of agricultural production. This is chiefly the result of taking in more fallow land for the cultivation of cash crops, but it may also result from increased output per acre. Thus the increased groundnut production from Northern Nigeria in the early sixties is attributable to both increased acreage and to the widespread use of fertilizers. What has yet to come is increased output per worker.

The third is increased specialization in a single crop and the emergence of large areas under pure stands, in contrast with the traditional agricultural systems. Thus we find extensive areas under pure stands of cocoa in the cocoa belts of Ghana and Nigeria, oil-palm groves of Porto Novo (Dahomey), and the maize in the environs of Accra, Lagos and Lomé.

Cash cropping may involve the utilization of indigenous crops already part of the agricultural system, such as the oil-palm, though the resultant intensification will modify that system. It may also involve the introduction of exotics, the most notable example being that of cocoa. We can also distinguish between three types of cash cropping in order to assess the consequences upon the agricultural systems, namely, tree cropping, annual cropping for export and the raising of food crops for cash sale.

In attempting to assess quantitively the development of cash cropping there is the problem of data collection. While good studies of tree cropping do exist (Galletti et al., 1956; Hill, 1957; Brasseur-Marion and Brasseur, 1953; Tricart, 1957; Hiernaux, 1948) this problem is particularly acute where food crops are grown partly for sale and partly for subsistence (but note, Gold Coast, 1953). There is great need for further studies.

The proportion of land devoted to cash crops naturally varies very widely but, by increasing the land requirements of the community, it lowers the critical population density and at the same time hastens the same agricultural changes that are also contingent upon increasing population density. Thus *arabica* coffee was introduced into the Bamenda Highlands of West Cameroon between 1945 and 1950. The most favourable area was in the scarp foot zone of the High Lava Plateau, where the soils were particularly favourable (Gleave and Thomas, 1968). But this was already the most intensively cultivated and densely settled area of the Bamenda Highlands (Bawden and Langdale-Brown, 1961). We have already shown there was pressure on land resources in our reference to Kimbaw. This has been intensified with the widespread adoption of coffee cultivation.

At first small amounts of coffee were introduced into the compound land, but later expansion has taken place at the expense of both the food-crop land and the fallow. Further, many of the coffee plots are interplanted with food crops such as cassava, yam and sweet potato. The land now devoted to food crops is farmed for two years and then fallowed for two years. Thus the coffee plots are almost in permanent cultivation and the food-crop plots are on the verge of becoming so. Partly because of land shortage coffee bushes are spaced about 6 ft apart instead of the 9 ft said to be ideal. Yields are by any standard low, probably due partly to overcrowding and under-cropping (Gleave and Thomas, 1968). Thus cash cropping is an uneasy compromise in terms both of subsistence and cash crops, but the peasant farmers are much better off than they were before the introduction of coffee.

In the *Terre de Barre* of Togo and Dahomey the principal export crops are palm kernels and oil. The density of the palms varies very greatly, but usually it is sufficiently low to allow the under-cropping of annuals. These are chiefly maize, cassava and groundnuts. Cultivation methods are the normal bush-fallowing ones, the land below the palms being either under crops or in fallow. In places, however, especially around Porto Novo, densities of palms reach such levels that under-cropping is impossible (Brasseur-Marion and Brasseur, 1953). These groves extend up to the Nigerian frontier, but beyond they give way to an extensive area of *kola*. This emphasizes that economic conditions and governmental policies affect the choice and spread of cash crops just as much as physical and cultural factors.

In the same area the growth of the larger towns, Porto Novo, Cotonou, Lomé and even Accra, has led to an increased demand for maize. This has led to an intensification of maize cultivation made possible only by drastic reductions in fallow periods so that in some areas up to 60 per cent of the land is under maize and 40 per cent under fallow (White, H. P., 1965). Excessive monoculture of maize in the lower Densu Valley north-west of Accra has led to the forest being replaced by useless grasses of low nutritive value and to the almost complete degradation of the soil.

Cash cropping undoubtedly sometimes hastens the development of permanent agriculture and, along with population growth, the depletion of soil resources. But it may also open up hitherto little-utilized areas to cultivation. In the years after the war the introduction of swamp rice was encouraged as a much-needed cash crop in the north-east of Ghana. The area suffered from population pressure but the rice could be grown on land that was flooded during the wet season and therefore hitherto useless.

It has recently been shown that groundnut cultivation in Kano Province has affected the pattern of population growth. Some areas were not particularly favourable for the traditional cereal economy and therefore could not support a dense population. These have since been developed for groundnut production and this has made denser settlement possible (Mortimore, 1966).

But the best illustration of this point has been the opening up for cocoa production of the moist semi-deciduous forest of Ghana. The cradle of the cocoa industry was the Akwapim Range, north-east of Accra. In the 1890s, because of population pressure, the Akwapim and Krobo peoples began to move into the virtually empty areas of Akwapim to the east of the Densu and Akim Abuakwa to the west. Suitable lands were purchased in blocks on behalf of extended families from local chiefs. Later purchases were made on behalf of 'companies' of unrelated individuals and the land within the blocks subdivided into strips (Hunter, 1963).

Although the Ashantis developed their own cocoa farms centred on Kumasi, the migrant farmers from the south-east swept across the central forests until in the late 1940s and early 1950s they reached the Ivory Coast frontier in Brong-Ahafo. After 1940, population pressure was added to the effect of swollen-shoot disease in the older areas. Until control and re-planting measures became effective after 1960, the only solution in the worst-affected areas was to abandon cocoa-growing altogether. Many farmers on the devastated areas purchased lands in Brong-Ahafo and established new farms of healthy cocoa (Hill, 1963; Hunter, 1963; Biebuyck, 1963).

The devastated areas (White, H. P., 1959; Hunter, 1961) were the most highly developed rural areas of Ghana and had been deprived of their main source of cash income. However, the rapid growth of Accra and other large towns had created a demand for food crops such as maize, cassava and plantains. Thus a large quantity of the food crop hitherto raised for sub-sistence now began to be sold. In Akim Abuakwa it was shown in 1952-3 that 30 per cent of the food crops produced were sold to Accra and other towns (Gold Coast, 1953). Population growth and production of crops for sale have both led to intensification of land use and reduction of fallow periods in Akwapim and Akim Abuakwa until it can almost be said that permanent cultivation has emerged.

The man–land relationship

The argument thus far suggests a direct relationship between population density and agricultural systems which may not in fact exist. Indeed, there is considerable evidence for suggesting that in certain circumstances the relationship is in some measure indirect. This is particularly so when popu-lation pressure upon resources of cultivable land is reduced by part-time non-agricultural activities by the farmers. In the recent study of three village areas close to Kano such secondary occupations are apparently im-portant to most of the population. A distinction should be made between those members primarily dependent upon such non-agricultural pursuits and those who practise such pursuits in a secondary role to farming. Occupational categories cut across this distinction, weaving, for instance, being practised as a main occupation and also in a secondary role (Morti-more and Wilson, 1965). In the former case it appears to be associated with small-holdings so that weaving continues even through the wet season, the agricultural responsibilities of its practitioners being light enough to allow this. In the latter case it is largely a dry-season occupation, the main effort during the wet season being devoted to farming. There is a wide variety of strictly secondary activities within the small survey area studied.

These findings are confirmed in the study of three village areas in the congested districts of Katsina Province where practically all the farmers are occupied in some craft, trade or service occupation during the dry season, while some continue their activities through the wet season (Luning, 1961). Indeed, it is suggested that many adult males have two, three or even more such occupations. Further, the women are also engaged in earning cash especially by spinning and the preparation of foodstuffs and ingredients for sale. Even the children collect grass on the farms and sell it in the villages to the owners of sheep and goats. Most members of the household are therefore occupied in making some money in one way or another.

Both these examples are taken from the Hausa area of Northern Nigeria and lest it be thought that they are atypical another instance will be briefly considered. A wealth of information on family budgets was collected in 1948 among the Nsaw tribe in Kimbaw in the Bamenda Highlands. In 1948 coffee, now the main cash crop, had only recently been introduced into the area. Two features of the budgets are prominent. The first is the pitifully low cash incomes of those households entirely dependent upon agriculture except for the few already growing coffee. The second feature is the much greater cash income of those households engaging in some secondary pursuit be it tailoring, or sewing and selling clothes, or smithing, or building, or part-time employment with the Native Authority (Kaberry, 1952).

The relationship is similarly indirect where paid labour is an important aspect of the rural situation. This is characteristic of the village areas near both Kano and Katsina surveyed in the studies to which reference has already been made. In the Kano study area small-scale labouring on the farms of others during the wet season was fairly general, while a few men derived most of their wealth in this way. The labourers are drawn from those small farmers with insufficient land to employ them and there seems to be no shortage of such labour. This is equally true of the study area in northern Katsina. In Bugasawa Landanawa village 39 per cent of the farmers are also paid labourers. In Birnin Kuka, on the other hand, only 3 per cent of the farmers are also employed as labourers. The proportion of labourers varies from village to village. This variation is related to the incidence of small and large farms and to the importance of craft industries which also vary from village to village.

Subsidiary non-agricultural occupations and paid agricultural labour are two means of overcoming the imbalance between size of farms and avail-ability of labour but they are not the only ones. Migration is important in performing a similar role. Migrations are broadly divisible into long-term and short-term, and each of these can be further divided into movements for paid labour and for farmlands (Prothero, 1964). Short-term migrations for paid labour are probably the best known of the West African migra-tions. These take the form of a southward movement of labour from the northern portions – Mali, Upper Volta, Niger and from the northern parts of Ghana and Nigeria – to the more economically developed parts, usually nearer the coast and to the ports and major towns. The migrants leave after work on the harvest is finished and return with the onset of the rains and the new farming season. Whether migrating in search of money or of trade the economic motive for the migrations is outstanding, while the absence of large numbers of adult males during most of the dry season must help considerably to conserve food supplies in the home areas.

Of equal local significance are the seasonal migrations in West Africa to farmlands. In these cases both the pattern and the season of migration may be substantially different from the labour migrations discussed above. Seasonal migration to farmlands tends to be short distance migration. It is a feature of life in the congested areas of Iboland and Ibibioland and of some areas of medium population density in Eastern Nigeria (Udo, 1964). In the congested areas the population density is between 800 and 1,000 per square mile; in the areas of medium density it is between 600 and 800. The areas to which the migrants go, rarely more than 100 miles from the home area, vary considerably. Some are river floodplains, others are open hill country, but all are alike in having relatively low population densities and therefore reserves of unused farmland and palms. The fuller utilization of these under-used resources is important in providing surplus food to meet the increasing demands of both congested and urban areas and in contri-buting palm oil and palm kernel for export that otherwise would be wasted. Similar although longer-distance seasonal migrations of farmers also take place from the Mali Republic, Portuguese Guinea and Guinea to western Senegal and Gambia where they are both employed in growing groundnuts and also cultivate them for themselves (Jarrett, 1949). They thus increase export production and are responsible for fuller utilization of resources.

In a somewhat different way, permanent and long-term migration also may modify the man–land relationship. The migrant usually maintains close ties with his home area. Commonly he returns there for brief periods to attend ceremonies and festivals while working away, and usually he builds a house in his home village to which he retires. Large houses, clearly not the result of locally earned wealth, are to be found in the most unexpected places in West Africa. But more important than this is the remitting home of money either to keep poorer relatives or to assist in some communal project such as the building of a school. Capital and cash importation can be a valuable economic prop for communities in the less economically advanced areas. In Shaki, for instance, one of the Yoruba towns in the Guinea savanna woodland zone and with, therefore, the problem of finding a viable economic base in modern conditions, £50–£60-worth of postal orders from Ghana were cashed in one week in 1957, and Shaki migrants are to be found equally in the Nigerian cocoa belt to the south of the town (Mabogunje and Oyawoye, 1961).

Many farmers thus obtain additional income from occupations other than the cultivation of their own holding: by some local non-agricultural activity, by becoming hired labourers, or by short- and long-term migration, which renders the man–land relationship indirect. It becomes even more so when part of this cash income is used to make good deficiencies in self- and locally-produced food. This presupposes a market in basic food commodities. There are many studies by anthropologists, economists, geographers and sociologists that testify to the existence of such a market (White, H. P., 1956b; Bohannan and Dalton, 1962, *inter alia*). Suffice it here to point out the dependence of the Yoruba cocoa crescent for food upon the derived and guinea savanna zones to the west and north; of the Ibo concentration upon food brought in from neighbouring areas; of the more densely settled forest areas in Ghana on yams from Ejura and Wenchi; and of the forest areas of Ivory Coast on yams moved in from Bouaké Cercle.

The features just described as making the man–land relationship indirect are all associated with densely populated areas. They may also be found in the more sparsely settled parts of West Africa but not as a series or complex. Indeed, their increasingly wide adoption in an area is the result of population pressure increase.

Conclusion

G. V. Jacks (1956) envisaged three stages in the evolution of the soil under man's stewardship. The first stage is one of ecological balance, where shifting cultivation does not permanently deplete soil fertility. The second is one of soil exhaustion, during which man mines the fertility of the soil without replacing plant nutrients removed by cultivation or by erosion. The third is the conserving stage, when man returns as much or more to the soil than is removed by cultivation, so that it remains productive. These stages seem to go hand in hand with the three phases in the evolution of agricultural practices from shifting cultivation through rotational bush fallowing to semi-permanent and permanent cultivation.

Most of West Africa, especially the Guinea savanna woodland zone, is still either in the first or in the early part of the second phase of this evolutionary process. A large part, particularly where cash cropping is significant, is also in the second phase, but only where there are plantations or very limited areas of intensive cultivation has the final phase been reached.

Further, in most areas where permanent cultivation is already practised agricultural systems are marked by a dichotomy between arable and stock farming. Thus in many areas in West Africa, even in the final stage, the more efficient farming systems are based upon doubtful sources of fertility. A case in point is the permanently cultivated areas around the towns of the Sudan zone such as Kano or Katsina. Permanent cultivation is based here largely on town refuse and night-soil, together with rather spasmodic grazing of crop residues by nomadic cattle stock. With the dwindling supply and increasing cost of firewood, kerosene is likely to replace wood as the domestic fuel, thus reducing the supply of wood ash. Again, the health authorities and public opinion will demand other systems of refuse collection and sewage disposal. In these circumstances permanent cultivation as practised now is likely to be seriously threatened (Pullan, 1966).

The stimulus for advance from one phase to the next comes from population pressure. But to this have been added, since 1920, and to an increasing extent since 1945, the growing land requirements for cash cropping. The most crucial point comes with the end of the second phase, especially if entry to the third is delayed, for then the danger of land deterioration is greatest. It has been suggested that this point is reached in the Sudan zone of north-western Nigeria when densities rise to between 200 and 250 persons per square mile (Prothero, 1962), and in the forest zone of Yorubaland to between 350 and 400 per square mile.

To forestall permanent loss of fertility, there is a strong case for hastening passage into the conservation stage to preserve land that is capable of supporting greater numbers if a more efficient farming system is practised. But over much of West Africa, there is a lack of manpower to support such a system, while paradoxically, in restricted areas of over-population there has been a failure of resettlement schemes to transfer people to underpopulated areas. This failure bears witness to the difficulties involved in the planned redistribution of population.

If the long-term aim of agricultural development in West Africa can best be served by the encouragement of efficient systems based on permanent cultivation, i.e. the hastening of Jacks' conserving stage, then ecologically based techniques of assessing critical population density are open to serious question. Allan (1965) assumes that the aim of agricultural development is to maintain soil fertility by preserving the balance between loss of fertility during short cultivation periods and the fertility replenished under long fallows. His method is, therefore, dedicated to keeping agriculture in the first or early part of the second of Jacks' three stages. In the short term this is a sensible policy, but for the large parts of West Africa it is inappropriate since agricultural systems have already passed this phase. As a medium- and long-term aim it seems to be impracticable even for those areas in West Africa in the earlier stages, for even here population and economic pressures are likely to build up as development proceeds. The identification of the critical period of transition from rotational bush fallowing to permanent cultivation is, in our view, more important in the West African context than diagnosis of critical population density as defined by Allan. The key to advance is the establishment of permanent cultivation by suppressing fallows, by the introduction of green manure crops, the widespread pur- chase of artificial manures and by genuine mixed farming. But this key has remained out of reach so far.

Furthermore, Allan's argument is based on the assumption that agricul- tural systems in tropical Africa are static. Each agricultural system is defined by the land-use factor, the ratio between the length of time that the soil will sustain cultivation and the length of time that the land must remain fallow for the restoration of fertility. Neither of these two com- ponents is static since more demanding crops, in terms of soil fertility, are replaced by less demanding ones as soil fertility falls. Elsewhere agricul- tural systems have been changed by increasing the area under permanent cultivation at the expense of the bush-fallowing land, by the use of manure and compost or by the planting of nitrogenous shrubs in the fallow. African agricultural systems, far from being static both in pre-colonial and colonial days, have changed as occasion demanded. The intensive, advanced systems of certain upland areas and the very adoption of cash crops in many areas are particularly significant in this respect. Therefore, because the man–land relationship is dynamic, studies should be made over time as well as over geographical space. But the number of areas for which this is possible is few indeed because of the lack of historical data.

Finally, there is not a direct relationship between man and the land in the rural areas of West Africa, as Allan implies. We have tried to show this is sometimes the case even in areas of low population density and that it appears to become less direct as population density increases. But always, even in the Kano sample studied by Mortimore and Wilson, where pressure on resources seems to be as great as anywhere, there is a marked reluctance to relinquish rights in the land. This is widely true even in otherwise completely urbanized southern families. Perhaps it is because rights in land are even yet the only universal form of capital. But in any case agriculture remains the basic activity in West Africa. Because population densities are increasing so rapidly, the relationship between this increase and the systems by which agriculture is practised is of fundamental importance to West African studies.

AUTHORS' NOTE

The argument developed in this chapter bears considerable similarity to parts of that in E. Boserup, *The Conditions of Agricultural Growth*, London, 1965. In preparing early drafts of this chapter we were unaware of Boserup's work, and are grateful to A. D. Goddard, of Rural Economy Research Unit, Ahmadu Bello University, Zaria for drawing our attention to it. In view of the fact that we arrived at our ideas independently, it it seemed worthwhile to proceed with publication particularly as we have relied upon different portions of the literature in developing our argument.

References

AHN, P. 1958. The principal areas of remaining original forest in Western Ghana. *Jl. W. Afr. Sci. Ass.* **5**, 91–100.

AKENHEAD, M. 1944. *Agriculture in Western Dagomba* (mimeographed paper), Dept. of Agriculture, Accra.

ALLAN, W. 1965. *The African Husbandman*, Edinburgh.

BALDWIN, K. D. S. 1963. Land-tenure problems in relation to agricultural development in the Northern Region of Nigeria. In BIEBUYCK, D. (ed.), 1963, op. cit.

BAWDEN, M. G. and LANGDALE-BROWN, I. 1961. *An Aerial Photographic Reconnaissance of the Present and Possible Land-Use in the Bamenda Area, Southern Cameroons* (mimeographed paper), Department of Technical Co-operation, Directorate of Overseas Survey, Forestry and Land Use Section, Tolworth.

BIEBUYCK, D. (ed.), 1963. *African Agrarian Systems*, London.

BOHANNAN, P. and DALTON, G. 1962. *Markets in Africa*, Evanston.

BRASSEUR-MARION, P. and BRASSEUR, G. 1953. *Porto-Novo et sa Palmeraie*, Dakar.

BUCHANAN, K. M. 1952. Nigeria – largest remaining British colony. *Econ. Geogr.* **28**, 452–73.

BUCHANAN, K. M. and PUGH, J. C. 1955. *Land and People in Nigeria*, London.

COPPOCK, J. T. 1966. Agricultural developments in Nigeria. *J. trop. Geogr.* **23**, 1–18.

DAVIES, J. G. 1946. The Gyel farm survey in Jos Division. *Fm Forest* **7**, 110–13.

ENJALBERT, H. 1956. Paysans noirs; le Kabré du Nord-Togo. *Cah. d'outre-mer* **9**, 137–80.

FAO. 1962a. *Africa Survey: Report on the Possibilities of African Rural Development in Relation to Economic and Social Growth*, Rome.

1962b. *Report on the Land Use Survey of the Oyo-Shaki Area, Western Nigeria.* (mimeographed paper), Ibadan.

FIELD, M. J. 1943. The agricultural system of the Manyo-Krobo of the Gold Coast. *Africa* **14**, 54–65.

FRÉCHOU, H. 1955. Les plantations européennes en Côte d'Ivoire. *Cah. d'outre-mer* **8**, 56–83.

FROELICH, J. C. 1949. Generalités sur les Kabré du Nord-Togo. *Bull. Inst. Fran. Afr. Noire* **11**, 77–105.

1952. Densité de la Population et Méthodes de culture chez les Kabré du Nord Togo. *Compte Rend. Cong. Int. de Géogr., Lisbonne* 1949, Tome 4, Section 5, 168–80.

GALLETI, R., BALDWIN, K. D. S. and DINA, I. 1956. *Nigerian Cocoa Farmers: An Economic Survey of Yoruba Cocoa Farming Families,* London.

GHANA. 1959. *East Dagomba Agricultural and Livestock Survey, Northern Ghana* (mimeographed paper), Ministry of Food and Agriculture, Kumasi.

GLEAVE, M. B. 1965. The changing frontiers of settlement in the uplands of Northern Nigeria. *Niger. Geogr. J.* **8**, 127–41.

1966. Hill settlements and their abandonment in tropical Africa. *Trans. Inst. Br. Geogr.* **40**, 39–49.

GLEAVE, M. B. and WHITE, H. P. 1967. The West African Middle Belt: environmental fact or geographers' fiction? Unpublished paper read to Inst. Br. Geogr., Sheffield, 1967.

GLEAVE, M. B. and THOMAS, M. F. 1968. The Bagango Valley: an example of land utilisation and agricultural practice in the Bamenda Highlands. *Bull. Inst. fond. Afr. noire* Sér. B., xxx, 655–81.

GOLD COAST, 1953. Agricultural statistical survey of South-East Akim Abuakwa, 1952–3. *Stat. Econ. Pap.* No. 1, Government Statistician.

GROVE, A. T. 1951a. Soil erosion and population problems in South-East Nigeria. *Geogrl J.* **117**, 291–306.

1951b. *Land Use and Soil Conservation in Parts of Onitsha and Owerri Provinces,* Bull. geol. survey Niger. **21**.

1952. *Land Use and Soil Conservation on the Jos Plateau,* Bull. geol. survey Niger. **22**.

1961. Population and agriculture in Northern Nigeria. In BARBOUR, K. M. and PROTHERO, R. M. (eds.), *Essays on African Population,* London.

HARRISON CHURCH, R. J. 1961. *West Africa: Study of the Environment and Man's Use of it,* London.

HIERNAUX, C. 1948. Les aspects géographiques de la production bananière de la Côte d'Ivoire. *Cah. d'outre-mer* **1**, 68–84.

HILL, P. 1957. *The Gold Coast Cocoa Farmer: A Preliminary Survey,* London.

1963. Three types of Southern Ghanaian cocoa farmer. In BIEBUYCK, D. (ed.), 1963, op. cit.

HILTON, T. E. 1959. Land planning and resettlement in Northern Ghana. *Geography* **44**, 227–40.

HUNTER, J. M. 1961. Akotuakrom: a devastated cocoa village in Ghana. *Trans. Inst. Br. Geogr.* **29**, 161–86.

1963. Cocoa migration and patterns of land ownership in the Densu Valley near Suhum, Ghana. *Trans. Inst. Br. Geogr.* **33**, 61–87.

1965. Regional patterns of population growth in Ghana, 1948–60. In WHITTOW, J. B. and WOOD, P. D., *Essays in Geography for Austin Miller,* Reading.

INTERNATIONAL BANK, 1955. *The Economic Development of Nigeria,* Baltimore.

IRVINE, F. 1953. *A Text-Book of West African Agriculture, Soils and Crops,* London.

JACKS, G. V. 1956. The influence of man on soil fertility. *Advmt Sci.* **13,** 137–45.

JARRETT, H. R. 1949. The strange farmers of the Gambia. *Geogrl Rev.* **39,** 649–57.

JOHNSTON, B. F. 1958. *The Staple Food Economies of Western Tropical Africa,* Stanford.

KABERRY, P.M. 1952. *Women of the Grassfields,* Colonial Research Publications No. 14, London.

LABOURET, H. 1941. *Paysans d'Afrique occidentale,* Paris.

LUNING, H. A. 1961. *An Agro-Economic Survey in Katsina Province,* Kaduna.

LYNN, C. W. 1937. *Agriculture in North Mamprusi,* Department of Agric. Bull. 34, Accra.

1946. Land and planning in the Northern Territories of the Gold Coast. *Fm Forest* **7,** 81–3.

MABOGUNJE, A. L. and OYAWOYE, M. O. 1961. The problems of the Northern Yoruba towns: the example of Shaki. *Niger. geogr. J.* **4,** 2–10.

MABOGUNJE, A. L. and GLEAVE, M. B. 1964. Changing landscape in Southern Nigeria: the example of Egba Division, 1850–1950. *Niger. geogr. J.* **7,** 1–15.

MANSHARD, W. 1961. Land use patterns and agricultural migration in central Ghana. *Tijdschr. econ. soc. Geogr.* **52,** 225–30.

MIÈGE, J. 1954. Les cultures vivrières en Afrique occidentale: essai de leur repartition géographique, particulièrement en Côte d'Ivoire. *Cah. d'outre-mer* **7,** 25–50.

MORGAN, W. B. 1955a. Farming practice, settlement patterns and population density in South-Eastern Nigeria. *Geogrl J.* **121,** 320–33.

1955b. The strip fields of Southern Nigeria. In STAMP, L. D. (ed.), *Natural Resources, Food and Population in Inter-Tropical Africa,* Bude.

1957. Some comments on shifting agriculture in Africa. *Research Notes,* Dept. of Geog., Univ. of Ibadan, No. 9, 1–10.

MORTIMORE, M. J. 1966. Population distribution, settlement and soils in Kano Province, Northern Nigeria, 1931–62. Unpublished paper read at First African Population Conference, University of Ibadan.

MORTIMORE, M. J. and WILSON, J. 1965. *Land and People in the Kano Close-Settled Zone,* Ahmadu Bello University, Department of Geography, Occasional Paper No. 1.

NADEL, S. F. 1942. *A Black Byzantium: the Kingdom of Nupe in Nigeria,* London.

NETTING, R. M. 1965. Household organisation and intensive agriculture: the Kofyar case. *Africa* **35,** 422–8.

NYE, P. H. and GREENLAND, D. J. 1960. *The Soil under Shifting Cultivation.* Commonwealth Bureau of Soils, Tech. Comm. 51, Farnham Royal.

OKONJO, C. 1966. A preliminary medium estimate of the 1962 mid-year population of Nigeria. Unpublished paper read to First African Population Conference, University of Ibadan.

POGUCKI, R. J. H. 1950. Report on land tenure in native customary law of the Protectorate of the Northern Territories of the Gold Coast. *Land Tenure in Gold Coast* 1.

1952. Report on land tenure in customary law of the non-Akan areas of the Gold Coast Colony, Pt. 1: Adangbe. *Land Tenure in Gold Coast* 2.

1954. Report on land tenure in customary law of the non-Akan areas of the Gold Coast Colony, Pt. 2: Ga. *Land Tenure in Gold Coast* 3 and 4.

1956. A handbook of main principles of rural land tenure in the Gold Coast. *Land Tenure in Gold Coast* 5.

PROTHERO, R. M. 1957a. Land use at Soba, Zaria Province, Northern Nigeria. *Econ. Geogr.* 33, 72–86.

1957b. Land use, land holdings and land tenure at Soba, Zaria Province, Northern Nigeria. *Bull. Inst. fr. Afr. noire* 19, Sér. B., 558–63.

1962. Some observations on dessication in North-Western Nigeria. *Erdkunde* 16, 111–19.

1964. Continuity and change in African population mobility. In STEEL, R. W. and PROTHERO, R. M. (eds.), *Geographers and the Tropics*, London.

PULLAN, R. A. 1966. Tropical soils – are they a limiting resource in the development of agricultural production in tropical West Africa? Paper read to Section E, British Association, Nottingham.

SMITH, R. 1941. *Agriculture in North Mamprusi* (mimeographed paper), Department of Agriculture, Accra.

1945. *Agriculture in South Mamprusi* (mimeographed paper), Department of Agriculture, Accra.

TRICART, J. 1957. Le café en Côte d'Ivoire. *Cah. d'outre-mer* 10, 209–33.

UDO, R. K. 1964. The migrant tenant farmer of Eastern Nigeria. *Africa* 34, 326–38.

1965a. Problems of developing the Cross River District of Eastern Nigeria. *J. trop. Geogr.* 20, 65–72.

1965b. Disintegration of nucleated settlement in Eastern Nigeria. *Geogrl Rev.* 55, 53–67.

VINE, H. 1954. Is the lack of fertility of tropical African Soils exaggerated? *Second Int. Af. Soils Conference, Kinshasha (Leopoldville)*.

WHITE, H. P. 1956a. Avatime: a highland environment in Togoland. *Malay. J. trop. Geogr.* 8, 32–9.

1956b. Internal exchange of staple foods in the Gold Coast. *Econ. Geogr.* 32, 115–25.

1958. Provisional agricultural regions of Ghana. *J. trop. Geogr.* 11, 90–99.

1959. Mechanised cultivation of peasant holdings in West Africa. *Geography* 45, 269–70.

1965. Terre de Barre – the basis of a West African agricultural region. *Bull. Inst. fond. Afr. noire* 27, Sér. B., 169–82.

1966. Dahomey – the geographical basis of an African state. *Tijdschr. econ. soc. Geogr.* 57, 61–67.

WHITE, S. 1944. Agricultural economy of the hill pagans of Dikwa Emirate. *Fm Forest* 5, 130–4.

WILLS, J. B. 1962. The general pattern of land use. In WILLS, J. B. (ed.), *Agriculture and Land Use in Ghana*, London.

11 The zoning of land use around rural settlements in tropical Africa

W. B. MORGAN

Given certain favourable circumstances, notably high cost of movement or ransport of goods, a tendency for patterns of land use to be arranged in rings around rural settlements may arise from the basic economic principles governing the location of agricultural activities (Chisholm, 1962). This is particularly the case among peasant communities, for, although wages may tend to be low by world standards, the relative costs of movement between the farmer's home and his fields may be high compared with his other costs, or estimated effort, involved in production. This arises from the fact that most such journeys are performed on foot, or at best nowadays on a bicycle. Implements, seed and manures or composts must be carried out to the fields. Where rural settlements are nucleated and large, say over 5,000 people, distances of 5 miles or more to the farthest fields have to be regularly traversed by some of the farmers, whether holdings are fragmented or not. Where fragmentation occurs, as is frequently the case, similar distances may have to be walked, even where the farms are evenly dispersed. The limit to such regular journeys from the home is set by the farmer's estimate of the amount of time he feels he can afford to lose in journeying during a day, against the return he obtains by such journeying. Where the rewards are high, the effort may be correspondingly great. Beyond the limiting distance, commonly of the order of 4–7 miles, the farmer may find it worth while to cultivate providing he sets up a subsidiary or seasonal homestead; in Africa sometimes only a temporary shelter. Here he may reside for brief periods of concentrated effort, and his choice of enterprises in this temporary location will be, in part at least, governed by the amount of time and its seasonal incidence thus available to him.

Because the farther fields cost more to reach in terms of effort or wages, it follows that unless they have other special advantages of soil, moisture or site, the costs or effort involved in production there will be correspondingly higher, and the returns for the total of the farmer's labour will be thereby reduced. The farmer thus finds it more difficult to achieve on the farther fields that level of intensity of effort which it is possible to achieve on the nearer fields. The factors governing the competition between the various enterprises available for the use of land thus differ according to location in relation to the settlement. The greater the effort involved in travel, or in transport (particularly of bulky goods of low profitability per unit weight, such as manures) the more marked this tendency must become. The relationship holds not only for the effort involved in actually producing the crop but also for the effort involved in gathering it and bringing it home. However, the relative importance of the different factors varies greatly between the effort involved in production and that involved in harvesting. In theory one might have supposed that highly profitable crops of low bulk were preferred on the more distant fields (Lösch, 1954).[1] In fact this is only true of certain cases. By and large, where zoning occurs in tropical Africa, the crops more distant from the homestead or compound are those which can most successfully be produced by extensive methods, even though their value, and their profitability, may not be high. However, where frequent access to a crop is needed, for culinary purposes for example, or where storage is best achieved by leaving the crop in the ground and gathering from it as needs arise, then location near the home must be preferred.

The model of location one may erect from the premises described is similar to that described by von Thünen (1826), and the zones have in certain instances been described as 'von Thünen rings'. However, this notation needs qualification. Although von Thünen did pay particular attention to the zoning of land use within the farm or estate, and although one of the features of particular importance to him in the relation of land-use zones to towns was the urban supply of manure, there are in fact some important differences between the model for the 'isolated state' and the model for the farm or rural settlement in tropical Africa. The 'isolated state' provides a market model, but the farmhouse or the farm village is the centre both of inputs and of consumption or disposal of produce. In many instances the input factors appear to be more significant in the creation of zones than the factors of consumption or disposal. This would appear likely in peasant agriculture when one considers that for most crops the effort involved in cultivation is very much greater than the effort involved in harvesting and carrying the produce to the storage bins in the farmhouse. It is not a market, but a production area with which, for the most part, we are concerned. The relative costs of production therefore are the chief governing factors, particularly where marked differences in the labour requirements of certain crops occur, or where important factors in production, such as a supply of manure, must be derived from a given centre. The

---

[1] It is the profit on a commodity, not its value, which pays the cost of transport. Ideally the choice of crop in any given location will be that which in combination with other crops chosen will yield the maximum profit or rent for the farm as a whole.

manure factor is particularly important, and frequently gives rise to a zone of higher fertility immediately around the settlement. In this zone high labour inputs realize greater returns than equivalent inputs on more distant unmanured lands. Apart, therefore, from the question of the distance factor, the disposal of refuse and night-soil immediately round the compound, together with the keeping of small livestock, particularly goats and fowls, close to the dwellings, creates an inner zone of high fertility, where permanent cultivation or very long rotation cultivation is possible. This, the compound, garden or kitchen-garden land, is of widespread occurrence throughout tropical Africa. The associated soils are not only richer in nutrients, but frequently retain moisture for longer periods, or even receive more moisture from the roofs of the dwellings. Crops needing richer soils, or longer periods to mature than are possible on the main cropland, find a more suitable environment here, sometimes in association with tree cultivation. A double land economy of this kind, compound land and main crop land, is a common feature not only in West Africa but also throughout Central Africa, particularly in the Congo Basin (Miracle, 1964). Such crops commonly include tobacco, sugar cane, bananas, various vegetable and leaf plants, peppers, pineapple and papaya. Some long-maturing varieties of plants otherwise regarded as main crop are also frequently found on compound land, particularly maize and, where the shade of trees is available, cocoyams. Similarly, crops which do badly on heavily manured soils by, for example, developing too much leaf or stem, find a more suitable environment in the outer zone.

Examples

One of the clearest examples of concentric zones or rings of cultivation is that provided by Prothero's studies of Soba village in the province of Zaria in Northern Nigeria (Prothero, 1953, 1957a, 1957b). Fieldwork was effected in the latter part, that is the early dry season, of 1952, using air photographs of February 1951. Prothero recognized four major zones of cultivation (Fig. 11.1):

  1. Land within the village walls, manured and permanently cultivated.
  2. Land outside the village walls, extending a $\frac{1}{2}$–$\frac{3}{4}$ mile, also manured and permanently cultivated.
  3. Land extending another $\frac{1}{2}$–$1\frac{1}{2}$ miles in radius from the village, cultivated without manure by rotational bush-fallow methods.
  4. An outer ring, extending another $1\frac{1}{2}$ miles, except in the south where its width was everywhere less than $\frac{1}{2}$ mile. This was the zone of shifting cultivation where the area of fallow grass was very much greater than the area under crops. In addition low-lying *fadama* or floodland

was also cultivated, on plots usually rented. Land within the village walls
was subdivided into:

(*a*) Land within the compound walls, containing various kitchen
garden crops, chiefly tomatoes, calabashes, papaya, loofah, rama, okra
and a little guinea corn.

(*b*) Land between the compounds, chiefly for tobacco, but also
planted to okra and guinea corn.

Category 2 land mainly produced guinea corn and cotton, with some
tobacco and groundnuts.

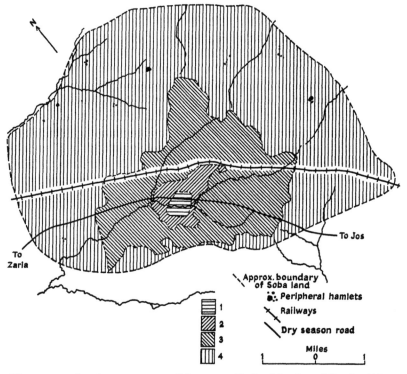

Figure 11.1 *Land use zones at Soba, near Zaria, Northern Nigeria* (After
Prothero, 1957a).

Category 3 and 4 land also produced guinea corn, cotton and groundnuts, so that apart from its lack of tobacco it differed hardly at all in choice
of crops from category 2. In certain locations it would have been difficult
to distinguish the boundaries between these categories of land use, particularly between 3 and 4. Farming practice varied considerably between one
farmer and another and there were normally many marginal cases, difficult

to distinguish in any system of classification. A further complication in the case of Soba was that while it was a common practice for individual farmers to have a 'farm' in each of the categories listed, not all farmers had their holdings divided in this way. Some farmers held only one farm. Other farmers cultivated, not from the village itself but from hamlets located towards the periphery of Soba's lands. The relationship of distance to the problems of manuring, choice of crop and choice of agricultural activity may thus have differed considerably from one farmer to another.

In northern Ghana, in Kusasi District, Manshard (1961b) distinguished land-use zones around a single farm:

1. Land immediately around the compound, well manured and with a radius of some 6–12 m.

2. Land only partly manured, extending a further 18–36 m.

3. Land near the farm, but not manured, extending an additional radius of some 10–30 m.

4. An outer zone of wooded savanna with scattered 'bush farms'.

Category 1 produced a great variety of vegetable crops, together with sweet potato, cotton, tobacco, maize and a fringe of sesame. Categories 2 and 3 were mainly bulrush millet, sorghum, melon and cowpea. Elsewhere in northern Ghana concentric ring systems of cultivation have frequently been distinguished (Wills, 1962), especially in what have been called the 'compound farming areas' of the extreme north-west and north-east. Allan (1965), also writing of northern Ghana, described three concentric zones of homestead garden: the kitchen garden, regularly manured; the compound garden, also regularly manured; and the semi-compound garden, with only occasional manure. The kitchen garden produced a variety of vegetable 'industrial' crops, the compound garden a mixture of sorghum and early millet, followed by early and late millet, and the semi-compound garden an alternation of sorghum and late millet with some cowpeas and groundnuts. Beyond was the 'bush' garden, never manured, formerly enjoying long fallows, but today cultivated by short fallow cycles, and producing sorghum, bulrush millet, groundnuts and pulses. Allan compared these zones with those distinguished by Dumont (1957) for a cereal-growing village near the River Shari in Chad, showing their striking resemblance. A similar pattern has been noted among the Bobo of Upper Volta with their manured compound land permanently cultivated and producing maize, quick-growing bulrush millet, guinea corn and tobacco; their more distant manured land also permanently cultivated, receiving less manure, associated with *Acacia albida* trees and producing sorghum, cotton, maize and groundnuts, with rice and yams on the moister soils; and finally, the farmland proper with no manure, dependent on fallows, and producing mainly sorghum and maize (Savonnet, 1956, 1959). This was called an 'evolved' system by Savonnet, who compared it with the simpler or 'semi-evolved' system of the neighbouring Dagari, with only one category of manured land, and the 'archaic' system of the Lobi and Koulango with no manured land, although he noted the cultivation of tobacco around the dwellings. The Bobo kept livestock in some numbers, and the extent of their manured lands depended on the numbers of sheep, goats and cattle, for which the *Acacia albida* trees provided some dry-season fodder. Fences or permanent hedges were, however, essential to protect the crops from animals living in or about the compounds. Further examples have been summarized for West Africa by Sautter (1962), who has compared them with examples of infield–outfield cultivation in Western Europe, has tried to establish causes, and has endeavoured in certain cases to examine something of their evolution.

All the examples so far quoted occur in the savannas. In the rainforest such concentric zoning is rarely indicated except around very large towns, such as Ibadan, and then as a zoning of commercial crops in relation to market. In Ghana, Fortes, Steel and Ady (1947) noted the concentration of food farming around Kumasi with a zone mainly of cocoa beyond. Generally, rain forest villages do not have livestock in such large numbers as do villages in the savanna lands. Moreover, the tasks of clearing land and of weeding are so onerous that the extensive methods characteristic of the peripheral lands of some savanna villages could hardly pay. Even where fallows are fairly long, say seven years or more in West Africa, the cultivation practised depends on high yields per acre when compared with savanna yields. Holdings may not therefore need to be as large, and the distance problem may be diminished. However, the use of compound land, fertilized by household refuse, producing kitchen-garden crops and associated with a tree cultivation, chiefly of oil-palms, has been noted among the Ibo, both in the forest and in the savanna (Bridges, 1938; Grove, 1951; Morgan, 1955). Outside this *ani uno*, town land or compound land, is a zone of nearer land, the *aguolie* into which settlements tend to expand as they grow, and, beyond this, the *ani agu* or field land, part of which in northern Iboland was formerly reserved for thatching grass. Subdivisions of *ani uno* are recognized. In Awgu, for example, there are two such subdivisions approximately in concentric zones around the settlement, including *okpuno* or town land proper and *isiagu* or the land nearest the town land. These distinctions, however, are of only small importance in relation to field crops, and are more significant in relation to settlement siting and rights in tree crops.

Some problems of classification and distinction

In the study of actual cases, the distinctions between different kinds of land and their locations are not always as clear as some of the examples quoted would indicate. Local names for near, middle distance and farther fields may become elevated to the status of precise terms, and general notions of some very approximate crop relationship become identified with a rigid spatial system. The greater the distances involved, the more clearly should patterns be revealed. However, the greater distances occur around the larger settlements, and the larger the settlements, the greater the number of farmers, and the greater the possibility of variation in agricultural practice on the different holdings. The relationship of distance with manure is not too difficult to assess, but the relationship with crops is, in most cases, not at all easy. Frequently local soil variations, for example, are a more potent factor. The author made three traverses in August 1966, each of nearly 2 miles in length, from the edge of the town of Eruwa in the savanna of Western Nigeria, across its farmlands, which lie in a fairly compact circle round it. Each field, on each of the traverses, was recorded, its contents listed, and its mean distance from the town estimated. Only three crops occurred sufficiently often to make statistical analysis worth while, and for each a rank graph to demonstrate the effect of distance has been prepared (Fig. 11.2); the farther apart the consecutive mean distances recorded, the steeper the gradient of the graph. Cassava clearly has a fairly uniform distribution throughout the cultivated area examined. At over 3,100 yds from the town the first three cassava fields have big gaps between them, and in fact occur in an area mainly of grass fallows. Other smaller gaps correspond with patches of streamside forest crossed by the traverses. The graph of maize shows larger breaks, with a more pronounced tendency to nearly horizontal grouping at 2,000–2,500 yds, corresponding with an area of suitable light loams, less satisfactory for alternative crops such as yams. Yams generally require higher labour inputs than either cassava or maize, and more care in cultivation. One might therefore expect to see them concentrated much more on the nearer lands. To some extent this appears to be the case. There are few yam fields at beyond 1,800 yds, and those that do occur at the farthest locations are on fairly heavy and moister soils where yam cultivation is more worth while. The yam occurrence at just over 2,000 yds is again on moister soil, but in valley-bottom land. There is no use of manure or compost at Eruwa. Few livestock are kept and most farmers would not think the effort of collecting their dung worth while. None of the minor crops showed any pronounced tendency to be located near the settlement. Most oil-palms, bananas, plantains and cocoa occurred near the town, but so did the most suitable, moisture-retaining soils.

Tobacco was an ordinary field crop, grown fairly extensively on a commercial basis, and not restricted, as in previous examples, to compound land. Field sizes showed something of the distance factor, perhaps reflecting a greater demand for nearer land. Thus on one traverse field sizes sampled at a $\frac{1}{4}$–$\frac{1}{2}$ mile from the town averaged less than a $\frac{1}{2}$ acre, while at $1\frac{3}{4}$–2 miles the

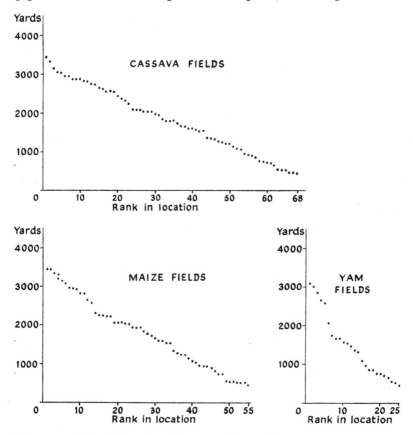

Figure 11.2 *The relation of crop location to distance from a town – traverses at Eruwa, Western Nigeria.* Fields ranked in order of occurrence from Eruwa Post Office on three selected routes. Fields consist of all occurrences of a crop, whether sole or mixed. All distances in yards from Eruwa Post Office, August 1966.

sample averaged $1\frac{3}{4}$ acres. Welldon (1957) noted around the nearby town of Ilora a distinction between the *oke etile* or nearer cropland and the *oke egan* or farther cropland. The air photograph of Ilora (Plate 1) shows near the town and at its southern end a narrow zone of very small fields, each less than $\frac{1}{2}$ acre, with larger fields farther out. In Dahomey Hasle (1965)

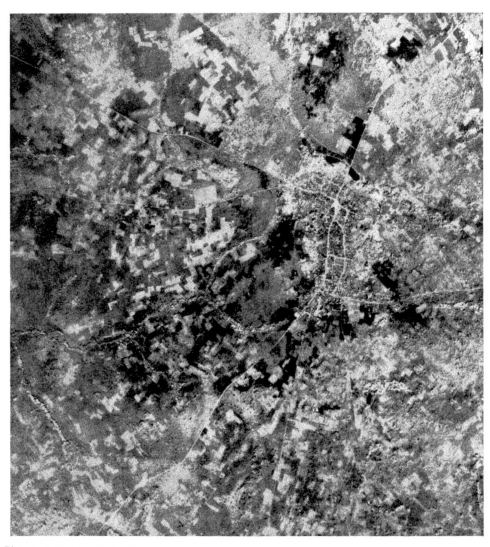

Plate I LAND USE PATTERN AROUND ILORA TOWNSHIP, WESTERN NIGERIA
The town is surrounded by grass, burned areas and very small fields, especially to the south (reproduced with the permission of the Director of Federal Surveys, Nigeria).

recorded the occurrence of small fields near Somba villages, and larger fields at beyond a 10 km radius.

A further complication is in certain instances provided by the livestock, which are normally the source of the manure necessary for intensive cultivation on compound land. These are mostly goats and poultry, but can include cattle, sheep, pigs and domestic animals. Where the numbers of livestock are sufficiently large household scraps and refuse are inadequate to feed them, and the innermost zone of heavily manured cropland is replaced by a zone of grass. This is frequently the case in villages in Gambia, and is clearly shown on the land-use maps of Gambia (Gambia, 1959) which provide the most detailed survey of land use available for any country in West Africa, and indicate the existence of zonal rings around the villages. The cost of fencing and hedging may be high, and cultivation near the village may in consequence be impracticable. This was the case with several farm villages in the Eruwa district of Western Nigeria, where there was no systematic collection of manure and no inner zone of high intensity of cultivation. In consequence, no attempt to protect fields from livestock was thought worth while, and the innermost zone consisted either of forest or woodland.

Distribution

The lack of suitable examples from East and Central Africa of the zoning of cultivation with distance from the compound or village is at first surprising. Almost all the examples are from the West African savannas, and appear best developed around long-established settlements. It is tempting to correlate this with the apparently better-developed traditional overland trade of West Africa and a tendency to greater population growth, together with the development of larger and more stable settlement units. It would, however, be too simple a solution and would accord badly with much of the so far limited evidence available. Certainly the lower population densities of East and Central Africa, the common occurrence of settlement dispersal, and the general preference for shifting cultivation, with fallows long enough to permit the regeneration of woodland, are important factors. However, nucleated settlements of some size and stability do exist in East and Central Africa, but in only a few cases are they associated with rotational bush fallow. Mostly, especially in the case of the smaller villages, they are associated with a very limited zone of permanent manured cultivation and a very broad zone of shifting cultivation. Even within West Africa concentric rings are frequently absent, even around well-developed villages, and in certain cases entirely different solutions to the problem of developing an efficient land-use system have been preferred, as for example the strip or *huza* system of the Krobo (Field, 1943; Manshard, 1957, 1961a). In

L

Central Africa increased demand upon the land appears to have been met in several cases, not by zones but by the *citemene* system, that is by increasing the over-all supply of ash to the soil, by importing wood to the cleared area from the surrounding district (Richards, 1958). The Lamba and Lala cut trees right down for this purpose, but the Bemba employ the more sophisticated method of pollarding, which permits faster timber regrowth. The Bemba also rotate crops, chiefly finger millet, groundnuts and beans, and use less land than the Lamba and Lala for equivalent production. Near the village they have a mound cultivation for the intensive garden production of maize, sorghum, cassava and pulses.

Such mound cultivation employs a great variety of crops, very much greater than in the main crop outer fields. It is in such cultivation that crop mixture technique has been most fully perfected, in order partly to raise the total output on the small area of manure-rich land adjacent to a settlement, and in order to suppress weed growth, thereby reducing labour needs on land otherwise heavy in its labour demand. Such mixture is facilitated by the fact that so many of the crops concerned are peppers and herbs, each satisfying only a small requirement. The more remote communities, thrown back almost entirely on their own productivity to satisfy their requirements, need to grow as great a variety of crops as possible, and in some cases depend greatly on the production of their compound land or kitchen gardens. Thus the Azande on the northern margin of the Congo Basin have courtyard gardens supporting okra, sweet potatoes, maize, melons, tomatoes, roselle, aubergine, peppers, mangoes, papaya, groundnuts, bananas and yams (de Schlippe, 1956). Elsewhere in Central Africa the Alur, Bututsi, Beni-Lubero, Bangu, Songe, Unga and Tabura all have village or compound gardens, supporting an equally great variety of small crops, with occasional staples such as bananas, and occasional commercial crops such as tobacco (Miracle, 1964).

Advantages of multiple land-use systems

Despite the effectiveness of *citemene*, *huza* and other systems of cultivation, it is clear that multiple land-use systems, not necessarily in concentric rings, hold a very large number of advantages for peasant agriculture with limited labour and nutrient resources. For their successful operation they need a great variety of crops, differing from one another in their labour requirements and incidence for tillage, sowing, weeding and harvesting, in their tolerance for pests, diseases and weeds, in their rates of growth, and in their tolerance of soil moisture and nutrient status. Apart from the kind and size of market open to the African peasant farmer, which is perhaps the chief problem limiting the expansion of his productivity, his most fundamental problems are labour inputs, and the maintenance of nutrient levels in the soil. Although a peasant farmer works an average of only a few hours each day, in many cases only four (Clark and Haswell, 1964), frequently he does so because of a bottleneck in the labour requirements of his agricultural operations. Any form of crop specialization makes temporary labour essential in order to keep permanent labour fully employed. This is true even of a crop such as cassava, the cultivation of which has a relatively low labour requirement, when it is grown on any large scale for sale in urban markets. The alternative is crop and land-use variety. Clearing forest, woodland or tall secondary bush is a particularly demanding operation. Phillips (1964) has estimated for southern Nigeria a normal requirement (without stumping) of 39–55 man days, each of eight hours, per acre. As the clearance of low bush regrowth takes less than half that time, of planted fallows (tephrosia, pigeon pea, mucuna) only 8–15 man days, and of low bush and grass only 2–4 man days, rotational bush fallow methods are much less demanding in clearance labour than shifting cultivation. The possibility of lower yields on recleared land than on newly cleared land is offset by the reduction in labour time involved. Thus a rotation of six years only requires clearance of one-sixth of the cropped area each year, whereas a regular abandonment of cultivated land every one or two years needs the regular clearance of the whole or half of the cropped area. Some crops do much better on freshly cleared land, others do better in some later shift in a rotation. The combination of different cycles of cultivation, made possible by employing different zones of land use, gives a variety of clearance demands, spreading the seasonal use of labour, and enables the peasant to produce many different crops in a single season. A common simple combination is of permanent cropland near the compound with more distant shifting cultivation. In the former there is little dry season labour demand, but a considerable rainy season demand for weeding (4–10 man days per acre in southern Nigeria according to Phillips) and for manuring (about 10 man days per acre for digging, carrying and spreading 3 tons). The cultivation and planting techniques are all more elaborate, more careful and more time-consuming. In the shifting cultivation zone the maximum labour demand is in the dry season for clearing and burning (often a complementary activity to hunting); tillage techniques are normally cruder, and little or no weeding is done. The crops grown need to be fairly tolerant of weed infestation and, where temporary residence in the fields is impossible, to have some resistance to attack by pests, particularly by birds.

Concentric land-use zones due to innovation

One might explain the different zones in terms of the creation of a complex to satisfy the various demands of labour and of cropping in relation to distance. One might view the zones as evidence of an agricultural system in a state of near equilibrium. For some observers, however, the zones are evidence not of equilibrium but of instability due to changes in economic pressure on agricultural techniques and land resources. For example, Boserup (1965) criticizes the 'static geographic theory of land use', arguing that population increase should produce changes in agricultural technique in order to raise productivity. Thus an agricultural community might go through several stages in the transformation of its techniques and its land use, none of which need necessarily be complete for the entire area occupied, and all of which could therefore co-exist in a single system of operations.

> At every stage of this process of transformation the fields first chosen for more frequent cropping would be among those relatively well suited for the next step in the development towards more intensive patterns of land use. This would be flat fields at the stage when plough cultivation was to be introduced, it would be land near rivers or wells when annual cropping with irrigation was to appear, and at the point of transition to intensive manuring, fields close to the centres of settlement would be chosen for the conversion (Boserup, 1965).

One may, therefore, create a model of zones of ever greater intensity, developing from the centre of farming operations and moving outwards as demand increases. Such a model assists our understanding. Nevertheless, an equilibrium situation can exist, where population growth is nil or where population densities remain nearly constant, due to the export of the surplus. Again, innovation can take place for a variety of other reasons, and does not necessarily, even apart from local ecological considerations, have to begin at the centre.

Something of the effects of population pressure on land use patterns has been shown in southern Iboland, where the more densely populated districts have a higher proportion of permanently cropped compound land and of oil-palm groves (Morgan, 1955). Permanently cropped land is virtually owned land. With increasing population pressure, demand for land increases and the need, where possible, to establish ownership rather than use-right arises. Raulin (1963) has demonstrated similar features at Madarounfa, a Hausa village in Niger. Around the village for some 4–5 km radius there is no vacant land. The 'free' lands form an outer zone where people with insufficient land move to create new farm hamlets. In the inner zone, as the demand for land has increased, the patrimonies of the extended families have become divided into fragmented holdings by households and by individuals. Fields are kept in cultivation for ten to fifteen years, and rested in fallow for only two or three years. Apart from the direct effect of pressure on the fallow, there is an indirect effect through fear of encroachment. People prefer to keep land in cultivation for as long as possible in order to maintain their claim on it. In the hill-lands of both West and East Africa, or on island sites such as Ukara in Lake Victoria, the defensive advantages attracted small groups of virtually refugee cultivators whose increasing numbers were for long strictly confined to their hill-lands. In such situations very high densities of population have been achieved, and these have been supported by some of the most intensive forms of agriculture found in tropical Africa. In more peaceful times hill cultivators could farm on the neighbouring plains, their outer zone, but under attack they were forced to retreat to their heavily manured, often terraced, uplands (Gleave, 1965; Allan, 1965). Wilmet (1963), writing of agriculture in Central Africa, has claimed that the garden of permanent cultivation marks an evolutionary stage in the development of a cultivation system. Its absence often coincides with economies in which gathering plays an important role. Presumably, if clearance should become very extensive and gathering decline in importance, more attention to garden cultivation could be expected.

However, innovation may come to the outermost lands first. Social and political stability have enabled many hill peoples to establish new farms in the plains. In many cases this is not just a relief from pressure on land resources but the creation of a superior economy in terms of its returns for labour expended, compared with the old intensive economy. The extensive methods of shifting cultivation have frequently proved more productive for farmers limited to hand tillage. At the same time they have made possible an extensive commercial cultivation, encouraged by the creation of new markets and roads. Thus the Kofyar of Northern Nigeria, described by Netting (1965), have developed a commercial cultivation of yams, sorghum, bulrush millet and cowpeas on the plains, supplemented by a traditional permanent garden in the hills, producing staples, and supporting tree crops and livestock. The heavy labour demands of shifting cultivation for clearance and tillage, however, have encouraged the development of group labour in large work parties, and the hiring of labour from elsewhere.

Grove (1961) has suggested that concentric land-use zones are likelier to occur in areas of rather low density of population than in high-density areas. Where there is considerable population pressure on the land, differences in soil quality should become more and more significant in the land-use patterns. Thus in Northern Nigeria concentric rings occur in areas with less than 150–200 people per square mile, and above that density the rings meet, and the land-use systems become much more soil selective.

A similar feature may be seen in Gambia where villages located on the most favourable terrace lands are so close to one another that the cropping zone rings around the villages, while still recognizable, assume a more linear, almost elliptical form, and some of the features exhibit a close relation to soils, drainage and slope (Gambia, 1959; Haswell, 1953, 1963). This does not invalidate previous arguments, for there is enormous regional variation, although it is possible that there is in any one area where the circumstances favour the formation of concentric land-use rings, an optimum population density, so that further increases in population, instead of intensifying the ring system, destroy it.

Some other factors

Burning

Wherever grassland is extensive, burning has a special significance in relation to land-use zones. Study in the forest–savanna mosaic areas of Western Nigeria reveals a distinction in the timing of burns in relation to distance from the settlement. Very early burning at the beginning of the dry season takes place wherever patches of grassland occur immediately next to a settlement. Such patches are normally used for grazing livestock. With the removal of old grass, a fresh growth appears even during the dry season, and provides some fodder. At the same time the burn is much lighter than later on in the dry season, with less risk of sparks which may set fire to housing. Burning in the outer zone of grassland takes place throughout the dry season, but more especially in the first half. Such burning is intended mainly to drive wild animals into traps, or flush them from cover. Study from air photographs indicates that burns stop abruptly at the edge of cropland, or wherever grass gives way suddenly to forest. In the cases examined few of the burned areas were more than 2 square miles in extent. Most burns were tear-shaped scars, made irregular in outline by cultivated land, roads, streams or rainforest (Fig. 11.3) (Morgan and Moss, 1965; Moss and Morgan, forthcoming). Theoretically the early burning of the inner zone favours shrub regrowth, but no doubt the heavy demand on the vegetable resources of that zone by village livestock, especially goats, effectively prevents this. The rather later burning of the outer zone appears to favour grass, a fresh growth of which begins during the dry season. End of dry season burning takes place in effect in the 'middle' zone of croplands where trash and weeds are piled up and burned, and the ashes dug into the soil either immediately, or when the soil has been made easier to work with the first rains. This spatial pattern helps to minimize the risk of fire damage to the settlement, and encourages the development of grassland in an area where hunting is enjoyed, or where the cultivation favoured is one of low labour inputs, even though the resultant yields may also be low. One serious problem is that crops next to the grassland must either be harvested before burning begins or must be able to avoid damage by being green, or with the important part of the crop protected by being underground, as for example cassava and groundnuts. Cotton would appear particularly liable to damage, and it would be worth examining whether burning is better controlled in cotton-growing areas or has a different timing, or whether the cotton is kept well away from the larger patches of grassland. Exposure of the soil as an aftermath of clearance and burning results in a rise in soil temperatures due to insolation (Pitot and Masson, 1951; Pitot, 1953). On the soils with a fairly high clay content this can result in baking and the development of a surface very difficult to break. Late burning and clearance of rubbish in the fields tends in Western Nigeria to be very late, minimizing the period of exposure to insolation. The work of digging and planting in the second and later shifts of a rotation must thus be rushed into a quite short period wherever planting is to be early. The remainder of the area is worked well into the rains. In the grassland the earlier burn for hunting appears to be followed by a fresh flush of grass, thus reducing the effect of insolation for a period before the grass is pulled and burned again on clearing fresh patches for cultivation. In Western Nigeria the rainy season is long, but wherever the season is a mere five months or less, the problems of timing, of burning and tillage in the various lands must be acute, and must restrict the area available for cultivation.

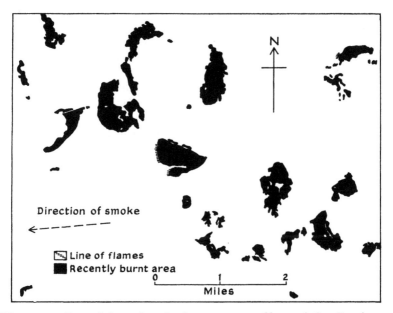

Figure 11.3 *Recently burned patches in savanna near Shepeteri, Oyo Province, Nigeria, December 1962* (drawn from aerial photographs CN/82–100 of the Canada–Nigeria programme of Canadian Aero Service Limited).

Settlement pattern and land tenure

Some comment on the effects of settlement pattern on the zones of land use
has already been made, but needs a little enlargement. Clearly in a situation
of complete dispersal of small family farm units, each with its own lands
close at hand, the likelihood of distance having an effect on land-use zoning
is small. Wherever the lands are fragmented and scattered, however, zoning
may take effect even with dispersal, but will not be apparent from a land-
use survey as such. Only a survey by holdings, identifying the owner or
use-right holder of each piece, would be able to discover whether such a
relationship existed. Where settlements are agglomerated into large units,
and the rights to land around the settlement are more or less uniform for
most of the farmers concerned, then concentric rings of distinctive land use
may arise. In theory the larger the settlement the more marked the effect,
but, as explained in the first paragraph, beyond about 4–7 miles distance
subsidiary or temporary settlements may be set up, providing new nuclei of
cultivation. These may, in effect, provide their own distinctive land-use
zone, but otherwise, in a land-use study based on simple cropping criteria,
they may complicate the pattern. Frequently, therefore, the apparent optimum development of land-use rings, with reference to a single settlement,
takes place where settlements are of the order of 2,000–3,000 population,
cultivating over an area some 2–5 miles in diameter (allowing a considerable
area in waste, woodland, forest and fallow). However, even in the case of
the ideal size of settlement, with an apparently compact area of cultivation,
land tenure differences between the farmers may easily obscure the pattern.
Thus the problems of the farmer with the bulk of his land in an inner zone
are different from those of the farmer with most land in the outer zone or
from those of the farmer with an even distribution of land in all zones. Each
may, equally correctly, make a different choice of enterprise in similar
locations. Differences in holding size may affect choice even where distributions by zones are similar, for the farmer on the very small holding may be
forced, in order to scrape a living, to employ intensive methods even at
some distance from the settlement, while the farmer with the very large
holding may find that intensive methods are only profitable on the nearest
lands, if at all. Equally, differences in soil moisture, consistency and
nutrients may be much more significant than distance, so that while increasing distance may raise costs of movement of manure or crops, its
operation as a factor is so obscured by other factors that it appears of little
significance in the operation of a given agricultural system.

   Another complicating factor is the tendency for agriculture to develop
not in uniform circles but in sectors. This may arise for convenience in
clearing and cultivating jointly operated field units; for social reasons, e.g.

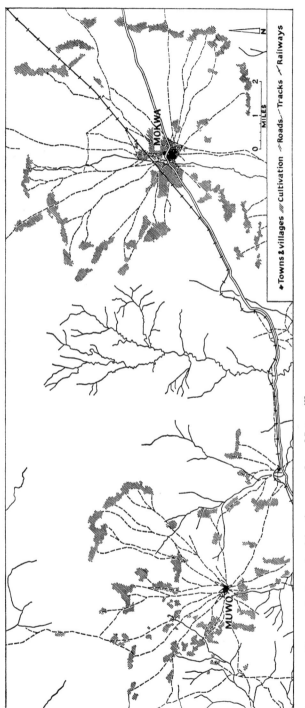

Figure 11.4 *Concentric rings of cultivation around two Nupe villages.*

where all the land in one direction is held by one kinship group, and all the land in another direction by another group; or because of convenience in movement by systems of major and minor paths out of the fields from the settlement, so that isopleths of equal access would tend to be star-shaped rather than circular, and the systems operate by so many of the sectors thus defined by the main paths in turn. In the last case the farther out cultivation proceeds, the more the 'units' break up into scattered fragments, or apparently conform to linear zones along the paths, instead of circular zones around the settlement. In some northern Ibo communities this linear tendency is quite marked. In other cases, however, the concentric ring tendency has been encouraged by clearance from existing fields, in great arcs, instead of from footpaths. Around some Nupe villages in central Nigeria, for example, the cultivated lands form remarkable, almost complete circles, with the fields curving with their long axes parallel with the arc (Fig. 11.4). They provide some of the most perfect examples in tropical Africa of concentric rings of distinctive land use, but as such are somewhat exceptional.

References

ALLAN, W. 1965. *The African Husbandman*, Edinburgh, 193–206, 242–4, 247–9.

BOSERUP, E. 1965. *The Conditions of Agricultural Growth*, London, 57–9.

BRIDGES, A. F. 1938. *Report on Oil-Palm Survey, Ibo, Ibibio and Cross River Areas*, unpublished MS.

CHISHOLM, M. 1962. *Rural Settlement and Land Use*, London.

CLARK, C. and HASWELL, M. R. 1964. *The Economics of Subsistence Agriculture*, London, 111–138.

DUMONT, R. 1957. *Types of Rural Economy*, London, 65–69.

FIELD, M. J. 1943. The agricultural system of the Manya Krobo of the Gold Coast. *Africa* 14, 54–65.

FORTES, M., STEEL, R. W. and ADY, P. 1947. Ashanti Survey, 1945–6. *Geogrl J.* 110, 149–79.

GAMBIA. 1959. *Land Use*, 1 : 25,000, Directorate of Overseas Survey.

GLEAVE, M. B. 1965. The changing frontiers of settlement in the Uplands of Northern Nigeria. *Niger. Geogr. J.* 8, 127–41.

GROVE, A. T. 1951. Soil erosion and population in south-east Nigeria. *Geogrl J.* 117, 291–306.

GROVE, A. T. 1961. Population densities and agriculture in Northern Nigeria. In BARBOUR, K. M. and PROTHERO, R. M. 1961, *Essays on African Population*, London, 125–6.

HASLE, H. 1965. Les Cultures vivrières au Dahomey. *Agron. trop.* 20, 725–46.

HASWELL, M. R. 1953. *Economics of Agriculture in a Savannah Village*, Colonial Research Study No. 8.
1963. *The Changing Pattern of Economic Activity in a Gambia Village*, Overseas Research Publication No. 2.

LÖSCH, A. 1954. *The Economics of Location*, New Haven, 60–3.

MANSHARD, W. 1957. Agrarische Organisationsformen für den Binnen-markt bestimmter Kulturen im Waldgürtel Ghanas. *Erdkunde* 11, 215–24.

1961a. Afrikanische Waldhufen- und Waldstreifenfluren. *Die Erde* 92, 246–58.

1961b. *Die geographischen Grundlagen der Wirtschaft Ghanas*, Wiesbaden.

MIRACLE, M. P. 1964. *Traditional Agricultural Methods in the Congo Basin*, Food Research Institute, Stanford University, Stanford.

MORGAN, W. B. 1955. Farming practice, settlement pattern and popula-tion density in south-east Nigeria. *Geogrl J.* 121, 320–33.

MORGAN, W. B. and MOSS, R. P. 1965. Savanna and forest in Western Nigeria. *Africa* 35, 286–94.

MOSS, R. P. and MORGAN, W. B. forthcoming. Soils, plants and farms in West Africa. *Symposium Volume of the British Ecological Society.*

NETTING, R. M. 1965. Household organisation and intensive agriculture: The Kofyar case. *Africa* 35, 422–9.

PHILLIPS, T. A. 1964. *An Agricultural Notebook*, 2nd edition, Ikeja, 173.

PITOT, A. 1953. Feux sauvages, végétation et sols en Afrique occidentale française. *Bull. Inst. fr. Afr. noire* 15, 1369–83.

PITOT, A. and MASSON, H. 1951. Quelques données sur la température au cours des feux de brousse aux environs de Dakar. *Bull. Inst. fr. Afr. noire* 13, 711–32.

PROTHERO, R. M. 1953. Land use at Soba, Zaria Province. *Research Notes*, Department of Geography, University College, Ibadan, 2, 3–10.

1957a. Land use at Soba, Zaria Province, Northern Nigeria. *Econ. Geogr.* 33, 72–86.

1957b. Land use, land holdings and land tenure at Soba, Zaria Province, Northern Nigeria. *Bull. Inst. fr. Afr. noire* 19B, 558–63.

RAULIN, H. 1963. Cadastre et terroirs au Niger. *Étud. rur.* 9, 58–79.

RICHARDS, A. I. 1958. A changing pattern of agriculture in East Africa: The Bemba of Northern Rhodesia. *Geogrl J.* 124, 302–14.

SAUTTER, G. 1962. À propos de quelques terroirs d'Afrique occidentale, essai comparatif. *Étud. rur.* 4, 24–86.

SAVONNET, G. 1956. Système d'occupation du sol dans l'ouest de la Haute Volta. *Symposium de Geographie, Institut Français d'Afrique Noire*, Dakar, 27–31.

1959. Un système de culture perfectionnée, pratiqué par les Bwaba-Bobo-Oulé de la région de Houndé (Haute Volta). *Bull. Inst. fr. Afr. noire* 21B, 425–58.

SCHLIPPE, P. DE, 1956. *Shifting Cultivation in Africa: The Zande System of Agriculture*, London.

VON THÜNEN, J. H. 1826. *Der isolierte Staat in Beziehung auf Landwirt-schaft und Nationalökonomie*, Rostock.

WELLDON, R. M. C. 1957. *The Human Geography of a Yoruba Township in South-West Nigeria*, B.Litt. thesis, Oxford; quoted in MITCHEL, N. C., Yoruba Towns. In BARBOUR, K. M. and PROTHERO, R. M. 1961. *Essays on African Population*, London, 287–8.

WILLS, J. B. 1962. *Agriculture and Land Use in Ghana*, London, 214–18.

WILMET, J. 1963. Contributions récentes à la connaissance de l'agriculture itinérante en Afrique occidentale et centrale. *Bull. Soc. belge Étud. Geogr.* 32, 51–63.

12 Colonial administrative policies and agricultural patterns in tropical Africa

J. A. HELLEN

'Agricultural development must go beyond its colonial framework' – so wrote the French agronomist René Dumont in his survey of the New Africa (Dumont, 1962). Increasingly the African states as well as the external development agencies are reconsidering these legacies, which so tenaciously influence institutions in many developing countries, and are seeking to modify or eradicate them. Rural economic development in Africa encounters many of its greatest difficulties in a single factor of development – land. But because of the complex pressures the African husbandman has to accommodate, the agrarian landscape is more than a simple expression of current allocation of means to a desired end. The belief that low productivity in the traditional sector is due to disregard of land as a capital asset has led to the erroneous view that development must follow if the situation can be altered. In consequence the morphological interpretation of the cultural landscape as an indicator of underdevelopment demands not only the evaluation of the more obvious cultural and ecological factors but also detailed investigation of colonial, legal, social and administrative determinants which can contribute to an understanding of the landscape chronology. From a methodological standpoint this approach to the inner *Raumstruktur* of the developing countries owes much to developments in the general field of landscape analysis and is considered as one of four main analytical subjects by Troll (1962; and see also 1950). If, as can be argued, agrarian landscapes in the tropics are becoming even more finely differentiated in the geographical sense, as a consequence both of population increase and of better adjustment to potential biological productivity, it would seem that aspects of colonialism as a causal factor in such areal differentiation deserve re-examination.

That the colonial powers have left an imprint on the institutions of their former dependencies is undisputed, but the relatively short duration of *effective* administration in most parts of Africa has meant that dualism persists; the co-existence of statute and customary law, of monetized and subsistence economy, of westernised élite and tribal majority is evident. Decolonization cannot realistically be taken at its face value. Furthermore, the interaction of the various metropolitan powers with the mosaic-like web of tribes produces visible land-use patterns of subtle distinctiveness and any attempt at elucidation must reject outright the theories of economic imperialism alone. This chapter will therefore examine aspects of those circumstances and policies which are implied by trading relationships, export patterns and land-use patterns on a crude territorial basis and appear to be important in having caused significant changes of cultural landscape. Such changes will be those associated with African agricultural patterns which serve to differentiate certain classes of modification or transformation.

The independence of many states in Africa has opened up a challenging field of inquiry into this particular colonial legacy which demands early investigation if images are not to be blurred. Of particular value would be numerous sample surveys of rural settlements to provide exact details of land utilization under specific agricultural systems; it might then be possible to establish a datum line against which to study changes implicit in records of administrative and legislative action. Very often accounts of landscape change rely on oral tradition or the reports of early colonial administrators. The nature of components of the African cultural landscape, be they dwellings, field boundaries or any other man-made feature, is such that they are erased permanently by the regenerating vegetation, by the scouring rains and devouring termites.

By colonial administrative policies are to be understood those decisions arrived at in metropolis, colonial capital, province or district alike, which serve to enforce, regulate, alter, encourage or otherwise influence the population's actions or the colony's appearance and functions. Such policies may be indirectly effective by sanctioning the policies and actions of third parties whether they be commercial, missionary or of any other organization. They may lead to complete transformation of a landscape by a single crop, to the displacement of an entire tribe, or merely to construction of a contour ridge. Such policies may be put into effect by a single administrator or an army of local government personnel; they may be centrally promulgated but moulded to local conditions or they may be inflexibly applied. In all instances they express an intention to interfere, for better or worse, with the *status quo* and must record a measurable effect which is of interest to the geographer. Any resulting agricultural patterns may therefore be examined on a variety of scales subject to the availability of the appropriate detailed studies. Many will indicate substantial change arising from new crops, new techniques and new agrarian structures; others will indicate features of an induced decay or degradation and others again will demonstrate gradual adaptation from within or compensating moves from without. Few situations are likely, on close examination, to have remained substantially unchanged.

The colonial frontier, for all its evident artificiality, embraces problems far greater than the merely tribal by preserving dissimilar administrative and legal infrastructures with associated linguistic, monetary and other features antagonistic to regional economic development. But within these frontiers colonial administrative policies appear to have induced landscape change chiefly through the operation of land policy. Such policy is variously motivated and may have had direct effects as evidenced by alienation of land, encouragement or even enforcement of cash cropping or the implementation of conservation ordinances, planned peasant colonization and the like. It may function indirectly through the operation of certain fiscal or social policies and non-agricultural development and lead to such features as recruitment of alien African labour leading to displacement of indigenous peoples, as in the plantations of West Cameroon, or to the exodus of a major tribal group in response to unpopular land policies, as in the eastward drift of Mawiko peoples from Angola to Zambia. 'In African conditions', Lord Hailey has written, 'the attitude of the Administration in matters relating to land policy affords one of the most significant indications of the relations it seeks to establish with the African population' (Hailey, 1957). In High Africa, at least, the land problem has epitomized the conflict of opposing interests and appears to have provided the dynamic for the movement towards self-determination. Those colonies or protectorates which succeed in attracting and retaining a settler element seem therefore to be characterized by general problems largely unknown in Low Africa, yet the conflict probably owes its origins as much to new ideas as to racial antagonism. It is sometimes argued that any degree of development implies a change of environment. If this is so, then the agrarian reform forced on newly autonomous states by internally generated demands for higher agricultural production will necessitate the resolution of these inherited conflicts which have hitherto remained quiescent in the normal state of social and economic dualism typical of colonialism.

Land and indigenous agriculture

The effect of colonial rule on the land has been as varied as the differing European cultures would suggest. In some instances the former freedom of tribal movement has been thwarted by the designation of tribal reserves and consequent spatial restriction of numerous groups, whereas in others manifestations of tribal identity have been suppressed. Common to most colonialists, however, has been a misunderstanding of the African material and ritual relationships to land and therefore of the patterns of occupance. Confusion has arisen particularly in the interpretation of land tenure of a communal and individual sort and in the definition of land 'ownership' as such.

On their first arrival in Africa, Europeans encountered a system very different from their own and this led them to make mistakes of interpretation – by speaking of *communal* ownership they implied that every tribesman had equal rights to every part of the tribe's land and they regarded the chief as the owner of the land. In fact, as Gluckman has shown, the chief commonly acted as trustee of the land for the tribe and his ownership gave certain *rights*. Important among these were the chief's right to demand allegiance from persons using his land, to hold land not taken over by his subjects or to repossess land allocated but not used, to demand produce in tribute, to control the settlement of his people and to legislate in matters of land holding or land use (Gluckman, 1945). The system of chiefly control distinguished by the *hierarchy of estates* and typical of such tribes as the Zulu and Tswana is, however, only one of the four main systems of land holding in Africa. *Feudal* systems with landlords and tenants exist and that of the Baganda is well known. A third type may be classified as a lineage or *unilateral descent* system under which land is held by a family descended from a common ancestor, for example, among the Yoruba. A fourth type embraces those systems where residence is sufficient to invest the individual with rights over land. Throughout the great majority of diverse systems, however, runs the common feature of access to land through membership of a community; but that this community is of greater importance than the sum of its individual members has been demonstrated by de Schlippe in the case of the Sudanese Azande where 'the pattern of behaviour inherent in the customary system of agriculture constitutes the force of adhesion of the group to the natural environment' (de Schlippe, 1956).

Indicative of the gulf between the European and tribal culture is Richards' comment on the Bemba shifting cultivator in Zambia who did not 'measure land, assess its value in size or productive capacity, or conceive of this value as a figure to be permanently maintained at a given level by the expenditure of effort and capital. He views the country round him as all one unit, all accessible to him, and all ready to supply his needs' (Richards, 1939). Such attitudes were no defence against settlers or concessionnaires who took land by right of conquest, treaty or grant and who, to quote Bohannan's phrase, would have accepted the white man's pattern of an 'astrally based grid map in terms of which people are, by a legal mechanism, assigned rights to specific pieces of earth-pieces which maintain their integrity even when their owners change' (Bohannan, 1963). But if this disparity was true of some areas, and particularly of those of low density worked by shifting cultivators, it appears to have led to modification in attitudes to land values among more densely settled African peoples at an early stage and these moves towards systematic individualization of land tenure must also be considered.

The common attitude of the African community towards the question of individual land needs would be one of satisfaction if the individual could maintain himself and his family on a subsistence basis; the production, consumption and storage of food were effectively controlled until the surplus of cash crop became desirable. Accordingly, under forms of subsistence agriculture it is reasonably easy to calculate the land requirements of the 'garden family' and the carrying capacity of the land itself. Allan has expressed this in a numerical relationship of *population–land factors* under conditions of shifting agriculture (i.e. bush fallowing): land-carrying capacity is estimated in terms of the fertility of given soil types, the percentage of cultivable land, the number of gardens needed to give a complete regeneration cycle (termed the *cultivation factor*) and the mean area *per caput* under cultivation at any given time (Allan, 1949). Thus under many systems and in particular habitats *optimum* population densities are found to be of the order of two or three persons per square mile where no artificial fertilizers are used. Since over the greater part of Africa shifting agriculture is operated in thinly populated areas where rainfall agriculture is possible, this concept of optimum man–land ratios is particularly valuable in understanding field patterns. The question of the optimum size of land holding relates to the subsistence demands placed on it and the capacity of the available family labour to clear and cultivate a certain area predominantly of seasonal crops such as cereals and legumes and to a lesser degree, except locally, of perennial crops such as bananas, cassava and yams.

Exact information on the amount and disposition of land worked by any community is less easily obtained than population data for a given territory. Just as registration of births and deaths among the African population has been the exception rather than the rule, so registration of land title has been in most cases a localized and recent experimental innovation; indeed, the U.N. Economic Survey of Africa has noted that 'information on the size of holdings is lacking. Since, however, traditional farming is almost entirely dependent on family labour and primitive technology, it may be presumed that production is predominantly on a small scale – a few hectares per family' (UNO, 1959). Aerial photography may reveal relationships between the broad land-use patterns and ecological zones down to the dimensions of the landscape cells, or ecotopes which together constitute a landscape mosaic, just as a ground traverse may establish the nature of the soil, groundwater or plant communities along a catena; but however helpful these aids may be in assessing potential biological productivity, they explain little in terms of African agricultural production or productivity. In consequence, any quantitative information is drawn from a variety of local field studies in preference to the estimates of administrations. That primitive techniques impose severe limitations on the extent of land used in any

one year by the immensely varied group of 'shifting' cultivators is borne
out by the roughly comparable size of holdings in many tropical countries,
and in Africa a figure of about one acre per head under main crops seems
reasonable on present evidence, the figure increasing over weak soils. Even
where the shift to settled agriculture does occur and the high labour input
of bush clearance is eliminated after the first season, total areas worked
appear to rise only slightly if at all. The important next stage as far as land
requirements are concerned has followed on the introduction of cash crops,
and except where techniques have been improved the critical population
densities have moved towards ecological imbalance of man and land
(Allan, 1960).

Agricultural and associated rural settlement patterns which derive from
tribal subsistence economies are not yet sufficiently well understood to support firm generalizations, although it may be noted that certain of the
principles of location operate even though costs and profitability are not
overtly considered. Particularly under soil selection systems of shifting
agriculture, newly colonized gardens may be located at ever-increasing distances from the major settlement and necessitate the erection of temporary
garden huts until the village elders decide to transfer the entire community.
It has been argued that 'under adverse conditions of physical and poor
technical accomplishments, a combination which betokens a low population
density, the distance certain shifting cultivators shift their main abodes is
very similar to the figures which have been quoted for normal, as against
exceptional circumstances' (Chisholm, 1962). Whereas it may often be true
that a crop zonation is to be perceived within a radius of cultivation
orientated towards a single village, two highly significant factors have to be
weighted: first, the ecological soundness of much of African agriculture,
dependent as it is upon certain bioclimatic indicators of fertility, tends to
support catenary or mosaic distribution of crops, and secondly, the psycho-
logical factor may often impel members of an established group to break
away and found their own village for non-economic reasons.

Quite often the net result is a complex pattern of land use made more so
by dependence on a wider hinterland, an extensive fringe belt from which
forest products such as rare timbers, thatching grass, medicaments, game
and other items may be obtained and which ought to be counted as integral
with the over-all land requirement. It might be argued that in Africa as a
whole there is a major distinction in this diversity between those subsistence
economies which have existed with and without the mechanisms of exchange usually epitomized by markets. This would appear, on present
evidence, to distinguish West Africa, where traditional markets were
established long before European intervention, from southern and south-
eastern Africa where markets have for the main part come with the

European (Hodder, 1965). The wider significance of these characteristics, which accord in part with the division of tropical Africa into *high* and *low* regions, lies in their reflection of response to external forces such as trade links along routeways and consequent specialization of production where resources, population density and the form of political organization were favourable for the development of market foci. Land under such a circumstance might appear to move increasingly towards a state in which it assumes economic as well as social and political importance for the Africans concerned and favours assimilation into the world monetized exchange economy.

Land and the Europeans

Of all the landscape features attributable to European action and influence, the cultivated crops are the most widespread geographically, and probably the most significant culturally and economically. Some plants may have arrived fortuitously whereas others have been deliberately, even forcibly, introduced and have transformed not only the subsistence sector but also formerly negative areas whether by casual infiltration of crops like maize and groundnuts or by the regimented and wholly man-made plantation landscape of sisal, tea or coffee. Unlike those of the Egyptians or Azanians, for example, the European introductions have been notable not so much for the mobility they have permitted but rather for their introduction to Africa of large-scale production of a single crop to satisfy a market demand as epitomized in commercial plantation tillage. They have set in motion great fluxes of migrants, circulating or seasonal labour rather than *Völkerwanderungen*, and they have deprived Africans in some areas of substantial blocks of land but have made possible colonization of semi-arid or highland habitats. There is no opportunity to discuss here the respective contributions of Arabs, Portuguese and later colonists, building on the indigenous crop range which has been so extended that Africa has become a major commodity producer in a few decades. But it is important to note that the introduction of such crops has had notable side-effects on relations between tribes and on tribalism itself. Administrative paternalism has tended to favour the sedentary cultivator at the expense of the once-dominant pastoralists, and the stateless group without a centralized authority at the expense of the centralized state. Within the tribe new crops demanding new techniques have often shifted emphasis from communal to individual activity and at the centre there has been decay of tribal politics and their replacement by mere tribal administration. An equally important outcome of the new cultivated crops has been the degree to which the administrative policy has aided their spread in order better to exploit land resources. In some cases this has been through compulsory native cash cropping, in some cases by peasant settlement schemes and in some cases by the extension of European-owned and directed farms and plantations.

Europeans' attitudes to land would appear to separate into two significant categories in this context: the one concerned with their own cultural experience, the other with mercenary considerations towards land acquired in the specifically colonial context by seizure of political power. The earliest were essentially of the *plantation* type, operated for profit or prestige. Land was seen as one of the factors of production, to be exploited in association with labour, capital and the necessary technical expertise, imported as the need arose. Europe was and is an area characterized by a high proportion of land given over to agriculture, as compared with other continents, and was typified by a situation in which *property* and *contract* rather than *status* were important. The etymology of the term colony implies cultivation or habitation, but the European colony in Africa very early differentiated between colonies of exploitation and colonies of settlement, linked with motives such as expansion of trade or emigration. Following Maunier, colonization requires occupation and legislation, an element of fact and an element of law (Maunier, 1949). The quality and extent of occupation may vary considerably and the interests of the colonists show more or less importance in shaping legislation and policy locally or in the metropolis. Similarly, colonial status demonstrates a cycle of development in relations between the metropolitan country and the dependency, both passing through phases of domination, partnership and emancipation, and it is important to view changed patterns of occupance and legislation as response to change in the mother country, to international pressures and to internal forces.

From a synoptic view Hance's map of source areas of export production in Africa is most valuable in emphasizing the discrete 'islands' of production which are characteristic of the continent. In its location of 95 per cent of total exports by value, the symbols each representing 0·01 per cent, the geographer has at his disposal a quantified synopsis of major agricultural patterns operating within or in association with the money economy (Hance *et al.*, 1961). The pattern of these agricultural producing areas – ignoring in this context the mineral realm – emerges as one of *coastal* and *peripheral* islands, of *highland* islands, of *irrigation* islands and *special* islands. The year chosen – 1957 – was a high-water mark of colonialism in Africa and seems in retrospect singularly appropriate for the purposes of this study. Exports in that year showed 64·9 per cent of the total value deriving from vegetable products, of which the principal categories were stimulants and spices (28 per cent), vegetable oils (16·6 per cent) and fibres (10·2 per cent). Conforming with these production islands, one might also stress the importance of land alienated and settled by Europeans in High Africa who make use of the labour pools of the intervening tribal land. These areas have their counterparts in Low Africa where land was set aside for agricultural concessions or has from an early date seen rapid increase in small-scale African production of certain traditional plantation crops.

The agricultural patterns associated with the large farms and plantations are in general related to certain obvious morphological features such as the arrangement of the crop, processing plants, roads and labour compounds but their effects are much more extensive than the immediate farm-complex. They usually generate high labour demands and create a *rural exodus* of employable males, enforced or voluntary. Degradation of tribal lands as a result of degenerating indigenous agricultural systems is as much a hazard at a distance of hundreds of miles as in the immediate vicinity. Because of belated intervention by administrations, conservation ordinances, resettlement and colonization schemes were often instituted long after perceptive agriculturalists had warned of incipient rural slums and gullied tribal lands. An additional factor in determining the major patterns of agricultural activity has been the growth of other enclaves of non-traditional employment where urban areas, due to administrative, industrial, transport and other functions, have expanded and accelerated the modification of traditional agriculture by creating markets, however small. Often the eccentric position of capital or primate cities has created in effect core areas with distinct pioneer fringes beyond which the agricultural patterns have experienced little direct incentive to alter; the existence of so-called 'Cinderella' provinces as in the case of the North Western Province in Zambia and the Ouadaï Province in Chad support this generalization. Yet there has often been a modification of this core area phenomenon in the line of rail or main road effect whereby native settlement and cash cropping have been attracted to a linear zone parallel to the routeway with resultant increase in land 'values'. This is evident in the examples of the Bas Congo–Katanga Railway where a remarkable diffusion of the Baluba of Kasai and associated intensification and increase in total area of land under cultivation has occurred; as Nicolaï states: 'le rail a simplement mis en contact la civilisation européenne et la civilisation indigène' (Nicolaï and Jacques, 1954). This extension of European attitudes to land via communications networks may be plausible in the case of the Dakar–Niger or Livingstone–Ndola lines of rail, but as O'Connor has significantly demonstrated for Uganda it may no longer be true that railways can promote economic development if a flexible road transport system operates (O'Connor, 1965).

Perhaps the most significant contrasts between African and European attitudes to the land may be summed up in the statement that natural resources are effectively cultural appraisals. By whatever means the process of acculturation was directly or indirectly prosecuted, it was inevitably disruptive because the colonialists had too little time to appreciate their ignorance of the African milieu or African society.

Policies and patterns: aspects of action and reaction in the African Colonies

The Portuguese territories

Since 1951 when Portugal converted her colonies to overseas provinces, that country has further strengthened its determination to seek its own solution in Africa. Post-war conditions, which fortified more liberal policies in France or Britain and led to the more direct development of the African sector in various economies, have not ameliorated the lot of the non-assimilated in overseas Portugal. Although in combined areas Angola and Moçambique occupy 779,082 square miles and dwarf the 13,948 square miles of Portuguese Guinea, each has been overshadowed by the economic demands of Brazil. Although there are now probably 400,000 whites in Portuguese Africa, it has proved impossible to establish a large white-settler community in the face of the New World's greater attraction and for this reason the emergence of a distinct policy was but little earlier than those of other European powers arriving centuries later.

Portuguese policy operates on the basis of certain fundamental principles, of which the concept of political – and hence social and economic – unity between Portugal and her overseas provinces is particularly significant. This is expressed in strongly centralized administrative control, either at the territorial or Luso–African level and a domestic political situation not altogether typical of other European ex-colonial powers. Two juridical classes are recognized – the citizen and the native – and this explicit hierarchy underpins the philosophy behind the social and economic development by simplifying the issues involved in the contact between the institutions of the Western and of the 'primitive' culture. Recent concentration on African and *petit-blanc* peasant farming schemes would suggest that the view that the majority are simply instruments of policy having the right to develop toards *evolué* status may be changing.

Whereas Brazil was not neglected after its discovery in 1500, the interior of Moçambique was not effectively administered until after the military campaigns of 1895–9, Angola until 1907–14 and Portuguese Guinea until 1916. Until that stage feudal conditions characteristic of an earlier Europe had been preserved by the granting of land title in the form of *prazos* (estates) in Moçambique from the sixteenth century with the aim of stimulating European colonization in southern Africa. Exploited by their owners for slaves and tribute, these *prazos* degenerated into petty empires which were able to defy Lisbon despite their being outlawed in 1832. It was not until the attention of other European powers was drawn to southern Africa that Portugal was forced to reform the system in 1890. Estates were sold to the highest bidder who in turn paid the government a proportion of tax revenue and undertook to cultivate and develop the land and to protect resident Africans. With this onset of modern development profitable operation became essential and by 1900 over two-thirds of Moçambique had passed under the control of the chartered concessionary companies. Chief of these were the *Companhia de Moçambique* which received its charter in 1891 and continued to exploit and administer the land westward from Beira between Zambezi and Save until 1942, and the *Companhia do Niassa* which was granted a charter in 1894 and sought to exploit 96,500 square miles between the Rovuma, the Rio Lùrio and Lake Nyasa on a thirty-five-year concession; the *Companhia do Zambézia*, granted a concession but no charter in 1892, commanded an area of about 60,000 square miles in the lower Zambesi area (Quelimane and Tete districts) until the title expired in 1929. This last company, unencumbered by administrative costs and benefiting from the enterprise of its sub-concession holders such as the Sena Sugar Estates Company, was more successful than the other concessionaires in developing areas vastly greater than metropolitan Portugal. Writing of the Moçambique and Zambézia companies in particular, Duffy has noted that during their short existence they 'had greater significance for the colony than had the *prazo* in three centuries. To a large extent they initiated the transformation of the interior of the province from a wilderness to an economically productive region . . . The benefits the Africans derived from their presence seem less than negligible' (Duffy, 1961).

The interest of the Portuguese territories lies in no small way in their remarkable failure to develop as settler colonies for a dominantly agricultural and overpopulated mother country. In the case of Moçambique, 44 per cent of which lies below 330 ft (100 m) and only 13 per cent above 3,300 ft (1,005 m), not only has white settlement proved to be disappointingly slight but also dependence on indigenous labour has proved acute and has brought the dual sectors of the economy into conflict through the operation of labour laws which in the past excited condemnation. Portuguese insistence that the African had an obligation to labour led to excesses in their native policy exemplified in Angolan recruitment for the São Tomé cocoa plantations and the Moçambique recruitment for the Witwatersrand. Changes in native labour policy were stimulated by foreign opinion and today labour is generally operated within the voluntary or contract category rather than the obligatory class, but the harmful effects of rural depopulation persist.

In Moçambique the indigenous agriculture is essentially of a subsistence character, the population (1960 Census – 6·6 millions) of such Bantu groups as the Maravi, Tonga and Shona, being relatively thinly settled away from the coastal areas and exploiting some 12·35 million acres principally under variants of shifting cultivation (Guichonnet, 1965). The internal division between native and non-native areas is not altogether clear since decrees setting aside reservations for the exclusive use of Africans have been fairly numerous since 1900. The 1955 *Native Statute* for both Angola and Moçambique stipulated that 'natives who live in tribal organizations are guaranteed, in conjunction, the use and development, in the traditional manner, of land necessary for their villages, their crops, and for the pasture of their cattle' (Duffy, 1961). In the absence of native courts, Portuguese civil and criminal law is operative and there is no recognition of individual rights in the native reserves. In Moçambique there were 2,501 separate European-type holdings in 1962, extending over 3·95 million acres, of which 988,400 acres were actually cultivated using a labour force of 134,416 Africans and a mere 1,302 Europeans (Guichonnet, 1965). Probably about 60 per cent of the agricultural exports originate from the European sector despite this great disparity in the cultivated areas and labour engaged – in Angola it is nearer 80 per cent – and a specialized pattern has emerged with plantations alone producing sugar, sisal and most of the tea. Certain crops, among which copra and maize are good examples, are grown by both Africans and Europeans and others solely by Africans. Of these cotton was forcibly introduced in 1932 and is extensively produced from what were originally compulsory African plots of about 5 acres per family operated under the aegis of the *Junta do Algodao*. Under this system the Africans received the seeds from concession holders who bought back the crop at guaranteed prices. Rice was similarly first introduced as an obligatory crop and, like cotton, culture led to serious ecological and social disequilibrium. As with native taxation, so obligatory crops were a feature of many colonies in a local endeavour to make them self-financing and thus to avoid a politically unpopular charge on the metropolitan budget.

In Angola the population is very thin in terms of total area, and the great Bantu groups such as Ovimbundu, Chokwe and others still practise extensive agriculture. Despite the much more common *planalto* areas climatically favourable to white settlement, Angola, which was effectively pacified much later than Moçambique, seemed primarily to serve as a labour pool. More than any other cultigen, coffee, first introduced to Angola from Brazil in 1837, dominates agricultural production and is of particular importance north of the Cuanza River; it is now the most important agricultural export commodity but this situation has brought about the implicit hazards of a monoculture as well as worsening the already great labour shortages of plantation labour. Van Dongen has commented that 'labour demands on the part of the coffee growers have led each year to extensive displacement of African able-bodied men from their tribal villages, causing family hardship and depressing local birthrates' (van Dongen, 1961). Although the number of white farmers is only about 3,000, the Angolan economy derives a disproportionate benefit from their lands, between 70 and 80 per cent of agricultural exports stemming from this quarter. Coffee plantations, predominantly in northern Angola, required 85,647 African workers in 1956; sisal plantations another 29,000 (von Gersdorff, 1960). As in Moçambique, cotton was an obligatory crop after 1932, although it had been exported as early as 1856. So unpopular was it that it precipitated population drift to territories to the east and African families continue to cultivate holdings of about 2·5 acres under a similar concessionary arrangement to the detriment, particularly in Malanje District, of normal subsistence cropping.

The important and heavily capitalized plantation companies of the Portuguese territories stand in contrast to the settler policy of the *colonatos*, conceived nearly fifty years ago but not put into effect as modern enterprises until 1949. In Moçambique the Limpopo scheme of fourteen villages in which some 1,000 white and 500 black families had been settled by 1964 in an irrigated area of 37,100 acres is well known. In Angola the setting is distinct from the latter which lies at an altitude of about 330 ft (100 m), for it occurs at a height of between 3,200 ft (975 m) and 6,250 ft (1,905 m) in the *planalto* zone and extends through 38 *colonatos* in nine districts and involves about 2,900 farming families (Niemeier, 1966). The attempt to create a Lusitanian peasantry in her Commonwealth would appear to have reversed the *laissez-faire* policy of Portugal which in spite of *de jure* unification of black and white interests and opportunities has favoured the localized transformation of her territories into plantation landscapes, African cash cropping playing a hitherto minor role. The argument is often paraded that Portugal has, with the exit of the other European powers, passed out of the era of traditional colonialism and entered upon a new phase of mutual benefit to all Portuguese citizens, white and black; placed in a monopoly position vis-à-vis other European countries she can now, without her traditional antipathy towards foreign capital, press ahead with development of industry as well as agriculture orientated towards domestic markets (Neumann, 1965). If this be the case, Portuguese administrative policies may yet influence a more objectively implemented land-use pattern, benefiting from the mistaken initiatives lacking the historical perspective which has been so costly elsewhere in Africa and evolving the hybrid institutions which have eluded others in the search for acceptable development forms.

The Belgian Congo

That the Belgian Congo, a vast mosaic of landscape and ethnic components, should have passed into the control of a metropolitan country
which had not even sought to be a colonial power yet inherited a colony of
exploitation *par excellence* is not without its irony. In effect, by the state's
reluctant annexation of the Congo Free State and the proclamation of the
Charte Coloniale in October 1908, the Congo passed from nominal independence to colonial status. After 1885 and in accordance with the conditions imposed by the Act of Berlin, Leopold II's private 'estate' had in
effect been an *international* colony. After 1908 the area passed through a
second phase in which it was influenced by a unique combination of administrative, missionary and commercial forces, finding expression in the
economic and social sectors to the ultimate detriment of political institutions. Concerning the decolonization process in the Congo, Young writes
of this colonial trinity that 'each fashioned a formidable organizational
structure in its own sector of activity' and that the ensuing colonial system
was 'unparalleled in its penetration into the African societies upon which it
was superimposed' (Young, 1965). The same author notes apropos of the
administration that 'no Congolese, rural or urban, could have failed to
perceive that he was being administered. In the urban centres this is hardly
surprising, but what differentiated the Belgian system from others in Africa
was the extent of its occupation and organization of the countryside'. Such
an effective administration was achieved by a tight network of *secteurs*,
composed of amalgamated native chiefdoms and given legal sanction as
circonscriptions indigènes from 1933, within the purely administrative network of provinces, districts and territories.

   The Free State had been characterized by a monopoly system in association with concessions which collectively extended over vast areas of 'vacant
native lands: of three types, railway concessions covered 22·5 million acres,
the Katanga concession 112·5 million acres and the forest concessions an
immense undisclosed area' (Hailey, 1957). Under Belgium the system was
regularized. Native land law recognized no individual land ownership, this
being invested in the clan and allocated by the *chef de terre*, custodian of the
land, as the need arose. The early administration recognized as tribal land
only that which was in actual occupation, but this was modified in 1906 by
a decree permitting a threefold increase on this inadequate allocation, and
subsequent legislation, particularly that of 1934 and 1935, sought to protect
native lands and rights of cultivation. The situation regarding the exact
nature of such land under native occupation, however, remained ill-defined
in the absence of specific tribal reserves following inconclusive attempts at
delimitation. Thus those lands vested in the state could be disposed of on

a freehold or leasehold basis by the Government and this function was further delegated to certain concession-granting authorities. The broad effect of this change was a swing away from the crude exploitation of such forest commodities as wild rubber to a situation in which freehold was ceded to a concessionaire when specified conditions of development had been achieved.

Although the move towards the systematic individualization of native land tenure had begun as early as 1933, it was another twenty years before the first decree was issued. Two major phases of colonial policy can be discerned in this connection. In the initial phase the inflow of capital was encouraged by the grant of concessions, such as those of the Unilever associate *Huileries du Congo* (1911), which led to a great upsurge in palm product exports from rural areas. The formation of essentially urban-based private enterprises such as *Union Minière* in Katanga and *Forminière* in Kasai five years earlier had led to pressure on labour supplies and forced the entire country towards a state of imbalance between capital and labour. Subsequently there was a reversal of policy whereby the *classe paysanne* was fostered as a logical extension of the 'educational process' implicit in the obligatory cultivation programmes enforced since 1917. From a geographical standpoint the Congo can conveniently be considered as having suffered a progressive rural depopulation in tandem with a marked success in urban development; initially facilitated by labour recruitment, this largely male exodus subsequently received tribal sanction. After 1924 the administration's policy turned against excessive recruitment and set a ceiling of 5 per cent on the number of able-bodied Africans permitted to leave the village for long-distance migration, with provision for a total of not more than 25 per cent to be absent within specific work zones (Anstey, 1966). In fact the situation deteriorated rapidly in some areas with over half the able-bodied male corps being absent at any one time and this had predictable effects on the material and social fabric of the family, village and agricultural systems as well as side-effects on associated ecosystems. The Belgians went a stage further in emphasizing the dualism of Congolese society by providing areas in which those Africans who wished to live outside tribalism could exist in urban *milieux de liberté* and rural *centres extra-coûtumiers*. So successful was this active detribalization process that between 1935 and independence this sector increased in numbers from 600,000 to 3 million.

The *paysannat indigène* policy, which had first been mooted in 1933, was not put into effect until the first resettlement schemes were set up in 1942. This official policy marks the second phase of major change in the Congo when serious and costly remedies against rural depopulation began to operate. By independence about 200,000, or 8 per cent of the rural population, had voluntarily joined these schemes and the Belgians began to register solid achievement not only in raising living standards but also in controlling traditional shifting agriculture. None the less Europeans continued to play an important role in agriculture until Independence in 1960. In 1958 there were 113,000 whites in the Congo, of whom 9,621 were engaged as colonists. Although the amount of land ceded to private persons or companies by 1958 was only of the order of 11·6 million acres, or slightly more than 2 per cent of the total Congo, the million acres actually under cultivation represented 40 per cent of the total area under cash crops (INFORCONGO, 1960). With over 300,000 Africans in wage employment on farms and plantations, the marketed production of European holdings demonstrates the overriding areal importance of palm products, coffee and rubber, and confirms the prominence of this type of organization at the close of the colonial rule.

In some ways the late emphasis on African rural development parallels experience in the Portuguese territories, both administrations having decided to attempt to recreate in Africa the European peasantry. In an admirable study of the Congolese Kwilu, over one million of whom occupied one of the most densely populated areas of the territory, Nicolaï's conclusions are of great interest (Nicolaï, 1963). He wrote that 'L'administration n'a donc introduit aucune reforme importante, aucune culture nouvelle. Sa seule intervention effective consista en cultures d'ordre éducatif, c'est à dire en cultures imposées . . . L'administration n'a guère réagi (ou mal ou trop tard) contre l'exode rural.' The consequences of a policy expressed in what Anstey terms 'a whole series of achievements, principally material and most related to unreflecting economic purpose', are etched into the Congolese landscape but in fairness to the Belgian administration must be set against the scale of the territory and the practical difficulties of controlling the whole rather than sectors of the economy and society.

French territories

France, with over 4 million square miles of territory at the zenith of her colonial power, presents features of marked distinctiveness in her colonial policy, not least those which derive from Napoleonic centralization in the nation state, her atomized domestic politics, or her own loss of independence in 1940. Brunschwig has assessed the course of French colonialism up to 1914 demonstrating the effect of defeat in Europe in 1870 and France's search for renewal in colonial expansion (Brunschwig, 1966). Comparisons between Britain and France in Africa are particularly valuable since French domestic policy failed to stimulate the emigration characteristic of the British situation. Slower industrialization distinguished her from a Britain more concerned with calculated commercial gain than prestige and permitted her generals to carve out what were effectively military rather than settler colonies. The over-all impression of the French colonial period is one of remarkable success achieved in integrating the African intelligentsia with metropolitan French culture, thereby emphasizing the contrasting patterns of acculturation between anglophone and francophone Africa. The French colonial policy aimed, to some degree idealistically, at assimilation with the mother country and although local differences in the colonies and trust territories existed, the central legislature and the Colonial Ministry in Paris, despite distinctions between metropolitan and colonial law, imparted a marked unity in the colonial empire, contrasting with the British Empire, each of whose territories were viewed as separate entities.

European settlement of the French African territories was in many instances hampered by their very vastness combined with relatively small population, difficult terrain and unpromising resources. Moreover, the closeness of such North African territories as Algeria and Tunisia to France understandably made them of greater practical interest. Algeria, in particular, acted as an experimental theatre where the conflict between French *colon* and native land-holder was played out and legislative and administrative machinery was developed to resolve it. Apart from those in North Africa, the French territories in Africa consisted of two great blocks, French West Africa (A.O.F.) and French Equatorial Africa (A.E.F.) based on Dakar and Brazzaville respectively, together with Togo and Cameroon in the West and Côtes des Somalis in the East. Malagasy (Madagascar) and the Comores may be regarded as distinct from this French tropical Africa. Even at the pre-Independence census of 1956, the number of French and other foreign citizens in A.O.F. did not exceed 90,000, of whom more than half were in Senegal, and A.E.F. totalled only 25,000, two-fifths of whom were in Moyen-Congo; Togo had about 1,300 whites, Cameroon about 17,000 and Djibouti less than 5,000. Judged by the criteria of effective settlement in tropical Africa, France would appear to have met with conspicuously little success in some areas, although the peculiar demographic shadow effect of the North African settlement goes far to explain this and the alternative stress on concessionary régimes and native agriculture. Another significant feature of the French tropical African territories involved their land-locked positions, isolating them in terms of exports and making their economic development heavily dependent on improved communications, particularly railways. This interior position of much of French Africa in the Saharan and Sahelian zones meant that as a culture area it was predominantly Muslim, and consequently much of the local law and custom was influenced by Muslim principles. Not only was the majority of population on many of the now autonomous states Muslim – Guinea, Senegal, Mali, Niger and Chad being examples – but, more significantly, the élites were Muslim. As in the case of the British in Nigeria, so

the French encountered and reacted to clearly differentiated social environments straddling the natural life zones; the pagan forest peoples were more
subject to Christian missionaries and education and the Muslim savanna
peoples retained institutions which were not necessarily amenable to
changes implicit in an alien cash economy or modern techniques, for all
their long-standing trans-Saharan trade links.

Founded in 1895, A.O.F. was administered from Dakar and Senegal became the core area for an expanding hinterland, Upper Volta being
amalgamated as late as 1919 and Mauritania in 1920. The port of Dakar
provided an outlet for railways converging on it from Mauritania and Mali,
and the rail network relates strikingly to present areas of groundnut production both inside Senegal and in western Mali. Similar features are
evident in the case of railways leading inland from the coast at Conakry,
Abidjan and Cotonou. Indeed, it has been noted that 'the economy of West
Africa has been completely reversed by the railway links between the
Islamic states of the Niger basin and the old trading ports of the Gulf of
Guinea . . . the new groupings in French-speaking Africa form themselves
on the railways' (Carrington, 1960). The economic development of A.O.F.
has been a slow process, Lord Hailey's *Survey* commenting that 'there was
little pressure to secure land for colonists. An attempt to start a plantation
system in Senegal in the nineteenth century had proved abortive, and
henceforward economic policy was directed mainly to the development of
the peasant economy' (Hailey, 1957). French land policy in the four territories of Moyen-Congo, Oubangui-Chari, Gabon and Chad, which had
been federated in 1910 to form A.E.F., manifested a concessionary character which took as its model Leopold's Congo but avoided direct intervention by the state. After the granting of exclusive rights over agriculture,
forestry and industry to various companies by the Commission of Colonial
Concessions from 1899, more and more of equatorial Africa was ceded in
return for cash payments and performance of certain administrative functions. Normally African reserves totalling one-tenth of the ceded areas
were allocated, but in effect practically the whole of A.E.F. passed into the
hands of concession-holders, and such was the frantic exploitation that
French liberal opinion obliged the Government, when excesses were made
public, to intervene. By 1910 the apogee had been passed and financial mis-
management sent the concession system into decline. Land which had been
taken out of native hands when originally alienated was not, however, reallocated to the tribes but reverted to the state in terms of metropolitan
decrees of 1899 covering *terres vacantes et sans maître*.

Progressively, exports changed from the uncultivated types such as
rubber and timber to a widening range of cultivated crops such as cotton,
coffee and cocoa, to which were later added groundnuts, tobacco, sisal and

other plants. At one extreme in Gabon, much of which was shrouded in rain forest including *okoumé*, export production has remained markedly localized as in the case of oil-palm, cocoa and rice, and the *paysannat* schemes seem to be in their infancy. At the other extreme Chad, a country of remarkably different physical conditions, strikingly illustrates the effect of introducing a new crop. Under the cotton monopoly *Cotonfron*, cotton production was commenced in 1928 and by the time of Independence extended to about 600,000 acres of largely peasant holdings and accounted for 80 per cent of exports by value. For the traditional cultural landscape such schemes were revolutionary in the long term, leading as they did with implementation of the *paysannat indigène* policy from 1952, to wholesale regrouping of villages. The history of cotton in A.E.F. has been a somewhat chequered one since its inception in 1924, despite the million and a half people in Chad and Ubangui estimated to be wholly or partially dependent on the crop. Thompson and Adloff state unequivocally that 'for nearly three decades the African farmer in northern A.E.F., was more a victim than a beneficiary of the official pressure to grow ever larger quantities of cotton' (Thompson and Adloff, 1960).

French administration was aided by a hierarchy of land, village and canton chiefs which facilitated operation of their policy which initially resorted to the *prestation* system of taxation whereby liability could be absolved by prescribed labour terms in lieu of cash payment. In the early years the concessionary companies perpetrated many abuses and even when labour legislation improved the situation the shortage of manpower on the plantations continued and necessitated contract labour by migrants from distant parts of both federations. Inevitably there were deleterious effects on native agriculture. Throughout French tropical Africa most land continued, in the absence of settler competition, to remain under traditional systems of ownership and in consequence when changes came they concentrated on converting *de facto* ownership to *de jure* forms, with individualization dependent on registration (*immatriculation*) of title. But by introducing French laws of possession and in particular the *jus utendi et abutendi*, which Dumont has criticized, the originally sound intention of developing a peasantry of proprietors rather than occupiers brought with it the seeds of future conflict between individual and national interests.

Seen as a whole, French tropical African territories demonstrate, nevertheless, a continuity of agricultural emphasis. De Carbon has calculated that four major commodities (groundnuts, coffee, cocoa and timber) accounted for three-quarters of exports in 1958 and 1938 (de Carbon, 1964). In many important respects the French territories under consideration suffered from a deficiency of capital investment up to the outbreak of the Second World War, as did the Portuguese possessions, when compared with the British territories, although investment in *per caput* terms qualifies regional advantages; Frankel's estimates of total overseas investment in Africa up to 1936 suggested that the British Empire territories received £941 million (of which £523 million were invested in South Africa), the Belgian Congo £143 million, the Portuguese territories £67 million, A.O.F. £30 million and A.E.F. £21 million (Frankel, 1938). Certainly the attitude of the metropolitan power to foreign capital investment facilitated or hindered certain developments; France discriminated by compelling a majority French holding in any investing company whereas Belgium had an 'open door' policy, and as early as 1928 France had begun to consolidate her trading empire by a customs union and tariff system.

Post-war change in French African policy found its expression in the 1946 constitution which recognized that progress towards the original goal of either political assimilation with metropolitan France (i.e. overseas departments) or development of the status of associated state within the French Union might not be speedily achieved, and accordingly a transitional status of associated territory was envisaged, necessitating extension of French citizenship to hitherto juridically dualist colonial populations, albeit retaining a split franchise (Robinson, 1955). Development in the interests of local populations was accelerated by the formation of an investment fund, commonly known as *FIDES*, which drew on French governmental subventions committed to expenditure on agriculture, research and social infrastructure. In the case of A.E.F. the French credits had to concentrate particularly on an infrastructure which was, to quote Thompson, 'strengthened beyond its productive capacity' (Thompson and Adloff, 1960). In the case of A.O.F. high-cost new or continuing development schemes, such as the massive irrigation colonies of the *Office du Niger* and elsewhere, have transformed the agricultural patterns but, with the exception of the Ivory Coast and Guinea, the concessions never attained great size or importance and have not inflicted damage which has first had to be made good. Consequently the Africans in A.O.F. would appear to have taken greater advantage of the opportunities of participating in the cash economy, where natural conditions permitted this, than in most other parts of Africa.

Former German territories

It would be wrong to ignore totally the effects of German colonization, even though it was a mere episode, extending effectively only from 1884 until 1914, which gave way to the B Mandates of Tanganyika, Ruanda-Urundi, Togoland and the Cameroons, together with the C Mandate of South West Africa. With Bismarck's annexation of Lüderitz Bay in 1884, the contributions to African exploration of Barth, Schweinfurth, Nachtigal and von Wissman gained tangible advantages for Germany and led to a colonial empire of 700,000 square miles in Africa. Settlement was confined almost wholly to East Africa where the German East Africa Company, founded in 1891, operated for the entire period. Plantation agriculture was established in Kamerun (chiefly coffee) and Togo (cotton). The East African settlement concentrated on the Dar-es-Salaam and Usambara Mountains regions and white settlers numbered little more than 5,000. A breakdown of population indicates that there were only 685 planters and farmers in East Africa in 1911, 111 in Kamerun and 5 in Togoland (Henderson, 1962). Nevertheless the 80-year-old German-founded commercial plantations of west Cameroon retain their importance to the present day and Bederman has noted that 'except in the realm of human relations, the Germans were undoubtedly the most progressive of colonial rulers. Many of their problems were solved by scientific methods' (Bederman, 1966). Since many Germans privately repurchased their former plantations and farms or moved into the Portuguese territories, their effect on agricultural patterns in certain areas has not been entirely negligible.

The British territories

If France was characterized by centralization and a rational approach to her colonies, then Britain might be stereotyped as decentralizing authority by the indirect rule through existing political institutions and preferring the empirical to the theoretical approach. Certainly this simplification would serve to explain the great variations both between and within the former British territories. This variability is further complicated by geographical heterogeneity as such which differentiates territories on a basis of their suitability or not for white settlement. Their population was in some cases already dense and their institutions well developed; in other cases there existed a positive incentive to create settler colonies where climatic conditions so disposed and the consequent influx not only of Europeans but also Asian settlers served to sharpen land conflicts.

Early penetration of the ultimately British areas had been accomplished by traders and missionaries who were assisted in pacification rather than administration by the British Government. Early administration was a charge on various of the great chartered companies which followed in the footsteps of the private Sierra Leone Company of 1800, and the aloofness of the ruling caste vis-à-vis commerce tended to place administrators in the position of adjudicators between native and European interests. The British African Empire grew essentially from the amalgamation of territories hitherto administered by the Foreign Office, but the evolution towards a unified Colonial Office policy is a complex matter best left to historians.

M

Private enterprise laid the foundation of British spheres of influence and of these the Royal Niger Company (1886), the Imperial British East Africa Company (1888) and the British South Africa Company (1889) were particularly significant in this regard. The contrast between possessions in High and Low Africa is marked and the question of alienation of land is crucial, especially where the companies handed over administration to Whitehall and the concessions granted to them by local chiefs became matters of legal dispute. Large tracts of land were alienated for white colonization in both eastern and southern Africa and native reserves were set aside as early as 1894 in Southern Rhodesia and 1904 in Kenya. The obviously impending clash of interests in High Africa – in Low Africa concessions generally concerned specific mineral, forestry or other rights – was partially resolved by a Privy Council decision in 1918 on the Matabele Concession in Southern Rhodesia, which determined that unalienated company lands passed to the Crown. This decision implied forfeiture and to some extent enabled the Crown to redress the balance upset by wholesale confiscation of tribal lands. Pacification had led not only to spatial restriction of tribes but also to an increase in their numbers in areas of locally inferior soils and the new ruling facilitated extension of native reserves and creation of *native trust lands* – that is, areas administered for the benefit of the native inhabitants. The consequence of this complex division of certain territories into blocks of land alienated for white settlement, unalienated Crown Land, native reserves, native trust lands, native purchase areas, game and forest reserves and various other categories, have obviously been far-reaching in legal, social, economic and other terms. In the case of South Africa, where 89 per cent of the land was alienated to Europeans, and Southern Rhodesia, where the figure was about 49 per cent, the concept of native reserves has come nearest to the Roman practice of absolute separation of native and colonist and is effectively a segregation (*apartheid*). Elsewhere the marked dualism of society was nowhere so harshly discriminatory as in this region of Romano-Dutch law and could neither in scale of land holdings nor size of immigrant population, whether in Kenya or Zambia, support more than temporary white nationalism.

Private enterprise in Low Africa took on a different form and British West Africa was spared massive expropriation of lands. Cultural segregation was shared in some measure with settler colonies but for reasons of institutional conservatism. Lugard's 'invention' of *indirect rule* whereby the power of the existing authorities was consolidated, was born of necessity and realized by the gradual adoption of local government through the native authority. The dominant factor in this tribal organization, as Carrington suggests, 'was always land tenure and the West African policy intended that land should not be alienated. There were too many parts of the world ... where recognition of tribal authority had merely enabled chiefs to sell the land to concession-hunters. Under the Lugard system the alienation of land was prevented' (Carrington, 1961). Peasant agriculture rather than plantation agriculture was the means by which traditional 'plant..tion' crops were produced on the encouragement of the Colonial Office and Udo has argued that this anti-plantation policy was, despite the sound intention to protect indigenous tribal life, a retarding influence in the case of southern Nigeria's economic growth (Udo, 1965).

In the case of Nigeria, its division into a colony and protectorate and the existence of some 300 tribes extending from the coast to the northern emirates has led to considerable diversity of land tenure and variations in agricultural patterns consequent upon cash cropping and steady population increase. Meek has investigated the complexities of land tenure and land administration in Nigeria and it would be wholly futile to attempt here any generalizations from this formidable survey save to note the gradual shift towards individualization of holdings both in Muslim and pagan areas and the creation of a free peasantry (Meek, 1957). Of great importance has been the role of native authorities since 1945 in disposal and control of land, and sales or pledging of land have increased to a point where fragmentation, sub-tenancies and other features, which may lead to the emergence of a landless proletariat, must cause concern. Nevertheless land utilization data collated by Meek show only 30,723 square miles under farm crops, and 3,935 square miles under tree crops, while 220,402 square miles from a total area of 339,169 square miles in Nigeria remain actually uncultivated bush or waste. The incipient pressures consequent upon increasing commercial value of land may therefore tend to encourage coalescence of the present nodes of agricultural cash cropping centred on the southern cocoa-palm zone and the cotton–groundnuts zone north from Kaduna.

British colonial policy, like that of other colonial powers, has undergone significant changes which have found expression in the visible landscape. From a *laissez-faire* economic policy, indirect rule spread with the strength of dogma throughout British tropical Africa and it only received a severe jolting when democratization of indigenous institutions began to be effected and the fragility of tribal institutions in the face of agricultural or any other change became manifest. Policies of the paramountcy of native interests – from 1930 – and the preparation for colonial self-government began to stimulate development in economic and social spheres as well as the political, and it became obvious that in place of annual budgeting, co-ordinated development plans were essential. The basis for accelerated development was laid by the Colonial Development and Welfare Act of 1940 and the series of colonial development plans which were formulated after 1945. The British Overseas Food Corporation and the Colonial Development

Corporation inaugurated numerous projects, both pilot and major in scope; some of these achieved success while others, despite large-scale, often mechanized operation, did not. As in other colonial areas, Britain appears to have swung increasingly towards creation of a 'middle class' African peasant, modelled in part on the success of his fellows on *paysannat* communal farms in the French territories. *Group farming* schemes involving relatively low capital investment achieved success in Nigeria and the *peasant farming* schemes introduced into Northern Rhodesia in 1948 were symptomatic of changed emphasis. Significantly there seems to have been ready transfer of successful concepts within the British sphere as well as a willingness to learn from Belgian and French experience. Given time this trend might well have led to a *convergence* of cultural landscape types both within the British sphere and, making due allowance for variance in ecological conditions, within colonial Africa as a whole. Dispersion of settlements and progressive enclosure in advance of *de jure* ownership became increasingly common.

In Uganda the British Commissioner, Sir H. Johnston, had already in 1892 begun to 'encourage the more intelligent among the natives to exchange their present communal holding which checks individual enterprise for the acquisition of personal titles'; in the Gambia the Governor introduced a form of native leasehold in 'waste lands' in 1895, and a comparable development occurred in the Gold Coast two years later with the publication of the Public Lands Bill (Branney, 1959).

This movement towards personal ownership, of either a freehold or leasehold character, has, however, remained slow throughout most of the colonial period, retarded by uncertainty of purpose in economic and social policy and the apparent antipathy of many African societies towards any promotion of individual interests at the expense of the group. In tropical Africa as a whole it remains the interesting exception to the rule although its accelerated application in the post-war years promised radical changes on the territorial rather than the local scale.

Conclusion

For any comparison to be made between former colonies or groups of colonies, the most appropriate method of analysis of agricultural patterns as expressions of administrative policy would seem to be not the sectoral approach, but rather the genetic classification of distinct types of *agrarian landscape* and the assessment of their areal and socio-economic importance. In examining this *Landschaftswandel* it would seem appropriate to consider particularly the three-dimensional aspects of any distributional patterns, since the *inversion* of the cultivated areas of which Pierre Gouru has written, whereby intensive techniques are, as in the Asiatic tropics, applied to the lowest soils instead of clinging to the plateau surfaces or hill slopes, has often been initiated by the European plantation or government agriculturalist or more prosaically 'in the interests of good administration'. Similarly, the search for 'temperate' conditions has caused the European to extend greatly the altitudinal range formerly available to African agriculture. Stress has been laid upon dualism in society and economy and very largely this has been shown to hang on attitudes to land as a resource where rights *in rem* and rights *in personam* have seemed incapable of resolution. The selection of criteria for categories of agricultural patterns can lead to confusion if account is taken of such factors as bioclimatic regions, population–land factors and institutional features, with a resultant multiplicity of classes which clouds perception. Most agricultural patterns in tropical Africa would, however, seem to fall into one of a number of major categories if zones of overlap are disregarded, and the following provisional classification is suggested. Some explanation and exemplification of this landscape classification is called for, but it must be stressed that the categories are tentatively formulated in the interests of reducing the self-evidently multifarious landscapes of reality to simpler proportions. The suggested groupings derive from experience and evaluation in depth of a single colony made possible by former active involvement in provincial administration. Shorter visits to other African territories, together with selective reading of the literature, suggest that the classification may deserve a wider application.

1. *Archaic:* A group comprising unmodified shifting, semi-permanent or permanent traditional systems under customary tenure.

2. *Transitional:* A group showing features of regression or progression. The *degenerate* forms, developing as a consequence of over-population, male depopulation or related causes and displaying such features as erosion, excessive fragmentation and neglect; the *regenerate* forms display adaptation by adoption of new techniques and cultigens voluntarily, by direct but limited European intervention through soil conservation measures, welldigging, seed-farms, etc., or by such direct actions or agencies as road construction and private commercial operations.

3. *Ephemeral:* A special group including illegal squatters, displaced political refugees and other short-lived occupancies such as those associated with tsetse-fly or game clearance schemes, dam construction, etc., where agriculture nevertheless persists.

4. *Colonial:* This group comprises patterns arising from essentially *exogenetic* or *endogenetic* causes.

The *plantation* landscape is essentially a European introduction dependent on large-scale alienation or concession of land. The *large-farm* landscape, unlike the plantation, is usually confined to areas reserved for white settlement. Both tend to develop laboursheds far beyond their immediate environs.

African *settlement scheme* landscapes may be differentiated into irrigated and rainfall agriculture types and represent policies aimed at intensifying cash-crop production, diversifying existing patterns or widening employment opportunities. Some show evidence of developing from an existing base and are conventionally of the *peasant farm* type, with or without individual freehold or leasehold tenure. Others have developed from uneconomic disposition of holdings and merely represent reorganization by consolidation and the rationalization of communal land use. Others are on a scale often dependent on mechanization and strongly centralized control, co-operative marketing, etc., and approximate to the state farm or syndicate with share cropping or leasehold tenures.

Although in the strict sense no tribe in an African colony could entirely escape the notice of a European administration, and all were therefore subject to exogenous factors of change, in practice many were so remote from core areas and so beset by adverse environmental or locational factors as to continue virtually unchanged in their forms of agriculture. It cannot be over-emphasized how lightly administered were certain areas, particularly in the larger but more thinly populated territories, and it has been in such areas that survivals – in the shape of fossilized land-use patterns – are most significant. Policies of social and economic development were universally modified by the exigencies of two world wars and the retrenchments of the depression years and any detailed study of official reports covering single districts reveals how much of the colonial *pax* was in fact an enforced stagnation with mere interludes of activity. Since farmers are by reputation conservative, the term *Archaic* does not seem misplaced in describing the antiquated systems which exist where static factors outweigh the dynamic and an absence of land hunger is typical, although such stagnant situations may, nevertheless, quickly pass into the *Transitional* phase. Characteristically the areas covered by such archaic agrarian landscapes will be almost entirely self-sufficient in economic terms and denied access to commercial markets because of remoteness, inadequate communications or discriminatory marketing levies and tariffs; administrative and agricultural priorities at central government level would tend to the preservation of the *status quo*. The Kaonde, a group of shifting cultivators of the Congo–Zambezi watershed, were such an example. Their isolation was verified by the author when working as an administrator in a district of over 16,000 square miles (the size of Switzerland) yet populated by a mere 32,000 people, for most of whom the annual tour by the district officer was the sole contact with the white man's administration. As early as 1938, some 33 per cent of taxable males were absent on labour migration, and by 1961 this figure had increased to about 60 per cent; yet this remote participation in the money economy had been compensated for by internal changes in the division of labour, rather than easily visible signs of progress or decline in the agrarian landscape as such, because the physical resources have had to support only low population densities. Another tribal case in which a traditional and archaic pattern has been retained would appear to be in the Jie area of Karamoja in north-eastern Uganda. The British ended large-scale movements of the essentially transhumant Jie, and the effect has been to differentiate the fixed homesteads (*ngireria*) at which older men and the women maintain the dairy herds and carry on agriculture, from the peregrinating stock camps (Gulliver, 1955). Writing of the Fur people in the Sudan, Lebon has noted how these sedentary cultivators, whose terraced slopes are remarkable, preserve 'an economy almost entirely self-sufficient owing to the inaccessibility of most parts of the Mountain (Jebel Marra) from the markets' (Lebon, 1965). While survivals of this sort are without doubt decreasing in number and extent, it seems probable that most larger territories embrace examples.

Throughout most parts of Africa the *Transitional* form is dominant among agricultural patterns since traditional systems have only been capable of maintaining food supplies without deterioration of the land where demand has remained under control. Pressure on land clearly arises from a steep rise in population, the introduction of cash crops (in addition to subsistence crops), the conversion from hoe to plough, and, in certain cases, direct competition with immigrant groups. Kraal manure is frequently unavailable for overworked soil and fallow periods decline as decreased yields set in. Where no increase in the farm area is available, over-cultivation and fragmentation of existing holdings usually appear and the resulting 'cultivation steppe' will be commonly exposed to excessive runoff, desiccation and ultimate degradation to bad lands topography. Alternatively, the breakdown of traditional agricultural systems may result from excessive wage migrations where able-bodied males are no longer available for such tasks as heavy bush clearance. Often this transition has led to degeneration in traditional systems which may be characterized by neglect of subsidiary crops and a consequent fall in the quality of the local diet. Degeneration has been a relatively common feature in many African territories and has commonly needed drastic governmental remedies, often involving the wholesale transfer of population or rehabilitation of landscape and thus involving the substitution of a *Transitional* landscape by one within the *colonial* category. The case of the Ngoni tribe, spatially restricted by the arrival of the British and the subsequent alienation of land to white settlers in eastern Zambia, is an example of an accelerating decline (Hellen,

1962). Similar land hunger arose in the Kikuyu areas of Kenya as well as in the Usambara and Uluguru Mountains of Tanzania, while in West Africa there are many degenerating areas adjacent to towns or cash-cropping land. The plain of Barotseland which Livingstone has described as bearing a close resemblance to the Nile Valley and whose people he described as never being 'in want of grain, by taking advantage of the moisture of the inundation they can raise two crops a year', has experienced a decline (Livingstone, 1857). This followed the ending of slavery (1906) and tribute labour (1926) which had made possible the construction and maintenance of a 400-mile network of drainage and navigation canals under the paramount chiefs and the cultivation of extensive peat seepage soils. Since that time the productive area has declined and difficulties have arisen over a high water-table re-established by this neglect; Barotseland has ceased to be self-sufficient in foodstuffs in a situation where land is still plentiful (Hellen, 1963). In certain highland areas the neglect of terracing or box-ridging has had a similar effect and parts of the Matengo area of south Tanzania demonstrated this deviation from good conservation prac-tice (Stenhouse, 1944, quoted in Allan, 1965).

Transitional but regenerate forms are commonplace and may coexist alongside examples of glaring degeneration. One tribe may respond to change by conscious adaptation and another may languish in an identical habitat. Central government agencies may inject limited financial, material and technical aid with a similarly sporadic outcome. Unlike certain forms of colonial landscape pattern, the transitional but regenerate form does not represent a wholesale reorganization of agrarian landscape or society but rather a selective, albeit scientific, intervention of modest proportions. Mandatory rules have often had a greater effect than self-help, and soil conservation legislation has been particularly important in halting degrada-tion of African areas. Prohibition of steep-hill-slope cultivation and the widespread adoption of contour ploughing, bench terracing and afforesta-tion of interfluves, has proved beneficial. One of the classic areas is exempli-fied by the Kabale area of Kigezi in Uganda where steep slopes of red loam were supporting population densities in excess of one person per acre by 1943. The Babika were said to cultivate slopes up to 35° and to graze slopes of up to 45° and their numbers were further increasing through immigra-tion and a high birth rate (Purseglove, 1946). Subsequent conservation measures following survey and controlled resettlement have supported the inherently good conservationist practices of the people and the unwitting decline occasioned by congestion has been halted. In a less striking manner the compulsory introduction of certain crops has had a widespread re-generative effect, as has voluntary acceptance of crop rotation to overcome shortage of land.

Ephemeral patterns are, as implied by the term, short-lived occupances due to extraordinary events or short-term expedients. Areas were set aside in the immediate pre-war period for the resettlement of landless Kikuyu labourers from European farms and the Mau Mau emergency of the 1950s saw the so-called 'villagization' for security reasons of an otherwise typically dispersed Kikuyu settlement habit. Similarly, ephemeral patterns exist where, for example, refugee groups have moved into neighbouring countries (e.g. the movements from the Congo, Rwanda and the Sudan) or where squatters occupy overspill zones from which they can expect eventual expulsion. Illegal squatting appears to have been a recurrent problem both in European farming and plantation areas and persists into the independence phase. Illegal compounds and villages with their associated subsistence agriculture taking the form of peri-urban fringe-belt areas are a common feature of expanding towns and by their very nature must be regarded as ephemeral by the authorities. During major projects such as tsetse-fly clearance it has often been found necessary to concentrate a generally thin and scattered population into settlement corridors until rehabilitation has been completed.

Colonial landscapes are probably the most clear-cut of categories because of the scale or intensity of intervention of the imperial power. The plantation or large-farm landscapes are clearly distinguishable by the nature and scale of their organization and, notably, by concentration on a few generally exotic crop types which dominate vast areas. In general such landscapes represent an extreme expression of exogenetic forces made possible by an almost total disregard of the indigenous social environment. Settlement scheme landscapes represent, by contrast, a local balance between major European policy involving such aims as the development of export commodity production and the rational distribution of population or reorganization of land tenure within the limits imposed by the expectations and capabilities of the native groups involved. Scant attention has been paid to indigenous society in certain cases such as the major irrigation projects, even to the extent of importing alien African colonists; in other cases the pattern may represent a highly effective modernization of existing husbandry effected with the full co-operation of those people involved. In nearly all cases the settlement schemes and peasant farms have been well chronicled by colonial governments and their histories, if not their true costs, recorded in countless ministerial reports. Such zones or foci are, together with the areas retained as plantations and large farms, the key to further transition and transformation of the cultural landscape. The stage of active competition in land use and land ownership must inevitably place archaic and transitional categories under increasing pressure to change. In effect, with certain modifications, the *colonial* group of agrarian landscape patterns becomes the post-independence *political* landscape, in so far as this latter type is an expression of state policy, which admits the inadequacy of traditional agriculture, and has the power to effect change.

Under whatever form of land tenure, however, many of the African territories are faced with the urgent necessity of providing non-agricultural occupations lest they be overwhelmed by population increase and frustrated by the inefficient political expedient of handing over land to the landless. The nature of the nationalist political platform of pre-independence days was such as to invest the slogan 'people without land deserve land without people' with the weight of dogma. If in retrospect much of the colonial legacy seems inappropriate to present African realities and the evolution of land policy, to borrow Dumont's description, 'anarchique', many of the forms imposed or experimented with by Europeans remain guide-lines in an area of limited choice. With a fresh start in Africa the administrations concerned have a unique opportunity to study the relative merits of external models – whether they be smallholder co-operative, state syndicate or commune – and to decide whether land reforms and hence landscape change should be imposed autocratically or democratically. In retrospect, decreasing returns may well prove to be the Achilles' heel of peasant agriculture, and hence the absence of an inherited prior commitment to peasant proprietorship in many countries will permit the introduction of systems offering the greatest productivity from land and people in a global situation where food deficits rather than economic growth demand immediate attention.

References

ALLAN, W. 1949. *Studies in African Land Usage in Northern Rhodesia*, Rhodes–Livingstone Pap. 19.
  1960. Changing patterns of African land use. *Jl R. Soc. Arts* **58**, 612–29.
  1965. *The African Husbandman*, Edinburgh.
ANSTEY, R. 1966. *King Leopold's Congo*, London.
BEDERMAN, S. H. 1966. Plantation agriculture in Victoria Division, West Cameroon. An historical introduction. *Geography* **51**, 4, 349–60.
BIEBUYCK, D. (ed.). 1963. *African Agrarian Systems*, London.
BOHANNAN, P. 1963. 'Land', 'tenure' and land-tenure. In BIEBUYCK, D. op. cit., 101–15.
BRANNEY, L. 1959. Towards the systematic individualization of African land tenure. *Jl afr. Admin.* **11**, 4, 208–14.
BRUNSCHWIG, H. 1966. *French Colonialism 1871–1914*, London.
CARBON, L. B. DE. 1964. Problems of economic development of French-language countries and territories. In ROBINSON, E. A. G. (ed.), *Economic Development for Africa South of the Sahara*, London, 138–83.
CARRINGTON, C. E. 1960. Frontiers in Africa. *Int. Affairs* **36**, 4, 424–39.
  1961. *The Liquidation of the British Empire*, Reid Lectures of Acadia University 1959, London.

CHISHOLM, M. 1962. *Rural Settlement and Land Use: an Essay in Location*, London.

COLSTON. 1950. *Principles and Methods of Colonial Administration*, (Colston Papers), London.

DONGEN, E. VAN, 1961. Coffee trade, coffee regions and coffee ports in Angola. *Econ. Geogr.* **37**, 4, 320–46.

DUFFY, J. 1961. *Portuguese Africa*, London.

DUMONT, R. 1961. *Afrique noire: développement agricole. Réconversion de l'économie agricole: Guinée, Côte d'Ivoire, Mali*, Paris.

1962. *L'Afrique noire est mal partie*, Paris; English trans. 1966, *False Start in Africa*, London.

FRANKEL, S. H. 1938. *Capital Investment in Africa*, London.

GERSDORFF, R. VON, 1960. Angola. *Die Länder Afrikas*, Deutsche Afrika-Gesellschaft, Bonn, 14.

GLUCKMAN, M. 1945. African land tenure. *Rhodes–Livingstone J.* **3**, 1–12.

GOUROU, P. 1956. The quality of land use of tropical cultivators. In THOMAS, W. L. (ed.), *Man's Role in Changing the Face of the Earth*, Chicago, 336–49.

GUICHONNET, P. 1965. Le Mozambique: esquisse géographique. *Globe*, Geneva **105**, 35–96.

GULLIVER, P. H. 1955. *The Family Herds. A Study of two Pastoral Tribes in East Africa. The Jie and Turkana*, London.

HAILEY, LORD, 1957. *An African Survey*, London.

HANCE, W. A., KOTSCHAR, V. and PETEERC, R. J. 1961. Source areas of export production in tropical Africa. *Geogrl Rev.* **51**, 4, 487–99.

HELLEN, J. A. 1962. Some aspects of land use and overpopulation in the Ngoni reserves of Northern Rhodesia. *Erdkunde* **16**, 3, 191–205.

1963. Barotseland. *Geogr. Rdsch.* **15**, 4, 149–60.

HENDERSON, W. O. 1962. *Studies in German Colonial History*, London.

HODDER, B. W. 1965. Some comments on the origins of traditional markets in Africa south of the Sahara. *Trans. Inst. Br. Geogr.* **36**, 7, 97–105.

INFORCONGO, 1959–60. *Belgian Congo*, 2 vols, Brussels.

LEBON, J. H. G. 1965. *Land Use in Sudan*, World Land Use Survey, Monograph No. 4, Bude, Cornwall.

LIVINGSTONE, D. 1857. *Missionary Travels in South Africa*, London.

MAUNIER, R. 1949. *The Sociology of Colonies. An Introduction to the Study of Race Contact*, 2 vols, London.

MEEK, C. K. 1946. *Land Law and Custom in the Colonies*, Oxford.

1957. Land tenure and land administration in Nigeria and the Cameroons. *Colonial Research Series* 22, London.

NEUMANN, H. 1965. Portugal's policy in Africa. The four years since the beginning of the uprising in Angola. *Int. Affairs* **41**, 4, 663–75.

NIEMEIER, G. 1966. Die moderne Bauernkolonisation in Angola und Moçambique. *Geogr. Rdsch.* **18**, 10, 367–76.

NICOLAI, H. 1963. *Le Kwilu, étude géographique d'une région congolaise*, Brussels.

NICOLAI, H. and JACQUES, J. 1954. *La Transformation des paysages congolais par le chemin de fer. L'exemple du B.C.K.*, Institut Royal Colonial Belge, Brussels.

NYE, P. H. and GREENLAND, D. J. 1960. The soil under shifting culti-
    vation. Commonwealth Agricultural Bureaux, *Technical Communi-
    cation* **51**.
O'CONNOR, A. M. 1965. New railway construction and the pattern of
    economic development in East Africa. *Trans. Inst. Br. Geogr.* **36**, 21–30.
PURSEGLOVE, J. W. 1946. Land use in the over-populated areas of
    Kabale, Kigezi District, Uganda. *E. Afr. agric. J.* **12**, 3–10.
RICHARDS, A. I. 1939. *Land, Labour and Diet in Northern Rhodesia*,
    London.
ROBINSON, K. 1955. French Africa and the French Union. In HAINES,
    C. G. (ed.), *Africa Today*, Baltimore, 311–36.
SCHLIPPE, P. DE, 1956. *Shifting Agriculture in Africa: the Zande System
    of Agriculture*, London.
STENHOUSE, A. S. 1944. Agriculture in the Matengo Highlands. *E. Afr.
    agric. J.* **10**
THOMPSON, V. and ADLOFF, R. 1960. *The Emerging States of French
    Equatorial Africa*, Standford.
TROLL, C. 1950. Die geographische Landschaft und ihre Erforschung.
    Studium Generale 4/5, 163–181 (reprinted in 'Okologische Landschafts-
    forschung und vergleichende Hochgebirgsforschung'. *Erdk. Wissen*,
    vol. 11, Wiesbaden 1966).
    1962. Die geographische Strukturanalyse in ihrer Bedeutung für die
    Entwicklungshilfe. *Basler Beitr. Geogr. Ethnol.*, Geographische Reihe
    5, 25–52.
UDO. R. K. 1965. Sixty years of plantation agriculture in Nigeria: 1902–62.
    Econ. Geogr. **41**, 356–68.
UNO. 1959. *United Nations Economic Survey of Africa*, New York.
YOUNG, C. 1965. *Politics in the Congo. Decolonization and Independence*,
    Princeton.

PART III

Case studies

13 Natural resource survey in Malawi: some considerations of the regional method in environmental description

A. YOUNG

Over the past twenty years it has increasingly become accepted that natural resource surveys should form a basis for agricultural development and land-use planning in the tropics. Recognition of this need arose, perhaps more than for any other single reason, from the lessons of the 1947 East African Groundnut Scheme, the failure of which was attributable in part to inadequate knowledge of the conditions of the physical environment (Phillips, 1959). The period since the Second World War has seen a considerable expansion in the soil-survey organizations of many tropical countries. It is now a normal requirement by international organizations such as the FAO and the World Bank that environmental surveys should precede land development projects to which international aid is to be given, and firms of consultants on agriculture and land-use planning are active in many parts of the tropics. In Africa natural resource surveys have been carried out by the British Directorate of Overseas Surveys, the French Office de la Recherche Scientifique et Technique d'Outre-Mer, the Belgian Institut National pour Étude Agronomique au Congo, as well as by various national soil-survey organizations. In many land-development projects, resource surveys are combined with engineering and economic investigations; but it may be noted that the need for corresponding studies of the effects of changes in settlement patterns, land tenure systems and other socio-geographical aspects is not yet widely recognized.

A distinction may be made between soil surveys and natural resource surveys. In a soil survey other aspects of the physical environment are investigated in less detail, and are presented as ancillary to a soil map; in a natural resource survey the mapping units are based on the physical environment as a whole, or on more than one environmental factor. There is a further division between detailed surveys, mapped on scales of 1:50,000 or larger, covering local areas and usually carried out with a specific development project in view; and reconnaissance surveys, presented on scales of 1:250,000 or smaller, covering the whole or large parts of a country and intended for general planning purposes.

In East and Central Africa the earliest examples of reconnaissance natural resource surveys were those of Zambia by Trapnell (1937, 1943, 1948). Surveys have subsequently appeared covering Moçambique (1955), Rhodesia (Vincent et al., 1961), Kenya, Uganda and Tanzania (Russell, 1962), and eastern Botswana (Bawden and Stobbs, 1963; Bawden 1965).

In Malawi resource surveys[1] carried out prior to 1959 consisted mainly of detailed soil and vegetation surveys of small areas (Nyasaland Department of Agriculture, 1954–61); a notable early soil map by Hornby (1938) was limited by inadequate base maps, while a vegetation classification applicable to the whole country had been established, and a preliminary vegetation map produced by Jackson (1954, 1959). In 1959 a reconnaissance natural resource survey was initiated, with the objects of assessing the agricultural resources of the country, mapping their distribution, and providing a classification framework to which detailed surveys could be related. For the northern and central parts of Malawi the results of this survey have been published as maps and memoirs (Young, 1965; Young and Brown, 1962, 1965[2]).

It is the purpose of this chapter to examine certain general questions that arose in the course of the Malawi survey. These questions relate to the areal differentiation of the physical environment, the relations between environmental factors, and the use of the concept of natural regions in presenting the results of a natural resource survey. The possibility of analysing relations between environmental factors by means of their mapped distributions is also discussed. Some general features of the environment of Malawi are described, primarily to illustrate the nature of regional differentiation involved; for detailed environmental descriptions reference should be made to the publications cited above.

Distribution patterns of environmental factors

Geology, geomorphology, climate, hydrology, soils, vegetation, fauna and disease may be termed the main factors of the physical environment. Of these, fauna and disease present special problems of distributional study, and are frequently omitted from natural resource surveys. The main environmental factors may each be subdivided; for example, hydrology includes groundwater conditions, river régimes and soil moisture.

Each factor varies over the land surface in a different manner. The characteristics of these distribution patterns, as represented on maps, depend partly on the intrinsic nature of the factors and partly on criteria

---

[1] Apart from geological survey, which is not discussed in the present paper except in so far as it relates to surveys for agricultural development.

[2] The survey of southern Malawi is being undertaken by Mr A. R. Stobbs, of the Directorate of Overseas Surveys.

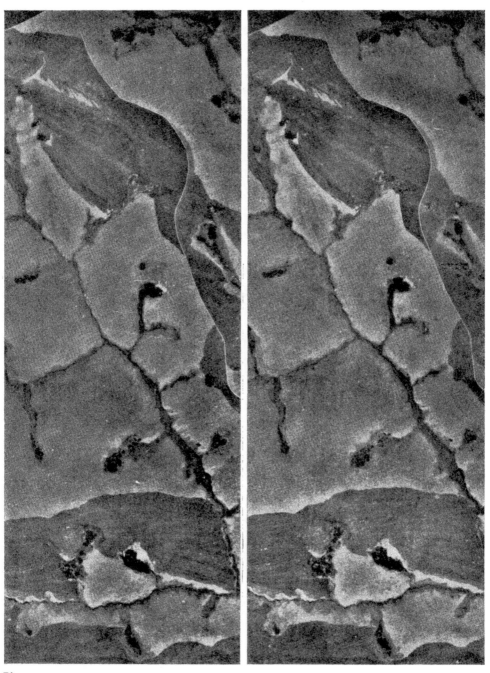

Plate 2 MAJOR NATURAL REGIONS OF MALAWI – I. THE HIGH ALTITUDE PLATEAUX: THE
NYIKA PLATEAU

Formed of granite, with the plateau surface here at an altitude of 8,000 feet (2,500 m.). A moderate
drainage density, convex slopes, with narrow strips of marsh in valley floors. Grassland, with patches of
montane evergreen forest in valley heads. The dark patches result from burning, and the fine lines are
game tracks.

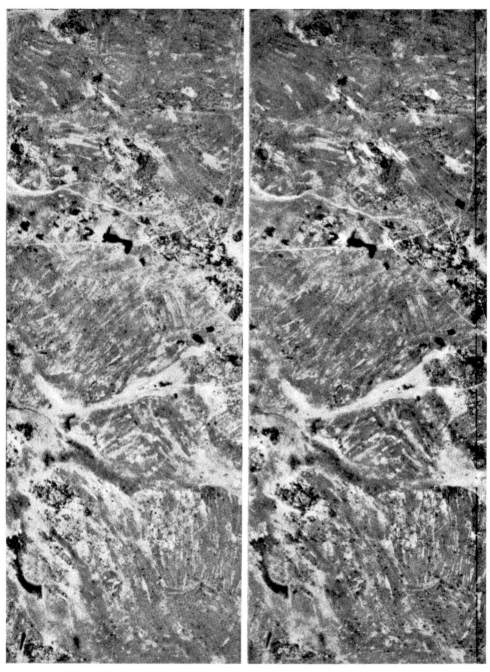

Plate 3 MAJOR NATURAL REGIONS OF MALAWI – II. THE PLAINS OF MEDIUM ALTITUDE: THE LILONGWE PLAIN

Formed of Basement Complex rocks, lying at an altitude of 3,500 feet (1,100 m.). A broadly-spaced valley system, gentle slopes, with level areas on interfluve crests. Almost entirely cultivated. Broad marshy valley floors, expanding at valley heads, under grassland. Villages sited adjacent to valleys; white areas result from overgrazing and cattle trampling.

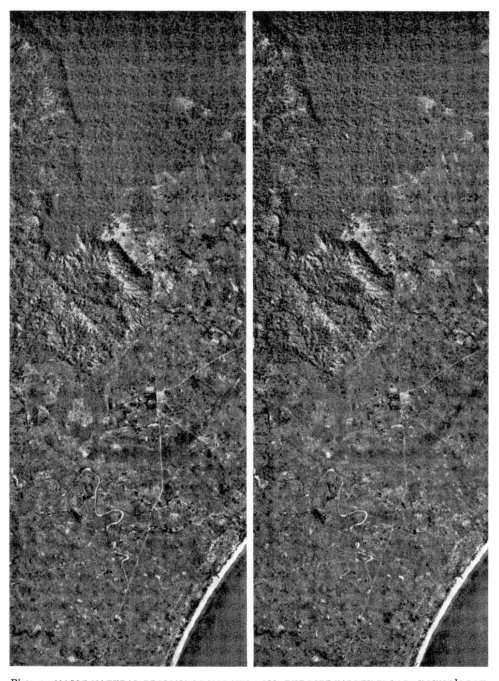

Plate 4   MAJOR NATURAL REGIONS OF MALAWI – III. THE RIFT VALLEY FLOOR: YOUNG'S BAY,
PART OF THE KARONGA LAKE-SHORE PLAIN
In the south a level depositional plain, mainly of alluvial origin but with lacustrine sand bars also
visible. The better drained areas under cultivation, the poorer drained carrying marsh grassland with
trees along levees. The wooded area in the north is a raised beach, which in the centre has undergone
badland dissection.

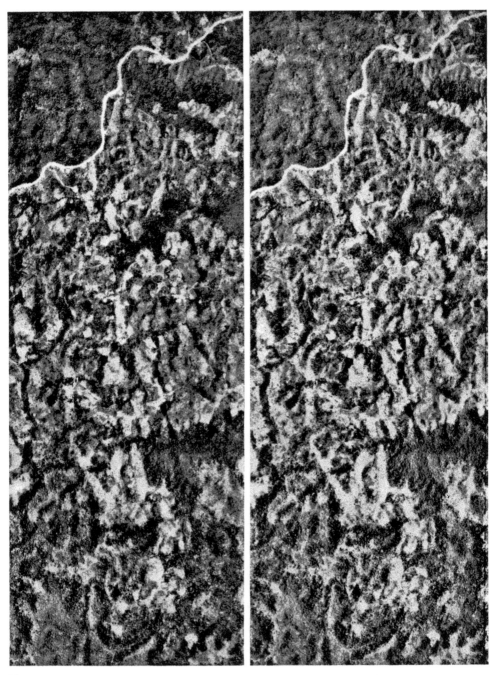

Plate 5 MAJOR NATURAL REGIONS OF MALAWI – IV. THE HIGH RAINFALL AREAS: THE EAST VIPYA SCARP ZONE

Showing close dissection typical of this region. Shifting cultivation has caused extensive removal of the moist woodland cover. The road follows the drainage divides. (Plates 2–5 are reproduced by permission of the Director of Surveys, Malawi.)

related to the survey: the classification system adopted, the level of classification mapped, the scale of the survey and the limitations of the available information.

Geological maps exhibit sharply-defined boundaries, between which are relatively homogeneous units. This is to some extent a consequence of the classificatory basis of such maps, which is primarily on the age of rocks and not their lithology, and in cases of gradual facies change may give a misleading impression; nevertheless as a generalization, areas of relatively uniform properties separated by abrupt changes over short distances are a feature of the distribution of rock types. The boundary pattern on maps of solid geology is frequently characterized by a banded, sub-parallel nature, reflecting the rock strike. Climatic distributions represent the opposite extreme, in which the variation in properties is in reality gradual and continuous; the boundaries shown on climatic maps are parameters selected to some extent arbitrarily, although Borchert (1953) has pointed out that there are areas of slight climatic gradients separated by certain natural boundary zones, across which there is a relatively steep gradient of one or more properties of climate. Climatic maps are also greatly over-simplified owing to limitations imposed by the available data. Because they are based on widely spaced recording stations the mapping units are necessarily large, and boundaries are usually drawn smoothly without regard for relief; yet the distribution, for example, of evapotranspiration varies considerably between valley and interfluve, and with slope aspect. In some cases mesoclimatic variations of this order of scale may be greater than those differentiating macroclimatic mapping units.

Geomorphological features vary at several levels of scale. Of particular significance is the repeated pattern of interfluve crest and valley floor; this causes a corresponding variation in groundwater and soil moisture conditions, which in combination with denudational processes gives rise to the soil catena. The latter is one of two main elements in the distribution of soils mapped at the level of the series, the other being that caused by rock type. Thus detailed soil maps show the dendritic pattern formed by valleys superimposed on the banded pattern of rock-type distributions; a third, more irregular, element may be added if drift deposits are present. On soil maps at a reconnaissance level, generalization is achieved by omitting intracatenal variation; the catena itself may be used as the mapping unit (e.g. Soil Map in Russell, 1962), it may be represented by bands of colour proportional in width to the component soil series (e.g. Milne, 1936; Soil Map in Kenya, Survey of, 1962), or the upper or predominant member of the catena only may be shown. On a smaller scale soils show the variation in properties caused by the effects of climate; zonal soil differentiation, which has the continuous and gradual characteristics of climatic distribution, is then added to the two main elements, slope- and rock-determined, of intra-zonal variation.

Vegetation, mapped at the level of the association or society, shows some comparable distributional features to soils, including intra-catenal variation and a degree of relation to rock type; over larger areas, associations and formations are influenced by progressive differences in climate. Vegetation differs from soils and other non-living environmental factors in that similar environmental conditions in different areas may not give rise to the same communities; if the areas are separated by a migration barrier there will be a different species composition leading to some degree of dissimilarity in life form and vegetation structure, although the process of convergent evolution tends to lessen such differences.

Natural regions and the relations between environmental factors

The non-coincidence of the distributions of each environmental factor gives rise to problems in presenting the results of a natural resource survey. Separate maps of each factor are necessary, but there is also a practical need to identify and map areas over which all environmental conditions are relatively uniform. It is desirable in presentation to stress the interaction of the different aspects of the environment; if the causal relations between factors can be demonstrated then the complexity, and apparent disorder, of their distribution patterns can be rationalized.

In the Malawi survey the concept of the natural region was applied. A natural region is here used to mean a part of the earth's surface within which the physical environment possesses the same major features, and in which individual environmental factors have a limited range of variation. The natural region is a basic unit for land-use planning, in which the whole environmental complex must be taken into account. The unity of a region results in part from a balance in the relations between environmental factors.

In some cases the relation between two factors is one of cause and effect operating in one direction only; for example, rock type influences soil, but there is no converse effect. In other cases the relations are mutual; for example, the soil moisture régime affects the vegetation, and conversely vegetation influences soil moisture conditions, through its effects on evapotranspiration and on the infiltration/run-off ratio. The main directions of influence between the factors of the environment are shown in Figure 13.1. Only the major effects are shown; causal interrelations to some slight degree exist between other pairs of factors. Systems of the type shown in Figure 13.1 have been discussed by Melton (1958); their equilibrium is maintained in part by feedback systems, represented as circular paths in the direction of causality, e.g. the climate–soils–geomorphology loop.

To determine the boundaries between regions, one possibility is to super-impose maps of the separate factors and select lines where these coincide or are concentrated in a narrow zone; but the practicability of this, as a means of environmental description, has yet to be demonstrated. The normal method is to select one factor and assemble the other data according to its boundaries; in some cases, different factors are used for subdivision at successive levels in the classification of mapping units.

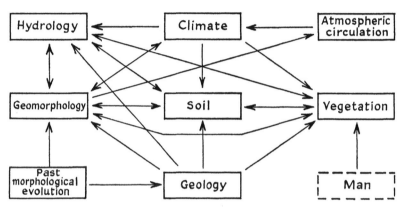

Figure 13.1 *Principal directions of influence between environmental factors.*
Man is shown only as an agent acting upon the environment.

In selecting a factor to use as the basis for natural regions, vegetation has the advantage that it is influenced by all the other factors (Fig. 13.1), and in this respect represents a response to the totality of the physical environ-ment; but it has the disadvantage that it may be considerably modified by anthropogenic influence, as has been demonstrated for Malawi by Hursh (1960). In the survey of Zambia Trapnell (1948) employed two-factor, vegetation–soil, mapping units. Climate was used as the main factor in the classic exposition of natural regions (Herbertson, 1905) and in most sub-sequent accounts on a world scale (Stamp, 1957; Hartshorne, 1961); it can-not easily be employed in detailed mapping owing to the transitional nature of the boundaries. In the map of Natural Regions and Areas of Rhodesia (Vincent *et al.*, 1961) the basis for the primary divisions was nominally effective rainfall, but in practice the latter was assessed from vegetation.

Geology forms an excellent basis for regional description in areas of gently dipping sedimentary rocks, but is less easily applied on the Base-ment Complex shield areas that underlie much of the African tropics. The use of soils for regional division in reconnaissance surveys has the dis-advantage that their boundaries cannot be directly distinguished on air photographs.

The use of geomorphology in determining natural regions has a number of advantages. First, the form of the land surface has substantial effects on climate, hydrology, soils and vegetation, while being itself strongly influenced by geology; it is therefore a suitable basis for analysing the causal relations between factors. Secondly, the boundaries between geomorphological mapping units are precise. Thirdly, related geomorphological criteria can be used to obtain a classification at different levels, giving a hierarchical system of regions of differing areal extent. A fourth considerable advantage is that geomorphology can be directly mapped from air photographs, with less control by ground observation than is necessary with vegetation, and without the element of indirect interpretation involved in air-photograph mapping of geology or soils.

Geomorphological mapping units

Until recently the intention of W. M. Davis, that geomorphology should serve as a basis for geographical description, had not been widely developed, and most geomorphological research was of a genetic nature. The past ten years have seen a change in this emphasis, and a number of techniques of descriptive geomorphology have been developed (e.g. Lewis, 1959, 1962; U.S. Corps of Engineers, 1959; Strahler, 1958; Thrower, 1960).

In describing the geomorphology of Malawi, units at four levels of scale were used:

1. *Slope units:* These are the divisions of an individual slope. They may be rectilinear, and described by their angle, or convex or concave, and described by their curvature and bounding angles. They are commonly of the order of 10–100 m in extent. Slope units form the basis of the technique of detailed morphological mapping, on scales of the order of 1:10,000–1:25,000 (Savigear, 1952, 1965; Waters, 1958; Young, 1964).

2. *Landforms:* Examples of these are a valley, scarp, pediment and floodplain. They are commonly of the order of 100 yds to 1 mile (100 m to 2 km) in extent, and can often be seen in large part from one viewpoint. In humid regions landform description can frequently be based on the characteristics of valleys; the main parameters are width, depth, maximum slope, predominant slope, and the relative extent of convex, plane and concave slope units on valley sides.

3. *Relief units:* Examples of these are a plain, range of hills, and scarpland zone. They are commonly of the order of 10–100 miles (20–200 km) in extent. They cannot usually be seen as a whole from a ground viewpoint, but form a readily distinguishable unit on air photographs. They are described in terms of the types and relative extent of landforms within them, and the proportions of slopes at different angles.

4. *Major relief units:* This term is here used for the main geomorphological divisions of a country or similar area, such as may be represented on scales of 1:1,000,000 or smaller. Each major relief unit is characterized by certain types of landforms, but with a wider range of detailed landforms and slope properties than that found within a relief unit; in this respect they are comparable with the 'classes of land-surface form' mapped for the United States by Hammond (1964).

At all of the above levels of scale the units are descriptive. They contrast with the systems of geomorphological mapping developed in France and Poland, in which a substantial genetic element is present (Klimazewski, 1963). The slope unit and the relief unit correspond in scale with the land facet and land system respectively as defined by Brink *et al.* (1966), with the difference that the latter units refer to the physical environment as a whole.

The distribution of slope units is related to that of intracatenal soil variations, or soil series (Bridges and Doornkamp, 1963). A landform is normally the geomorphological scale at which one complete soil–vegetation catena occurs. A relief unit commonly covers the whole area over which a particular soil–vegetation catena is found. Major relief units correspond to soil and vegetation differences at a higher order of classification; they may also show a degree of association with climate, through the influence of altitude on temperature and through orographic effects on rainfall.

The major natural regions of Malawi

There are five major relief units in Malawi (Young, 1962), illustrated schematically in Figure 13.2. These comprise three units of plateaux and plains, at altitudes of 5,000–8,000, 3,400–4,500 and 1,550–2,000 ft (1,500–2,500, 1,000–1,400 and 470–600 m), separated by two units of dissected

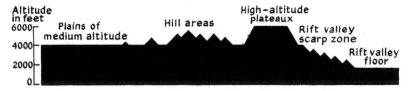

Figure 13.2 *Schematic cross-section of the major relief units of Malawi.*

land with steep slopes. The major relief units form a basis for the primary regional division of the country; each is characterized by distinctive landforms and by a limited range of other environmental conditions. One exception to the correspondence between geomorphological and environmental units is described under Region VI below. The term natural region was used in the Malawi survey for environmental regions of the order of scale of the relief unit, therefore the higher-order regions described here will be termed *major natural regions*. The distribution of the major natural regions is shown in Figure 13.3. Their characteristics are as follows:

1. *The high-altitude plateaux* (Plate 2): These are isolated plateau remnants at altitudes of 5,000–8,000 ft (1,500–2,500 m) surrounded by high and steep scarps. The most extensive are the Nyika and Vipya Plateaux; in southern Malawi they are represented by the Mlanje and Zomba Plateaux. Their landforms consist of broad valleys up to 200 ft (60 m) deep, with long gently convex slope units predominant on the valley sides; moderate slopes[1] are typical. The plateaux are preserved partly on intrusions of syenite and granite including, in southern Malawi, ring-complexes (Bloomfield and Young, 1961), but also occur on Basement Complex rocks. Climatically they have lower temperatures than the other major natural regions, and a mean annual rainfall mainly in the range 40–60 in (1,000–1,500 mm). The soil remains above wilting-point for all or most of the year, and the larger plateaux act as sources of perennial rivers. Using the classification of the Soil Map of Africa (d'Hoore, 1964), the soils of the high-altitude plateaux belong to the humic ferrisol and humic ferrallitic groups; soil formation will be discussed in the following section. Much of the vegetation is montane grassland or a shrub savanna of *Protea madiensis*; this is believed to be in whole or part a fire sub-climax, having replaced a former cover of montane evergreen forest which survives as patches in valley heads and on a few other isolated sites (Jackson, 1954).

The main differentiating factor within this region is altitude. The soils of the higher plateau remnants have more strongly developed ferrallitic properties, e.g. lower values for subsoil exchangeable cation saturation, than those at lower altitudes.

2. *The hill areas:* These comprise all areas with predominantly steep slopes that rise above the level of the medium-altitude plains (see §3 below). They include the scarps of the high-altitude plateaux, a number of relief units formed of hilly country, and numerous isolated hills that rise abruptly from the plains. Rivers rising in the hill areas are non-perennial, owing to a six- to seven-month dry season. The soils are mainly lithosols. The vegetation consists of *Brachystegia* savanna woodland or woodland, distinguished from that of the plains by the greater frequency of species common on shallow and stony soils, especially *Brachystegia boehmii* and *Uapaca kirkiana*.

The hill areas are differentiated according to the relative predominance of steep and moderate slopes, and the extent of soil types other than litho-

[1] The slope categories used here are: 2°–5°: gentle; 5°–15°: moderate; 15°–35°: steep.

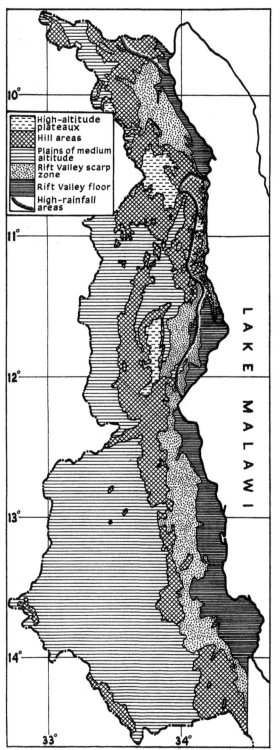

Figure 13.3 *Major relief units and major natural regions of northern and central Malawi.*

LAKE MALAWI

High-altitude plateaux
Hill areas
Plains of medium altitude
Rift Valley scarp zone
Rift Valley floor
High-rainfall areas

sols. A relation between slope angle and soil depth characterizes this region.

3. *The plains of medium altitude* (Plate 3): These are gently undulating to almost level plains, lying at altitudes of 3,400–4,500 ft (1,000–1,400 m); in southern Malawi they occur at 2,000–3,500 ft (600–1,100 m). They are the most extensive of the major natural regions, and include much of the better agricultural land in Malawi. The landforms consist of broad, gentle valleys, ¼–1 mile (400–1,600 m) wide and with maximum slopes varying from 2° to 8°. Valley floors of the smaller tributaries frequently contain no defined stream channel but are gently concave in cross-section, with a cover of swamp grassland. The climate of the plains belongs to the tropical dry-winter type (Köppen's Aw), with a single rainfall maximum and a long, seven- to eight-month dry season. Mean annual temperatures are 65°–70° F (18°–21° C), and mean annual rainfall 30–40 in (750–1,000 mm), of which 80–90 per cent falls in the four months December–March. During the dry season the soil moisture content falls below wilting-point to a depth of several feet. Most rivers, other than those rising on the high-altitude plateaux, cease to flow for one or more months in the late dry season. The soils on freely drained sites belong to either the ferallitic or ferruginous groups; horizons of nodular laterite occur locally, but become a normal feature only in the Kasungu region. A well-marked soil catena is character-istic of the plains: between interfluve crest and valley side the soil normally becomes paler or yellower; lower on the valley side a mottle appears in depth; a belt of sandy soils frequently occurs at valley-floor margins, with hydromorphic clays in valley centres. The vegetation over most parts of the plains consists of savanna woodland of short trees, in which species of *Brachystegia, Julbernardia* and *Isoberlinia* are frequently dominant; most of this woodland is in a degenerate condition as a result of burning, intensive use for fuel and other domestic purposes, and in some areas intermittent clearance for shifting cultivation (Hursh, 1960).

The main differentiating features within the plains are rock type and slope, both of which influence soils. Rocks of acid composition are usually overlain by ferallitic soils, and intermediate to basic rocks by ferruginous soils; while areas in which the maximum valley-side slope is only 2°–3° tend to have more highly weathered, almost structureless and agriculturally poorer soils than those where valley slopes reach 4°–8°.

4. *The Rift Valley scarp zone:* Lake Malawi (Nyasa), and its outlet the River Shire, lie along the African Rift Valley near to its southern ex-tremity. In northern and central Malawi its western margin forms a region of steeply sloping country lying below the altitude of the plains. Between Livingstonia and Nkata Bay, and also east of Dedza, this consists of a well-defined high scarp; elsewhere it forms a belt of deeply dissected country. Lithosols are the predominant soil type, and occupy a higher proportion

of the total surface than in the hill areas. The vegetation type is similar to that of the hill areas, but with different species becoming common at lower altitudes, including *Pterocarpus angolensis* and bamboo (*Oxytenanthera abyssinica*).

5. *The Rift Valley floor* (Plate 4): This differs from the plateaux and plains regions in being formed mainly of depositional surfaces. In northern and central Malawi it consists of a discontinuous lakeshore plain at 1,550–2,000 ft (470–600 m). In southern Malawi it is represented by the floors of the Bwanje Valley and the upper and lower Shire Valley; the upper Shire Valley lies at 1,500 ft (460 m), and is separated by a fall of 1,250 ft in 50 miles (380 m in 80 km) from the lower valley at 100–250 ft (30–75 m). The lakeshore plain is formed mainly of river alluvium but over limited areas by sands of lacustrine origin. Mean annual temperatures are close to 75° F (24° C), while rainfall may be from 25 to 60 in (650–1,500 mm) (approaching in its driest parts the boundary between Köppen's Aw and BS climates). The water-table lies close to the surface, giving imperfect site drainage over much of the region. The soils are mainly calcimorphic. Vegetation is distinct from that of other regions, with *Acacia* spp., baobab (*Adansonia digitata*), *Sterculia* spp. and palms (*Hyphaene ventricosa*) common.

Within this region small differences in elevation lead to considerable changes in environmental conditions. Slight depressions have poor site drainage, hydromorphic soils and a vegetation of swamp grassland. There are also areas of dissected surfaces formed by Tertiary and Recent sediments, probably corresponding to former lake levels and now standing at 50–350 ft (15–100 m) above the present lake surface; these are freely drained, with ferallitic or ferruginous soils.

6. *The high-rainfall areas* (Plate 5): The correspondence between major relief units and major natural regions breaks down in the case of areas with a mean annual rainfall exceeding 60 in (1,500 mm); these form a separate region, which fulfils the criterion of having environmental homogeneity but which transgresses the major relief units. High rainfall is found in three areas where the ground rises steeply north-westwards or westwards, giving pronounced orographic influence on rainfall from the prevalent south-east and east winds: the northern end of Lake Nyasa, the belt extending from Livingstonia to south of Nkata Bay, and, in southern Malawi, the south-eastern slopes of Mlanje Mountain. In addition to the higher total rainfall, the seasonal concentration becomes less marked, with 65–75 per cent falling between December and March and 5–10 in (125–250 mm) in the six driest months. Landforms in this region consist typically of a succession of V-shaped valleys with moderate to steep slopes separated by narrow convexities on interfluves. Soils are mainly red ferrisols of the type usually

associated with the equatorial forest zone. In the Nkata Bay region the
deeper soils carry semi-evergreen forest; beneath a deciduous canopy of
tall *Brachystegia spiciformis* there are evergreen strata that include trees,
shrubs and creepers. Elsewhere the region is covered by a *Brachystegia*
woodland taller and with a denser canopy than that of the hill areas and the
Rift Valley scarp zone.

The relations between environmental factors show different features
from those found in other regions. The rock composition/soil type relation
of the plains appears to be less marked, and lithosols are less frequent on
moderate and steep slopes. Within the forested area the undergrowth is
denser on deep soils than on shallow (Young 1960).

Areal associations and the relations between environmental factors

The descriptions of the major natural regions indicate a feature additional
to that implied by their definition of environmental homogeneity. It is that
each region is characterized by a distinct set of relations between environmental factors, or more precisely that the values (e.g. of rainfall, slope
angle, rock composition) governing such relations differ from one region
to another.

This leads back to the general problem, encountered in a wide range of
geographical studies, of isolating cause and effect among a system of interacting factors. Examining this problem in relation to determinism Martin
(1951) suggested that the concept of partial causes was applicable; this
approach has been clarified by the application of the techniques of multiple
and partial regression to show the extent to which one environmental
distribution can be accounted for in terms of others (Gregory, 1965), and
by the construction of correlation matrices between environmental variables
(Melton 1958). A similar problem has been encountered by soil scientists,
leading to discussions of the relative importance of different soil-forming
factors (e.g. Jenny, 1941 and, with respect to tropical African soils,
Paton, 1961).

The description of the environment on a regional basis reveals *areal
associations* between factors, that is, distributional patterns that are wholly
or partly coincident. The interpetation of such areal associations is similar
in principle to the explanation of a correlation between two variables:
while one distribution may be the cause of the other, at the opposite extreme the two distributions may have no mutual interaction but result from
some further factor affecting both. This difficulty is resolvable only by
reference to the likely causal processes involved.

Subject to this reservation, areal associations provide a means of studying
the relations between environmental factors. This will be illustrated by
reference to soil formation in Malawi.

Soil distribution and soil formation in Malawi

Taking the classification of the Soil Map of Africa (d'Hoore, 1964) as a basis, the soils of northern and central Malawi fall into ten classes. *Lithosols*, stony or shallow soils, are probably the most extensive of all soil types in the country. The *regosols* are developed from sands of lacustrine origin. The *hydromorphic* soils are black or mottled clays with a very coarse blocky to prismatic structure, waterlogged for all or part of the year. The *calcimorphic* soils are grey to greyish-brown, approximately neutral in reaction, with calcium carbonate concretions usually present in depth; they are mainly alluvial in origin.

The remaining types belong to the broad group defined by Kellogg (1949) as *latosols*: the red to yellow soils of freely drained sites in tropical climates in which rainfall exceeds evapotranspiration, acid in reaction and leached of exchangeable cations. The *ferruginous soils* are dark red to reddish-brown, with a strongly developed fine to medium blocky structure in the subsoil and visible clay skins on ped surfaces. The *ferrisols* are clays to sandy clays, deep, and with little textural differentiation in the profile; there is strong structural aggregation, giving the soil the physical properties of a loam despite a high clay content; these soils are strongly acid and highly leached of exchangeable cations. The *humic ferrisols* have similar properties, but with a topsoil organic matter content exceeding 5 per cent, compared with 2–3 per cent in the ferrisols.[1] The *ferallitic soils* are the most highly weathered of the latosol groups; they are yellowish-red to red, with a moderately sandy topsoil overlying a compact, heavier-textured subsoil; the subsoil is massive or with a weakly developed blocky structure with no clay skins. Over much of the area covered by ferallitic soils, laterite is uncommon. There are limited areas where it becomes a normal feature, and these have been mapped separately as *ferallitic soils with laterite*. The laterite usually consists of hard nodules $\frac{1}{10}-\frac{1}{2}$ in (2–12 mm) in diameter, varying from non-cemented to strongly cemented; the laterite horizon is typically 1–2 ft (30–60 cm) thick, occurring at a depth of 18–48 in (50–150 cm) and passing downwards directly into iron-impregnated weathered rock. *Humic ferallitic soils* have a similar humic topsoil to that of the humic ferrisols, but are more weakly structured and have a bauxitic weathering tendency (Young and Stephen, 1965).

There is a substantial degree of areal association between soil types and

[1] In applying to Malawi soils the definitions of the ferrisol class in the Soil Map of Africa a conflict exists between field and analytical characteristics. The red clays are classed as ferrisols on the basis of a moderately developed blocky structure and the presence of clay skins and weatherable minerals in depth; they are, however, more acid and have lower values of subsoil cation saturation than the soils classed, on the basis of profile characteristics, as ferallitic soils.

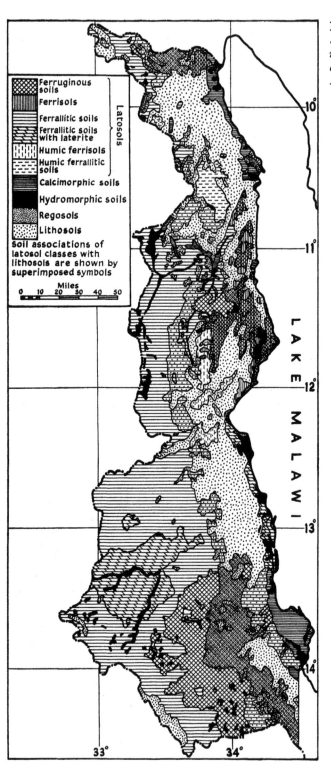

Figure 13.4
*Soils of
northern and
central
Malawi.*

the major relief units (Figs. 13.3 and 13.4). Humic ferallitic soils and humic ferrisols occur on the high-altitude plateaux. The lithosols predominate in the two relief units characterized by steep slopes, the hill areas and the Rift Valley scarp zone. Regosols occupy small areas on the Rift Valley floor, adjacent to the shore of Lake Malawi (Nyasa). The plains are mainly covered by the ferallitic and ferruginous soil groups. There is a partial association between calcimorphic soils and the Rift Valley floor. Hydromorphic soils and ferallitic soils with laterite occur on the two relief units with the most gently sloping landforms, the plains and the Rift Valley floor. The distribution of ferrisols, however, is unrelated to relief units but is associated with the areas of high rainfall.

A basic problem in the geographical analysis of soils is to explain why a given soil type occurs at a certain place, or over a certain area, and not in other places. The lithosols will be considered first. Lithosols in Malawi show an areal association with steep slopes; geomorphology, specifically slope angle, will be termed the *dominant factor* in lithosol formation. Except in the high-rainfall region, lithosols give place to one of the latosol types at angles below approximately 15°, while at over 40° rock cliffs occur; 15° and 40° are *limiting values* of the dominant factor. This relation is only found, however, within areas possessing a certain range of *associated conditions*. The relation occurs on a wide range of rock types, but not on quartzites, which are frequently overlain by lithosols on gentle slopes, nor on unconsolidated sediments. Rock type is an *independent factor*, since it has no necessary causal dependence on the dominant factor. With respect to hydrology, the slope/lithosol relation occurs under conditions of free drainage; this is a *dependent factor*, since free site-drainage is necessarily present on steep slopes. The climate is a tropical dry-winter type, with a single four- to five-month wet season. The mean annual rainfall is between 25 and 50 in (650 and 1,250 mm), but a distinction must be made between these two values: 25 in is the lowest rainfall in the area covered by the observations, but possibly the slope/lithosol relation remains the same under drier conditions hence it is not a limiting value; above 50 in, however, slopes of up to 25° may carry ferrisols, so a 50 in mean annual rainfall is a limiting value for the relation under discussion.

Some further examples will be given more briefly. The direct cause of the formation of hydromorphic soils is poor site-drainage; this condition occurs on gently concave valley floors and in enclosed depressions. With an indirect causal relation of this type either hydrology or geomorphology may be taken as the dominant factor.

In the formation of humic ferallitic soils the dominant factor is the relatively low temperatures. The limiting value is a mean annual temperature of 63° F (17° C), which occurs in Malawi at approximately 6,000 ft

(1,800 m); soil organic matter content shows a sharp increase above this altitude, suggesting that a relatively rapid change in the balance between humus formation and decomposition occurs (Young and Stephen, 1965).

The areal association between calcimorphic soils and the Rift Valley floor might appear to be related to temperature, through its effects on the balance between rainfall and evapotranspiration, but certain exceptions show that climate is not the essential cause. On moderately sloping areas on the Rift Valley floor, where raised erosion surfaces have been dissected, latosols occur; while on some river terraces of the plains, e.g. the North Rukuru, calcimorphic soils are found. These exceptions suggest that the dominant factor in calcimorphic soil formation in Malawi is a particular soil moisture régime, found on depositional surfaces, in which neither strong leaching nor frequent waterlogging occurs. This landform/hydrology/soil relation is found under a wide range of rainfall values.

Ferallitic and ferruginous soils both occur on the plains under almost identical conditions of relief, climate and hydrology. The vegetation which they each carry is distinct, but this leaves the differentiation of the two soil–vegetation units to be accounted for. The cause lies partly in the parent material: basic rocks, with a high content of ferromagnesian minerals, are overlain by ferruginous soils, while acid rocks are associated with ferallitic soils. This relation is directly demonstrable in a few areas, although its confirmation over much of the country waits upon detailed geological survey. Among the associated conditions is a mean annual rainfall of 50 in (1,250 mm), above which the soil properties on both basic and acid rocks tend towards those of ferrisols. This rock/soil relation agrees with findings from both Zambia and Rhodesia, that parent material is the dominant factor in soil formation on undulating erosion surfaces under a limited range of climatic conditions (Ellis, 1958), but that this dominance decreases where relief becomes more varied and under a high rainfall (Webster, 1960; Paton, 1961).

Ferallitic soils with laterite occur on the Kasungu Plain, a very gently undulating relief unit within the plains region, and near Kota-Kota on the Rift Valley floor, where they occur on an erosion surface standing 200 ft (60 m) above Lake Malawi (Nyasa). In both areas the laterite is found on level undissected surface remnants but is frequently absent from valley sides. This relation suggests it is a relict feature, formed as a ground-water laterite on a poorly drained, nearly level surface, prior to the cutting of the present valleys.

As a final example, the factors in the formation of ferrisols are shown in Table 13.1. Rainfall is the dominant factor. Among the associated conditions for this rainfall/soil relation, limiting values are a mean annual temperature of 63° F (17° C), below which humic ferrisols or humic

TABLE 13.1 *The formation of ferrisols in Malawi*

‖ = Limiting value

Dominant factor	*Climate*	Rainfall
Limiting values		Mean annual rainfall
		‖ 50–100 in (1,250–2,500 mm)
ASSOCIATED CONDITIONS		
Independent factors	*Geology*	Rock type: Acid to basic, igneous or metamorphic
	Geomorphology	Site: Valley side or interfluve crest
		Slope: 0°–25° ‖
	Climate	Type: Tropical dry-winter (Aw)
		Mean annual temperature: ‖ 63°–75° F (17°–24° C)
		Rainfall distribution: ‖ 85–95 per cent in 6 months
Dependent factors	*Hydrology*	Free site drainage (dependent on geomorphology)
	Vegetation	Rain forest (dependent on climate)

Soils formed when limiting values are passed

Mean annual rainfall	< 50 in (1,250 mm):	ferallitic or ferruginous soils
Slope	>25°:	lithosols
Mean annual temperature	< 63° F (17° C)	humic ferallitic soils
Rainfall distribution	< 85 per cent in 6 months:	no data from area surveyed

ferallitic soils are found, and a slope angle of approximately 25°, above which lithosols tend to occur even under high rainfall conditions. The table can be rewritten to show the influence of other factors; for example slope angle is related to the occurrence of ferrisols by a limiting angle of 25°, taking the rainfall as an associated condition.

This raises the question of which factor should be selected as dominant. The criterion proposed is the degree of areal association. If soils are being considered, the dominant factor is that which shows the greatest areal association with the soil distribution. Thus in the above example ferrisols occur on only a small proportion of the area of Malawi having slopes below 25°, but on a large proportion of that with over 50 in (1,250 mm) rainfall, hence rainfall is taken as the dominant factor in ferrisol formation.

This method of analysis has a bearing on the discussion of the relative importance of the soil-forming factors. One factor may appear to have a dominant effect on soil formation, provided that other factors remain within certain limits. For each of the associated conditions there is a range of tolerance. If the variation in the values of one factor within a particular country, or area, does not exceed certain limiting values, then the importance of that factor may be underestimated in studies based on that area. For example, over large parts of Rhodesia the relief and climate are relatively uniform, hence there is a considerable degree of areal association between geology and soils. Malawi has a wide range of relief, and consequently in analysing distribution within it geomorphology, both on the scale of the major relief unit and that of the landform, is most frequently the dominant factor.

The concept of limiting values in environmental interrelations helps to account for the feature noted above, that each natural region, in addition to having a limited range of environmental conditions, is characterized by particular relations between factors. This will result if the boundaries between natural regions coincide with limiting values for one or more factors.

Problems of regional description

In applying regional methods to the description of the physical environment a further problem arises from the fact that environmental analogues, that is, areas with closely similar conditions, may occur in isolated, widely separated places. For some purposes, for example the wider application of agronomic experimental results obtained at a particular site, this is no disadvantage. But for planning purposes it may be desirable to recognize, at some level in the regional classification, units that are spatially continuous or at least confined to one part of the country. In some cases these will be locally recognized as distinctive regions, and a local name available.

These two systems of regional subdivision may conflict at the same level of scale. For example, in Malawi within the major natural region of the plains of medium altitude it is possible to recognize, on the basis of environmental uniformity, classes comprising all areas with ferruginous soils, or all those within a given range of rainfall values. The alternative is to distinguish localized units, such as the Lilongwe Plain in central Malawi and the Fort Hill Plain in northern Malawi (Fig. 13.6). Each local unit has certain predominant environmental characteristics but these are not wholly exclusive; thus the principal soil types on the Lilongwe and Fort Hill Plains are ferruginous soils and ferallitic soils respectively, but exceptions occur in both cases. This conflict presents an obstacle to the concept of environmental homogeneity as the criterion for natural regions, a problem that has been recognized for the general case of regional method by Grigg (1965). It is partly resolved by the recognition that, whereas the range in properties of each environmental factor shows much overlap between regions, any particular combination of factors, or environmental complex, is more likely to be unique to a given locality.

The system of regional ordering used for Malawi arose in part from the method of survey. During air-photograph interpretation boundaries were drawn between all areas with a relatively uniform appearance, termed *photo units*. By comparison with observations from ground traverses the photo units were related to field characteristics; where field data were insufficient to support the distinction between two apparently differing photo units these were combined.

The areas obtained in this empirical way formed the lowest unit of regional classification to be mapped, termed *natural areas*.[1] To avoid numerous divisions, two mapping units excluded from the natural areas were recognized, namely hills and marshes; these were applied only to isolated inselbergs, small groups of hills and the smaller valley-floor marshes, similar but extensive features being classed as natural areas.

Many of the natural areas are made up of a number of spatially separated parts. In combining these into *natural regions* the primary aim was again that each region should have internal environmental homogeneity, to the lesser degree appropriate to the changed scale. But this aim was modified by adopting, at this intermediate level in the regional hierarchy, the principle that each region should wherever practicable form one continuous unit. Therefore groups of natural areas with closely similar properties but widely separated from each other were incorporated into different natural regions.

The distributional pattern of natural regions and areas is illustrated in Figures 13.5 and 13.6. The natural regions are largely areally continuous, but a few 'outliers' occur, for example in regions 4 and 8. The natural areas are in many cases separated into several non-contiguous parts. The natural regions are of the same order of size, and have a generally similar type of distribution, as the land units mapped in eastern Botswana by Bawden (1965).

In the 22,970 square miles mapped, 33 natural regions were distinguished, divided into 169 natural areas. The regions have a mean size of 710 square miles, with a range of 125–3,930 square miles. The natural areas average 137 square miles, and with a small number of exceptions range between 100 and 500 square miles (extreme values are 7 and 1,520 square miles). Regions and areas both tend to be substantially larger on the plains of medium altitude than on the other major relief units, leading to an overlap in areal extent between the two levels of the classification. This

---

[1] The terminology follows that of the Agro-Ecological Survey of Southern Rhodesia (Vincent *et al.*, 1961), which employs natural regions as the highest units, subdivided into natural areas.

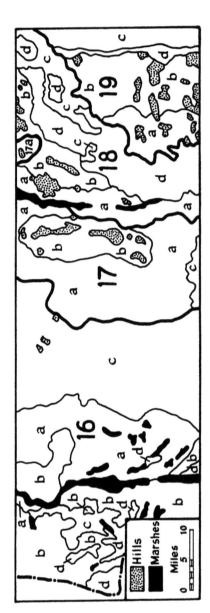

Figure 13.5 *Extract from the map of Natural Regions and Areas of Malawi* (based on Malawi, Natural Regions and Areas. Sheet 2, Central Malawi. Malawi Government, 1965).

Natural Regions	Area	Mean Annual Temp. (°F.)	Mean Annual Rainfall (in.)	Soil Parent Materials	Landforms	Vegetation	Soils	Present Land Use
16 Upper South Rukuru Valley	a	69–70	25–30	a–c: Basement Complex gneisses and schists d: Alluvial clays	a: Gently undulating plain, gentle and very gentle slopes b: Plain, level to very gentle slopes, broad marshy valley floors c: Broad, level surface remnants on interfluves, separated by widely spaced, broad, gently sloping valleys d: Level marsh with large termite mounds	a, c: *Brachystegia-Julbernardia* woodland b: *Brachystegia-Julbernardia* woodland, and *Acacia-Combretum* thicket of plateaux d: Marsh grassland with scattered trees	a: Weak ferallitic and sandy ferallitic soils; Mpherembe, Kapemba, Rumpi series b: Sandy ferallitic, possibly with weak ferallitic soils; Nkamanga Rumpi, Kapemba, Jalire, ?Mpherembe series c: Weak, ferallitic soils, with sandy ferallitic soils on valley sides; Loudon, Mpherembe, Bulala series d: Hydromorphic soils	a, b: Sparse cultivation with patches of modern cultivation near river c: Largely to entirely cultivated on interfluves, sparse cultivation on valley sides d: Uncultivated. Principal crops in all areas: maize, groundnuts, tobacco
	b	69–70	25–30					
	c	68–69	30–35					
	d	69	25–35					

No. / Name		Altitude (ft)			Geology	Physiography	Vegetation	Soils	Land use
17 Central Mzimba Hills	a	4500–4900	66–67	30–35	a–c: Basement Complex gneisses and schists	a: Gently dissected plain; broad valleys, low to moderate relief, gentle slopes	a, c: *Brachystegia-Julbernardia* woodland with *Cryptosepalum pseudotaxus, and Brachystegia* hill woodland	a: Sandy ferallitic soils; Kafukule series	a, b: Moderate cultivation; maize, groundnuts, tobacco
	b	4600–4900	66–67	30–35		b: Gently sloping pediments	b: *Brachystegia-Julbernardia* woodland	b: Weakly ferallitic and sandy ferallitic soils; Jandalala, Kafukule series	c: Sparse cultivation; including Perekezi Forest Reserve
	c	4400–5300	65–68	30–35		c: Low hills and ridges; mature dissection, moderate relief, moderate and gentle slopes		c: Lithosols and sandy ferallitic soils; Kafukule series	
18 Upper Kasitu Valley	a	3700–4200	68–70	30–40	a–d: Basement Complex gneisses and schists	a: Gently sloping valley sides	a–c: *Brachystegia-Julbernardia* woodland	a, b: Weakly ferallitic and sandy ferallitic soils; ?Jandalala ?Kafukule series	a–c: Moderate cultivation; maize
	b	4100–4400	68–69	30–40		b: Gently sloping pediments	d: *Brachystegia* hill woodland	c: ?Ferrisols and lithosols	d: Uncultivated
	c	4000–5000	66–69	34–45		c: Deep valleys amid hills, moderate and steep slopes		d: Lithosols	
	d	4000–6000	63–69	35–50		d: Steep, high scarp			
19 Vipya Plateau	a	5400–6000	63–65	50–55	a–c: Basement Complex gneisses and schists	a: Gently undulating high-altitude plateau; old-age topography, widely spaced valleys, broad marshy valley floors, gentle slopes	a, b, d: Montane grassland	a, b: Humic ferallitic soils; Vipya series	a, b, d: Uncultivated, partly forest reserve
	b	5000–5800	63–66	50–55		b, c: Dissected high-altitude plateau; mature dissection, moderate relief, broad convexities, moderate and gentle slopes	c: Montane grassland, with small areas of moist *Brachystegia* woodland	c: Ferrisols, humic ferallitic soils, and lithosols; Mazamba, Vipya series	c: Partly uncultivated, partly with estate cultivation; tung, coffee
	c	4100–4500	67–68	50–55		d: East-facing scarp of moderate steepness		d: Lithosols	
	d	4000–5500	64–69	50–55					

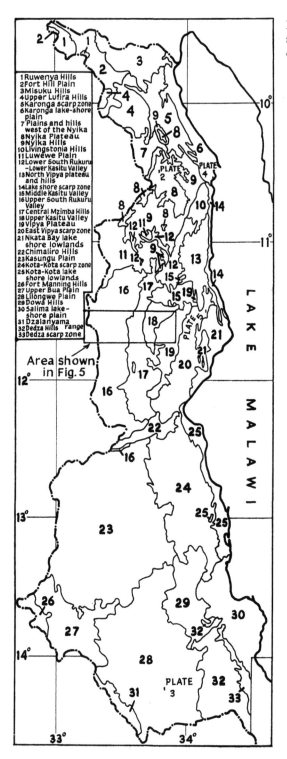

Figure 13.6 *Natural regions of northern and central Malawi.*

1 Ruwenya Hills
2 Fort Hill Plain
3 Misuku Hills
4 Upper Lufira Hills
5 Karonga scarp zone
6 Karonga lake-shore plain
7 Plains and hills west of the Nyika
8 Nyika Plateau
9 Nyika Hills
10 Livingstonia Hills
11 Luwewe Plain
12 Lower South Rukuru –Lower Kasitu Valley
13 North Vipya plateau and hills
14 Lake shore scarp zone
15 Middle Kasitu Valley
16 Upper South Rukuru Valley
17 Central Mzimba Hills
18 Upper Kasitu Valley
19 Vipya Plateau
20 East Vipya scarp zone
21 Nkata Bay lake shore lowlands
22 Chimaliro Hills
23 Kasungu Plain
24 Kota-Kota scarp zone
25 Kota-Kota lake shore lowlands
26 Fort Manning Hills
27 Upper Bua Plain
28 Lilongwe Plain
29 Dowa Hills
30 Salima lake-shore plain
31 Dzalanyama range
32 Dedza Hills
33 Dedza scarp zone

Area shown in Fig. 5

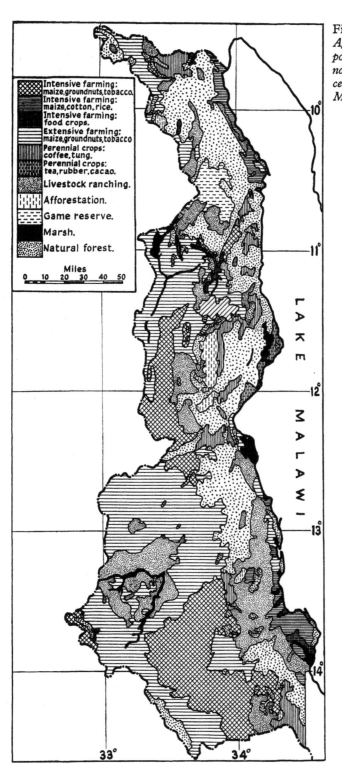

Figure 13.7
*Agricultural
potential of
northern and
central
Malawi.*

Intensive farming:
maize,groundnuts,tobacco.
Intensive farming:
maize,cotton,rice.
Intensive farming:
food crops.
Extensive farming:
maize,groundnuts,tobacco
Perennial crops:
coffee,tung.
Perennial crops:
tea,rubber,cacao.
Livestock ranching.
Afforestation.
Game reserve.
Marsh.
Natural forest.

Miles
0 10 20 30 40 50

LAKE MALAWI

10°
11°
12°
13°
14°

33°   34°

overlap is a necessary consequence of the nature of landscape distributions, if the principle of maintaining an equivalent degree of environmental homogeneity throughout each level of regional classification is adhered to. The major natural regions, including in this case their extension into southern Malawi, have a mean area of 6,100 square miles.[1]

At the scale of reconnaissance survey, soil series cannot be mapped in detail. Series were provisionally identified and defined in the course of the Malawi survey. Within most of the natural areas, however, more than one series was observed. The method adopted was to list for each natural area the series known, or believed likely, to occur in it (Fig. 13.5); in the survey memoirs regional keys to soil series identification are given.

To assess the resources for land use the natural area was taken as a basis. The agricultural and other land-use potential was assessed from the descriptions of each area. On the resulting map (Fig. 13.7) the distributions indicate the influence of the major relief units, as well as showing a measure of association with the natural regions. In applying the results of agronomic experimental work to the survey the soil series was employed. The series were identified at stations for which experimental data were available. Having established one or more type sites for each soil series, the agronomic results were reassessed in relation to these. The survey memoirs contain, following the soil series descriptions, accounts of their potential land use, nutrient status, responses to fertilizers, and suitable crops.

Comparison of regional units

The units most widely employed in natural resource surveys are *land systems*, a concept independently evolved in Australia, Britain and South Africa. The land system was originally defined as 'an area or group of areas, throughout which there is a recurring pattern of topography, soils and vegetation' (Christian and Stewart, 1953). The land system has more recently been defined as 'a recurrent pattern of genetically linked land facets' (Brink, *et al.*, 1966), the *land facet* being a part of the land surface which is, for practical purposes, environmentally homogeneous. Land systems are employed with a local connotation; similar land systems occur-

---

[1] The G scale proposed by Haggett *et al.* (1965) provides a useful means for comparison of areas. $G = \log (Ga/Ra)$ where $Ga =$ the area of the earth's surface $= 1.968 \times 10^8$ square miles, and $Ra =$ the area under investigation; a difference of $G = 1$ represents a tenfold scale difference. In these terms the natural areas have a mean of $G = 6.1$, the natural regions $G = 5.4$, and the major natural regions $G = 4.5$. Approximating the figures, the scale ranges are of the order of $G = 6.5$–5.5 for natural areas, 6.0–5.0 for natural regions, and, if the high-altitude plateaux are excluded, 5.2–4.2 for major natural regions. Thus the three levels of classification are separated by approximately $G = 0.5$ (i.e. each unit is subdivided into 5 at the next lowest level) but have a scale range of $G = 1$, or tenfold; consequently the units at different levels overlap in size.

ring in different areas are grouped into a single *abstract land system*; areally grouped land systems are classed as *land provinces* and *regions* (Brink *et al.*, 1966).

The regional classification employed in the Malawi resource survey was developed independently of the above work, and it did not include surveys, even on a sample basis, at the degree of detail of the land facet. The natural areas of the Malawi survey appear to correspond in degree of environmental homogeneity to land systems. The smaller average size of natural areas as compared with most mapped examples of land systems may be a result of the highly diversified relief of Malawi, rather than a narrower range of environmental conditions. In the survey of eastern Botswana Bawden (1965) defined *land units* as 'areas within a land system in which one particular factor differed from the general pattern', and from the descriptions given the Malawi natural areas more nearly correspond with the land units then with the land systems; the mean size of the Botswana land units is 800 square miles, similar to some of the largest natural areas in Malawi.

The areal extent of mapped examples of land systems varies considerably. For example, in the classic account of the Katherine–Darwin region, Northern Territory, Australia (Christian and Stewart, 1953) the mean area is 1,420 square miles, with the two smallest 42 and 190 square miles and the largest 2,750 square miles; whereas in a survey of part of Papua by Perry *et al.* (1965) the mean is only 160 square miles and there are five land systems of less than 10 square miles. The causes of such variations in extent may be partly subjective, different workers selecting smaller environmental differences as defining criteria for a land system. But the large differences in size encountered within a single uniform survey in Malawi suggest that the main cause lies in intrinsic variations in the degree of differentiation as between different types of country. Some systematization of the nature of these variations, for example in terms of macro-units of geological structure, relief, or climatic zone, may be possible.

The major natural regions distinguished in Malawi appear to be representative of environmental regions common to substantial areas of East Africa. The plains of medium altitude correspond to the '*Brachystegia–Isoberlinia* woodlands on Plateau soils' of Zambia (Trapnell *et al.*, 1948). The types of ferallitic soils that are most extensive on the plains, always on gently undulating landforms, represent the 'plateau soils' of central Tanganyika (Milne, 1936) and Zambia (Trapnell *et al.*, 1948), and the 'leached, pallid soils' noted by Watson (1962) as characteristic of African plateau areas. The Rift Valley scarp zone is represented in Zambia by the '*Brachystegia–Isoberlinia* woodlands on Escarpment Hill soils' which form belts bordering the Luangwa trough, and in Rhodesia by the 'very broken' country, with shallow soils, forming belts along the southern margin of the Zambezi Valley (Vincent *et al.*, 1961). The valley floors of the Luangwa in Zambia and the Zambezi and Sabi in Rhodesia are generally similar in land-forms and soils to the Rift Valley floor region, although they correspond more with the lower Shire Valley of southern Malawi than with the lake-shore plain. The Inyanga Mountains of Rhodesia are 'topped by a plateau of rolling grassland' at 7,000 feet, on which the rainfall, soils and vegetation are similar to those of the high-altitude plateaux of Malawi.

The studies by King (1962) of the distribution of erosion surfaces in Africa and on a world scale have a bearing on the possible wider develop-ment of areas with environmental conditions similar to those represented in Malawi. Three of the major natural regions of Malawi are geomorpho-logically based on surfaces: the plateaux, plains and Rift Valley floor regions. These surfaces have been discussed from the point of view of their genesis by Dixey (1937). From King's descriptions of the landforms characteristic of the Gondwana and African Cycles there is a clear corres-pondence with the Malawi high-altitude plateaux and medium-altitude plains respectively; the Rift Valley floor represents the Congo cycle. King considers that these surfaces are widely represented both in Africa and in other continents, on Basement Complex shield areas. If this view is correct, similar geological and geomorphological conditions are widely distributed; substantial parts of these shield–plateau areas have in addition tropical dry-winter climates. On this basis it is suggested, as an untested hypothesis, that environmental analogues may be present between the shield areas of Africa, Brazil and the Indian Deccan; and that the similarities between them in geology, major relief, landforms and climate will lead to closely similar soil types and to physiognomically similar vegetation communities.

Conclusions

In a natural resource survey, each of the factors which contribute to the environmental complex may to a certain extent be examined in isolation. Problems arise if an attempt is made to synthesize these factors into some form of regional description of the physical environment as a whole. The above discussion of these problems is based on the experience of a natural resource survey of part of Malawi. Although some of the results may be of wider application, they are based primarily on the environment of the tropical dry-winter climatic zone. Owing to resources available, the survey was confined to work on a reconnaissance scale, without the support of detailed studies of sample areas. A further limitation is that statistical methods have not been applied to the results, although it is recognized that a number of the questions discussed could be further examined by such methods.

The conclusions are therefore regarded as tentative, and subject to modification when applied to areas with different ranges of environmental conditions. It is suggested that:

1. The factors of the physical environment each show a distinctive distribution pattern when the variation in their properties is represented on maps. These distribution patterns are related to the classification system employed, the level of classification mapped, the intensity of survey, the scale of mapping and the intrinsic nature of areal differentiation of the factor concerned. Climatic zones, relief units, geological outcrops and landforms are four elements present in areal differentiation of environmental conditions; in humid areas the valley is normally the basic landform, giving rise to the catena pattern which appears in the distributions of hydrological conditions, soils and vegetation.

2. In natural resource surveys, the presentation of results should include maps and descriptions of the physical environment as a whole. The concept of natural regions provides a means for doing this. Within a natural region, each environmental factor possesses a limited range of conditions; a region identified on this basis is further characterized by certain causal relations between factors.

3. Geomorphology provides one possible basis for determining natural regions and for analysing the relations between environmental factors. Descriptive geomorphological units exist at four levels of scale, termed slope unit, landform, relief unit and major relief unit. Each of these units is associated with distributional features of other environmental factors at different levels of classification.

4. A comparison of the distribution patterns of individual factors of the environment shows partial similarities of distribution, termed areal associations. Such associations may be used to isolate the relation of one factor to another; consistent relations between pairs of factors occur over areas within which all other factors do not lie outside a certain range of properties, termed limiting values. For each factor there are certain critical limiting values, at which different features appear in the system of factor interrelations. The boundaries of natural regions frequently follow such critical values; the unity of a region results in part from the existence within it of a particular system of causal relations between factors. Under the tropical dry-winter climate of Malawi an important limiting value occurs at a mean annual rainfall of 50–60 in (1,250–1,500 mm), above and below which rock/slope/soil relations are substantially different.

5. The distribution pattern of natural regions, defined on the basis of relative environmental homogeneity, is areally discontinuous. A conflict

arises if it is considered desirable that spatially continuous regions, with a local connotation, should be recognized.

6. The distribution of any one environmental factor may be analysed on the basis of its areal associations with other factors. In relation to the distribution of the main soil types of Malawi, geomorphology is most frequently the dominant factor, that is, the factor having the closest areal association with a given soil type. Geomorphology/soil relations occur both on the scale of the relief unit and on that of the landform.

7. Six major natural regions occur in Malawi, five defined on the basis of major relief units and the sixth based on rainfall. The regions differ from each other in their main environmental features, the ranges in properties of individual factors, and the relations between factors. Regions with similar environmental conditions are believed to be extensive in southern Africa, and possibly occur in other continents.

Acknowledgements

Figure 13.7 is based on an agricultural assessment by Mr P. Brown, of the Ministry of Natural Resources, Malawi. Figure 13.5 and Plates 2–5 are reproduced with the permission of the Malawi Government.

References

BAWDEN, M. G. 1965. A reconnaissance of the land resources of eastern Bechuanaland. *J. appl. Ecol.* **2**, 357–65.

BAWDEN, M. G. and STOBBS, A. R. 1963. *The Land Resources of Eastern Bechuanaland*, Tolworth, Surrey.

BLOOMFIELD, K. and YOUNG, A. 1961. The geology and geomorphology of Zomba Mountain. *Nyasal J.* **14** (2), 54–80.

BORCHERT, J. R. 1953. Regional differences in the world atmospheric circulation. *Ann. Ass. Am. Geogr.* **43**, 14–26.

BRIDGES, E. M. and DOORNKAMP, J. C. 1963. Morphological mapping and the study of soil patterns. *Geography* **48**, 175–81.

BRINK, A. B. *et al.* 1966. *Report on the Working Group on Land Classification and Data Storage*, Christchurch, Hants.

CHRISTIAN, C. S. and STEWART, G. A. 1953. *Report on a Survey of the Katherine–Darwin Region, Northern Territory, 1946*, Canberra.

D'HOORE, J. L. 1964. *Soil Map of Africa, scale 1 to 5,000,000*, Lagos.

DIXEY, F. 1937. The early Cretaceous and Miocene peneplains of Nyasaland, and their relation to the Rift Valley. *Geol. Mag.* **74**, 49–67.

ELLIS, B. S. 1958. Soil genesis and classification. *Soils Fertilizers* **21**, 145–7.

GREGORY, S. 1965. *Rainfall over Sierra Leone*, Liverpool.

GRIGG, D. 1965. The logic of regional systems. *Ann. Ass. Am. Geogr.* **55**, 465–91.

HAGGETT, P., CHORLEY, R. J. and STODDART, D. R. 1965. Scale standards in geographical research: a new measure of areal magnitude. *Nature*, London, **205**, 844–7.

HAMMOND, E. H. 1964. Analysis of properties in land form geography: an application to broad scale land form mapping. *Ann. Ass. Am. Geogr.* **54**, 11–19.

HARTSHORNE, R. 1961. *The Nature of Geography*, revised edition, Pennsylvania.

HERBERTSON, A. J. 1905. The major natural regions: an essay in systematic geography. *Geogrl J.* **25**, 300–312.

HORNBY, A. J. 1938. *Soil Map of Southern Nyasaland*, Zomba.

HURSH, C. R. 1960. *The Dry Woodlands of Nyasaland*, Salisbury, Rhodesia.

JACKSON, G. 1954. Preliminary ecological survey of Nyasaland, *Proc. 2nd Inter-Afr. Soils Conf.* I, 679–90.

— 1959. Nyasaland: major plant communities. Map in JACK, D.T. *et al.*, 1960, *Report on an Economic Survey of Nyasaland 1958–59*, Salisbury, Rhodesia; reproduced, with revised legend, in PIKE, J. G. and RIMMINGTON, G. T. 1965, *Malawi: a Geographical Study*, London.

JENNY, H. 1941. *Factors of Soil Formation*, New York.

KELLOGG, C. E. 1949. *Preliminary Suggestions for the Classification and Nomenclature of Great Soil Groups in Tropical and Equatorial Regions*, Harpenden.

KENYA, SURVEY OF, 1962: *Atlas of Kenya*, Nairobi.

KING, L. C. 1962. *The Morphology of the Earth*, Edinburgh.

KLIMAZEWSKI, M. (ed.), 1963. *Problems of Geomorphological Mapping*, Warsaw.

LEWIS, G. M. 1959. Some recent American contributions in the field of landform geography. *Trans. Inst. Br. Geogr.* **29**, 23–36.

— 1962. Changing emphasis in the description of the natural environment of the American Great Plains area. *Trans. Inst. Br. Geogr.* **38**, 75–90.

MARTIN, A. F. 1951. The necessity for determinism. *Trans. Inst. Br. Geogr.* **17**, 1–11.

MELTON, M. A. 1958. Correlation structure of morphometric properties of drainage systems and their controlling agents. *J. Geol.* **66**, 442–60.

MILNE, G. 1936. *A Provisional Soil Map of East Africa*, London.

MOÇAMBIQUE, JUNTA DE EXPORTAÇÃO DO ALGODÃO. 1955. *Esboco do reconhecimento ecológico-agrícola de Moçambique*, 2 vols, Lourenço Marques.

NYASALAND DEPARTMENT OF AGRICULTURE. 1954–61. Sections on Ecology in *Annual Reports of the Department of Agriculture (Part II)* for the years 1954–5 to 1960–1.

PATON, T. R. 1961. Soil genesis and classification in Central Africa. *Soils Fertilizers* **24**, 249–51.

PERRY, R. A. *et al.*, 1965. *General Report on Lands of the Wabag–Tari Area, Territory of Papua and New Guinea, 1960–61*, Melbourne.

PHILLIPS, J. F. V. 1959. *Agriculture and Ecology in Africa*, London.

RUSSELL, E. W. (ed.), 1962. *The Natural Resources of East Africa*, Nairobi.

SAVIGEAR, R. A. G. 1952. Some observations on slope development in South Wales. *Trans. Inst. Br. Geogr.* **18**, 31–51.

— 1965. A technique of morphological mapping. *Ann. Ass. Am. Geogr.* **55**, 514–37.

STAMP, L. D. 1957. Major natural regions, Herbertson after fifty years. *Geography* **42**, 201–16.

STRAHLER, A. N. 1958. Dimensional analysis applied to fluvially eroded landforms. *Bull. geol. Soc. Am.* **69**, 279–300.

THROWER, N. J. W. 1960. Cyprus – a landform study. *Ann. Ass. Am. Geogr.* **50**, Map Suppl. 1.

TRAPNELL, C. G. 1943. *The Soils, Vegetation and Agricultural Systems of North-eastern Rhodesia*, Lusaka.

TRAPNELL, C. G. and CLOTHIER, J. N. 1937. *The Soils, Vegetation and Agricultural Systems of North-western Rhodesia*, Lusaka.

TRAPNELL, C. G., MARTIN, J. D. and ALLAN, W. 1948. *Vegetation-soil Map of Northern Rhodesia*, Lusaka.

U.S. CORPS OF ENGINEERS. 1959. *A Technique for Preparing Desert Terrain Analogs*, Vicksburg.

VINCENT, V. *et al.* 1961. *An Agricultural Survey of Southern Rhodesia*, Salisbury.

WATERS, R. S. 1958. Morphological mapping. *Geography* **43**, 10–17.

WATSON, J. P. 1962. Leached, pallid soils of the African plateau. *Soils Fertilizers* **25**, 1–4.

WEBSTER, R. 1960. Soil genesis and classification in Central Africa. *Soils Fertilizers* **23**, 77–79.

YOUNG, A. 1960. Soil surveys: Chinyakula-Chombe, Nkata Bay District. *Ann. Rep. Dept. Agric. Nyasal. 1958–9 (Part II)*, 147–9.

1962. Ecology of experiment stations. *Ann. Rep. Dept. Agric. Nyasal. 1960–1 (Part II)*, 7–27.

1964. Slope profile analysis. *Z. Geomorph.*, Supplementband 5, 17–27.

1965. *Map: Malawi, Natural Regions and Areas*, Sheet 1, Northern Malawi; Sheet 2, Central Malawi, London.

YOUNG, A. and BROWN, P. 1962. *The Physical Environment of Northern Nyasaland*, Zomba.

1965. *The Physical Environment of Central Malawi*, Zomba.

YOUNG, A. and STEPHEN, I. 1965. Rock weathering and soil formation on high-altitude plateaux of Malawi. *J. Soil Sci.* **16**. 322–33.

14 An ecological approach to the study of soils and land use in the forest zone of Nigeria

R. P. MOSS

In Chapter 8 a broad sketch was given of the ecological background to the study of land use in tropical Africa. Some of the basic relationships between plants and soils, animals and plants, men and plants, and men and animals were described, and to a certain extent systematized. If, in fact, these relationships are as important as was suggested, then it follows that approaches to land-use study different from the traditional geographical morphological approach are not only desirable but absolutely necessary. This is especially true if the study of land use is to be directed towards the planning of agricultural development.

The purpose of the present chapter is to outline the theoretical basis and to describe the practical implications of an approach which has been adopted in a relatively densely populated area of the closed forest zone of south-west Nigeria. Towards its northern margins, the region also embraces an area of forest–savanna mosaic, and an area of savanna with forest galleries. It thus affords a wide variety of basic vegetation conditions into which the patterns resulting from human activity must be fitted.

The location of the area is shown in Figure 14.1. It covers some 1,500 square miles, and is underlain by a wide variety of soil types derived from Tertiary sediments, which range in lithology from ferruginous sandstones with mudstone bands to clays and clay shales with interbedded limestones. To the north of the scarp which marks the limit of the sediments, there is a large area of coarse-grained acid igneous and metamorphic rocks, often very high in quartz content. The detail of the soil pattern is markedly affected by the erosion history (Moss, 1965), which has been associated with the development of hard or indurated plinthitic deposits. These have developed into breakaways which are associated with well-defined soil toposequences.

Climatically the principal variation in the area is in length of dry season. This is indicated by the fact that there is only one month with less than one inch of rain in the south-east, but four in the north-west. This is associated with a decrease in annual total from about 70 in (1,790 mm) to

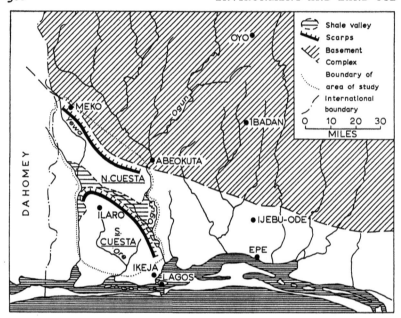

Figure 14.1 *Location of the area studied in south-western Nigeria.*

45 in (1,125 mm). These gradients in macroclimatic characteristics cut at a marked angle across the soil pattern, as it is related to lithology, and locally across the boundary between the closed forest and the savanna.

Basic concepts

In Chapter 8 the major plant–soil systems were viewed in terms of their present relationships as exemplified in water use, microclimate, nutrient cycles, organic matter and decomposition relations, periodicity and their animal component, especially in relation to pests and diseases. It was suggested that the contrast between forest and savanna was so great in these respects that it was necessary to view the forest–savanna boundary as perhaps the most basic divide in West African ecology. This was not to deny the fact of considerable variation both in the forest and, even more strikingly, in the savanna, but rather to insist that this was variability within the limits set by a major system of relationships. In the context of the more local variation with which the present study is concerned the question then is, whether the same basic viewpoint may be validly applied there also.

The basic elements of the land-use/vegetation pattern consist of individual plots, parcels and areas of land with a definable plant cover. This cover is the result of the coincidence on that plot of a wide range of disparate groups of forces – biological, physical, social and economic – acting both in space and in time. Nevertheless, these individual plots are not haphazardly related to one another, since the forces producing them also operate in discernible patterns. Furthermore, each plot, and each distinct complex of juxtaposed plots of varying character, represents a contemporary ecological situation, which may be relatively homeostatic, in process of rapid change, or delicately balanced between a steady state and dis-equilibrium. The basic problem therefore is the delimitation of the areas of operation of particular complexes of ecological relationships, whether strongly conditioned socially and economically or not, and the evaluation of the kinds of relationships between them.

It is important to notice that no distinction is made between 'vegetation' and 'land use'. In areas of dense population in the tropics both are in fact expressions of the same complexes of agricultural activity, and both have strong ecological implications (Ross, 1954; Clayton, 1958; Moss, 1961, 1963). The distinction is therefore in the present context at best useless, and at worst quite specious and misleading.

The focus of attention is thus the understanding of particular ecological complexes or systems, and their delimitation in respect of three different aspects. *First*, attention must be given to the way in which they 'work' as coherent units, that is their *internal function*. This facilitates delimitation of the areas in which they operate, since it gives meaning to their morpho-logical characteristics in terms of 'land use' or 'vegetation type' on the small individual plots of land of which these areas are composed. *Second*, it is necessary to investigate their relationships with other contiguous com-plexes or systems of land use and vegetation, that is their *external function* in relation to phenomena of a similar kind. This makes it possible in some measure to assess the character of the boundaries drawn, and the gradients and zones of transition, between adjacent complexes. *Finally*, assessment must be made of their interaction with other systems of quite different character (which also vary in time and in space) that is their *environmental relationships* or their interaction with phenomena of a different kind. In the context of this study the environmental system of principal interest is the soil; but this category might equally well include macroclimate, or the socio-economic factors. Thus, it is convenient to consider each of these sets of relationships in turn.

Internal function

This includes the characteristics of the husbandry system, its expression in terms of particular plant groups, whether planted or self-sown, and its association with particular problems of animal or plant populations in terms of weeds or pests. The description and delimitation of these *land-use/ vegetation systems* is achieved by the convergence of three lines of approach.

In the first place an evaluation must be made of existing knowledge concerning the agricultural systems employed in this, or in similar areas, and its translation into terms applicable to the situation under examination. The principal sources for this information are the studies of Buchanan (1953), Morgan (1960) and Moss (1960, 1963), together with a number of Agricultural Department Reports (Moss, 1953, 1955, 1957b). Numerous other articles are also important, though not directly concerned with this area (Clayton, 1958; Tinsley, 1964, *inter alia*).

Secondly, the available aerial photographic cover must be examined (for 1953–54, on 1:25,000 approx., and for 1962–63, on 1:40,000 approx.), and characteristic patterns of individual *land-use units* distinguished. These are then interpreted in terms of the knowledge already available, and further elucidated by selective work in the field. It is important to point out that there is no attempt to make any kind of land-use map based on the morphological approach. The focus of attention is always the general ecological significance of the collective pattern, not the morphological significance of particular photo images which can be related to field cultivation or other discrete plot patterns.

Finally, there must be an evaluation of the ecological relationships implied in the patterns studied. By this means the significance of the function of the individual land-use units may be assessed, and the characteristic ones selected as an indicator by means of which the area of operation of the particular set of relationships which constitute the particular system may be delimited from the aerial photographs and by selective field study.

These three lines of approach converge to facilitate description and understanding of the land-use/vegetation systems.

External function

Understanding of the relationships between the different systems may be attempted by comparing the internal relationships of adjacent systems. This generally poses particular questions which may be answered by further planned programmes of observation, specialized study of particular problems, or by analogy with similar situations elsewhere. It must be emphasized that the present study is merely a reconnaissance, and that the chief value of the ecological approach suggested is perhaps that it pinpoints precise problems which need further investigation, rather than providing full answers to the questions which may be posed. It is this fact that makes the approach so useful in agricultural planning.

Environmental relationships

Having thus elucidated the broad internal and external functions of each system it is then possible to examine their relationships with other systems of a different kind. The advantage of first concentrating upon the land-use/vegetation systems themselves is that it facilitates selection of those aspects of land variables within the environmental systems which are of demonstrable importance to land use and vegetation. It may in fact be argued that this is the only meaningful way to approach 'environmental' study (Andrewartha and Birch, 1954; Morgan and Moss, 1965a). Mere correlation between already classified morphological land-use units and soil types yields little or no *information*, even if it were logically and statistically valid. Correlation between *variables* within the plant system and within the soil system is both valid and informative. Hence in the present study concern with the soil is restricted to examination of particular soil properties and their variation, and not with soil groups as such, however classified.

The application of this basic thinking to the area of study will perhaps serve to clarify the issues involved.

The land-use/vegetation systems and their internal function

A broad classification of the individual land-use units of importance in the area is shown in Table 14.1. This is not intended to be a complete grouping, since the degree of subdivision employed could be extended almost indefinitely, as is the tendency in conventional land-use survey (cf. the First with the Second Land Use Survey of England and Wales). In this case, in view of the decreasing usefulness of each further level of subdivision, the classification is carried only far enough to facilitate adequate characterization and delimitation of the systems in the area of study. Furthermore, the categories distinguished have been based as far as possible on their ecological characteristics.

Table 14.2 shows the principal systems present in the area of study, together with the land-use units which serve as indicators of the extent of the operation of these systems. It also summarizes their basic ecological characteristics. Four points need to be emphasized. *First*, it is necessary to point out the importance of the population imbalance of the mirids (*Sahlbergella singularis*, especially) in System IIIB. The importance of this in relation to soil character will be examined later. *Second*, reference must be made to the different complexes involving both grass-with-trees and closed forest types (IVA, B, C), each of which implies a different kind of ecological system. This also will be amplified later when soil factors are introduced into the discussion. *Third*, it must be emphasized that within each system variability is chiefly to be seen in the degree to which the human animal dominates the available area of the system. This is reflected in a variable density of those units, especially plantations and cultivation patches, which reflect direct human influence. It is also possible to

TABLE 14.1 *Classification of individual land-use units*

MAJOR GROUP	SUB-GROUP	TYPE	UNIT
I. **Controlled or Semi-Controlled** Units actually involved in crop production, or in producing useful materials	1. *Standing crops:* herbs or shrubs cultivated to the virtual exclusion of others; but with some self-sown plants, often deliberately preserved	(1) *Annual crops:* crops which occupy only one year in the agricultural cycle (either by virtue of their life cycle or by husbandry practice)	(i) *Roots:* cassava, yams, cocoyam, etc.
			(ii) *Cereals:* maize, rice, guinea corn, etc.
		(2) *Perennial crops:* crops which occupy several years in the agricultural cycle	(i) usually pineapple, and sometimes cassava. Also cotton
		(3) *Vegetable or garden crops:* herbs or shrubs, often perennial, producing accessory foods, and technological or medicinal requirements, on a more or less permanent basis, usually close to settlements	(i) usually peppers, tomato, *Solanum* spp., etc.
		(4) *Cleared land:* land cleared for cultivation but without standing crops	
	2. *Tree plantations:* trees, or occasionally large herbs, planted or preserved for the harvesting of useful materials. Other large species usually mostly excluded	(1) *Fruit or seeds*	(i) *banana:* including plantain, when in groves
			(ii) *cocoa:*
			(iii) *kola:*
			(iv) *oil palm:*
			(v) *coffee:*
			(vi) *citrus:*
		(2) *Sap or bark:*	(i) *rubber:*
		(3) *Wood:*	(i) *Timber:* distinguish between pure and mixed stands
			(ii) *Poles or firewood:* usually exotic species, such as teak, cassia and gmelina

II. Rotational and Successional: Units dominated by self-sown plants, transient in character, and part of the normal successions to closed forest or grass-with-trees

1. *Woody:* with dominant woody plants and non-graminaceous herbs

   (1) *Early regrowth:* immediate post-cultivation regrowth less than 2m high, with or without relic forest trees
       (i) *Herbaceous regrowth:* with herbaceous weeds dominant
       (ii) *Woody regrowth:* with regenerating stools, coppiced stumps and tree seedlings dominant

   (2) *Thicket growth:* woody plants dominant, 2–6 m high, with developing tree seedlings, and oil palm (usually)
       (i) *Dominant shrubs and small trees:*
       (ii) *Climbers dominant:*

   (3) *Low secondary forest:* woody growth 6–10 m high, with crowns of young trees beginning to emerge from general canopy level, with oil-palm (usually)

2. *Grassy:* with grasses very important or or without trees or thicket clumps

   (1) *Grass regrowth:* immediate post-cultivation regrowth
       (i) *Low dense mat:* low mat of grasses and herbs less than 1 m high
       (ii) *Herb-rich:* regrowth over 1 m high, with high proportion of herbs, especially *Indigofera* spp., and *Tephrosia* spp.
       (iii) *Low grass:* grass regrowth less than 1 m high, usually dominated by *Imperata cylindrica*
       (iv) *High grass:* grass regrowth over 1 m high, with *Andropogon* spp., *Hyparrhenia* spp., and/or *Pennisetum* spp.

   (2) *Savanna types:* grasses with trees of varying density and frequency
       (i) *Open savanna:* scattered trees now here forming a closed canopy; distinguish between species-rich and species-poor units

TABLE 14.1 *Classification of individual land-use units*—cont.

MAJOR GROUP	SUB-GROUP	TYPE	UNIT
			(*ii*) *Farm savanna:* woody species dominated by non-fire-tolerant species, forest climbers often conspicuous
		(3) *Transition woodlands:* more or less close canopy of trees, with both savanna and forest species, usually with *Anogeissus leiocarpus*	
III. Relatively permanent Units: units involved in the late stages of the succession, and more or less stable climaxes	1. *Forest types:* dominated by woody species in open or closed canopy; savanna grasses do not occur	(1) *High secondary forest:* woody growth with developing tree canopy 10–20 m high, dominated by secondary forest species, but with high forest species of increasing significance	
		(2) *Broken forest:* an irregular mosaic of canopy heights produced by the break-up of the main forest canopy, usually by windblow	
		(3) *High forest:* tall forest without oil palm, usually 25 m or more in height of main canopy, with developing layered structure, and high forest tree species dominant	

(4) *Swamp forest:* tall forest in seasonally or permanently swampy areas along watercourses, or on adjacent flats, with dense understorey of lianes and shrubs

(5) *Detached forest areas:* areas of forest separated from the main closed forest zone by savanna types and their associates

    (i) *Forest galleries:* closely associated with watercourses

    (ii) *Forest outliers:* not associated with watercourses; two main types, those associated with rocky hills and steep slopes, and those not

May be subdivided on floristic grounds

2. *Savanna types:* trees, with grass important or dominant in herb layer

    (1) *Savanna woodland:* more or less closed canopy of savanna trees, with grasses of variable importance; canopy height 10–20 m

    (2) *Open savanna:* grass layer more or less continuous, with numerous trees of 6–10 m high or more, but without closed canopy; not easy to differentiate from II.2. (2).(i) above

3. *Grass types:* trees virtually absent, with grass layer dominant, though not necessarily continuous

    (1) *Low sparse grass:* discontinuous cover of low grasses with very few stunted low trees

    (2) *Grass thicket:* grass cover with occasional small thicket clumps; usually found in seasonally swampy areas only

TABLE 14.2 *Land use–vegetation systems in the area of study*

SYSTEM GROUP	SYSTEM	NOTES	INDICATOR UNITS
I. Closed forest	A. *Secondary forest*, with thicket regrowth	Usually in forest reserves only, but also on some steep slopes; stable	Low and high secondary forest, with very occasional high forest
	B. *Swamp forest*, with associated thicket regrowth	Always associated with valley bottom sites and areas of high water table	Swamp forest, with climbing palms and *Mitragyna ciliata*
II. Closed forest cultivation	A. *Rotational bush fallowing*, with annual crops, and woody regrowth	Intensity of cultivation very variable; subdivisible on basis of crop associations; some unused land; stable	Annual crop cultivation, without plots of permanent tree crops
III. Closed forest tree crops	A. *Cocoa plantations*, with some rotational bush fallowing, and cultivation of annual crops preparatory to the planting of cocoa	Variable density of plantations; much young cocoa; mirid infestation not killing over heavy clay soils, but stability on light soils dependent on intensity of mirid infestation	Cocoa plantations, no derelict plantations; no kola plantations
	B. *Kola plantations with cocoa plantations*: with some rotational bush fallowing, and no young cocoa plantings	Variable density of kola plantations; mirid infestation on cocoa severe, with many derelict plantations; unstable and transitional to IIIC	Kola plantations with cocoa plantations
	C. *Kola plantations*: with a few areas of rotational bush fallowing	Kola plantations often very dense or almost continuous; food crop cultivation very limited in both area and scope; stable	Kola plantations

IV. Savanna	A. *Rotational bush fallowing*, with grass or grass-herb fallow, *with forest galleries*	Forest galleries may be savanna woodland, though some rainforest galleries may be present. Cultivation not extensive; stable	Annual crop cultivation: with savanna type fallows and successional types; forest galleries
	B. *Rotational bush fallowing*, with grass fallow, but without forest galleries	Very variable density of cultivation and considerable variability owing to edaphic factors; stable	Cultivation with grassy fallow but no forest types
	C. *Rotational bush fallowing*, with dominantly grassy fallow, but *with some patches of woody fallow*, and some patches of woodland or forest	Very variable density of cultivation with some unused land; woodland may be closed forest or savanna woodland; may be unstable in some locations	Cultivation with grassy fallow and patches of woody fallow
V. Complex	A. *Flood plain complex* with islands of rotational bush fallowing with woody or grassy fallow on drier parts and swamp forest along water-courses	Very complex pattern, essentially a mixture of IB, IIA, and IVB, too intimately mixed to be mapped separately on the present scale; stable	

distinguish different degrees of maturity in the patterns according to the extent to which the current situation represents a stable state with respect to human influence.

Finally, it is important to notice that the function of a particular land-use/vegetation unit within any system is to be seen principally in its relations with the other units present and not in any inherent quality viewed in isolation. Thus 'cultivation' in System IIIA, even though it may contain similar crop plants to the similar unit in IIA, fulfils a different purpose in relation to the system as a whole. Furthermore, in all systems it is impossible to decide on purely morphological grounds whether well-developed thicket growth is the end stage of a fallow period or the indication of land now unused for cultivation owing to its being in title to an absentee tenant (Mabogunje, 1959).

The idea of the land-use/vegetation system thus facilitates the recognition of related groups of land-use and vegetation units which represent a coherent, though not discrete, ecological unit. Furthermore these units can be described in functional terms which facilitate understanding of the working of the factors which produce the pattern revealed by morphological study. The reverse is not true.

The external functions of the land-use/vegetation systems

Figure 14.2 is a simplified map of the land-use/vegetation systems described from the area of study. They have been delimited from aerial photographic mosaics using the criteria noted in Table 14.2, and the field data of the Soil Survey of the area made in 1955 and 1956 (Moss, 1957a, b). Two important sets of relationships between different systems are clearly illustrated by this map: first, those which link the three tree-crop systems, and second, those which bring together the systems including grassy fallows. These cannot, however, be fully discussed until soil factors have been introduced, and it is therefore sufficient at this stage simply to point out that the three tree-crop systems are related to one another by a process of change from cocoa plantations through mirid-blasted cocoa plantations with kola plantations, to pure kola plantations. The reasons for this sequence, which is clearly seen in its spatial expression on the map (Fig. 14.2), are complex and related to both ecological and economic factors

(*a*) Secondary and high forest and thicket   (*b*) Freshwater swamp forest and floodplain complex of rotational bush fallowing with swamp and thicket along watercourses   (*c*) Rotational bush fallowing with woody fallow   (*d*) Grass with scattered trees and occasional rotational bush fallowing with grassy fallow   (*e*) Grass with trees and rotational bush fallowing with grassy fallow   (*f*) Grass with trees, thicket and rotational bush fallowing, usually with grassy fallow, but with some woody fallow   (*g*) Cocoa plantations, sometimes with new plantings, with some rotational bush fallowing with woody fallow, and some secondary forest   (*h*) Cocoa plantations with capsid and considerable dereliction; little new planting; with increasing areas of kola plantations. Some rotational bush fallowing   (*k*) Kola plantations with some rotational bush fallowing.

Figure 14.2 *Land-use systems in the Otta–Abeokuta area.*

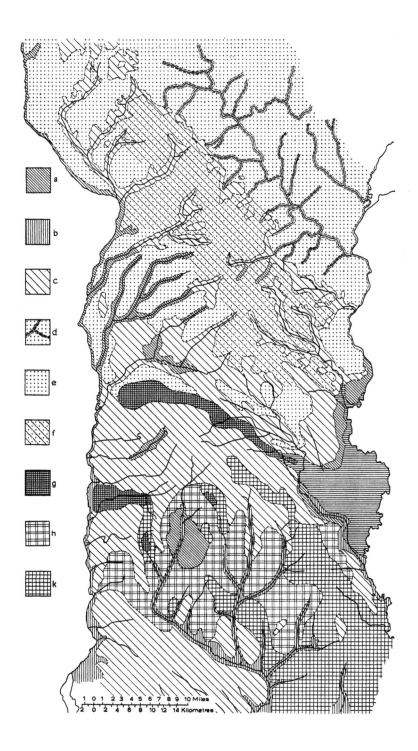

a

b

c

d

e

f

g

h

k

1 0 1 2 3 4 5 6 7 8 9 10 Miles
2 0 2 4 6 8 10 12 14 Kilometres

(Moss, 1968; Hopkins, 1966). The grass-with-trees complexes present an even more complicated situation, but it seems that edaphic factors, especially those related to the soil moisture system, are very important. Thus the following sequence seen on the map may not represent a simple time-sequence of change from closed forest conditions to savanna:

> rotational fallowing with woody fallow → patches of rotational fallowing with woody fallow within rotational fallowing and grass-with-trees → grass-with-trees with forest galleries.

Finally, it is relevant to observe that, even when two systems are contiguous, it does not necessarily follow that the boundary or transition zone represents the same kind of relationship throughout its length. Thus the boundary between the closed forest systems and the savanna systems may in part be a fire boundary, in part related to soil moisture conditions and in part related to cultivation practices (Moss and Morgan, 1968). Similarly, the boundary between the cocoa plantations system and the rotational fallowing with woody fallow may be a boundary of extending cocoa cultivation or of abandonment of cocoa plantations in the face of pest attack (Tinsley, 1964).

The environmental relationships of the land-use/vegetation systems

In this chapter the principal concern is with soil conditions. In relation to plants and the land-use/vegetation systems, soil properties may be considered in two distinct categories. First, there are those properties, such as nutrient status, organic matter functions and the activity of micro-organisms, which are intimately related to the plant and animal components of the ecosystem. These properties change with the alterations, both inherent and induced, in the plant–animal component, and may be termed *ecological* soil properties (Moss, 1968). Secondly, the soil body possesses another category of properties which are more directly related to weathering processes, geomorphological history and regional hydrological characteristics. This group includes such features as soil depth, presence and kind of pan or indurated layers, particle size distribution, clay character and soil water-table levels as affected by regional groundwater levels. These properties are not directly related to the organic component of the ecosystem, do not change rapidly (though groundwater levels obviously show seasonal and some shorter-term fluctuations), and cannot readily be altered by suitable agronomic practices, except at considerable expense. These properties thus contrast strongly with the *ecological* group, and have been termed *morphological* (Moss, 1968).

It is with these morphological properties that the present section is concerned. Since they are dominantly related to various aspects of earth history and contemporary hydrology, they often vary with the form of the land surface (Moss, 1965). Thus on a local scale it is possible to map from aerial photographs facets of the land surface which differ from one another in their morphological properties and which therefore constitute areas which present different limitations with reference to plant growth. These units have been termed '*habitat-sites*' (Moss, 1968). On a more general scale it is possible to map the extent of particular limiting factors which can then be related to the land-use/vegetation systems which occur. A map of limiting factors for the area of study is shown in Figure 14.3.

Comparison of Figures 14.2 and 14.3 immediately reveals two significant relationships between land-use/vegetation and limiting factors. The first is related to mirid attack, and the second to the relationships between the closed forest areas and those characterized by grass-with-trees. In both cases the influence of soil depth, texture and, subsidiarily, of nutrients, is seen. Thus mirid attack produced plantation dereliction only on soils with 12 in. or more of sandy layers and not on the montmorillonitic clayey soils derived from the calcareous shale. Furthermore, the closed forest systems are dominant on the deeper, moister sites in the savanna areas, (especially on colluvial slopes below breakaways) and savanna types do not extend to the south of the main band of shale. In more detail tree density is related to soil depth, being least on the shallowest soils over hard plinthite and greatest on the deeper soils towards the centre of interfluves.

Discussion

Each of the land-use/vegetation systems which have been distinguished represents a coherent and distinct set of ecological relationships expressed in terms of a characteristic pattern of land use or vegetation on the individual plots of which it consists. The relationships thus expressed are, first, *internal*, expressing the organization within the system; second, *external*, characterizing the relationships between the systems; and third, *environmental*, concerned with the relationships of the plant–animal systems to other systems, notably the soil. These three sets of relationships are, however, not independent but act together to produce the patterns which are observed. The relationships characteristic of the systems in the area of study are shown in Figure 14.4. The diagram is more or less self-explanatory, but the following points require emphasis. *First*, it is necessary to point out the existence of *thresholds* which are ecologically conditioned, in connection with certain aspects of human activity. The most important ones are those relating to cultivation pressure in lowering the vigour of the woody fallow, and intensity of burning in relation to the change from woody to grassy fallow. These have clear implications for the idea of *critical*

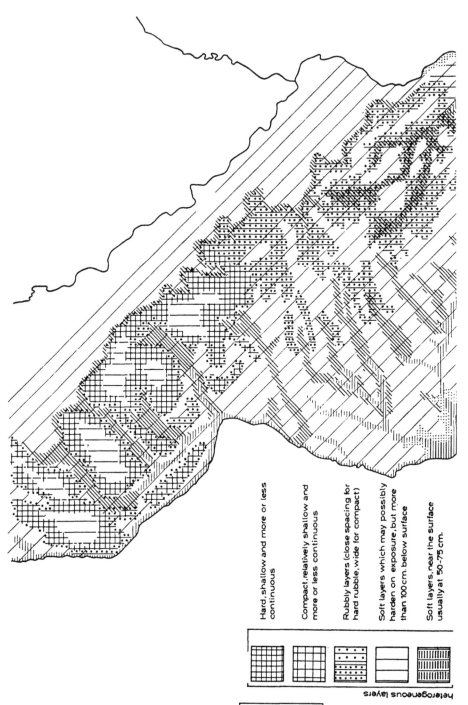

Hard, shallow and more or less continuous

Compact, relatively shallow and more or less continuous

Rubbly layers (close spacing for hard rubble, wide for compact)

Soft layers which may possibly harden on exposure, but more than 100 cm. below surface

Soft layers, near the surface usually at 50-75 cm.

heterogeneous layers

LIMITING FACTORS

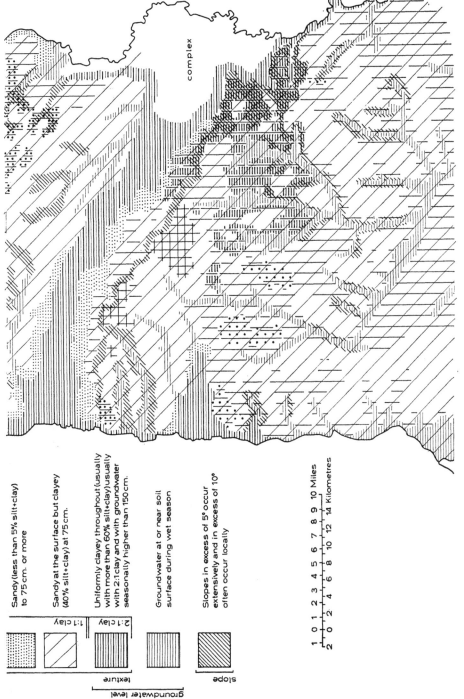

Sandy (less than 5% silt+clay) to 75 cm. or more

Sandy at the surface but clayey (40% silt+clay) at 75 cm.

Uniformly clayey throughout (usually with more than 60% silt+clay) usually with 2:1 clay and with groundwater seasonally higher than 150 cm.

Groundwater at or near soil surface during wet season

Slopes in excess of 5° occur extensively and in excess of 10° often occur locally

texture

1:1 clay

2:1 clay

groundwater level

slope

complex

Figure 14.3 *Generalised map of habitat-site factors in the Abeokuta area.*

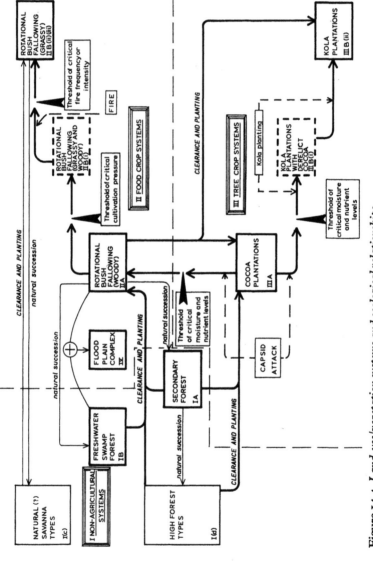

Figure 14.4 *Land-use/vegetation systems and ecological relationships.*

population density put forward by Allan (1965). *Second*, the significance should be noted of the intimate interrelation between those changes effected by direct human intervention, and possible ecological changes occurring indirectly as a result of such intervention. *Third*, attention must be drawn to the complex interactions which may collectively produce an almost irreversible ecological change. This may best be illustrated by reference to the two major problems already referred to, namely the change from cocoa to kola, and the relations between closed forest and woody fallow and grass-with-trees and grassy fallow.

The first of these is related to the ecological complex described by Tinsley (1964), in which, by means of the connected action of mirids, fungus infection and swollen-shoot virus, a sequence of changes is brought about which results in the development of grass-with-trees from the original plantations. Two important differences are, however, apparent: *first*, the absence of the swollen-shoot virus in the mirid-affected areas at the time of the investigation, and *second*, the presence of an alternative tree crop in this area of somewhat higher rainfall than the location described by Tinsley.

Most of the cocoa in the present area of interest was planted in the early years of the present century (Galetti *et al.*, 1956; Hopkins, 1966), and planting was encouraged immediately after the First World War. Dereliction of many plantations set in and now there is little or no cocoa left in the original area of densest plantings, around Agege, in the south-eastern corner of the area. Virtually all has been replaced by kola plantations which are still spreading northwards and westwards through the zone of derelict cocoa plantations. More recent plantings of cocoa are in the west and north-west, near Ilaro and Igbogilla, on heavy montmorillonitic soils. Dereliction over the lighter, poorer soils was once attributed to nutrient deficiencies but other factors now appear to be more important, especially the break-up of the cocoa canopy as a result of damage to the growing points of the trees by mirid bugs, and subsequent infection with the fungus *Calonectria rigidiuscula*. Such canopy break-up results in disruption of the water and nutrient cycles of the healthy mature plantation. Litter additions are considerably reduced or completely prevented, the sandy upper layers of the soil dry out, and the feeding roots of the cocoa die. Due to the absence of feeding roots lower down in the solum and increasing water deficiency in the upper layers of the soil, the trees die. On heavier soils better supplied with mineral nutrients the trees would survive such an attack, and regenerate even after coppicing.

In the absence of an alternative crop such derelict plantations would revert to food crop cultivation and fallow or, if other factors such as swollen shoot entered the complex, might be degraded to grass-with-trees and grassy fallow. The presence of a profitable alternative crop, however, notably kola, facilitated the change to the new plantation system.

The relationships between the systems, including woody fallow and those with grassy fallow, are less easy to define. Nevertheless, recent work in this and adjacent areas has revealed a number of important facts (Moss and Morgan, 1968; Morgan and Moss, 1965b). *First*, the close correspondence of the areas of woody fallow with the sites having characteristics favourable to a water régime sympathetic to the plant–soil system must be emphasized. *Second*, despite a considerable increase in population density, evidenced by a higher proportion of land under cultivation and the proliferation and enlargement of settlements, there is no evidence from the two sets of photographs that the areas of woody fallow and closed forest have diminished. In fact, in some favourable situations they have increased in extent. *Third*, it must be continually remembered that burning is a very variable factor. Some areas are only rarely burned; others are affected by fire at least once a year, and sometimes twice. Furthermore, fires almost always stop quickly on reaching a woody area, and the effect of the fires is often markedly accentuated by the presence of soil factors, such as shallowness or sandiness, which adversely influence the moisture properties of the solum. *Finally*, it is often forgotten that indigenous cultivators frequently take quite deliberate steps to prevent fires spreading from savanna into forest and woody regrowth, such as clean weeding, or the planting of teak along the boundary, which effectively eliminates grasses when its canopy develops.

In view of these considerations it is suggested that the present situation is fairly stable, despite the increasing cultivation pressure produced by an increasing population in rural areas as well as the demands for food by the large urban centres near by, such as Ibadan, Abeokuta and Lagos. It may be that great changes have taken place in the past, but the present situation is at least in a state of delicate equilibrium of homeostasis.

Conclusion

It is suggested that the ecological approach to the study of land use sketched in this chapter offers a number of advantages over conventional land-use study based on detailed morphological description, at least in the tropical African situation. In the first place, an appreciation of the ecological

---

(A) Areas where present land-use system is ecologically unstable   (B) Areas where present land-use system is conditionally stable   (C) Direction of movement of encroaching land-use system boundary   (D) Boundaries between land-use systems. Boundaries between areas of differing ecological stability   (E) Boundaries changing by direct extension of particular crops into new locations.   (*Land-use systems*) Lower case letters refer to Map of land-use systems Figure 14.2.

Figure 14.5   *Stability of land-use systems in the Otta–Abeokuta area.*

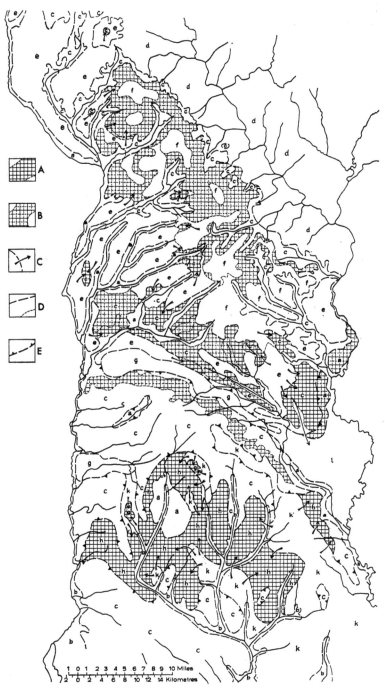

A

B

C

D

E

1 0 1 2 3 4 5 6 7 8 9 10 Miles
2 0 2 4 6 8 10 12 14 Kilometres

O

relationships operating in the context of the social and economic forces which create and mould the need for agricultural production facilitates understanding and generalization about the land-use patterns and their variability. As a corollary to this, it reduces considerably the amount of detailed study that is necessary to further this understanding and validate the generalizations.

Secondly, this approach seeks to treat the study of what is essentially a dynamic situation by means of a dynamic model which recognizes and evaluates the constantly changing situation. To this task the static model which is the basis of the conventional land-use map is conceptually inappropriate. Ecological study itself facilitates the construction of a map (Fig. 14.5) which attempts to show the direction and character of the changes which are taking place at the present time.

Thirdly, an approach of this kind makes possible the selection of the significant environmental relationships from the mass of possible influences and effects. This, conventional study can never do, since it is concerned with correlating *one* classification based on observed properties with *other* classifications of the environment based on different sets of properties. Concentration of attention on the variables in each system facilitates *evaluation* of the *relationships* of *variables* in different systems to one another.

Finally, the approach focuses attention on the basic questions requiring answers if an adequate understanding of the working of the land-use systems is to be understood. Such understanding is necessary if agricultural planning is to be based upon clear, precise prognosis rather than upon factually based speculation, however informed it may be. Clearly much more detailed understanding is essential; in particular the basic conceptual model needs to be developed by mathematical and logical analysis in order to facilitate valid experimentation and prediction. Nevertheless the concept, unlike the traditional model, is capable of such development and has already made possible informative reconnaissance. Furthermore, it has made possible cartographic expression of the relationships involved. It is therefore suggested that an ecological approach of this kind affords a more fruitful approach to land-use study in tropical Africa than the alternative morphological approach, and could have far-reaching consequences for the planning of agricultural development.

References

ALLAN, W. 1965. *The African Husbandman*, Edinburgh.
ANDREWARTHA, H. G. and BIRCH, L. C. 1954. *The Distribution and Abundance of Animals*, Chicago.

BUCHANAN, K. M. 1953. The delimitation of land-use regions in a tropical environment. *Geography* **38**, 303–7.

CLAYTON, W. D. 1958. Secondary vegetation and the transition to savanna near Ibadan, Nigeria. *J. Ecol.* **46**, 217–38.

GALETTI, R., BALDWIN, K. and DINA, I. O. 1956. *Nigerian Cocoa Farmers*, Oxford and London.

HOPKINS, A. G. 1966. Personal communication.

MABOGUNJE, A. L. 1959. *The Changing Pattern of Rural Settlement and Rural Economy in Egba Division, S.W. Nigeria*, unpublished M.A. thesis, University of London.

MORGAN, W. B. 1960. Agriculture in southern Nigeria. *Econ. Geogr.* **35**, 138–50.

MORGAN, W. B. and MOSS, R. P. 1965a. Geography and ecology: the concept of the community and its relationship to environment. *Ann. Ass. Am. Geogr.* **55**, 339–50.

1965b. Savanna and forest in Western Nigeria. *Africa* **35**, 286–94.

MOSS, R. P. 1953. *Report on the Soils of Cocoa Plots near Ilaro, Abeokuta Province*, Res. Div., Ministry of Agriculture, Western Region, Nigeria, Ibadan.

1955. *Report on a Visit to the Eggua and Ohumbe Forest Reserves, Abeokuta Province*, Res. Div., Ministry of Agriculture, Western Region, Nigeria, Ibadan.

1957a. Some notes on the soils of the Western Region, with special reference to plantations of exotic trees. *Exotic Forest Trees in the Western Region of Nigeria (Commonw. Forestry Conf. Pap.)*, 19–29, Government Printer, Ibadan, Nigeria.

1957b. *Report on the Classification of the Soils Found over Sedimentary Rocks in Western Nigeria*, Res. Div., Ministry of Agriculture, Western Region, Nigeria, Ibadan.

1960. Land use mapping in tropical Africa. *Niger. Geogr. J.* **3**, 8–17.

1961. *A Soil Geography of a Part of South-western Nigeria*, unpublished Ph.D. thesis, University of London.

1963. Soils, slopes and land use in a part of south-western Nigeria: some implications for the planning of agricultural development in inter-tropical Africa. *Trans. Inst. Br. Geogr.* **32**, 143–68.

1965. Slope development and soil morphology in a part of south-west Nigeria. *J. Soil Sci.* **16**, 192–209.

1968. An approach to the study of the relationships between land use, vegetation, and soil factors in tropical West Africa. *Pacific Viewpoint* (October 1968).

MOSS, R. P. and MORGAN, W. B. 1968. Soils, plants and farmers in West Africa. In GARLICK, J. P., (ed.) *Human Biology in the Tropics: Proceedings of a Joint Symposium of the Society for the Study of Human Biology and the Tropical Group of the British Ecological Society*.

ROSS, R. 1954. Ecological studies in the rain forest of Southern Nigeria. III: Secondary succession in the Shasha Forest Reserve. *J. Ecol.* **42**, 259–82.

TINSLEY, T. W. 1964. The ecological approach to pests and disease problems of cocoa in West Africa. *Trop. Sci.* **6**, 38–46.

15 Man–water relations in the east central Sudan

ANNE GRAHAM

This chapter attempts to show the relationship between man and water in a semi-arid environment. It deals with the way in which limited supplies of drinking water affect the distribution of settlement, the way of life of the people concerned and consequently the use that can be made of the land. It attempts to show too something of the human response to changes in the water supply and how this in turn can result in a radical alteration in the economic exploitation of the area. While it is essentially a study of an area of some 7,000 square miles in the eastern Sudan it is thought that at least some of the points and problems discussed may have relevance to other parts of the continent.

In much of Africa development is hindered by lack of water. Parts of the continent will probably never be developed because the provison of adequate water supplies would be uneconomic. Certainly it is unlikely that much of the true desert will ever bloom like the rose. It is rather in the marginal environments of semi-desert and dry savanna that modest improvements in water supply, which are both feasible and economic, can make possible the fuller utilization of large areas of land. Over much of these areas at present imperfect use is made of the resource base because water supplies are scarce and scattered.

The problem of improving rural water supplies has attracted comparatively little academic attention. Rather, interest has been focused on large-scale irrigation projects. These have certain drawbacks. Many of the more obvious sites have been developed already and many of those that remain are likely to be costly to exploit and involve the country concerned in heavy foreign debt. The schemes even when completed bring direct benefit to only a small part of the country, a fact which affects the balance both of internal political power and of regional economic development. Moreover, implementation of these projects frequently involves human problems of adjustment to new techniques and new environments and often a considerable physical movement of population is necessary.

Schemes to improve the rural water supply are, by contrast, very different. They are usually much cheaper. They attract little attention because the governments concerned are unlikely to seek foreign capital for their implementation. The benefits of better rural water supplies can often be spread over a large area, thus minimizing the political problems of discrimination in favour of a particular region. By their very nature such changes do not uproot the existing way of life of the people concerned but rather open the way for its physical extension with opportunities for economic betterment.

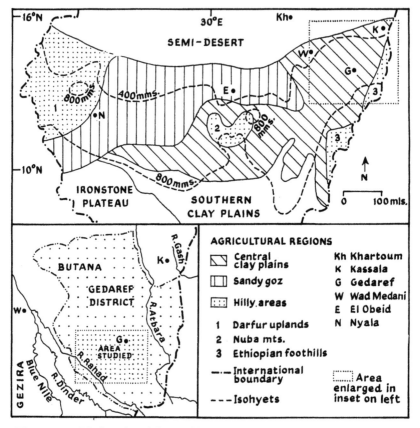

Figure 15.1 *The location of Gedaref in the central Sudan.*

In probably no other African territory have improvements in rural water supplies been so marked as in the Sudan in the last twenty years. Although the Sudan stretches from the Libyan desert in the north to tropical forest within a few degrees of the equator in the south, the main beneficiary of these improvements has been the savanna lands of the central Sudan (Fig. 15.1). Here the annual rainfall which varies from about 15 to 30 in (400–800 mm), has always been sufficient for a certain amount of grazing and seasonal agriculture. In the past, however, much of the area was under-utilized because of the lack of drinking water; in the sporadic patches where water supplies were adequate and settlement concentrated over-exploitation was the norm. Recently two new methods of exploiting the potentially available water have profoundly altered the provision of rural water supplies. First, deep bores have been sunk in the rocks of the Nubian sandstone series. These Mezozoic sediments underlie superficial deposits over quite large areas of the northern savanna and semi-desert and yield moderate quantities of underground water. While the function of deep bores is to tap a hitherto unused source of water, lying some 500–800 ft (155–244 m) below the surface, the other new method of obtaining water is merely a modernization of a previously known technique. Much of central and eastern Sudan consists of a gently sloping plain of aggradation. The clay of which it is composed is impermeable. It is therefore possible to conserve water by excavating basins or *hafirs* in the clays, in such a position that they are filled by surface run-off. Traditionally, these tanks were excavated by hand; the innovation of recent years has been the use of machinery for this purpose. The resulting 'mechanized' *hafirs* form water points with rather different characteristics from their hand-dug predecessors. The effects of this new type of water point on the economic and social structure of the area have been profound; the consequences for part of the area are described later in the chapter.

It will be clear from the foregoing that the whole area of the central Sudan provides an admirable setting for the study of man–water relationships. Gedaref District, which lies in the eastern part of the area between the Blue Nile and the River Atbara, is eminently suitable for a study in depth. The traditional sources of water in this area are very varied, not only in their origins but also in their adequacy and their reliability. The correspondingly varied modes of existence of the inhabitants illustrate the intricate impact of water on man. But the area also contains deep bores and numerous mechanically excavated *hafirs*; thus some analysis of the way in which man responds to changes in water supplies is possible. First the basic data of the area – its peoples, its physical characteristics, and in particular its rainfall – must be furnished.

The people
The peoples of the area practise a wide variety of ways of life. They range from nomads who occasionally plant a crop to settled cultivators with little interest in animal husbandry. The human scene is further complicated by the fact that Gedaref is an area of economic opportunity. For historical reasons it was underpopulated even before the new bores and *hafirs* increased its capacity to absorb new people. Land is still generally available

and rainfall is usually adequate for cultivation. Since rights to land are usufructarian, settlers are able to obtain holdings with the result that large numbers of immigrants have been, and still are, attracted to the area. The population now comprises three main groups of approximately equal size. As well as the indigenous Arabs there are people from the less prosperous regions of the western Sudan. While they generally have a veneer of Muslim culture they also display cultural traditions which are non-Arabic in origin. Some have come as genuine settlers. Others are young men who plan to return home after several years with enough money to marry. They often work as agricultural labourers. In addition Gedaref lies on the main east–west route across Africa, the route traversed by pilgrims to Mecca. Many West Africans pause here for several years to earn money by farming both on their way to and on their way back from the Holy City. Some eventually decide to remain permanently in Gedaref. Each of these very different peoples has its own concept of what constitutes an adequate water supply for its own mode of life and this in turn is reflected in its choice of settlement sites.

The physical environment

The dominant physical feature of the area is the clay plain, which slopes gently to the River Rahad in the west. Its monotony is broken by two groups of low hills: in the west the granite and serpentine rocks of Qala' en Nahl, in the east the basalt Ridge of Gedaref which overlies Nubian Sandstone (Fig. 15.2).

The clays of the area are alkaline and montmorillonitic. Their most striking feature, from the standpoint of water supplies, is their complete impermeability which prevents percolation and makes possible the use of *hafirs* to conserve water. The depth of the clay mantle varies considerably from east to west. Close to Gedaref Ridge it is too shallow for *hafirs*, while in contrast very adequate depths can be obtained close to the Hills. *Hafirs* are sited so as to be filled either by direct run-off from a hill or by a *khor* – a seasonal stream. But the drainage system is poorly developed owing to the absorptive qualities of the clay, the comparatively low rainfall and the general absence of relief. This often means that only a short section of a stream's course can be relied upon to flow, a factor which severely limits the scope for siting *hafirs*.

Again because of their impermeability the clay plains yield no subsurface water supplies. Well-water can only be obtained from the pediment zone around rocky outcrops where percolation takes place. The types of well found in Gedaref vary greatly according to the way in which the rocks have been weathered.

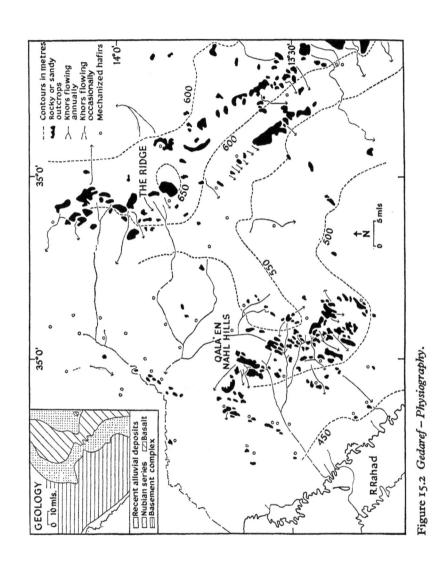

Figure 15.2 *Gedaref — Physiography.*

The Qala' en Nahl Hills are formed of the oldest rocks in the area, those of the pre-Cambrian Basement Complex. Most are granitic but there are two groups of serpentine outcrops, one in the south and one in the north. The serpentine gives rise to long, low, debris-covered hills. The weathered rock is very porous and percolation is accelerated along numerous shear-lines and faults. Here ample well-water is available throughout the year but because the rocks are so permeable the wells are the deepest in the whole area, reaching on average a depth of 130 ft (40 m). The granitic outcrops form residual hills of typical inselberg shape. Rising abruptly out of the plain, they are boulder-clad or dome-shaped with steep, smooth surfaces. After rain, water from the bare rock surfaces percolates into the sandy pediment. Here shallow wells tap water at an average depth of 58 ft (19 m). Compared with other wells in the area those in the granite have very small unconsolidated catchments. Not surprisingly most of them regularly peter out in the dry season, usually in January.

The Gedaref Ridge consists mainly of a series of Tertiary basaltic silts, but here and there are found small outcrops of underlying Nubian Sand-stone. The basalt is by far the most important source of well-water in the area. Almost 80 per cent of all the wells are found here. They vary enormously in adequacy and in depth from about 10–65 ft (3–20 m). The best and shallowest are usually near the crest of the Ridge where streams concentrate run-off from large catchments in flat-topped hilly country. The Nubian Sandstone also has wells but it is a less important source of water because outcrops are fewer and often consist of only low rises. The wells here do, however, yield water throughout the year, albeit slowly and mostly from depths of over 98 ft (30 m). Furthermore, the Nubian Sandstone is a source of underground water which has recently been tapped by bores. The fact that fewer than ten bores were still yielding adequately after three years of use would seem to suggest that the supplies of potable water in the area are limited.

The River Rahad flows through the west of the area studied and is a tributary of the Blue Nile. Although it normally floods its banks when it rushes down in spate in the rainy season, it ceases to flow for about six to seven months each year. Nevertheless it probably represents the largest and most permanent source of water in the area. Even in the dry season the southern part of the river has pools, while in the north, although the river-bed may be dry, water can be obtained by digging holes between $6\frac{1}{2}$ and 10 ft (2–3 m) deep in the silt.

Rainfall

Virtually all sources of water in Gedaref depend on local annual rainfall for replenishment, the exception being the Rahad which rises in the

Ethiopian Highlands. Rainfall decreases from south to north, or rather from south-south-east to north-north-west, for the proximity of the Ethiopian Scarp modifies the isohyet map which is otherwise typical of the African savanna zone (Fig. 15.3). In the north annual average rainfall is about 19 in (480 mm); in the south it probably exceeds 27·5 in (700 mm). The acacia grassland vegetation which covers the area varies accordingly. In the drier regions of the north it consists of short annual grasses, with scattered thorn-bushes which concentrate to form dense thickets along the meandering lines of seasonal watercourses. Southwards there are tall perennial grasses with open deciduous woodland. Owing to differences in rainfall totals water points in the north are generally less adequate than those in the south. Moreover, rainfall in this area is marginal for agriculture, and people consequently tend to be migrants and to rely to a comparatively large extent on animal husbandry. (North of the area under study there are virtually no more villages except along the Rahad. Nomadism is general and such cultivation as does exist is either confined to depressions, or to fields which have banks on the downslope sides in order to concentrate run-off.) In the south, by contrast, better water points and greater agricultural potential provide opportunities for a more settled way of life.

Gedaref's rainfall is seasonal. Rain falls mainly between June and early September and is connected with the northward movement of the intertropical convergence zone. During the long dry season many water points become inadequate. This may cause whole villages to become transhumant. In other instances villages are able to remain settled but are forced to spend the dry season fetching water from a more adequate source nearby. This preoccupation with collecting water in the dry season may be one of the reasons why there are so few crafts among the indigenous Arabs and why dry season occupations in the area as a whole are comparatively unimportant.

Occasional showers occur in April and May. These are of great importance for two reasons. First, good early rains replenish water points and enable those cultivators who are transhumant to return to their villages in sufficient time to prepare their fields properly before the cultivation season. Secondly, when the previous rainy season has been poor early rains are of a more general importance, for they can do much to terminate the hardship of villagers who are eking out the last of their rapidly declining and sub-average water supplies. The dates at which the first rains occur can, however, vary by several months not only from season to season but also between neighbouring places in the same year.

Total rainfall is also very variable. In the last fifty years the rainfall of Gedaref Town has varied from 64 to 160 per cent of the mean. Moreover, since the mode lies 1·0–1·6 in (25–40 mm) below the mean the chances of

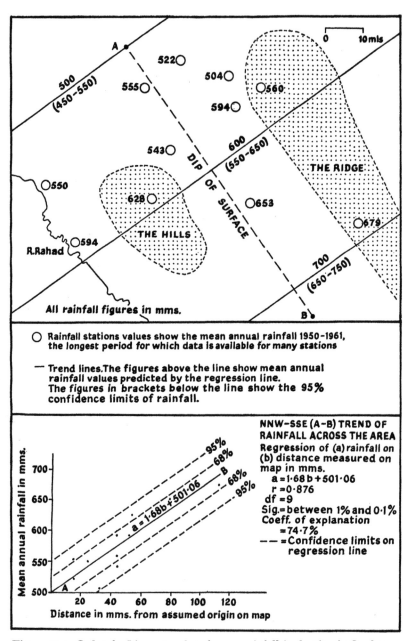

Figure 15.3 *Gedaref – Linear trend surface on rainfall* (reduction in Surface Slope = 74 per cent).

obtaining a rainfall less than the mean are considerable. Rainfall is more variable in the north than in the south. But since the northern ways of life are generally more flexible it is the comparatively settled people of the south who may be the more inconvenienced in a bad year. 1961 was a typically poor season, with rainfall everywhere about 25 per cent less than normal. Wells dried up five to six weeks early. About half the mechanized *hafirs* failed to fill completely, and many of the villagers in the Hills which normally relied on making no more than local movements to fetch water were forced to migrate. That year transhumance increased by 20 per cent.

Total rainfall can also vary by as much as 50 per cent over distances of only a few miles. Thus even in a good year some villages will borrow water from more fortunate neighbours.

Relationships between water and man in Gedaref

Before embarking on a detailed description of man–water relationships in the Gedaref area it is important to realize that the concept of a precise balance between man and water is purely theoretical. Man is influenced not only by water supplies but also by the whole of his physical and social environment. Thus while man's ability to use such water as is potentially available limits the location of settlements, the detailed siting of a village is the result of political, tribal or individual considerations. Similarly, while ways of life are considerably influenced by the seasonality and unreliability of water supplies they vary according to the traditions and economic requirements of the peoples concerned. But within these limitations it will be useful to consider the relationship between water supplies and human activity as reflected in settlement pattern, agriculture, land use and ways of life.

Settlement

In Gedaref the settlement pattern reflects the very uneven distribution of water supplies. There are no sub-surface water supplies in the impermeable clay plain and percolation is confined to zones around the rocky outcrops. Villages therefore cluster tightly at the foot of the Qala' en Nahl Hills or the Gedaref Ridge, where well-water is available, or are strung out along the semi-permanent course of the Rahad. These three nuclei of settlement were long separated by vast stretches of uninhabited plain. Only with the recent construction of deep bores and *hafirs* has some colonization of the plain become possible (Fig. 15.4). The settlement pattern is not, of course, completely determined by the water-supply situation. Water supplies are generally more adequate in the south than in the north because rainfall is higher, but the south is nevertheless comparatively underpopulated. The reason for this discrepancy lies at least partly in the fact that the south has more dense vegetation and is unhealthy and unpleasant in the rains.

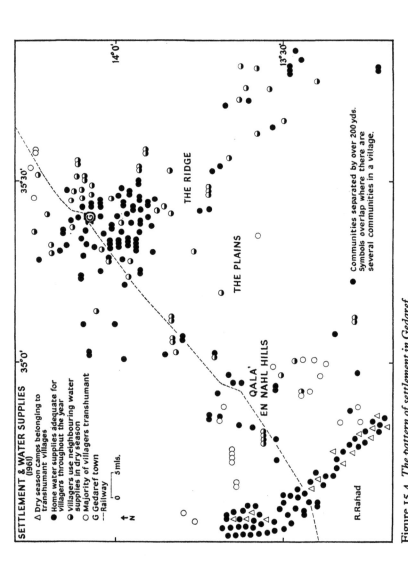

SETTLEMENT & WATER SUPPLIES
(1961)

△ Dry season camps belonging to
   transhumant villages
● Home water supplies adequate for
   villagers throughout the year
◑ Villagers use neighbouring water
   supplies in dry season
○ Majority of villagers transhumant
G Gedaref town
--- Railway

● Communities separated by over 200 yds.
  Symbols overlap where there are
  several communities in a village.

N
0   5 mls.

35°0'
35°30'
14°0'
13°30'

THE RIDGE
THE PLAINS
QALA'
EN NAHL HILLS
R.Rahad

Figure 15.4 *The pattern of settlement in Gedaref.*

Permanent settlement is therefore avoided by Arabs and those with large numbers of livestock who find the northern environment more suited to their way of life. Northern herds, however, move south to graze the area in the dry season. Many of the villages that do exist are fairly new. They are often inhabited by recent immigrants, West Africans or West Sudanese, who are attracted by the high rainfall which ensures rich and reliable harvests. They are unhampered by livestock and are not deterred by the hard work of clearing the bush or the generally rugged conditions. The underpopulation of the south is not completely explained by the difficulties of the environment. Written history testifies to a dense Arab population in the area at an earlier time when it lay on the main trade route into Ethiopia. Remnants of this Arab population survive today in only a few villages. The south is now a backwater and remote from markets for the important lines of communication pass through the north. Until land and water started becoming scarce in the north, immigrants had no need to tap the possibilities of the south.

Agriculture

Virtually everyone in the area obtains his basic livelihood from agriculture. The montmorillonitic clays provide a soil which is remarkably uniform[1] as well as being exceptionally fertile. Both because of its physical characteristics and because of the small average gradient of only fifteen minutes found in the plain, the soil is resistant to erosion. Its heavy texture, however, restricts the range of crops which can be grown. Thus, immigrants entering Gedaref from the west must of necessity abandon their former dependence on staple crops such as *dukhn* or pennisetum millet and groundnuts, which thrive on the generally lighter soils of their homelands, and adopt the agricultural practices found in their new environment.

The major factor causing variations in farming within the area is rainfall. In the north-west aridity is such that it is only possible to grow one crop – dura or sorghum millet. Yields are liable to suffer severely from fluctuations in rainfall. Drought conditions occur one year in four.[2] Better yields can be obtained by cultivating in depressions, where run-off accumulates, but in the flat Gedaref plains there are very few sites where this natural flush irrigation can be used. Hence people tend to practise a dual

[1] The actual clay content of the soil averages 70–74 per cent, but samples tested ranged from as low as 45 per cent in the north to 81 per cent in the south. However, this soil can contain appreciable quantities of coarser material without its physical properties being affected.

[2] About 18 in (450 mm) of rain are regarded as a minimum necessary to mature a crop. This figure may seem high but an appreciable proportion of the rainfall is absorbed into the structure of the clay and is therefore not available for crop growth. In the sandy soils of western Sudan as little as about 12 in (300 mm) may suffice for grain.

economy relying on animal husbandry both as a means of obtaining a cash income and as an insurance against crop failure. Southwards, as rainfall increases, it becomes possible to grow sesame as well as dura. Fluctuations in rainfall are less severe in this part of the area. Moreover, since dura prefers a modest amount and sesame a good deal of rain, one of these crops is likely to be a success whatever the rainfall. In the very far south rainfall is such that a wider range of crops begins to be found and a few farmers grow small amounts of cotton or even of other grains and pulses.

Sesame is tolerant of poor soils. Therefore, although everyone in the area would like to grow sufficient millet to satisfy their family requirements, sesame actually predominates on the stony soils near the Ethiopian foot-hills or on imperfectly cleared new land in the pioneer zone.

The shortness of the agricultural season results in a considerable degree of rural unemployment during the rest of the year. Crops are not usually planted until July when the rainy season is well advanced. Sesame is harvested in September and *feterita*, the normal variety of dura grown in the area, in December. Traditionally grain is not broadcast but planted in rows, using a *seluka* or digging stick. Between-row weeding is necessary about a month after planting. Using these methods and with some assist-ance from his family a farmer can cope with about 10 feddans of land.[1]

Most holdings are rather smaller, the average size being just over 8 fed-dans, but there are marked variations. Over 65 per cent of the farmers have less than 8 feddans and 40 per cent less than 4 (Baptista, 1960). In the south, where rainfall ensures high yields but where the clearing of fields is arduous work, farmers are often content with about three feddans. Con-versely, in the north farmers require a larger holding to compensate for low yields. About 70 per cent of the holdings consist of a single plot. A further 20 per cent have two plots. While this leads to compact farming units, in the north at least, the possession of several scattered plots can insure to some extent against the uneven distribution of rainfall.

Fields are cultivated continuously to the point where a farmer would normally expect to abandon his land. There is no appreciation of the need to maintain soil fertility by the use of manure, although animals do in fact graze the stubble after harvest. Where dura and sesame are both grown there will probably be a simple dura/sesame or dura/dura/sesame rotation, but almost half the farmers grow only dura and a sixth only sesame. After about ten years of continuous cropping the decline in yields becomes appreciable. On new land dura yields 2·5–5 *ardebs*[2] per feddan and sesame 1·0–1·5 *ardebs* (depending on rainfall); on 'old' fields the figures fall to

---

[1] A feddan is the unit most often used for measuring land in the Sudan and Egypt and equals 1·038 acres or 4,200 sq. metres.

[2] An *ardeb* equals 198 litres. An *ardeb* of *feterita* dura weighs approximately 146 kilos, an *ardeb* of sesame about 119 kilos.

1·5 and 0·3 *ardebs* respectively. Despite this about a third of the plots in Gedaref have been regularly cropped for over ten years, a third for five to nine years and only a third, many of which are in pioneer areas, for less than five. The main reason for retaining a worn-out field is the difficulty of obtaining fresh land sufficiently close to water to make agricultural operations feasible. Thus under the traditional systems of farming the area is imperfectly exploited.

Uneven distribution of water supplies has produced an irregular pattern of land use in an area of otherwise uniform physical characteristics (Fig. 15.5). Fields, or *bildat*, extend outwards from the village avoiding areas close to watering places and animal paths. This periphery of cultivated land seldom extends more than 5 miles beyond the village water point, for this is about the distance a farmer can travel daily to work. The proportion of cropped land within the peripheral zone or the degree of over-cultivation found is a function of the size and age of the settlement. This in turn tends to reflect the water supply situation, for areas where water is easily available were usually settled at an earlier date and those where water supplies are abundant have been able to support large villages. Only occasionally did enterprising farmers open up fresh land by constructing temporary water points farther out in the plains and camping beside them during the cultivation period. Even then these fields sometimes had to be abandoned if the scant water supplies dried up before harvest.

An extensive system of agriculture called *harig* enabled parts of the waterless plain, the open grassy areas, to be put under occasional cultivation. Under this system little labour is required and such water as cultivators require can therefore be brought with them. (The land is cleared after the rains have started. Newly sprouted vegetation is burnt off along with the old dry grass, thus both eliminating the need for later weeding and enriching the ground with ash. After a couple of crops have been taken the area is abandoned for about three years until a sufficient grass cover has again accumulated.)

There is also some extraction of gum arabic from the *hashab* (*Acacia senegal* Willd.). This tree occurs widely in the southern part of the area where rainfall exceeds about 23 in (600 mm). It is found in mixed stands seldom exceeding 20 per cent of the forest species. Although it grows wild, it is preserved even in cultivated areas because of its value. People are eager to obtain rights to extract gum in allocated areas of forest 20–100 feddans in extent. The trees are tapped in the dry season at monthly intervals from the end of November until March. Other forest products are also extracted. There is some gathering of perennial grasses and timber for building and a certain amount of charcoal-making for fuel. But in general the clay plains were but little used.

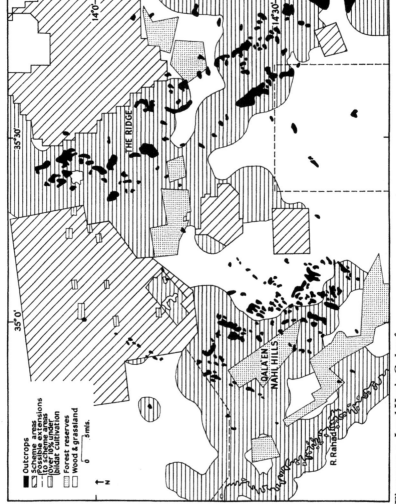

Figure 15.5 *Land Use in Gedaref.*

Legend:
- Outcrops
- Scheme areas
- Possible extensions to scheme areas
- over 10% under bildat cultivation
- Forest reserves
- Wood & grassland

THE RIDGE

QALA EN NAHL HILLS

R. Rahad

0   5 mls.

N

35° 0'   35° 30'   14° 0'   14° 30'

The advent of mechanized *hafirs* and bores has changed all this. Much of the once-wooded plain is now under cultivation. The new water points have in some cases led to an extension of peasant cultivation, in others to an extremely different form of agricultural exploitation. For the technical revolution in water supplies went hand in hand with the innovation of machinery for cultivation. Over a million feddans in Gedaref District are now devoted to the government-sponsored Mechanized Crop Production Scheme,[1] and the area under crops has been increased by 150 per cent. The landscape of the Scheme areas with its vast fields, treeless horizons and geometric grid of dirt access-roads contrasts sharply with the irregular, scattered plots and paths winding through the bush which form the traditional agricultural scene (compare Fig. 15.9 with Figs. 15.6, 15.7, and 15.8). Indeed, with its dumps of fuel oil and water towers its appearance is more akin to the Canadian wheat belt. The land-use map has been radically altered so that today extensive areas of uncultivated land are found only in the remote south.

Animal husbandry

Animal husbandry is the most important and widespread secondary occupation found in the area. Most families have at least one goat to provide domestic milk supplies. The large majority have a donkey, or more occasionally a camel, for transport purposes. These domestic animals are treated as part of the household for watering purposes and seldom move far from the village. Attitudes to large herds of cattle – or more rarely of camels or goats – are different. Among major stock-holders there is a close correlation between tribal origins and the number of stock held. In the case of the Arabs two factors have made for large herds – their nomadic tradition that cattle-rearing is a prestigious occupation and the opportunity afforded during their long settlement of the region to amass capital which, perhaps for want of other uses, has been invested in stock. The position of the western Sudanese is more varied: those with a tradition of cattle-rearing and who have been in the area for some time often have large herds, others with no such tradition or of more recent arrival have not. Most West Africans, regarding themselves as in transit, are reluctant to tie up their savings in cattle, but a few second- or third-generation villages have adopted Arab attitudes even though they come from tribes with no stock-rearing traditions.

It is uncommon for members of these groups to keep animals for purely commercial purposes. The tendency is to accumulate stock and to hoard it as a form of wealth. Only when ready cash is required are animals sold. This often occurs at the end of the dry season when food supplies run

---

[1] Referred to hereafter as the Scheme.

short. At this time the animals are in poor condition and the financial returns correspondingly low. This uncommercial attitude would make any
ordinary economic classification of the grazing patterns inappropriate.

Grazing patterns are influenced by the distribution and characteristics
of such water points as are available to animals. Their number is limited
by two factors. First, it is difficult to obtain large amounts of water quickly
from certain types of water points such as deep wells and these, because of
the large quantity of water required, are not considered desirable for
watering stock. Secondly, where water supplies are scarce men take
precedence over animals and animals over the requirements of irrigation.
(Thus in the Ridge at the height of the dry season, the market gardens turn
their pumps to watering nomadic livestock.) These factors mean that in
practice animals and men often obtain water from different water points.
In the dry season herds are often sent away to avoid exacerbating the water
shortage in the village. One consequence of this is that the villagers have
the benefit of animal products from their herd for part of the year only.
Another is that livestock migrations are on a much larger and more general
scale than human migrations. (Over 45 per cent of the village herds migrate
in the dry season compared with only 9 per cent of the villagers.) It follows
that the movement of animals provides a much more sensitive indicator of
water scarcity than the movement of men.

TABLE 15.1 *The amount of water needed by some domestic animals*

Animal	Amount needed (gallons)	Frequency of watering
Camel	22·7	Every 4 days
Cattle	10·0	Every 2 days
Cattle	6·5	Each day
Donkey	4·0	Each day
Goat	2·0	Each day

After Grabham, 1927.

Since water supplies differ markedly in the different settlement areas
certain traditional movements of cattle are especially typical of certain parts
of the district. There are four main types of grazing pattern. Where water
supplies are particularly good, stock are often at home throughout the year.
Over half the villages with animals of this sort are situated in the Ridge.
Where home supplies are less good, cattle may still be grazed locally
throughout the year if adequate water supplies are available in a neighbouring village during the dry season. As with the human population,
movements of this kind are most typical of the edge of the Ridge where

90 per cent of the villages whose cattle make this movement are found. Other herds are transhumant. They pass the rainy season at their villages but move to established dry-season camps when water supplies become inadequate. Over half the livestock from the Qala' en Nahl Hills with their highly seasonal water supplies make this movement, as do about a quarter of the herds from the most arid northerly part of the Ridge. Finally, there are those herds which travel most of the year in the care of nomadic herdsmen. Their movements are more complex. In the dry season they move south to graze the perennial grasses. Precise routes are determined by water availability and most herds are channelled along the banks of the Rahad in the west or the Atbara in the east. In the rains they move north away from the unhealthy riverine environment to a point safely beyond the cultivated areas, and then fan out across the Butana where the fresh annual grasses provide seasonal pasture. Occasionally only the rainy-season part of the nomadic movement occurs. Both cattle from the Scheme area and some from the densely cultivated parts of the Ridge migrate to Butana to avoid damaging standing crops. In the case of the Scheme area this exodus is obligatory. Large herds from the Rahad, on the other hand, are sent north to avoid the insects and disease so prevalent locally at this time.

The size of the area grazed by livestock in Gedaref depends on several factors – the season, the type of animal concerned, the frequency with which it needs watering and the distance it can travel between waterings. During the rains livestock can browse a wide area, for when the grass is verdant little drinking water is needed and such as is necessary can be found in temporary pools in the clay. The situation is very different during the dry season. The grass dries up and as the winter ends temperatures rise. Preferably all animals, except the camels, should be watered every day. Goats can manage a daily trip to water about 2·5 miles away, cattle go farther – up to 3·5 miles. Quite commonly, however, especially at the edge of the Ridge, cattle are watered every two days. They can journey up to 7 miles to the water point. Very occasionally it is necessary for goats to walk 5 or 6 miles to water every two days. They remain alive, but under these conditions they cannot provide the milk which is their *raison d'être*. Camels, requiring water every five days, can graze or fetch forest products from the remotest part of the area between waterings. The paucity of dry-season water points and the very limited range grazed by livestock results in severe over-use of areas adjacent to water supplies and the seasonal under-utilization of the rest of the pasture resources of the area. Particularly in the north where the annual grasses are especially susceptible to over-grazing, there are signs of a permanent deterioration around many water points.

The effect of water supplies on ways of life

Within the area of Gedaref water sources vary widely in their adequacy. The ways of life men adopt in response to these varying degrees of water shortage show a similar diversity. Where water supplies are plentiful men can, if they wish, remain in the same place throughout the year. Where water is short people must obtain supplies from elsewhere, even if this involves moving house. The only alternative – to reduce the amount of water used in time of shortage – is difficult, for with normal consumption at 3·6 gallons a day per head, housekeeping operations are already economical in the use of water.

In discussing ways of life in the area it is convenient to consider four categories – the nomadic, the transhumant, those who resort to local movements for their water supplies and those who are fully settled. In practice, these categories are not sharply defined. Between the way of life of the nomad and that of the settled agriculturalist there are many intermediate stages. Thus there is relatively little difference between, on the one hand, the nomad who takes up agriculture in the rainy season, and on the other, transhumant Arabs from the northern hill areas whose cattle spend most of the year elsewhere. In the same way where a settled village has many men who regularly obtain seasonal employment elsewhere, the whole population can become transhumant should unusual water shortage arise.

Nomadism

The essential difference between the nomads and the transhumants is that the former live in tents and spend much of their time moving, while the latter live in villages but spend several months each year in another area. Nomadism represents one of the most extreme adaptations that man can make in response to limited water supplies. In the past, when the techniques used to obtain permanent water supplies were more limited, a much higher proportion of the population was probably nomadic. Their mobility permitted them to make good use of the scattered grazing and water resources. And they avoided the fate of agriculturists who, tied to the few permanent water points that existed, fell an easy prey to bandits and armies in unsettled times. Today there are no nomads with rainy-season camps in the area, but the dry season brings many from the north. Some travel slowly, pausing for several weeks at each water point. Others move swiftly to an established dry-season camp, returning to the same place for four to eight months year after year. Most nomads, especially those with regular dry-season camps, spend the height of the dry season by a river, since few *hafirs* or wells can provide adequate water for large numbers of stock in May or June. Recently the number of nomads has probably increased as a

MAN–WATER RELATIONS IN THE SUDAN

direct result of the increase in water supplies suitable for use in the dry season. Some bores provide enough water to supply a dry-season camp. Moreover, the *hafirs* in the Scheme area sometimes contain enough water after the harvest to support animals while they graze the surrounding stubble.

Nomadism is the best way of utilizing a variety of temporary water points. Each point supports the herds for a short period at a time and the surrounding area is not therefore severely over-grazed. The nomads' willingness to move from place to place ensures that the best possible use is made of available pasture. Stock are maintained in reasonably good condition and the nomad has a reliable supply of animal products throughout the year. On the other hand, his standard of living is not high; because of the need for mobility he has few material possessions. It is almost impossible to provide nomads with social services, though in the dry season they may occasionally be able to use those supplied to settled villages near their camps.

Transhumance

Transhumance is traditionally a regular dry-season movement by the entire population of a village to an established camp. The movement is usually a response to limited water supplies but it is occasionally caused by economic factors. While the transhumant existence is less flexible than the nomadic it is nevertheless highly adaptable. It does not matter if rains are unusually poor or particularly heavy; the villagers can move earlier or later than usual. Fears of exceptional shortage cause no undue upheaval for families can be sent ahead to the dry-season camp while the men remain to complete the harvest. Where water supplies are regularly really short, sesame can be substituted for dura because this crop matures several months sooner and therefore allows an earlier move to the dry-season camp. Sometimes a migration made out of necessity is turned to economic advantage. People who would have been idle during the dry season are able, because water is easily obtainable, to take on dry-season occupations. Indeed, dry-season transhumance may be undertaken for *purely* economic reasons. Some people, most of them West Africans who want money for the pilgrimage, move right out of the area to pick cotton in the Gezira. The availability of water at home has little to do with their decision to seek employment of this sort.

Local movements

For the third category in the area local movements suffice to obtain adequate water. Typically, villagers travel a distance of some 4 miles for from four to six months in the dry season. Generally speaking, the longer the period in the year during which the village's water supplies are inadequate and the greater the distance to dry-season water points, the more prosperous the inhabitants must be if they are to continue living there. Where the neighbouring water point is more than 3 miles away, transport animals are needed to fetch water. Arab and old-established and fairly wealthy villages usually have sufficient livestock for this purpose, others may not. For them the only alternative to moving is to pay for water to be fetched. A donkey can manage a round trip of up to 12 miles daily and can carry enough water for a family. Where the nearest water point is more than 6 miles away, however, a village normally must become transhumant. Only very occasionally has a community sufficient camels with which to fetch water from farther afield or access to a truck which can serve the same purpose. Generally villages are keen to avoid the inconvenience and unpleasantness of moving if they can. Permanent settlement in villages of this third category, however, entails disadvantages. They seldom enjoy the additional benefits of public services. Schools and dispensaries require water on a scale which is found only in villages with ample supplies throughout the year. Moreover, journeys to fetch water are time-consuming and may effectively preclude family heads from engaging in any more profitable dry-season occupation.

Sedentarism

The advantages of living in a fully settled village, not least that of permanent water supplies, are obvious. But there are disadvantages here, too. These settlements are nearly always found in areas of dense population. Fuel may be expensive, the land over-cultivated and new fields obtainable only a considerable distance away. Grazing may be difficult to find locally. It is important to realize, too, that both people that are fully settled and those who make local movements to obtain water may have difficulty in adapting their ways of life to meet exceptional water shortages.

The physiographic regions

The four physiographic regions in the area contain water supplies so different in character that they deserve to be studied as individual water environments. Indeed, the detailed and varied responses of the population to the possibilities and limitations of these contrasting environments serve to throw into relief the immediacy and subtlety of relationships between man and water in this part of the Sudan.

The Gedaref Ridge

With its generally good water supplies the basalt ridge of Gedaref supports over 40 per cent of the villages in the area; the surrounding Nubian Sandstone supports a further 5 per cent. The region is also the site of Gedaref Town, the district's headquarters, though even the comparatively abundant water resources of the Ridge are hard put to support this settlement of 50,000 people. Water supplies are particularly good near the crest of the Ridge. Here there are often old Arab villages containing as many as several thousand families. Yet the quantity of water is such that nomads passing through the area on their seasonal migrations traditionally make use of certain of its wells. It is even possible, in response to urban demands, to irrigate small market gardens and maintain commercial dairy herds throughout the year.

Over two-thirds of the villages of the Ridge are able to use their home wells throughout the year. The remaining third, which runs short of water in the dry season, is situated mainly along the margins of the Ridge where wells are deep and low-yielding. However, these villages can often fetch water from better wells at the centre of the Ridge only a few miles away. This obviates the need for transhumance which is all the more fortunate in that the River Atbara, the nearest permanent river, lies a considerable distance away.

The good water supplies of the Ridge have attracted dense settlement. This in turn has led to an acute shortage of arable and pasture land. Nowhere else in the area are the problems of over-cultivation and over-grazing so clearly marked (Fig. 15.6). Over 15 per cent of the fields in the Ridge have been cultivated continuously for over twenty years; fewer than 20 per cent for less than five years.

The situation is particularly bad at the centre of the Ridge. Here villages contain few newcomers for, although water is ample, the size of settlement is restricted by the scarcity of agricultural land. These old communities are now often hemmed in by villages belonging to the more recent immigrants who in order to obtain farms were forced to settle by the poorer water points along the margins of the Ridge.

In some cases land has deteriorated so much that village sites in the centre of the Ridge have been abandoned. In others, population pressure has been so great that villages have been forced to split. While part of the village has remained at the centre of the Ridge an offshoot has formed a new community on the edge of the clays, obtaining fresh land at the price of a precarious water supply. In other cases men from Ridge villages move out into the clay plains to cultivate, living in temporary housing in their fields during the rainy season and supporting themselves with water collected in depressions or laboriously carried with them. Thus, in the Ridge the problem of over-cultivation is such that people are beginning to prefer fresh land to decent water supplies. In recent years the situation has improved at least for the wealthier families who have had recourse to tractors to cultivate extensive fields far out in the plains.

The general problem of over-population is clearly shown by the fact that when mechanized *hafirs* were established around the Qala' en Nahl Hills and in the plains, emigrants from the Ridge were predominant among the settlers.

The Qala' en Nahl Hills

The Qala' en Nahl Hills offer a much less hospitable environment for settlement. Because of poor water supplies the area was sparsely peopled for much of this century. Even today, despite modern improvements in water supplies, only 20 per cent of the area's villages are found here, and settlements are often small, isolated and lacking in amenities. As a result, in contrast to the Ridge, land is freely available. Often fields extend only 2–3 miles beyond a settlement. Not only is this the only part of the region where there are still extensive areas of *harig* but also much of it is within walking distance of villages. Moreover, even today its pasturelands tend to be under-utilized, for the fact that most of the mechanized *hafirs* are reserved for human use forces herds to remain migrant. No large herds of cattle remain in the Hills in the dry season and in the case of five villages even the goats are transhumant. Neither the wells in the serpentine nor those in the granite are highly regarded as water sources although both are of considerable interest.

The disadvantage of serpentine wells is that the water has to be hauled from a considerable depth. It seems that this source of water was unused before the importation of modern well-drilling tools and the wells date back only some sixty years. The effort required to draw water from them has been a deterrent to their multiplication. Even today there is only one village in the southern hills and, altogether, only 3 per cent of the settlements in the area under study are supported by wells in the serpentine. That there is ample water available is shown by the fact that the second largest town of the area is found in the serpentine zone. In this case, however, because of its administrative function, the town has been provided with a shallow bore and pumps to obviate the difficulty of raising water from a great depth. Where the need is sufficiently great, therefore, the well-

---

(a) *left* The centre of the Ridge with its typically dense settlement based on ample well water. Over-cultivation is indicated by the fragmented pattern of *bildat*, and indeed most of the non-cultivated land is scarred by abandoned fields. All local timber has long since been used. (b) *right* This shows a village which moved to the edge of the Ridge 13 years ago, obtaining new agricultural and grazing land at the price of a precarious water supply. Cultivated land is more extensive and plots much larger than in (a). There is still some woodland. The village well is supplemented by using hand-dug *hafirs* for livestock and by sharing the water resources of the neighbouring settlement.

Figure 15.6 *Contrasting landscapes on the Ridge.*

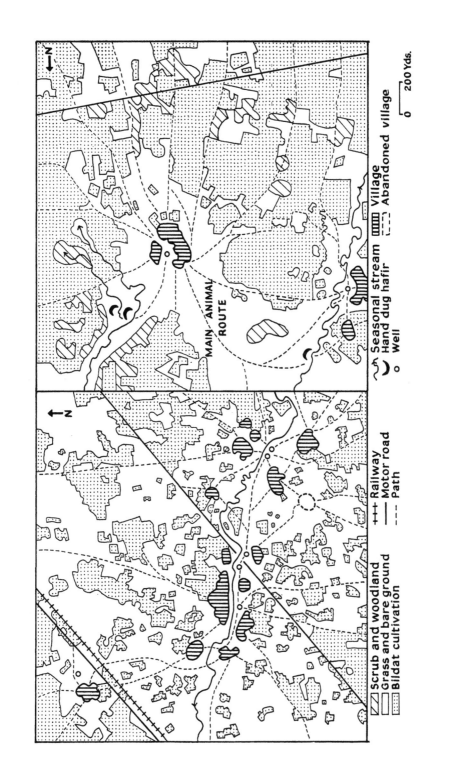

N

MAIN ANIMAL ROUTE

N

Scrub and woodland
Grass and bare ground
Bildat cultivation

Railway
Motor road
Path

Seasonal stream
Hand dug hafir
Well

Village
Abandoned village

0   200 Yds.

water resources of the serpentine can be exploited through capital investment in modern equipment and the serpentine areas in the south provide
a potential area for agricultural settlement.

The granitic areas are a case study in themselves. The extreme seasonal
fluctuations of well-water in this region and, indeed, the outright failure of
most wells in the dry season have produced communities which till recently
were in large part transhumant. Until the early 1950s over 50 per cent of
the villages in the Hills migrated to established dry-season camps along the
Rahad. While this way of life has been to some extent overtaken by modern
technical developments its partial survival provides valuable illustration of
how limited water supplies have influenced the economic exploitation of
the area.

The inadequacy of the wells in the earlier period is clear from the fact
that, even in order to achieve a semi-settled existence, supplementary
sources of water were needed. Nowhere else in the area was settlement supported by such a multiplicity of minor water points (Fig. 15.7). Not only
were hand-dug *hafirs* far more important than in the Ridge, but also the
granitic outcrops provided two additional sources of water. *Gallits* are
simply basins in the rock, created by chemical weathering along a joint plane.
Their function is to conserve run-off after a storm until it is evaporated
away. *Jamams* are holes scooped out at the very foot of the hill at a point
where percolating water seeps slowly from a joint in the rock above.

Each of these types of water point played its distinctive part in the traditional pattern of transhumance. In the rains, water supplies naturally presented no problem. People used whichever of the village water points were
most convenient as they went about their agricultural pursuits. Meanwhile, cattle grazing the plain beyond the fields drank from natural pools.
Later in the year livestock used hand-dug *hafirs* and, when these were
finished, were led off by herdsmen to the river. Families followed after the
harvest when wells became inadequate. Only a few householders would
remain behind as caretakers. They opened up *jamams* to supplement the
trickle of water that still percolated into the wells. With the first showers
of spring, the *gallits* filled. These provided sufficient water for the men
to return to prepare their fields. Then as soon as the main rains caused
the wells to rise the women, children and livestock would return swiftly
before the clays became a quagmire. Yet despite all this ingenuity in the
use of water some villages were actually abandoned because of inadequate
water supplies.

While the traditional transhumant life was readily accepted by the
Arabs, who were thus able to utilize two areas of grazing, it proved less
attractive to the other groups. With increasing population pressure in the
district as a whole, however, and the desire for new land, the number of

West African and West Sudanese settlers increased. Like the Arabs, some were able to turn the enforced transhumant movement to their profit. Some immigrants acquired livestock. Others, particularly West Africans, took up fishing. More recently some settlers from the Rahad have actually begun migrating to the Hills in order to practise a dual agricultural economy. They accept the seasonal nature of water supplies in the Hills in exchange for the benefits of good land, for grain production here is much better than in the ill-drained land near the river which is infested with weeds and birds. Then after the dura harvest they return, to spend the dry season cultivating their riverine gardens.

The advent of mechanized *hafirs* has profoundly modified the traditional patterns of existence in the granitic hills. Considerable depths of clay are found close to the inselbergs, and here mechanized *hafirs* have been used to supplement village water supplies. Only about a third of the villages are now transhumant and in some cases this transhumance is retained for reasons of custom or profit rather than necessity. Improved water supplies have accelerated immigration and over 40 per cent of the settlements in the area have been formed in the last fifteen years. Some are like the older villages in that they use wells, followed by mechanized *hafirs* in the dry season. Others rely entirely on mechanized *hafirs* and, apart from the fact that they are sited at the foot of hills with all the benefits that this implies, are little different from the new villages in the plains.

The River Rahad

The River Rahad provides water throughout the year but its main attraction lies in the large and readily obtainable supplies that it offers in the dry season. Thus this environment not only contains almost a third of the area's villages but also absorbs a vast seasonal immigration of men and livestock.

Notwithstanding the string of villages which lies on the well-drained sandy levées along the Rahad's course, there are considerable drawbacks to permanent settlement along the river. When the river floods, communications are cut and the area is unhealthy for man and beast. These drawbacks are much worse in the south than in the north, with the consequence that the general north–south variations in the settlement pattern of the whole area are accentuated. The age of villages decreases steadily from north to south. Since the early 1920s young West Africans have been emigrating from their parent villages in the north, where riverine alluvium was scarce and overcropped, to form pioneer agricultural colonies farther south. But despite the inducements of both water and land, permanent settlers have been attracted to the south only in the last few years, more particularly since government use of residual insecticides has helped to reduce the dangers of disease in the rains.

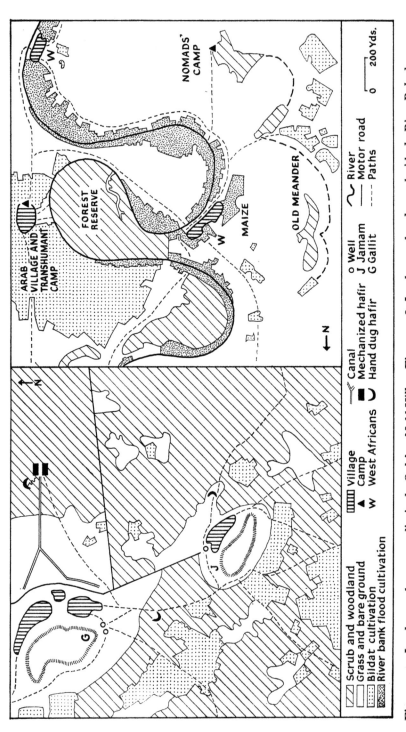

Figure 15.7 *Land use and water supplies in the Qala' en Nahl Hills.*   Figure 15.8 *Land use and settlement beside the River Rahad.*

Scrub and woodland
Grass and bare ground
Bildat cultivation
River bank flood cultivation

▦ Village
▲ Camp
W West Africans

Canal
Mechanized hafir
Hand dug hafir
○ Well
J Jamam
G Gallit

River
Motor road
Paths

0   200 Yds.

15.7 shows typical hill-foot villages and the wide variety of water resources traditionally used in the Qala' en Nahl Hills. Compared with the ridge a modest proportion of the land is cultivated. 15.8 shows the variety of settlements found by the River Rahad in the dry season. Notice the importance of river-bank cultivation near the two West African villages, compared with the Arab village which has made greater use of *bildat*.

The Rahad provides the most striking example of how different people find different parts of Gedaref best suited to their needs. It is, *par excellence*, an area of West African settlement, over 80 per cent of the villages containing peoples of non-Sudanese origins. This is reflected in the fact that over 40 per cent of the villages have no cattle. Moreover, where *bildat* land behind the levées floods sufficiently, food grains other than dura – maize, and even a little rice – have been introduced by the immigrants, although locally the Arabs make no effort to cultivate them. The riverine environment is particularly attractive to pilgrims because it provides ample sources of income even in the dry season. Most West Africans are skilled cultivators. Here, as the flood-waters retreat, they make use of the damp banks to grow a wide variety of high-price fruit and vegetables (Fig. 15.8). Tomatoes and peppers are the most popular, for these can be dried and sold as condiments, thus making rapid transport to the market unnecessary and avoiding a glut and consequent low prices at harvest time. The drying pools are extensively fished and even the least enterprising can gather and sell the fruit of the wild *sunut* tree which grows exclusively on land subject to annual inundation.

The attraction of the Rahad for the Arabs is that it supplies easy and ample water for both man and more especially beast in the dry season. Arab use of the environment is almost exclusively on a seasonal basis. There are admittedly a few permanent Arab settlements beside the river, but even these Arabs normally find it necessary to send out not only their livestock but often even their women and children to the healthier conditions of the northern interior in the rains. In the dry season there is a large immigration into the area. The dry-season camps belonging to the people from the Qala' en Nahl Hills have already been mentioned. The river also provides dry-season quarters for groups of nomadic Arabs from the Butana and performs a further function as a well-watered routeway for those moving south. Needless to say, as the dry season progresses the riverine strip becomes increasingly devoid of edible vegetation.

Thus it is at the Rahad that tribal differences in the economic exploitation of the environment are most marked. The growing of rain-fed dura and sesame, the main occupation common to everyone in Gedaref, is here of least importance. Dry-season pursuits assume predominance and it is in these that tribal variations are greatest. The differences are such that occasional racial clashes arise as a result of the conflicting uses to which the riverine area is put in the dry season. In particular the green riverside gardens (*gerf*) may block access to a watering place and provide a great temptation to hungry stock.

The clay plains

The clay plains have been the principal theatre for the most dramatic development in man–water relations in the history of this area. The coming of mechanized *hafirs* was greeted in the Sudan as a technical revolution which could transform the entire central rainlands. In the pages that follow an attempt is made to assess in a critical manner the nature of these new water points and the degree of change they have engendered.

First, mechanized *hafirs* must be compared with their hand-dug predecessors. Hand-excavated *hafirs* have always been important in Gedaref. They are usually constructed close to hills where natural run-off collects. They are horseshoe-shaped and open to allow the water to enter freely. With a shallow sloping floor and an average capacity of 3,000 cu m they hold water for two to three months after the rains. They are normally found in connection with existing settlements and are particularly convenient for watering animals. A few, however, are sited away from the villages at the margins of the Hills or the Ridge. They were mainly constructed in the 1930s when population pressure in traditional areas first became apparent. They were then used to provide water for temporary camps so that cultivators or, occasionally, grazing could spread a little way into the clays.

Mechanized *hafirs* are usually much larger with a capacity of some 15,000 cu m and a depth of at least 18 ft (6 m). With their small surface area, evaporation is low. They are not merely supplementary sources of water but also can support a permanent settlement of about 200 families provided they have only a few household animals. The essence of the new technique is its comparative independence of the physical environment. Many mechanized *hafirs* are fed by artificial canals and can therefore be sited out in the plains away from the obvious catchment areas. In particular short-lead canals are often used to link *hafirs* with the short seasonal streams which otherwise dissipate themselves in the clays. Thus the greater flexibility in the siting of the mechanized *hafirs* permits a greater penetration of the clays than is usually possible in the case of hand-dug ones. (Jefferson, 1952, 1954; Robertson, 1950.)

The Mechanized Crop Production Scheme

The effect of these *hafirs* on land use in Gedaref has been very great. While the areas opened up for peasant agriculture and animal husbandry have not been insignificant, it is the Scheme that has been really important. This Scheme has profoundly altered the economic exploitation of the area. With over 600,000 feddans annually under crops, the Scheme has produced over two-thirds of the District's millet and one-fifth of its sesame in recent years. In the early 1960s Kassala Province became the second most important grain-producing area in Sudan with over a third of the country's dura acreage. This prominent position is largely the result of the Scheme, for Gedaref District, apart from two relatively small irrigated areas, is the only part of the Province with enough moisture for extensive cultivation.

The Scheme was commenced in 1945 when it was hoped it would help to increase the then scanty supplies of food in the Middle East. Experiments in the use of heavy duty tractors to cultivate the clays and in the use of machinery to excavate *hafirs* occurred simultaneously, for without adequate water for cultivation large-scale agricultural developments in the plains would have been impossible. Initially the expansion of the Scheme was slow. At first the Government cultivated in its own right and employed its own labour. Later a participating-cultivator method was tried. The present organization has existed since 1955 and in 1957 the Scheme areas were greatly extended until they reached the boundaries shown on the map. Certain problems may limit the further expansion which the Government favours. The existing northern boundaries have adequate rainfall for cultivation only in seven years out of ten. In the south, costs of clearing the heavy bush have to be met by the scheme-unit holder, so few are ready to rent new scheme-units in this area. Moreover, the early rains here may be so heavy that the ground becomes impassable to tractors before planting is complete.

The scale of operations in the Scheme is vast. Thousand-feddan holdings are rented to entrepreneurs on an annual basis for a nominal sum. Scheme-unit holders are men of substance; they must possess sufficient capital to sustain them against crop failure (officially set at a bank balance of £2,000 Sudanese), provide their own machinery and recruit their own labour force.

Crops grown are similar to those found outside, with sesame rather less and cotton rather more important than among traditional agriculturalists. The land is prepared for cultivation by disc harrowing after the rains have caused the weeds to germinate. A second harrowing to eliminate later-germinating weeds is followed by sowing. Such additional weeding as is necessary, and all harvesting, is done by hand. Because of its uneven stand and drooping heads dura cannot be harvested by machine. (The American varieties which can be combined are not considered palatable by the Sudanese.)

The scheme-unit holders are primarily interested in quick returns; but these have not always been easy to achieve. As well as the difficulties faced by ordinary farmers the scheme-unit holder requires up to 200 workers for two months at harvest but labour is in short supply. It is increasingly costly and much of it comes from as far as the western Sudan. Tractors frequently break down on the heavy clays and adequate spares are seldom available. At least initially there was an absence of knowledge of the conservation techniques necessary for mechanized agriculture. Despite normally gentle gradients the great length of the uninterrupted ploughed land in these vast

P

fields has resulted in some soil wash. Since the Scheme has been extended
into the south and east where the land surface is more irregular, there have
been signs of more active erosion in places. Only after some holdings had
been cultivated for fifteen years and yields had fallen badly was a system of
fallowing evolved. Now after four years under crops, a scheme-unit lies
idle for four years, the former scheme-unit holder receiving another unit in
exchange.

The Scheme, a central government project, was imposed on Gedaref
from without. It is alien to the traditional way of life of the District. Most
scheme-unit holders are absentees, for only a few local merchants or tribal
sheikhs are wealthy enough to participate. Local farmers seldom work on
the Scheme; they are usually busy on their own farms at the time the
Scheme needs labour. Although the bulk of the people do not participate
directly in the Scheme, its very existence has had important implications
for the traditional agriculture of the District. Many villages lost *harig* when
the Scheme was established (indeed, the area in which the Scheme started
was chosen precisely because it consisted of grassland which did not need
clearing). Previously, enterprising farmers had obtained higher yields from
this land than they could from their older *bildat*. Villages bordering the
scheme-units have had their *bildat* consolidated so that it does not project
into the Scheme areas. In many cases this has meant that cultivation has
become more concentrated. Less land is available for new plots. Over-
cultivation and lower yields result. Not only have farmers lost land which
they used to cultivate but all have been faced with lower prices for their
crops as a result of competition from the Scheme. Peasant production of
dura, now a small proportion of the whole, has little influence on prices.
Increasing urbanization in Sudan, together with limited supplies of grain,
would have enabled the Gedaref producers to command high prices had
not the Scheme glutted the market. As it is, the home market is saturated
and only low prices can be obtained by exporting the surplus. Other
people lost their holdings of *hashab*, for mechanized cultivation, unlike
traditional methods, involved the complete clearing of the land. In view of
the substantial addition to income that local people receive from gum and
the very marginal profits of scheme-unit holders in recent years it seems
that the harm done by the cutting down of *hashab* has not been fully
justified.

Contact with the Scheme has, however, brought about some improvements in agricultural techniques. The large-scale undertakings have
inspired some farmers to extend their cultivation. The employment of field
labour, once on a small scale, has increased markedly and it is estimated
that local farmers employ over 7,000 workers each year. It is no doubt easy
to tap the labour force attracted to the Scheme. Others have bought trac-

tors, a modern alternative to investing in livestock. There are now over forty in the District but for every farmer owning a tractor tens of farmers hire them. (Over a third of the tractors owned by scheme-unit holders are hired to farmers outside the Scheme.) In villages both within and on the outskirts of the Scheme the use of tractors seems to be almost universal. The people of the Ridge have been particularly quick to make use of their new opportunities. Many have found it worth while to acquire distant fields for cultivation by machinery, local *bildat* being too small and worn-out for this purpose. Indirectly the Scheme has enabled new areas to be opened up for peasant cultivation, for access roads, primarily intended for scheme-unit holders, have also been used by local farmers, who if necessary buy water from passing vehicles, while they cultivate their roadside plots.

The Scheme has been responsible for the growth of Gedaref Town from a modest District headquarters and local market town to a major service centre. Its population more than trebled between 1955 and 1961. Some of the prosperity which the Scheme has brought to the middle-class merchants has spread to the ordinary people of the rural areas. Villages in the Ridge within a few miles of Gedaref have become suburban in character. The fragmentation of fields in this densely populated zone matters less if people are only part-time farmers. Pilgrims take dry-season portering jobs in town connected with the marketing of produce. The West Africans of the Rahad find a ready market for dried fish among Scheme labour. The demand for building materials both from the Scheme and from the growing population of Gedaref means that villages with camels can earn good money by extracting timber from the remaining areas of forest.

Settlement in the plain

In contrast to the huge expansion of agriculture, the spread of settlement into the plains has been much more modest, although villages have been established beside most of the new water points.

There is little settlement in the Scheme area. Very few villages existed there before it was started because of the paucity of water supplies. Some were established around bores or *hafirs* during the early phases but since the late 1950s all new water points within the Scheme have been reserved for scheme-unit holders' labour. Villages in the Scheme area, although in many ways similar to other pioneer villages in the plains, have additional disadvantages. Fuel and building materials (except dura straw) are not available locally and there are seldom shade trees. Villagers were originally issued with 20 feddan holdings (Fig. 15.9). These are now surrounded by scheme-units. Youths can only remain in the village or newcomers enter it if others leave for the town or are prepared to subdivide their holdings. The area around a *hafir* is often very marshy. Normally a village is sited farther

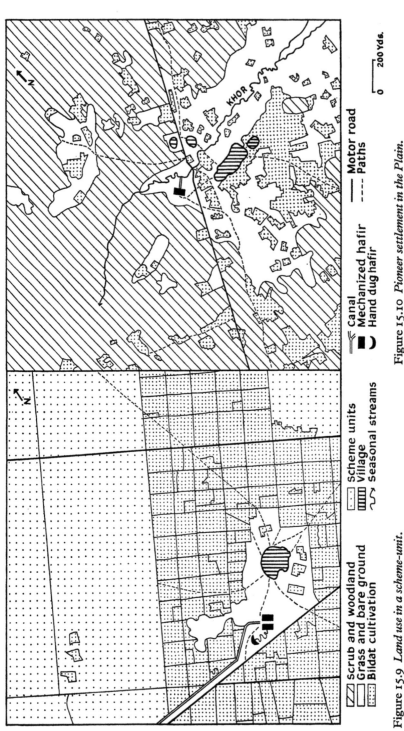

Figure 15.9 *Land use in a scheme-unit.*

Figure 15.10 *Pioneer settlement in the Plain.*

Figure 15.9 legend:

Scrub and woodland
Grass and bare ground
Bildat cultivation
Scheme units
Village
Seasonal streams

Figure 15.10 legend:

Canal
Mechanized hafir
Hand dug hafir
Motor road
Paths

0   200 Yds.

15.9 shows the agricultural landscape found in the Scheme area. The village, one of comparatively few within the Scheme area is based on a mechanized *hafir*. Because of pressure of population, over 17 per cent of the villagers own less than the standard 20 feddan plots, nearly all of which are cultivated by tractor. 15.10 shows a four-year-old pioneer settlement. Notice the contrast in settlement form with that of 15.9. Fields are still small and scattered. Only in the south-east is some consolidation of cultivated land taking place, particularly where the woodland is less dense.

off on better-drained ground but in Scheme areas villages are close to the
hafirs to allow more land to be cropped and they may therefore be extremely
unpleasant in the rainy season.

Altogether about a seventh of the villages studied are to be found in the
clay plains. Virtually all have mechanized *hafirs* or, very occasionally, bores
as their sole source of water supply. Settlement is spontaneous and dis-
organized, immigrants arriving very rapidly after the establishment of a
new water point, all of them motivated by the prospect of obtaining virgin
lands for agriculture. To begin with, patches of land are cleared haphazardly in the surrounding bush (Fig. 15.10). Only later as the cultivated
area increases does the traditional pattern of a peripheral zone of cropped
land around the village emerge.

The villages have a pioneer character. Services, apart from a rudi-
mentary general store, are usually lacking. They are very isolated, particularly when communications are cut in the rains. The settlement sites suffer
from lack of air movement and unlike the well-drained, slightly raised,
sandy sites found elsewhere are unpleasantly damp. Insects abound, particularly in imperfectly cleared areas in the south. Houses are often
inadequate, being constructed by men eager to get on with the job of
winning fields in the virgin bush. There are as yet few domestic animals
and standards of nutrition are low. Men considerably outnumber women.
Living conditions are poor and villages have a straggly, unkempt appearance. There is no established hierarchy, and village chiefs come and go
with remarkable rapidity. The *hafirs* have usually attracted colonists from
numerous individual tribes, although certain peoples predominate.

Most of the new settlers are of western Sudanese origin and members of
this group are found in all but two of the villages of the plains. In general
the western Sudanese will tolerate the hardship of the plains in exchange
for the chance to cultivate land so much better than that available in Darfur.
Young men unhampered by families or beasts are especially ready to
respond to the opportunities offered by the new water points. There are
also a certain number of West Africans, particularly in *hafir* villages within
the Scheme areas, for here, as well as possessing their own small holdings,
they can earn money by working as labourers on the scheme-units. Arabs
are seldom found in *hafir* villages. They are strongly attached to their
traditional settlement sites and normally inherit at least some of the land
around them.[1] They are reluctant to give up the amenities and comforts
available in their established home for a much more primitive existence in
a *hafir* village. With their mature family structure, a move would involve
transporting many dependents and severing other kinship ties.

[1] Although rights to land use are usufructarian, in practice land under cultivation
when the owner dies can be inherited.

Because mechanized *hafirs* are so often the sole source of water for a village it is vital that they should conserve and provide water throughout the year. Unfortunately mechanized *hafirs* on the plains are particularly susceptible to the vagaries of the rainfall, for unless rains are sufficiently heavy the streams which feed them may not flow at all. Moreover, because of their complex construction either the lead-off canal or the inlet mechanism may become damaged and prevent the *hafir* filling. It is often impossible to carry out emergency repairs in the rainy season and even annual general maintenance which is carried out by skilled repair teams from the Government is often inadequate because labour and transport costs are high. Badly maintained *hafirs* deteriorate rapidly, silting occurs and the *hafir's* capacity is gradually reduced. Meanwhile the *hafir's* population is still growing and eventually a point is reached where a precarious balance exists between men and water supplies. In a year of sub-average rains such a village is likely to experience considerable hardship, for the *hafir* will be inadequate. Most of the people regard themselves as permanently settled and have nowhere to go in the dry season if this should occur. The villages often lack animals to fetch water and are usually too far from neighbouring water points, anyway, to do this. They are, therefore, at the mercy of commercial vehicles which may bring them water at prohibitive prices.

Conclusions

Early in the present century an administrative official wrote that the development of the then sparsely peopled Gedaref District was entirely dependent on the development of its water supplies. Today, as this chapter has shown, this statement is scarcely less true. Yet there are certain problems inherent in man–water relations in this area the solution of which is likely to be critical for its future prosperity.

Over-exploitation of parts of the area is becoming a major problem. It is, perversely, often a consequence of ample rather than meagre water supplies. The capacity of the water points in question is the critical factor. Large water points attract dense populations and this leads ultimately to over-cultivation. Because water supplies are good, the accumulation of livestock usually proceeds unchecked and over-grazing results. In extreme cases dereliction of the land removes the means of earning a living and this, rather than scarcity of water, limits the size of the settlement and may even cause its decline. In contrast, where water supplies restrict the size of the settlement, the soil and vegetation resources of the surrounding areas tend to be preserved through less intensive utilisation. This problem is now fully recognized by the central government. Where it has a choice in the size of a particular water point, as is the case of mechanized *hafirs*, conservation factors play the largest part in its calculations.

It was originally hoped that mechanized *hafirs* and bores would ameliorate the problems of over-cultivation and over-grazing around the traditional water points by bringing new areas into use. It seems, however, that there is instead a danger of these problems being recreated over a larger area. This is because the settlement of new water points has normally been allowed to proceed spontaneously, and traditional methods of land use have been extended to the new areas without any advice from extension services on modern conservation methods. Even the agricultural operations of the Scheme are essentially extractive.[1] It is only due to the coherent structure of Gedaref soils that the effects of over-use around new water points have so far been less obvious than in the western Sudan where the lighter soils have become exhausted much more rapidly (Lebon, 1956).

A further problem is that most of the obvious sites for bores and *hafirs* have now been used; the expense and risk of developing the less obvious ones may limit the further exploitation of the plains. This is a particularly important limitation on the opening up of new land for grazing since the optimum use of pasture requires numerous small water points every few miles, each capable of supporting animals for a brief period. Agricultural development will not suffer a similar restriction since increasing mechanization means that areas farther away from water supplies can be cultivated.

It may be doubted whether the administrative structure, whatever its merits in other respects, is well fitted to solve these problems. Today most wells and hand-dug *hafirs* are constructed at the instigation and expense of the local authorities. The present rural councils are representative in structure. Consequently the siting of water points is influenced by local politics. Outlying pioneer villages, particularly immigrant ones, are at a disadvantage in relation to larger villages in the traditional settlement areas, which are able to exert powerful pressure on the councils. Thus a high proportion of new water points may go to areas already over-populated and serve only to accentuate local land-use problems. It should be said, however, that the initiative in the construction of deep bores and the mechanized *hafirs* lies with the central government which tends to site them in hitherto unexploited areas, thus to some extent counteracting the effects of local authority policy.

Private individuals and communities have not only lost their old initiative in creating water points but they have also abrogated much of their former

---

[1] In September 1968 the Sudan Government announced a World Bank loan for the development of a new Mechanized Crop Production Scheme in the clay plain between the Blue and the White Nile. This scheme is to be organised with far more attention to conservation considerations.

responsibility for maintaining them. The standards of maintenance of water points leave much to be desired. Villagers usually carry out routine maintenance on wells, but major repairs and the cleaning of *hafirs* (which requires a sizeable labour force) are usually left to the inadequate resources of the local councils. Yet, given the existing water-supply situation it is imperative that all water points should be maintained in the best possible condition. For only with constant vigilance and effort can this region of very marginal water supplies be most efficiently exploited.

All these problems become more acute when considered against a background of steadily increasing population. There is not only a high rate of natural increase but also a steady influx of immigrants. The new bores and *hafirs*, while permitting some spread of population, have been quite unable to absorb the increase. As a result the overcrowded areas of traditional settlement have become even more congested. Many established villages have now reached their maximum sizes in terms of the availability of either water or land. New settlements are being founded in areas of marginal water supplies. It is unfortunate that while bores and *hafirs* have made the economic exploitation of the plains possible they have done comparatively little to promote permanent settlement.

The inadequacies of the water-supply situation are particularly irksome at the present time. Economic development has brought financial prosperity but the achievement of better standards of living is hampered by poor water supplies. Often large villages lack public services such as schools and dispensaries because they would place too heavy a strain on their water points. In Gedaref Town seasonal water shortages and the absence of a proper piped water supply retard urban development.

Increasing prosperity is bringing a new demand for fully settled villages with permanent water supplies. At the same time it increases the amount of water considered necessary to maintain acceptable standards of health and hygiene. The need for mobility used to be accepted philosophically but it is unlikely that it will be so accepted in the future.

The acquisition of household property will make movement less easy. The secondary and tertiary sectors of the economy are expanding fast. A desire for the wealth to be gained from dry-season employment will make men reluctant to waste their time fetching water. Moreover, if trends in other countries are a reliable guide there will be an increased demand for a settled life *per se* as a symbol of progress quite apart from the benefits it may bring.

These aspirations come at the very time when, because of population pressure on existing water resources, any variation in water supply may be critical. The flexibility of life required to cope with fluctuations in water supply has increased rather than diminished, but it is more difficult to attain as life within the areas becomes more complex. What emerges very clearly is the fact that although modern technology may improve the means used to collect and conserve rainfall, water supplies are almost as dependent as ever on the vagaries of that rainfall. Given these circumstances, in the immediate future as well as in the past, the life and development of the area can only take place within the circumscribed framework permitted by its water supplies.

Acknowledgements

Fieldwork for this chapter was carried out in 1961 and 1962. The area was last visited in 1964. The use of the present tense refers to the period of the early 1960s. The author is grateful to the Goldsmiths Company for a scholarship to undertake this research, to the University of Khartoum and the Department of Land Use and Rural Water Development for information, facilities and encouragement, and to the Ministry of Agriculture for access to historical and statistical information about the Mechanized Crop Production Scheme. The author also wishes to thank the Survey Department (Air Photo and Cartographical Sections), the Geological Survey and the Meteorological Department, Khartoum, for providing material which assisted in the compilation of her maps.

References

BAPTISTA, J. G. 1960. *Report on the Sampling Survey to Estimate the Area under Dura and Sesame in Bildat Cultivation of Gedaref District*, Agricultural Economics Division, Department of Agriculture, Khartoum, Unpublished.

GRABHAM, G. W. 1927. *Water Storage in the Anglo-Egyptian Sudan*, Department of Geological Surveys, Khartoum, Unpublished.

JEFFERSON, J. H. K. 1952. A note on field experience of planning hafir excavation programmes. *Sudan Notes Rec.* **33**, 224–44.

1954. Hafirs or developments by surface water supplies in the Sudan. *Trop. Agric.* **31**, 95–108.

LEBON, J. H. G. 1956. Rural water supplies and the development of the economy in the Central Sudan. *Geogr. Annlr* **38**, 78–101.

ROBERTSON, A. C. 1950. *The Hafir – what – why – where – how*, Agricultural Bulletins, Khartoum.

16 Problems of land tenure and ownership in Swaziland

G. W. WHITTINGTON *and* J. B. McI. DANIEL

The election of Swaziland's first Legislative Council took place in 1964 and was followed in 1968 by full political independence. Attendant upon these events is the reappearance of problems of land tenure and land ownership which provided the British Administration of this former High Commission Territory with great difficulties in the past. In a country where urban-based political parties are emerging, policies contrary to those of the long-dominant chiefs are making their appearance. The power of the chiefs who control the distribution of tribal land is being challenged by the new, non-traditionalist political parties. The principal European political party representing the settler minority which is intimately concerned with land holding and ownership has so far given its support to the traditional African leaders. The clash of interests between the urban African politicians and European landowners has characterized the emergence of many African states which were formerly under colonial control. This was evident during the wide social, economic and political changes which took place in Kenya, and to a lesser extent in Malawi. The same trend is currently of vital importance in Rhodesia. This clash, which embraces a challenge of tribal authority, is aggravated where a European minority holds a disproportionate percentage of the land. These very features are at the root of an incipient problem involving land relations in Swaziland – a problem made more complex than in other parts of Africa because of the country's history and location.

The evolution of land tenure and ownership in Swaziland

Land tenure and ownership in Swaziland has had a complex evolution and the present situation has as its immediate cause events of the late nineteenth century. Penetration of Swaziland by Europeans was an inevitable result of the Great Trek and in 1845 and 1855 the Swazi chieftain, Mswati, ceded parts of his tribe's territory to the Lydenburg Republic. Although this marked the end of the cessation of actual territory, during the 1860s this same Swazi chief granted certain concessions within an area which accords quite closely with the present political unit of Swaziland. These concessions were made to European settlers who were interested either in grazing cattle

and sheep or prospecting for minerals. They were few in number, however, compared with the flood that was to follow in the twenty years after
the London Convention of 1884 by which Britain and the South African
Republic recognized Swaziland as totally independent. To the numerous
white concession seekers who thronged his kraal the reigning chief at that
time, Mbandzeni, granted rights over the larger portion of Swaziland for
such things as mineral exploitation, grazing, manufacturing iron, cutting
timber, and even selling insurance and advertising!

The confusion which this indiscriminate granting of concessions caused
in land relations was admirably summed up in the Swaziland Annual Report for 1907–8.

> 'Practically the whole area of the country was covered two, three or
> even four deep by concessions of all sizes, for different purposes and for
> greatly varying periods. In but few cases were even the boundaries de-
> fined; many of the areas had been subdivided and sold several times,
> and seldom were the boundaries of the superimposed areas coterminous.
> In addition to this, concessions were granted for all lands and minerals
> previously unallocated, or which, having been allotted, might lapse or
> even become forfeited. Finally it must be remarked that over these three
> or four strata of conflicting interests, boundaries, and periods there had
> to be preserved the national rights of the natives to live, move, cultivate,
> graze and hunt' (Swaziland, 1909).

It was not long before disputes arose over land rights where, for example,
the interests of pastoralists and mineral prospectors clashed. Furthermore,
the numerous disadvantages which the concessions brought manifested
themselves in disputes and conflicts between the Europeans and the Swazi
themselves. While Mbandzeni insisted that he still held sovereign rights
over all his land, having only granted the right to use it, many concessionnaires regarded the land as theirs. In 1903 Great Britain assumed direct
control of Swaziland when the Governor of the Transvaal was invested
with administrative powers over the country. At this time Swaziland was
still undeveloped and the question of land rights, involving land tenure and
ownership, was both complicated and unrelated to the actual distribution
of population as the Swazi continued to live on land granted to the concessionnaires. A small European community of pastoralists, prospectors
and traders was living mainly in the higher, wetter areas adjacent to the
Transvaal.

As the 1907–8 Annual Report said, 'It is hard to exaggerate the complexity of this chaos, or the difficulty of arriving at a solution which would
preserve to the Native his proper rights while securing to the concession-
naire some equitable enjoyment of the privileges and rights he had pur-

chased, and which had been confirmed to him by responsible courts' (Swaziland, 1909). One of the first acts of the British Administration was an attempt to bring some sort of order to the situation by organizing a division of the land between the white concessionnaires and the Swazi nation. As a first step £40,000 was expended on buying out monopolies

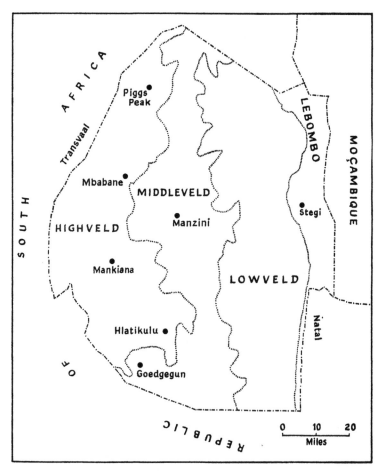

Figure 16.1 *The physiographic regions of Swaziland.* The Lebombo is altitudinally similar to the Middleveld.

and concessions, many of which had never been taken up. Secondly, in 1905 the concession areas were surveyed and their boundaries were demarcated. Thirdly, all concession holders, except those with mineral rights, were required to surrender one-third of their total acreage. This land, together with that purchased by the Government, was declared

'Native Area' and set aside for exploitation exclusively by the Swazi. By these means land, considered sufficient for all time for the needs of the people, was made available for return to the Swazi nation. Although the Swazi were against this reorganizing of land holding,[1] considering as they did that all the land was actually theirs, by 1914 they had been resettled in the reserves.

The partition of Swaziland established a threefold division of land: land held by the Swazi on a basis of communal tenure; individual holdings which were freehold or leasehold and occupied mainly by Europeans; and Crown Land. The third category, Crown Land, comprised areas which in 1907 had neither belonged to the concessionnaires nor had been occupied by the Swazi. The Government encouraged farming activities by selling Crown Land in small blocks of 60 to over 600 acres to ex-servicemen and to poverty-stricken Europeans (Poor Whites) who were virtually squatters at that time on Crown Land or even in the Native Area. Since 1946 Crown Land has also been given to the Swazi for occupation and certain portions of it are being retained for the urban development now beginning in Swaziland. It is, however, in the first two categories that the problems concerning land ownership and tenure are found.

Present categories of land tenure and ownership in Swaziland

Following upon the disentangling of the concessions and the modification of this pattern, two basic categories of land exist in Swaziland – Swazi Nation Land and land held under freehold or leasehold tenure, i.e. individual tenure holdings (I.T.H.).

1. *Swazi Nation Land*. This land is available for exclusive occupation by Swazi (Plate 6). All land in principle belongs to the Swazi Nation. It is apportioned by the Ngwenyama (or king) to local chiefs who, through their *indunas* or headmen, allocate the land to various families. As the land belongs to the nation it cannot be sold. The chiefs have the power to evict subjects from the land for treason or serious crimes or if they are no longer accepted as members of the community in which they live. Provided that the holding is effectively occupied families as a rule are not evicted. In these Swazi-occupied areas the land can be classified under three headings, viz. Swazi Area (S.A.) (i.e. Native Area), Lifa Land and Swazi Land Settlement areas[2] (S.L.S.):

(a) *The Swazi Area* is held in the name of the Ngwenyama in trust for the Swazi Nation and consists of those areas demarcated by the

---

[1] The Swazi deputation to England in 1907 to protest was unsuccessful and in 1926 the Privy Council judgment went against them in a case they brought to challenge the validity of the land partition.
[2] Formerly called Native Land Settlement areas.

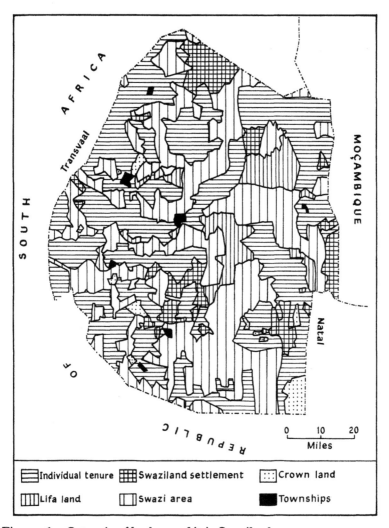

Figure 16.2 *Categories of land ownership in Swaziland.*

Swaziland Concessions Proclamation (No. 28 of 1907). The Swazi Area is scattered throughout the Territory in blocks varying in size from 1,300 to 268,000 acres and covering 37·6 per cent of the total area of Swaziland (Table 16.1).

(b) *Lifa Land:* 6·31 per cent of the total area of the country is land which has been purchased by the Swazi Nation with money paid into the Lifa Fund. The aims of the Fund are to reduce overstocking and to buy land. Provision is made for culling cattle in certain proportions from

all herds of more than ten head, once in three years. The cattle are sold
and a levy of from 25 to 40 per cent of the proceeds contributed to the
Lifa Fund. Adult males owning less than ten head of cattle have to pay
£2. The purchase of land has continued at regular intervals but de-
stocking has not emerged as an important objective. Lifa land which was

TABLE 16.1 *Swaziland: division of land* (Daniel, 1962)

	Total area (morgen[1])	*% of total*
I.T.H.	932,352	45·99
Lifa	127,933	6·31
S.L.S.	147,396	7·27
S.A.	762,302	37·60
Crown Land	57,349	2·83
Total	2,027,332	100·00

[1] One morgen equals 2·11654 acres.

previously held under freehold tenure now forms part of the Swazi
Nation land but its control is closely in the hands of the Ngwenyama and
not subject to subordinate chiefs as is the case in Swazi Area. Nevertheless, much of the Lifa land has been settled by persons moving in from
adjacent Swazi Area.

(c) *The Swazi Land Settlement Areas:* The Swazi Land Settlement
Scheme, initiated in 1946, aimed at making additional land available to
the Swazi. Freehold farms were purchased by the Government and over
130,000 acres of Crown Land were set aside for the implementation of
the scheme. On this land leasehold tenure was applied – 'once the Swazi
realizes that he or his heirs will be credited with the value of whatever
improvements he has effected on his allotment, a greater incentive to
progress and contentment will have been achieved' (Swaziland, 1945).
The Land Settlement Scheme was not a success (Daniel, 1966) and was
virtually abandoned after 1954. In practice, these areas are now subject
to the normal Swazi laws and customs regulating the holding and use
of land.

2. *Individual Tenure Holdings* (Plate 7). This land held on a freehold or
leasehold basis comprises 45·99 per cent of the total area of Swaziland. Very
few of the title-holders are Swazi and more land is owned by companies
than by individual farmers in this category. 15·5 per cent of the privately

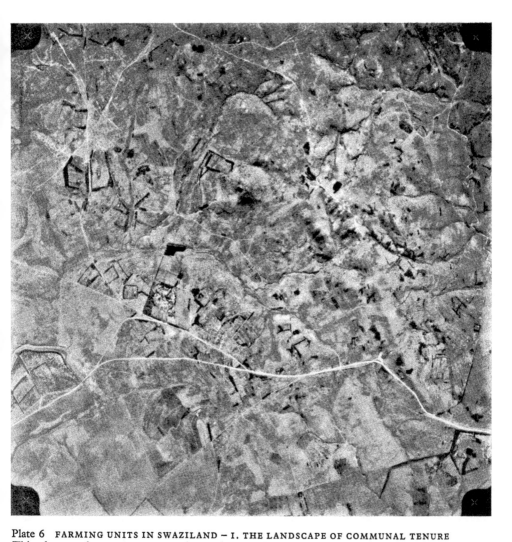

Plate 6   FARMING UNITS IN SWAZILAND – I. THE LANDSCAPE OF COMMUNAL TENURE
This photograph covers an area to the north-west of Manzini, the commercial capital of Swaziland. Most of the land here is Swazi Area on the western fringe of the highly dissected Maliaquma Hills. The land is held communally and there are no large fields. Due to the strongly accidented relief contour banks form the major component of the agrarian landscape. The land at the western extremity of the main road contrasts strongly with the Swazi Area. Here, in association with more level land and freehold tenure, is found a large individual block given over to cash cropping based on citrus fruit growing.

Plate 7   FARMING UNITS IN SWAZILAND – II. THE LANDSCAPE OF INDIVIDUAL TENURE
This photograph shows an area around the town of Goedgegun in the south-west of Swaziland. Most of the land here is in individual tenure, the main crops being wattle, tobacco and cotton. Such land is farmed in large blocks and in the main occupies areas of flattest topography. The attention paid to soil conservation is reflected in the large number of contour banks especially close to stream courses, e.g. in the zone leading north-eastwards from the town. (Plates 6 and 7 reproduced by permission of the Swaziland Government, Ministry of Works, Power and Communications.)

owned land consists of concessions held in perpetuity or on leases of more than ninety-nine years' duration. The owners of these concessions are being encouraged to exercise their option under a 1907 Proclamation and convert their title to freehold.

Although both the Swazi Nation Land and the privately owned farms are interspersed throughout the country, (Fig. 16.2 and Table 16.2) they contrast in two important aspects. The concept of communal tenure in the Swazi-held areas is very different from the idea of individual ownership of land, and there a type of subsistence farming is found whereas on the individual tenure holdings the emphasis is usually on commercial farming. The influence of land tenure on land use is therefore marked.

TABLE 16.2 *Swaziland: division of land within physiographic regions* (Daniel, 1962)

	Highveld %	Middleveld %	Lowveld %	Lebombo %
I.T.H.	57·78	35·56	41·14	52·82
Lifa	1·73	6·17	11·82	2·53
S.L.S.	3·63	9·46	6·67	16·42
S.A.	33·87	48·60	35·93	24·21
Crown Land	2·99	0·22	4·44	4·02
Total	100·00	100·00	100·00	100·00

Problems involving the different land categories

Certain aspects of the above land ownership categories pose problems which may well handicap the aim of continued progress in the important agricultural branch of Swaziland's economy. In Swaziland, as in many of Africa's other newly independent countries, land tenure and ownership is unfortunately intertwined with political attitudes (Jones, 1965).

Swazi Nation Land

Basically the Swazi-occupied areas are characterized by a subsistence economy. Over 90 per cent of the arable is devoted to the cultivation of maize, the basic food staple, and sorghum (Daniel, 1964), and great emphasis is placed on cattle keeping. Like the majority of Africans in neighbouring territories, the preoccupation with the social status afforded by cattle ownership has contributed to the failure of the Swazi to realize the full potential economically of their herds. Two salient features emerged from a study in 1960 (Daniel, 1964) of land utilization in the Swazi rural areas. The first was the extent of the maize shortage and the second was the extent of the overstocking. Fifty-five per cent of the homesteads had to buy maize and in absolute numbers livestock units should be reduced by 160,000 or 37 per cent. These problems are not new to Swaziland, but with the passage of time they have assumed larger proportions. The problem therefore is to raise the standard of farming in order to achieve several aims: to make the Swazi self-sufficient in maize; to bring these areas into the money economy by fostering cash-crop production; and to reduce the numbers and improve the breed of the Swazi cattle, and yet not destroy the balance, as has been done in parts of Rhodesia (Hamilton, 1965), between arable land-unit sizes and their present dependence upon numerous cattle for adequate manure. These aims bring the agricultural reformers into direct collision with the strongest feature of Swazi agriculture – conservatism (Hughes, 1964b). The weaning away of the peasant from age-old farming methods, tried and proved crops, and the keeping of too many inferior cattle is a difficult task but it is one that must be accomplished.

Low yields are often attributed to the fact that under the prevailing system of communal land the Swazi do not have security of tenure. On this issue there are two schools of thought. The one believes that it is virtually impossible to bring about improvement unless the prevailing form of land tenure (i.e. communal land tenure) is destroyed, or at the very least, modified into a co-operative or collective form. The second school of thought recognizes both the disadvantages of communal tenure and the need to make provision for the more progressive farmer, but does not believe that land tenure is necessarily the prime cause of low productivity or that sweeping changes in tenure, involving all homesteads, would be in the interests of the people as a whole. In any event, the question of land tenure will undoubtedly prove to be a most difficult problem in Swazi life, both in the agricultural and political spheres (Hughes, 1964b).

The disadvantages of communal tenure may be summarized as follows:

1. Security of tenure is threatened when aspects of commercial agriculture appear. These conflict with the traditional outlook and chiefs' fear that material rather than hereditary prerogative will become the criterion for leadership. If, for example, an individual becomes relatively prosperous as a result of adopting progressive methods of farming, he may be faced with an ultimatum from the chief to revert to traditional methods of agriculture or leave the community. Initiative is therefore stifled.

2. If a farmer sows new crops which ripen later than the traditional crops, they may be destroyed as all cultivated land is thrown open to communal grazing after the harvest. In the majority of Swazi-occupied areas there are strong objections to fencing of any kind.

3. Fragmentation of holdings, a feature of communal tenure, leads to inefficient farming and makes the introduction of machinery impossible.

4. The farmer cannot command credit raised on the security of his land.

In assessing the existing situation in Swaziland it must be admitted that in the overwhelming majority of cases customary law ensures security of tenure and that communal tenure has not prevented the introduction of soil conservation measures or the reorganization of scattered arable and grazing lands into consolidated units for the community as a whole. The continuing fragmentation of individual lands, even within the consolidated units, together with the practice of communal grazing have, however, hindered the quick and efficient implementation of these changes.

Progressive methods of farming were lacking on the 67 properties covering some 17,000 acres, owned by Swazi on an individual tenure basis when a survey was conducted in 1960 (Daniel, 1962). It cannot be claimed, therefore, that individual tenure automatically increases incentive and initiative. In addition to the problems of agrarian indebtedness, and those related to inheritance and subdivision which are characteristic of societies where individual tenure is found, two further considerations are pertinent to the question of land tenure in Swaziland:

1. On relatively small farm units where the emphasis is on dryland farming, and the majority of such units would be of this nature, the integration of crop and animal husbandry will be of paramount importance. This idea is foreign to the Swazi and it is by no means certain that mixed farming could be successfully introduced.

2. With the increasing demand for labour, associated with economic expansion in Swaziland and the Republic of South Africa, wage employment has proved to be a more ready source of cash than the cultivation of the soil (Hughes, 1954a). The failure of irrigation farming projects undertaken by the Swazi in the past is perhaps a symptom of this (Daniel, 1966). In the dual-based economy of the rural areas the incentive to participate in cash cropping has been inhibited and without this incentive any mass move towards individualized forms of tenure is unlikely to succeed.

Although the concept of leasehold which was tried in the Swazi Land Settlement Scheme was a failure, a very worthwhile experiment in land tenure, irrigation farming and cash cropping based on sugar cane is presently being conducted in the northern Lowveld. The Vuvulane Irrigated Farms Settlement Scheme, initiated in 1963, consists of three types of holding: major farms, farms of 60 acres and smaller units of 8–16 acres.

The larger farms can be purchased but the smaller holdings can only be leased. The long-term results of this experiment in which the progressive farmer cannot be held back by conservative chiefs could have far-reaching consequences in the future planning of agriculture in Swazi rural areas. Furthermore, the Government now plans to introduce intensive development programmes in selected rural areas where the emphasis will be placed on co-operative societies and community development in addition to agricultural improvements. The scheme is an attempt to build a sound rural society.

Whether one does or does not believe that agricultural improvement depends basically upon a reform of traditional land-tenure practice, there is clearly a need for further experiments, further research and an overwhelming need to make adequate provision for the more progressive farmer. Land tenure, involving social, economic and political considerations as well as the question of land utilization, is clearly one of the more important issues facing the newly independent Swaziland. Moreover, it is an issue which is now firmly established in the political arena.

The individual tenure holdings

The freehold and leasehold land of Swaziland is almost entirely in the hands of Europeans[1] (Fig. 16.3). Produce from it is of vital importance to the well-being of the country's economy but not all of the land is of equal value in this contribution. The most prosperous farms are those owned by large companies involved in the growing of sugar cane, timber and citrus fruits. These products are in the forefront of export cash earnings – in 1964 sugar earned 33 per cent of the total and timber products a further 25 per cent. These enterprises are a manifestation of white economic dominance which may well be resented in an independent Swaziland, particularly since the companies have strong South African connections. This could lead to nationalization on lines becoming common in Africa north of the Zambezi.

The privately owned farms show distinct differences. Some are well run and prosperous, growing a variety of cash crops which contribute to the export earnings. For such farms there is the same potential threat as for the company lands, and events in Kenya are very much in the thoughts of their owners. There are also farms, however, which cry out for attention. A low standard of farming, low productivity and the destruction of the soil – features which no country can allow to continue – are found particularly where the uneconomic subdivision of properties has taken place and where share cropping is practised as in the cotton areas of the south. But the most unsatisfactory feature of the individual tenure holdings is absenteeism among landowners.

[1] Eurafricans own approximately 100,000 acres.

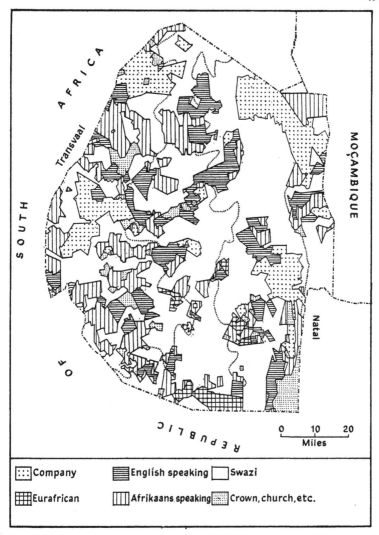

Figure 16.3 *Individual tenure land in Swaziland shown by commonly accepted social and institutional groups* (after Daniel).

Forty per cent of the farms in Swaziland, covering 35 per cent of the total area of the individual tenure holdings, are owned by absentee landlords (Fig. 16.4). Absenteeism, a factor which is preventing the full realization of the potential of land, is found throughout the country, but it is most marked in the Highveld and Middleveld where the land is used for the winter grazing of sheep owned by farmers who live in the Transvaal

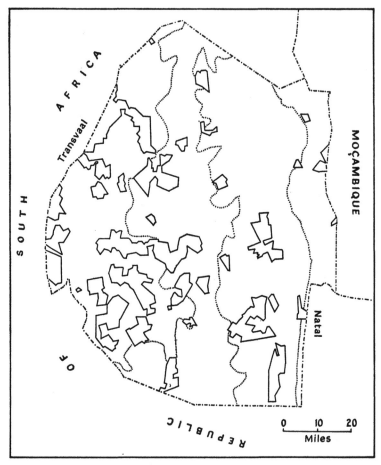

Figure 16.4 *The distribution of freehold farms in Swaziland owned by absentee landlords* (after Daniel).

Province of the Republic of South Africa (Fig. 16.5). The practice of burning the grass during the autumn permits the growth of new grass on which the sheep can feed during the winter months. These 'treksheep' farms which have been in existence for over a hundred years date back to the time of the concessions. In the first two decades of this century these farms were pasturing between 300,000 and 400,000 sheep each winter. During the thirties both the number of sheep and the acreage of the 'treksheep' farms (now 517,000 acres) decreased due to the spread of forestry as an economically attractive venture in the Highveld. In 1964, 87,104 sheep entered Swaziland from the Transvaal, the figure being the lowest for eighteen years.

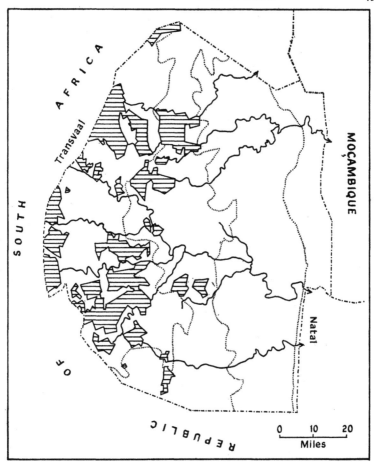

Figure 16.5 *The 'treksheep' farms of Swaziland.*

Apart from the danger of political friction over the ownership of the 'treksheep' farms the land use practised on them causes concern. Autumn burning of the grass increases the danger of soil erosion in the main watershed areas, thereby jeopardizing the future of the post-war irrigation and hydro-electric schemes in Swaziland. The hazard of soil erosion is further increased by Swazi squatters who practise poor subsistence farming which is unsupervised and without the high standard of soil conservation which has been introduced so successfully by the Department of Agriculture on the greater part of Swazi Nation Land.

In 1955 the Swaziland Government issued a proclamation making it unlawful to fire standing or uncut grass on any land at a shorter interval than twenty-four months or during the period from the beginning of May to the end of September. This provision, aimed at decreasing the burning on the 'treksheep' farms, applied throughout the country. The 'treksheep' farmers were greatly irritated by the proclamation, claiming that their farm management was not in any way detrimental to good land conservation as only half of the farmland is burned during the autumn and sheep are kept off the land for eighteen months after it has been grazed over the winter months. They also maintained that the grass was burned after the rains, thus ensuring that the ground was not left without a protective cover of vegetation during the wet season, and that autumn burning did not interfere with the replenishment of natural water resources.

The immediate problem at present related to the 'treksheep' farms is this matter of conservation. Whether or not the land could be used more economically is another question. The farms could be put under exotic timber, but this would commit most of the Highveld to a monocultural activity. The land is not suitable for agriculture as, despite its plentiful rainfall, it has many disadvantages – steep slopes, the majority over 8°, generally infertile soils and a high frequency of stream courses. The ownership of this land and its method of use do raise two other problems, however. In the first instance, Swaziland derives no benefit in income from the present form of land use. Secondly, the alien ownership of the land by absentee South African farmers is in itself a potential political danger and source of racial conflict.

Conclusion

Swaziland is potentially the only one of the three former British High Commission Territories in Africa that has any chance of leading a politically independent and economically viable existence. This chance was furthered by the opening in September 1964 of the rail link from Ka Dake in the western Highveld to Goba, thus giving Swaziland a direct route to Lourenço Marques (Whittington, 1966). If this economic viability is to be achieved it is vital that agriculture should play a full part. At present only the lands in company ownership and some of the properties held by individual farmers are really contributing strongly to the country's gross national product. Will these lands be the first to suffer the extremes of African nationalism or will the method of tolerance, prompted by the thought of financial gain, be practised in Swaziland on the Malawi pattern? In any event, the problems of absenteeism require urgent attention. The principal problem at the present time concerns the development of the Swazi Nation Land. It is not enough that this land should enable the Swazi to be nearly self-sufficient in maize. The arable potential of this land is greater than that. The incentive to raise productivity must be provided. This incentive is closely related to the dualistic structure of the rural economy and the Swazi system of land tenure. Until this problem is solved, the development of over half the total area of Swaziland will continue to lag behind that of the individual tenure holdings and thus hold back the contribution which agriculture should be making to the over-all growth of the country's economy.

References

DANIEL, J. B. MCI. 1962. The geography of the rural economy of Swaziland. *Institute for Social Research, University of Natal*, 70–72 and 159–69.
  1964. The Swazi rural economy. In HOLLEMAN, J. F. (ed.), *Experiment in Swaziland*, Cape Town, 204–50.
  1966. A review of certain government measures to improve African agriculture in the Swazi rural areas. *Geogrl. J.* **132**, 506–15.
HAMILTON, P. 1965. The changing pattern of African land use in Rhodesia. In WHITTOW, J. B. and WOOD, P. D. (eds.), *Essays in Geography for Austin Miller*, Reading, 255–6.
HUGHES, A. J. B. 1964a. Incomes of rural homestead groups. In HOLLEMAN, J. F. (ed.), *Experiment in Swaziland*, Cape Town, 251–69.
  1964b. Reflections on traditional and individual land tenure in Swaziland. *J. Local admin. Overseas* **3**, 3–13.
JONES, N. S. C. 1965. The decolonisation of the White Highlands of Kenya. *Geogrl J.* **131**, 186–201.
SWAZILAND. 1909. *Annual Report, 1907–8*, Cmd. 4448–5, 13.
SWAZILAND. 1945. *Report of the Department of Native Land Settlement.*
WHITTINGTON, G. 1966. The Swaziland railway. *Tijdschr. econ. soc. Geogr.* **57**, 68–73.

17 Agricultural change in Kikuyuland

D. R. F. TAYLOR

The Kikuyu Plateau is the traditional homeland of the Kikuyu people. It rises in altitude from about 4,000 ft (1,219 m) in the east to over 12,000 ft (3,657 m) in the Aberdare Mountains in the west, but above about 7,500 ft (1,981 m) the land is forest reserve and virtually uninhabited. The heavy relief-induced rainfall on the Aberdare Mountains has given rise to hundreds of small streams which have deeply dissected the volcanic rocks of the plateau into parallel ridges and valleys running west to east. The soils of the plateau are deep and fairly fertile and rainfall totals rise from 35 in (889 mm) per annum in the lower areas to over 100 in (2,540 mm) per annum in the mountains. Over most of the plateau the mean annual temperature is between 50° F and 60° F (10°–15° C) allowing a twelve-month growing season. The plateau is one of the most favoured agricultural areas in Kenya and supports over one million people. The 1962 census (Kenya Government, 1964) revealed average population densities of 400–500 per square mile, rising in places to over 1,000. Agriculture is the only significant economic activity and over the last fifty years it has undergone considerable change. The pace of this change has greatly accelerated in the last decade, leading to what might best be described as an agricultural revolution.

The traditional agricultural system

When Europeans first came into contact with the Kikuyu the traditional system of land tenure and land use, which has been called loosely the 'githaka system', was in operation. There is considerable controversy over this system, the main point at issue being whether it was one of individual or communal ownership of the land. Githaka means land and the term included all of the land owned by mbari, an extended family or sub-clan. Each member of the sub-clan cultivated an area of the githaka which was known as his ngũndũ. This plot of land was in a sense his private property although in theory it belonged to the mbari as a whole. The man could not sell or lease his land without the permission of the elders of the mbari, but otherwise he could do as he pleased. There were 'two apparently antagonistic principles constantly in operation – community (or kinship) and individual rights. It is the genius of the untouched native system that it preserves a stable equilibrium between the two' (Lambert, 1956). The position has been admirably summed up by Barlow:

'The sense of family ownership is so strong and the instinct to preserve the integrity of the family *githaka* is so deep-seated that the inquirer into the system of tenure may at times find difficulty in disentangling family rights and individual rights. Under normal circumstances, family control over the land remains inconspicuous, and the individual rights play the important part in the everyday life of the *githaka*. Every subdivision of the *mbari*, and every individual, down to the youngest son of the youngest wife of the most junior member of the family, have their indisputable rights in their respective portions of the land. And yet every transaction concerning any modicum of the land is preceded by consultation between the members of the *mbari* whose common interests are affected' (Barlow, 1934).

The power of the individual in the '*githaka* system', although limited to a certain extent by communal control, appears to have been fairly strong. Each man's *ngũndũ* was inherited by his own immediate family and was passed on from father to son for generations; in every aspect of land other than sale, the wishes of the individual farmer dominated. There has always been a strong element of individuality in the Kikuyu agricultural and land tenure systems. Kenyatta (1938) states that ' . . . the Kikuyu system of land tenure was never tribal tenure'. Leakey, in his testimony to the Kenya Land Commission of 1932, maintained, 'The Kikuyu do not recognize such a thing as a tribal tenure . . . the accepted forms of land tenure in Kikuyu country are individual tenure and family tenure. I would prefer to say individual ownership and family ownership' (Leakey, 1934). Beech (1917) goes even further and says, 'There is no evidence whatever to support communal tenure, either in practice or in tradition.' This element of individuality may well have hastened subsequent developments in land tenure.

The traditional land-tenure system was further complicated by individuals cultivating on the land of another farmer. There were a number of ways in which such cultivation rights could be acquired, the two most important being as a *muguri* and a *muhoi*.

A *muguri* was a man who made a redeemable purchase of land. The *muguri* was obliged to relinquish the land if the original seller or his descendants wanted to redeem it. The *muguri* had full rights of cultivation and building but had no right to dispense with the land he received.

The word *muhoi* (plural *ahoi*) comes from the verb *hoya*, meaning to ask for. A *muhoi* was a man who obtained permission to cultivate on the land of another purely on the basis of friendship. A *muhoi* had only cultivation rights and had to move when the owner asked him to, although he had the right to harvest standing crops. In return for cultivation rights, the *muhoi* usually presented his benefactor with beer and the first fruits of his crops.

Kenyatta maintains that each individual had the right to accept *ahoi* without consultation but the consensus of opinion seems to be that the elders of the *mbari* had to be consulted.

The traditional type of agriculture was entirely of a subsistence nature. Although each *mbari* may have had a fairly large acreage, only part of it would be under cultivation at any one time. Some form of shifting cultivation appears to have been practised but there is little reliable information on the form that this took. After clearance, cultivation on a *ngūndū* seems to have been continuous. Initially, however, the *ngūndū* may have been so extensive as to allow each cultivator to indulge in 'shifting agriculture'. The common land of the *githaka* seems to have been used for communal grazing, part of it being given to each male when he came of age. Once a man had been granted land by the elders that land was cultivated by himself and his dependents. There seems to have been no new land issued to a landholders of the *mbari*.

Just how much land a household needed to support itself would depend on the size of the family. Meinertzhagen's map of part of a *githaka* near Fort Hall in 1902 (Fig. 17.1) shows eight households and the land cultivated by each. Measurement of these holdings gives an average acreage under cultivation of 3·675 acres, the individual holding sizes being 2·42, 1·1, 3·81, 4·43, 4·25, 3·48, 4·85 and 5·06 acres respectively. Each area of cultivated land seems to be surrounded by an area of fallow land, but, as no boundaries are marked, it is difficult to decided whether this was fallow land held by the *mbari* or part of the *ngūndū* of the individual household.

Routledge, who was in the area at the same time as Meinertzhagen, writes of it: ' . . . all reclaimed land in Kikuyu, whether under cultivation or lying fallow, is private property' (Routledge, 1910). Beech's map of a part of Kiambu in 1915 (Fig. 17.2) shows most of the *githaka* carved up into *ngūndū* with definite boundaries. Unfortunately, this map give no indication of the extent of cultivation within each *ngūndū*. It appears that the 'shifting cultivation' merely amounted to a formless rotation practised on each *ngūndū*.

In the traditional Kikuyu agricultural pattern both seasonal and perennial crops played a part. The chief seasonal crops were millet, maize, beans and various other pulses. Millet was by far the most important cereal crop and was planted in the short rains. Millet was, in fact, so important that it gave its name (*mwere*) to the short rains which are called *kemera kya mwere*. Two types of millet were planted, bulrush millet or *mwaya* (*Pennisetum typhoideum*) in the lower areas, and foxtail millet (*Setaria italica*) in the higher areas. A local sorghum of a sweet white variety was also planted together with early types of maize, both yellow and black, but Humphrey (1945) considers these crops to have been of limited importance.

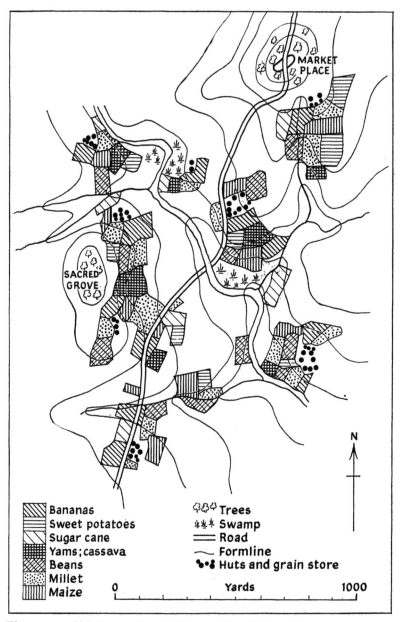

Legend:

- Bananas
- Sweet potatoes
- Sugar cane
- Yams; cassava
- Beans
- Millet
- Maize
- Trees
- Swamp
- Road
- Formline
- Huts and grain store

MARKET PLACE

SACRED GROVE

N

0   Yards   1000

Figure 17.1 Githaka *near Fort Hall, 1902* (after Meinertzhagen).

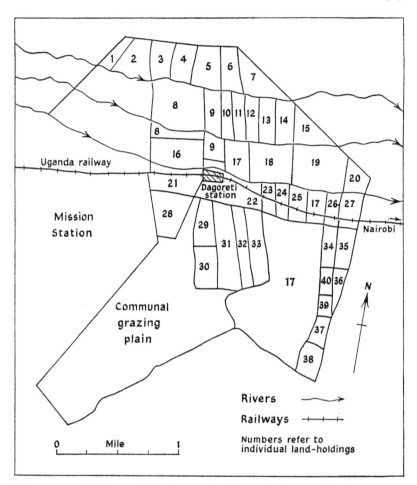

Figure 17.2 *A Kikuyu* githaka *in 1915.* This map is of an area in Kiambu District (after Beech).

The chief long-rains crop was the bean *njahi* (*Dolichos lablab*) which gave its name to the long-rains *kemera kya njahi*. *Njahi*, as well as being a food crop, appears to have had some sort of religious significance, as a small quantity of it has always been an apparently indispensable part of the bride-price.[1] Cow peas (*Vigna catiang*) and pigeon peas (*Cajanas indicus*) were also planted during the long rains.

The chief perennial crops were bananas (*Musa sapientum*), sugar cane

[1] In later years when *njahi* was replaced by the commercially more valuable French bean its price in the local markets rose to three times that of any other bean for this reason.

(*Saccharum officinarum*), sweet potatoes (*Impomea batatas*), colocasia and yams (*Diosorea* spp.). An analysis of Meinertzhagen's map (Fig. 17.1) gives the following results:

TABLE 17.1 *Fort Hall District: samples of crops and acreages, 1902*

Crop	Holding								Total Acreage
	1	*2*	*3*	*4*	*5*	*6*	*7*	*8*	
Millet	0·76	0·47	0·95	1·40	0·94	0·90	0·61	1·09	7·12
Bananas	0·43	0·45	1·01	1·03	0·66	0·69	1·10	0·60	5·97
Maize	0·37	—	0·14	0·53	1·43	0·37	1·00	0·68	4·52
Beans	0·47	0·18	0·29	0·55	0·61	1·00	0·44	0·55	4·09
Sugar cane	0·10	—	0·31	0·40	0·24	0·24	1·09	0·42	2·80
Sweet potatoes	0·29	—	0·27	0·21	0·33	1·14	—	1·48	2·72
Cassava and Yams	—	—	0·84	0·31	0·04	0·14	0·61	0·24	2·18
Total	2·42	1·10	3·81	4·43	4·25	3·48	4·85	5·06	29·40

Although only for a small area this shows the diversity of crops grown under the traditional agricultural system and illustrates the place of millet as the most important staple crop.

In the traditional Kikuyu economy sheep and goats were extremely important. They were herded communally on the common land of the *mbari* and provided a source of meat and clothing. They were perhaps even more important from a socio-religious viewpoint. Cagnolo (1933) records over thirty different kinds of ceremonies for which sheep and goats were required and there were indubitably many more. Sheep and goats were also an important part of the bride-price. Poultry were never important and there were definite taboos against the consumption of fowls, especially by men. Cattle, according to Humphrey (1945), came into importance only slowly, and for a long time their distribution was very uneven. Eventually they became a matter of great pride and the chief insignia of wealth. Killing of cattle for meat was extremely rare, although their milk was drunk.

Cultivation by hand was, and still is, virtually the only method of working the land, with much of the work being done by women. There seems to have been a fairly well-defined division of labour. Women were responsible for the collection of firewood, fetching water from the stream, various domestic duties and food storage. In the fields they were responsible for planting crops, weeding the land, harvesting, threshing and winnowing.

Men were responsible for the rearing and milking of cattle and most of the work connected with the livestock. Breaking virgin land and the care of certain crops such as tree crops and most perennials were the special concern of men.

Cultivation in the past was done with a digging-stick and the size of the area to be cultivated depended on the size of the family. Polygamy was the rule and as women did most of the cultivating they were a definite economic asset. Some important men had as many as forty wives but the average was probably two or three and thus the land which could be easily managed by the average family rarely exceeded 5 acres.

Although much of the work on the *ngũndũ* was done by the household itself, there was a certain amount of communal effort. Grazing was on a communal basis, the flocks of the whole *mbari* being put in charge of one or two men. There was the custom of *ngwatio* whereby groups of women cultivated the land together and the custom of *wera* by which a man could call together a group of people to help him with a particular task in return for beer and food.

Two types of manure were available to the Kikuyu; the cow dung accumulated in the cattle enclosures, and the compost made from waste collected in the homestead, the latter being further enriched by sheep and goat droppings. Although realizing the value of manure the Kikuyu do not seem to have used it extensively. Part of the reason for this may have been the difficulty of carrying the manure, but the chief reason was probably the taboos which were associated with the manure heaps. Dead calves and after-birth were buried in the cattle enclosures and for this reason the manure was rarely moved. There were even stronger reasons for not moving the compost heaps of the household: aborted and still-born babies were sometimes buried in the heap, and it was therefore left untouched.

The traditional Kikuyu system was very successful and therefore the apparent emptiness of the southern area of the Kikuyu Plateau at the close of the nineteenth century is somewhat puzzling. However, this area suffered from a swift succession of natural calamities at this time. A smallpox epidemic in 1898 was followed by an outbreak of rinderpest and a long drought. The famine caused was intensified by swarms of locusts which destroyed what few crops survived. Thus, the greatly reduced population moved northwards back into the Kikuyu homeland area of Fort Hall. This area had also been affected by these natural disasters but the drought had been less serious and more of the crops survived.

In 1903 the first attempt at a population estimate was made for Fort Hall District. A hut count was made, 27,000 huts being recorded (*History of Fort Hall*) and using a factor of five this would give a population of 135,000.

It is most unlikely that this population estimate was accurate for this would give a population density for Fort Hall District in 1903 of 240 per square mile. However, if the population was over-estimated by 100 per cent the density would still have been 120 per square mile. This is a considerable population density for what a contemporary anthropologist called a

Q

'prehistoric people' (Routledge, 1910), who a few years previously had undergone a series of natural disasters which had considerably reduced their total.

The impact of Europeans, 1910–50

The traditional pattern of land holding and agriculture began to change
when the Kikuyu came into contact with Europeans, if only slowly until
1920. The Kikuyu were confined to their own area and surrounded on all
sides by land alienated to Europeans. Within the Kikuyu Reserve a remarkable increase in population reflected the influence of the British administration in stopping inter-tribal warfare and in alleviating the problems of
famine and disease. In Fort Hall District, for example, the population in
1927 had risen to an estimated 162,890 and by 1940 it had climbed to
200,000, a population density of 352 per square mile.

The first result of this was that the common land in the *githaka* was
rapidly brought under cultivation as individual *ngŭndŭ*. This was accompanied by a fuller utilization of the land available to the Kikuyu. Until the
1920s the majority of the population had been concentrated in Middle
Kikuyu, the zone between 5,000 ft (1,524 m) and 6,000 ft (1,828 m). About
this time, however, there was a move into the colder, wetter areas of High
Kikuyu and the drier areas of Low Kikuyu, both of which were less
favourable for agriculture.

With the common land of the *mbari* exhausted and all of the land in the
Kikuyu Reserve occupied, the system of a sub-clan control of the land
began to decline rapidly and the individual began to play an increasingly
important part. The rapid breakdown of *mbari* control, probably facilitated
by the large measure of individualism existing within it, was reflected in
the great increase in local litigation from about 1930 onwards.

It is from this time also that fragmentation of land became an increasingly
important problem. By 1930 almost all of the common land had been used
up and, as there was no new land available, the only source of land a young
man had was from the subdivision of his father's holding. The eldest son
of each widow inherited that part of the *ngŭndŭ* she had cultivated. He in
turn divided this land among his brothers. At the same time each son
retained any land he had received from the *mbari* at his coming of age,
together with any land his father had given him on the occasion of his
marriage. Fragmentation was increased by certain other customs, such as
a father giving his daughter a small parcel of land on her marriage.

Redeemable sales to *aguri* complicated the picture and as land became
scarcer and scarcer the number of *ahoi* increased rapidly. The traditional
idea of a *muhoi* was extended to include a man who rented land for cash, a
trend which became increasingly evident in the 1940s.

In 1935, Fazan said of fragmentation, ' . . . these difficulties are not yet very acute but they threaten to become so in the next generation and are already a serious obstacle to progress'. By the 1940s, Fazan's predictions had certainly materialized. The District Commissioners' reports from all over Kikuyu country were full of references to the influence of fragmentation: ' . . . the fertility of the land decreases, the population increases and the fragmentation never ceases so that the economic return gets smaller to the family each year' (District Commissioner, Fort Hall, 1948).

This period also saw the introduction of new crops and the conception of selling agricultural surplus for cash. The most significant introduction was that of white maize (*Zea mays*) which rapidly replaced millet as the most important food crop. This was first introduced after the famine of 1917–18 and quickly 'attained an importance unjustified by its merits and disastrous to good farming' (Humphrey *et al.*, 1945). It became the staple food crop, being planted in both seasons. The traditional beans such as *njahi* (*Dolichos lablab*) were replaced by the commercially more valuable french bean and haricot bean (*Phaseolus vulgaris*). Other introductions included European vegetables and potatoes, plums and, as the first real cash crop, wattle, which soon rose to a position of great importance in the economy.

The agricultural economy slowly began to change from growing crops purely for subsistence to growing a surplus for cash with wattle as the principal cash crop. The taboos against the use of manure were slowly overcome and by 1938 the District Commissioners' reports were mentioning an increasing number of people using compost pits. The digging-stick was replaced by the *jembe*, a short-handled hoe with an iron head, and the *panga*, a machete-like knife, became the tool of almost every man in Kikuyuland.

Soil erosion became an increasing problem as the population density increased. As land became scarce, even slopes of 30° and more were utilized without any precautions being taken. The topography of the area and the growing of maize on the same land year after year made the situation even worse. The annual reports of the Agricultural Officers and the District Commissioners from the late 1930s onwards refer frequently to the steadily worsening problems of soil erosion: ' . . . most of the people have no apparent intention of saving themselves and their descendents, and are indeed continually breaking new steeply sloped land as soon as one's back is turned' (District Commissioner, Kiambu, 1943).

As the amount of land available was limited, the continual increase of population meant that the creation of a landless class and a class with insufficient land to support them was inevitable. Many of these people made their way to Nairobi and on to the European settlers' farms in search of a livelihood. Corfield (1960) has estimated that in 1912 there were approximately 12,000 Kikuyu outside the Kikuyu Reserve and by 1939 this number had risen to 200,000. Discontent over the land situation was the chief reason for the growth of political agitation. It is not surprising that the early Kikuyu political movements began in the 1930s, a time when the influence of increasing population pressure began to make itself really felt. Throughout the 1940s it became increasingly obvious that something drastic would have to be done about the chaotic land-tenure situation if any real agricultural progress was to be made. The idea of Land Consolidation was already being tried in Nyeri District when the discontent erupted into violence.

The Mau Mau emergency

A state of emergency existed in Kenya from 1952 to 1960 and the early period at least was one of virtual standstill in agricultural development. One benefit which arose from the emergency was that the people were forced by the administration to take action against soil erosion. This did much to combat the problem at the time, but it also engendered a political aversion to such control measures which has been making its influence felt ever since.

During the emergency a radical change was wrought in the settlement pattern in Kikuyuland. The people, for ease of administration and defence, were drawn into large villages usually situated strategically on a ridge top. Each family was given one quarter of an acre in the village, which was laid out with geometric precision. What had been a dispersed pattern of settlement with each man living on his own *ngŭndŭ* became a strongly nucleated pattern in a relatively short space of time.

The outbreak of violence during Mau Mau brought much trouble and grief to the people of Kenya. Thousands of people died and the economy was dislocated, but the Emergency did bring the problem of land into sharp focus and made its solution a problem of immediate and urgent concern. The result was the Swynnerton Plan which is revolutionizing agriculture in Kikuyu country. Whether or not funds would have been found for such a plan had Mau Mau not occurred is doubtful.

The Swynnerton Plan

The aim of this plan introduced in 1954 was 'to raise the productivity of the African lands, their human- and stock-carrying capacity, the income and standards of living of the people, while at the same time effecting a substantial increase in the resources and economy of the colony'. The plan in theory dealt with the whole of Kenya, but in fact most of the estimated £10,800,000 spent between 1954 and 1960 was spent in Kikuyuland.

Special attention was to be given to eight main points: the consolidation of fragments, security of land tenure, technical assistance to develop land on sound lines, the introduction of highly priced cash crops, ready access to water, the introduction of marketing facilities preferably of a co-operative nature, access to sources of agricultural credit large enough and flexible enough to meet the needs of a large number of small farms, and an agricultural bias to education. Land consolidation was the basis for development under the Swynnerton Plan. Economic farming on a large number of fragmented parcels of land is obviously difficult if not impossible. The extent to which fragmentation had progressed by the time of the inception of the Swynnerton Plan is well illustrated by Figure 17.3 which is of a *githaka* in Nyeri District. Fragmentation was bad in other areas of the plateau, the most extreme examples occurring around Fort Hall in Middle Kikuyu, the first area of the plateau to be occupied by the Kikuyu.

TABLE 17.2 *Fort Hall District: examples of fragmentation of holdings*

Area	Size of holding (acres)	Number of fragments
Location 3	3·5	24
Location 18	4·2	23
Location 18	3·6	17

Consolidation of these fragments and the reallocation of the land was the first and vital step in agricultural development. This was an extremely difficult task which would have been impossible without the co-operation of the people. Therefore it was not attempted unless the majority of the people wanted it, being in no case imposed from above but entirely based on the wishes, or rather the demands, of the people. Consolidation was certainly desired and sometimes it was used as a reward for political loyalty, those areas which were 'loyalist' during the Emergency being consolidated first.

The land unit chosen for consolidation was the *itura* which refers to a *githaka* and all its inhabitants. Once the decision was made to consolidate, the first step was the settling of all land disputes by the elders of the community in the traditional manner. The process of fragment gathering then commenced, the units of each individual land-holder being carefully measured and recorded. A period of thirty days was then allowed for any objection to the record of existing rights to be dealt with.

The next step was the measurement of the total acreage of the *itura* by air survey methods – this usually gave a different result from that achieved by ground survey. This difference was resolved by multiplying each man's land total by a factor achieved by dividing the air survey acreage by the ground survey acreage.

Figure 17.3 *Gatundu Sub-location, Nyeri before consolidation.* The holdings
of several landowners have been shaded (after Farm Planning Office,
Nyeri).

    The Land Consolidation Committee for each scheme consisted of a
majority of local people and a minority of representatives from the various
branches of the administration. After the programme of fragment gathering
had been completed, the Committee's first task was to detail the land needs
of the community as a whole – for roads, schools, village areas, hospitals,
dispensaries and burial grounds. The total average needed for this, both

for the present and for the foreseeable future, was expressed as a percentage of the *itura* acreage and an equivalent percentage was deducted from each man's land.

After final acreages were decided the new land-holdings were demarcated. This was done by a demarcation team in consultation with the elders and the landowner himself and every effort was made to ensure that the redistribution of land was equitable. An attempt was made to situate each new holding near the largest fragment of a man's original land and compensation according to a fixed scale was paid for standing crops lost as a result of redistribution.

Each landowner was given access to a road and water whenever possible and the new holdings tend to run in long narrow strips from the ridge top to valley bottom (Fig. 17.4). This ensures that wherever possible each landowner gets a share in the ridge-top land, the valley-side land and the valley-bottom land, as was the custom in the past. This division is based on the suitability of the different areas for different purposes. The ridge top is level with fairly deep soils and it provides the most suitable area for the building of homesteads. The roads and tracks of Kikuyuland also utilize the ridge tops and the communication network reflects this, being very well developed in an east–west direction and very poorly developed from north to south. The valley-side land is well drained and is the chief crop-growing area. It has gained an additional significance with the introduction of cash crops such as coffee and tea which require well-drained soils. The valley-bottom land has heavy black alluvial soils and is subjected to seasonal flooding. It is the only area of Kikuyuland on which sugar cane, which is used in beer-making, can be grown well and is therefore particularly valued.

Landowners whose holdings were not of an economic size, originally set at 3 acres but later reduced to one, were allocated a plot of land in the village. The remainder of their land was located near by so that they could move out and work it each day. The majority of landowners built houses on their own land, but some also elected to have a house and a quarter of an acre from their total area in the village.

A period of sixty days was allowed for any further objections and once these had been dealt with each man proceeded to develop his new farm. The first priority was the establishment of a boundary hedge to ensure that the demarcation lines were not lost. Once the hedges had grown the area was surveyed from the air again and a map prepared to show all the boundaries of the scheme. This map was used in drawing up the final title-deed which each landowner received, making him the full legal owner of the land. Consolidation was finally completed in Kikuyuland in mid-1965.

Consolidation has led to a marked change in the appearance of the landscape; a pattern of well-defined regular holdings is emerging all over

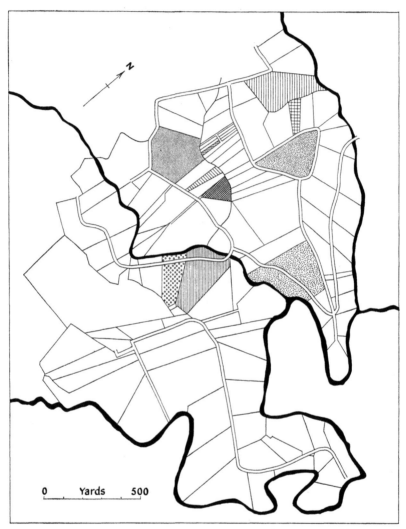

Figure 17.4 *Gatundu Sub-location, Nyeri after consolidation.* The discrepancy in area between this map and Figure 17.3 is caused by land transactions which took place during the consolidation process.

Kikuyuland. The old Emergency villages are shrinking as the people move back on to their land, but they are not disappearing entirely and room has been left for a possible future expansion. The emerging pattern is illustrated by Figures 17.4 and 17.5 and is in sharp contrast to that shown on Figure 17.3.

Originally the Swynnerton Plan aimed at a very careful planning of land

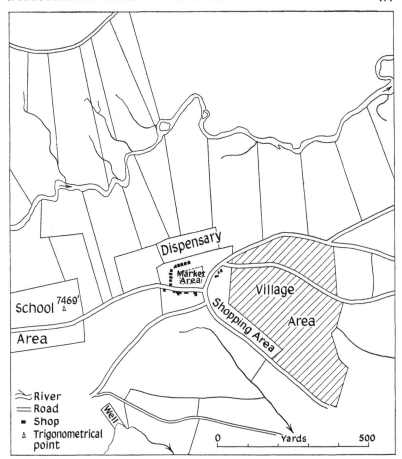

School 7469'
△

Area

Dispensary

Market Area

Village Area

Shopping Area

Well

≈ River
= Road
▪ Shop
△ Trigonometrical
   point

0   Yards   500

Fig. 17.5 *Consolidated holdings around Kinyona Village, Fort Hall District.*

use, to be carried out after consolidation in three stages: the Minimum Standard Layout, the Simple Farm Layout and the Farm Plan.

A Minimum Standard Layout has been adopted on all farms. Farm buildings are sited so as not to obstruct subsequent development. The farmer is advised to plant food crops on slopes of less than 20°, cash crops on slopes between 20° and 35° and trees and permanent grass on slopes greater than 35°. This establishes a basic land-use pattern which, once established, is difficult for the farmer to change.

The Simple Farm Layout builds upon and may coincide with the Minimum Standard Layout. It involves the siting and pegging out of areas for permanent cash crops and the division of the cultivable parts of the farm into a number of fields. These may be used for arable crops, grass or semi-permanent cash crops. These fields are all roughly the same unit size and may vary in number according to local farming conditions.

The aim of the Farm Plan is the fullest possible use of the land. The farmer is given detailed advice on the phased development of his holding and is assisted in obtaining superior quality livestock and seed. The aim is to create a self-sufficient unit which, after one or two years, will be independent of further loan capital. The emphasis in the first two years is therefore on cash crops and dairy cattle to give a quick cash return to finance further development.

Such careful land-use planning has proved beyond the resources of the Kenya Government, the principal limiting factor being the shortage of qualified staff. As a result less than 20 per cent of the holdings have reached the Simple Farm Layout stage and less than one per cent have benefited from the Farm Plan.

Although land use-planning is lagging, the introduction of cash crops has proceeded. At present *arabica* coffee is by far the most important cash crop. Coffee was one of the crops restricted to European farmers prior to 1950. The difficulty of controlling both the quality and the quantity of the coffee produced was the principal reason put forward to justify this restrictive attitude. Coffee has now been introduced on a co-operative basis and has been grown very successfully. The figures for the Fort Hall District give some indication of the rapid growth of coffee planting:

TABLE 17.3 *The growth of planting and value of coffee*

Year	Number of growers	Acreage	Value £'s
1953–4	109	14·2	Nil
1964–5	12,920	7,894·0	750,000

The quality of the coffee produced in Kikuyuland is high – the Kikuyu Co-operative Societies over the past few years have won the Kenya Coffee Board's Cup for the best-quality coffee produced. The Kikuyu Plateau is actually producing better coffee than most of the European estates, and although this is partially a reflection of the age of the coffee bushes, it indicates that the doubts expressed about quality maintenance were not fully justified.

The Government has also managed to control the quantity of coffee produced by the Kikuyu Co-operatives. This has not been achieved easily, and even with the restrictions on further planting introduced in January 1964, production, on the basis of existing acreages, is going to exceed the quota allowed to the co-operatives in Kikuyuland very soon. An over-dependence upon coffee has already appeared, for it provides 70–80 per cent by value of the Kikuyuland agricultural exports. The problems of selling coffee on the world market are manifold, and its price has been dropping fairly rapidly, now being well below the £500 per ton achieved by Kenya coffee in the late 1950s. The Kikuyu small-holder is at present a low-cost/high-quality producer and as such is in a highly competitive position, but on the basis of the present world market situation coffee production in the Kikuyu Plateau has been developed almost to its limit and it would be over-optimistic to expect an increasing monetary return.

Coffee was not the only cash crop introduced under the Swynnerton Plan. Attempts were also made to introduce pyrethrum, tea, pineapples and tobacco but although all of these crops are being grown the targets set by the Plan have not been realized. In Fort Hall District alone it called for 10,000 acres each of tea and pyrethrum; in June 1966 there were 1,100 acres of tea and 80 acres of pyrethrum.

Wattle bark was, prior to the Plan, a very valuable source of income. Competition from synthetic tanning has led to a rapid reduction in production in recent years. In some areas this drop offset the rise in coffee production.

TABLE 17.4 *Fort Hall District: value of selected cash crops*

Year	Wattle	Coffee	Total agricultural exports
1954	£359,422	—	£408,487
1961	£ 37,078	£307,652	£408,901

The Plan also envisaged a big improvement in the quality of the cattle in the Kikuyu area. This is being achieved mainly by upgrading local stock by artificial insemination. The Artificial Insemination Scheme began in the late 1950s and A.I. centres were established all over the plateau. Guernsey is the semen generally used, as the policy is to introduce high-butterfat-yielding breeds. Upgrading is meeting with some success in High Kikuyu which is the most suitable area for dairy cattle, but in other areas progress has been very slow.

Post-Swynnerton development and problems

There is little doubt that the Swynnerton Plan initiated an agricultural revolution in Kikuyuland and brought about a great increase in the wealth and standard of living of the people. Studies in Fort Hall District (Taylor, 1964) have revealed increases of up to 2,000 per cent in monetary income. All over the plateau there is tangible evidence of this in the form of well-built, hygienic houses and prosperous-looking farms, but there has perhaps been a tendency to regard the Plan as a complete panacea and there has been inadequate reappraisal of its progress and aims.

There is insufficient agricultural land in the Kikuyu Plateau to go round. The three administrative units of the Kikuyu Plateau cover 636,300 acres while the population total of these districts in 1962 was 1,005,800. Thus there was at that time only about six-tenths of an acre of land available per person. If the area of land unsuitable for either cultivation or grazing is subtracted, this figure is reduced to half an acre.

There is undoubtedly a large number of people with insufficient land and many with no land at all. An accurate total of landless people is practically impossible to obtain, but the 1962 Census revealed that out of 1,642,065 Kikuyu in Kenya, almost 40 per cent are found outside the Kikuyu home-land areas, and this gives some indication of the shortage of land. Not all of these 40 per cent are landless, nor did all of them leave the Kikuyu Plateau because they could find no land to farm, but the shortage of land has certainly been an important factor in causing migration from the plateau. Investigations within the plateau (Taylor, 1964) have suggested that, of the people remaining, somewhere between 6 and 10 per cent are genuinely landless and it is these people who pose a real threat to con-tinuing agricultural progress. Many of the landless people are *ahoi* who had to move off the land due to consolidation, and in this respect Land Consolidation has actually intensified the problem of landlessness.

If land or an alternative source of income is unavailable these people will appeal to relatives for a portion of land, or endeavour to rent land from large landowners. This will lead, and is leading, to refragmentation. The Government foresaw this danger and enacted legislation forbidding sub-division of land without the consent of Divisional and Provincial Land Boards. The Boards have granted very few requests, but all over Kikuyu-land an examination of the cultivated areas reveals the presence of tradi-tional boundary marks indicating that illegal subdivision is taking place. The dangers of such refragmentation are many (Allan, 1965) and could eventually destroy the consolidated pattern.

Allan noted this danger but saw the emergence of a new and distinct middle class of capitalist farmers acting as a stabilizing influence. In 1959 he observed, 'My impression was that, short of some political or economic catastrophe, the farmers as a class will survive and increase' (Allan, 1965). In 1965 this still held true but opposition to government land policy had grown to such an extent that it was being vented in the political arena by Kikuyu politicians like Bildad Kaggia, and investigations in the Fort Hall District (Taylor, 1964) revealed that in some areas over 25 per cent of the farmers were displeased with the changes that consolidation had brought.

As yet, although illegal subdivision has taken place, it does not seem to have greatly affected the land-use pattern. Most of the evidence of sub-division is seen in the parts of the holdings where food crops are grown, indicating that part of the food-crop area has been rented or given to another cultivator, who invariably grows the same crop as the landowner himself.

Many of the landless have also become a burden on relatives who feel bound to support them. An increased number of mouths to feed can mean an increased acreage of food crops, and a decreased acreage of cash crops. The average yields of food crops are at present low, being between three and four bags per acre for maize, and two to three bags per acre for beans. The Agricultural Department estimates that by the use of fertilizers, improved seed and rotation of crops the yields could relatively easily be tripled. The landless could therefore be supported temporarily without an increase in the acreage of food crops.

The number of the landless is bound to increase as population increases, and it is increasing at a rate in excess of 3 per cent per annum, so such a solution would be only temporary. One obvious answer to the problem is to obtain land for the landless. The Kenya Government began doing this in 1963 by buying land from European farmers in the former 'White Highlands' and settling African farmers on it. A number of settlement schemes are in progress, the eventual aim being to settle between 50,000 and 70,000 families on 1,500,000 acres.

There are three types of schemes: the High Density Small-holder Scheme, which is for Africans with limited capital and agricultural knowledge, aims at subsistence plus a minimum annual net income of £25-£40; the Low Density Small-holder Settlement Scheme for experienced farmers with substantial capital aims at subsistence plus a minimum annual net income of £200; and the Assisted Owner Scheme for experienced farmers with substantial capital aims at subsistence plus an annual income of at least £250. By June 1965 approximately 24,000 families had been settled, the majority of them on High Density Schemes.

Controversy rages over the economic merits of replacing large-scale European-run farms with small-scale African cultivators. Exponents of the scheme say that increased production will result, whereas opponents see in it the destruction of a large part of the country's production for export. The primary objective of the scheme was probably not economic but political. The 'White Highlands' were a symbol of colonialism, and settlement in them provided a means of dealing with the political problem of the landless.

The majority of the Settlement Schemes already established have not been conspicuously successful, probably because most of them are High Density Schemes, which have the least chance of economic success.

'The High Density Settlement Scheme has many doubtful aspects. Emphasis seems to be placed on the number to be settled rather than on the suitability of the size of the holding for efficient production. We think it is just as important for the African farmer that he should be able to farm on a reasonable scale as it is for the highly mechanized commercial farmer. Provision in farm size for the narrow margin of £25–£40 net income for a family over subsistence is in our view inadequate to cover fluctuations in the price of the main cash product and seasonal adversity. Much will, of course, depend on the extent of the effort of the family, knowledge of good farming and extension assistance' (International Bank, 1962).

As the International Bank Mission points out, an accurate assessment of the economic success or failure of the settlement schemes cannot be made for several years yet.

'The economic consequences of the scheme cannot be assessed reliably for some years. They will not be confined to the scheduled areas and many factors at present unknown to us (including the burden of loan finance) would have to be taken into consideration. It is not only a scheme for the transfer of land ownership. The deliberate replacement of large-scale with small-scale production may mean the destruction of capital assets and the adoption of less efficient forms of production. In some circumstances a small-holder could increase production on his holding over that previously yielded by large-scale farming by reason of a greater intensity of cropping and cultivation. Equally a fall in production can be visualized due to the loss of the benefits of large-scale farming and for instance to growing crops less suited to small-scale farming practices. It can mean the replacement of current production with a system of farming which would take time to attain its full potential. It would appear that no great increase in the number of people engaged in agriculture can be expected as a result of the settlement schemes because the farmers in the scheduled areas have been large employers of labour' (International Bank, 1962).

Some of the problems of the Settlement Schemes are illustrated by the Maragua Ridge Settlement Scheme, which is a High Density Scheme of 5,500 acres divided into 240 small-holdings. All the settlers are landless Fort Hall Kikuyu and the Scheme opened in June 1962.

This area is physically part of Low Kikuyu, being a ridge between the Maragua and Sabasaba Rivers, varying in altitude between 3,700 ft (1,127 m) and 4,300 ft (1,310 m). The plots vary in size from 10 to 130 acres, depending on site, but all are of the same value, 1,600 shillings. Each settler pays legal fees and stamp duty of 80 shillings but no deposit is required for the land. The price of the land is paid back in six-month instalments over thirty years at $6\frac{1}{2}$ per cent.

Twenty-four of the plots are classed as grazing plots, being either too steep or too dry for arable agriculture. Maize and beans are grown for subsistence and sunflowers, castor seed and fruit are being introduced as cash crops. Coffee was growing on some of the land when it was taken over and thirty-one settlers have some poor-quality coffee. Most of the land is of very low agricultural potential with poor, thin soils and a low and unreliable rainfall. It is marginal land for the intensive type of agriculture to which it is being subjected.

The quality of the settlers, both in terms of farming ability and the desire to farm, is generally very poor. A number have not developed their land at all, for, having land elsewhere, they are not genuinely landless. There are also a few settlers who, although nominally the owner of the land, are in reality the employees of rich land speculators and thus have little incentive to develop the land fully. The Settlement Authority has been very slow in evicting unsuitable settlers and this is giving rise to a decline in discipline and standards among the other settlers and to a general lowering of morale. This has been intensified by a delay in the introduction of public services such as schools and health centres. These problems are common to many of the settlement schemes and it may well be that social factors are of even greater significance than economic ones in explaining the lack of success of the schemes.

Even if settlement schemes were expanded to their maximum, they would only absorb a small proportion of the landless. They might be taken into other sectors of the economy, but at present not much relief can be obtained from this source as, apart from agriculture, there are few other economic activities in the Kikuyu Plateau. In the towns bordering the plateau, such as Nairobi and Thika, industrial development has taken place but at present the towns do not provide any real answer to the problem of the landless because there are already many unemployed people who cannot be absorbed by the existing industries. The development of industry, in the light of foreseeable economic conditions, is unlikely to be rapid enough to accommodate many of the landless.

The most encouraging prospect for solving this problem is perhaps to be found within the Kikuyu Plateau itself, rather than outside it. There are few areas of Kenya with greater agricultural potential; the area has what for Kenya is a reliable, adequate and well-distributed rainfall, the altitudinal range allows a wide range of crops to be grown and the soils are still basically fertile despite years of misuse. With Land Consolidation and the introduction of cash crops much progress has been made. The income of almost every family has at least doubled, and in many cases it has increased ten- or twenty-fold.

One result of this increased wealth is that many of the villages are beginning to acquire additional functions and are emerging as small urban centres. During and after the Mau Mau Emergency a number of small African-owned shops developed in the villages. Initially all of the shops were of the *duka* type: small general merchant shops with a limited stock to meet the limited demands of the community. The shop was almost always run by a landowner on a part-time basis. During consolidation space was left in the villages for the expansion of such shops and this growth is now taking place. Kandara Village in Location 1 of Fort Hall District is a good example of this. In 1958 when consolidation in this area was completed this village had five shops; by 1967 it had fifty-five.

The shops are not only growing in number but also beginning to specialize. The general merchant shop still exists, but in places like Kandara shops are being established which have a specific purpose such as baking, butchering, tailoring and shoemaking. In some of the larger centres branch banks and post offices have emerged. Some of the newer shopkeepers do not own land but are entirely dependent upon their business for their livelihood.

As demand increases more and more people will be fully employed in these developing 'central places' providing services for the surrounding area. The growth of such centres in the plateau will provide employment opportunities for the landless. They are already renting village sites from the County Councils and now make up the bulk of the village population.

Some of the landless in the villages are finding work as agricultural labourers. Clayton's work in Kikuyuland has shown that a 7-acre holding under intensive cash-crop cultivation needs a labour force of at least five adults. At present many of the holdings in Kikuyuland suffer from what Clayton (1964) calls a 'labour-scarce/land-abundant situation'. The normal Kikuyu family cannot supply the equivalent of a labour force of five adults and in order to be utilized to their full potential many small-holdings require additional labour. Although the average Kikuyu holding is only about 5 acres there is a considerable percentage of holdings larger than 7 acres; in Nyeri 14·3 per cent, in Kiambu 26·3 per cent, and in Fort Hall 25·2 per cent. The full development of these holdings could create a major employment opportunity for the landless.

The solution of the economic, social and political problems of the landless people in the Kikuyu Plateau would seem to lie in the development of agriculture within the plateau itself. The Swynnerton Plan initiated this development and it must now be continued and expanded.

The introduction and expansion of cash crops other than coffee would seem to be one profitable line of advance. The ecological conditions allow a wide range of products to be grown. High Kikuyu is ideal for the growing of tea, and vegetables of all kinds. Temperate fruits such as plums, pears, apples and strawberries grow well and their acreage could be considerably expanded. The area is well suited for dairy cattle, the high rainfall ensuring a good twelve-month growing season for grass. The production of 'baby beef' from calves is also a possibility and the conditions for the raising of sheep, pigs and poultry are very good.

In Middle Kikuyu ecological conditions are excellent for the growth of *arabica* coffee and a wide range of other crops is possible. The growing of pineapples and citrus fruits could be extended and vegetables like chillies, onions, garlic and capsicum grow well. This is the main maize-growing area of the plateau and cattle and pigs could be fattened on this crop.

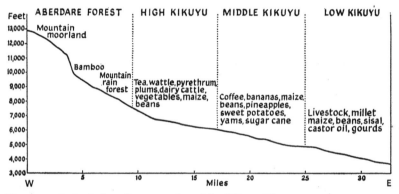

Figure 17.6 *Agricultural cross-section of the Kikuyu Plateau.*

The low parts of Low Kikuyu are really marginal land for arable agriculture and are climatically best suited to extensive stock-rearing. In the higher parts where rainfall is higher and the soils richer, tobacco and sisal grow well. Lentils, groundnuts, sesame, chillies, capsicum and castor seed can all be grown (Fig. 17.6). Favourable conditions exist for the growth of a wide range of crops but the successful introduction of new crops will be influenced strongly by the problems of marketing and by the attitudes and agricultural skills of the farmer.

Problems of marketing

There are three levels of marketing for the agricultural products of the Kikuyu Plateau: the local market, the Kenya market and the world market, each with its own special problems and potentialities. There is, however, one problem common to all of them – the problem of transport. A branch railway line and a main bitumenized road run north–south along the eastern edge of the Kikuyu Platea (Fig. 17.7). A large number of feeder roads run west–east down the ridges linking up with the main road and the railway line. These main routes are no more than 35 miles from any part of the plateau, yet for many months of the year they are extremely difficult to reach.

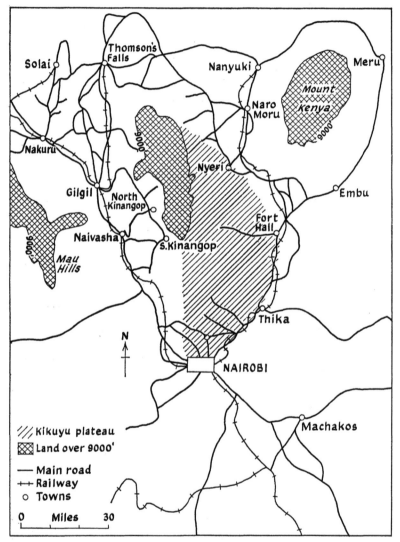

Figure 17.7 *Communications in central Kenya.* There is a great contrast between the transport network of the former 'White Highlands' and the Kikuyu Plateau.

The principal reason for this is the standard of the feeder roads. They are earth roads which only rarely receive adequate maintenance. Gradients are reasonable when the roads are on the ridge top but the roads have to switch ridge tops fairly often. This necessitates many steep gradients and many bridges which are often washed away during the rains. Thus in wet weather much of of the Kikuyu Plateau is inaccessible by road.

The local market

At the local market level there is considerable movement of produce between the three agricultural zones. High Kikuyu supplies vegetables, firewood, charcoal and milk; from Low Kikuyu comes gourds, sisal ropes, reed mats, charms and livestock; Middle Kikuyu supplies maize, sweet potatoes, bananas, beans and fruit. The buying and selling of the different products

Figure 17.8 *Distribution and size of markets in Fort Hall District.*

is centralized in local markets held in an open space in or near a village. The pattern of the markets in one Kikuyu District, Fort Hall, is shown on Figure 17.8.

The basic pattern of the trade is the exchange of various products from High and Low Kikuyu for the staple food crops of Middle Kikuyu which are more difficult to grow in the other two zones. The largest markets are therefore found in Middle Kikuyu, as can be seen in Figure 17.8. There are smaller markets situated along the border of High and Middle, and Middle and Low Kikuyu, the function of which is the exchange of produce between two zones, and these form secondary chains at about 6,000 ft (1,828 m) and 5,000 ft (1,524 m) respectively.

At present this trade operates largely on the basis of human porterage, the women carrying the produce to the markets on their backs. The volume of trade is already considerable and in 1963 this internal trade, which is not recorded in official statistics, was, in Fort Hall District, equal in value to the recorded agricultural exports. The principal barrier to the increase of this trade both in scope and volume is the poor road network.

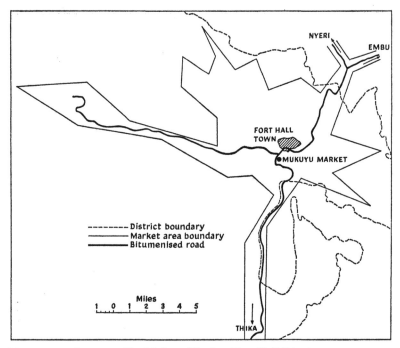

Figure 17.9 *Area served by Mukuyu Market, Fort Hall District.*

Where good communications exist the local markets begin to expand and to take on additional functions. An excellent example of this is Mukuyu market in Fort Hall District (Fig. 17.9). In 1960–1 a tarmacadam feeder road was built from the main road for a distance of about 12 miles. The effect on Mukuyu market was magnetic, it receiving an immediate surge of produce from all along the road. Being at the junction of the main north–south road with the only reliable east–west road into Fort Hall District, it became an important trade centre. Local entrepreneurs, using trucks, began to transport produce on a large scale along the road, especially bulky products such as firewood and charcoal, on which a good profit could be made. Round the market, stone-built native shops began to increase rapidly in both number and scope, even although the established town and administrative centre of Fort Hall was very close. The entrepreneurial activity might well develop along similar lines to the 'Mammy Lorry' trade of West Africa given an adequate road network.

The Kenya market

This market, particularly the urban area of Nairobi, offers a large potential to many of the products of the Kikuyu Plateau, especially those of High Kikuyu, such as fruit, vegetables and dairy produce. This market is already being partially exploited, especially by the farmers of Kiambu District. In 1962 food imports into Kenya totalled £6,890,000 and those of animal and vegetable oil and fats £1,041,000 (Government of Kenya, 1964). Many of these commodities could be produced within the Kikuyu Plateau. This would allow a diversification of the cash-crop economy of the plateau and, at the same time, result in a considerable saving in imports for the country. Although the demand exists the means of getting the produce to the market are inadequate.

The collecting and marketing of produce from a number of small producers poses many problems. It has been tackled in two ways: through statutory boards set up by the Government and through co-operative societies. 'Few underdeveloped countries can compare with Kenya in the magnitude of the organizational arrangements which have been provided by Government to further agricultural development and marketing' (International Bank, 1962). For the Kikuyu Plateau the main marketing organization is the Central Region Marketing Board which was established in 1959. Its main function is to organize the marketing of scheduled African produce. It acts as an agent for crops with guaranteed prices and as a trader for other crops. The various crops are bought for cash by traders who are licensed as agents of the Board. The produce is then sent by rail or road to Nairobi where the Board provides the important service of storage. In the case of products whose prices are not guaranteed by the Government the Board fixes prices in accordance with its own assessment of the market.

The growth of the co-operative societies, which deal direct with the Marketing Board, is also encouraged but apart from the coffee co-operatives there are few really active and successful societies in the Kikuyu Plateau. They are beset with many problems, the chief of which is that people do not seem to like a co-operative form of organization. The coffee co-operatives exist mainly because it is impossible to grow coffee without being a member of one. Thus the growth of co-operative societies does not suggest itself as the most effective way of improving the flow of produce to the Kenya market.

Some form of organized marketing is, however, required if the small producer is to get his goods to market. The Marketing Board's system of licensing traders to collect produce from the growers could be expanded. The encouragement of the local entrepreneurs' buying direct from the small-holder and reselling either to a market board or direct to the Kenya market would seem to be the most practicable system. In this respect the local markets might well be developed as collecting-points for produce which could be bought and transported by the traders. In this situation the traders would be competing for the produce and the possibility of unfair prices being paid would be reduced.

To cut out the middle man would be desirable but impractical as it would necessitate the collection of produce by the Marketing Board from the producers – a difficult and expensive process. With the significant lack of success of the co-operative system in the Kikuyu Plateau, the development of the local entrepreneur seems the most feasible alternative marketing method, and the one most suited to the present socio-economic situation.

The World market

This at present is of great importance, being the outlet for the most lucrative cash crop – coffee. Tea, wattle bark and pyrethrum are also sold mainly in the world market. The difficult world marketing situation for coffee is well known and as Kenya is a signatory to the International Coffee Agreement, coffee production must be controlled. The Economic Mission of the International Bank for Reconstruction and Development recommended that the Kenya Government take active steps to limit new coffee planting:

> The world supply situation for coffee calls for a reconsideration of the role of coffee in the future agricultural development in Kenya. With Kenya's present quota under the international agreement any surge in output in the next few years will face difficult marketing problems. Immediate action is therefore required to discourage new plantings.

The Kenya Government banned all new planting of coffee in January 1964. The world market for tea is much brighter than that for coffee.

> Unlike coffee there are no large stocks overhanging and threatening the world tea market. No sudden upsurge in production of a kind that would demoralize the world market is anticipated from the major producing countries but plantings and yields are increasing. On the whole there has been a reasonable balance between supply and demand for

many years and tea prices have shown greater stability than those of most other primary stable commodities (International Bank, 1962).

This is the cash crop which is being expanded most rapidly at present to offset the overdependence on coffee. The development of tea in the Kikuyu Plateau is in the hands of the Kenya Tea Development Authority established under the name of the Special Crops Development Authority in 1960. It is an independent commercial organization and is concerned only with what has been called 'small-holder tea' to distinguish it from estate production. By January 1965 there were 11,000 growers cultivating approximately 6,000 acres of tea in the plateau area. Two tea factories are now operating there and three new factories are planned for 1972. The Authority is a non-profit-making body responsible for providing the grower with planting material, supervising cultivation, collecting and processing the green leaf, and marketing the final product. Shares in the Authority are being made available to the growers, who will eventually control the whole organization.

Planting material in the form of two-year-old tea stumps is sold by the Authority to the small-holder at a subsidized price. Cultivation is supervised by a Tea Officer employed by the Authority, each officer supervising 500 acres. When the tea is ready it is plucked and carried to the nearest buying centre established by the Authority. These centres are distributed so as to ensure that no grower is more than 3 miles from one of them. The green leaf is collected from the buying centres by four-wheel-drive lorries and taken to the nearest tea factory.

Tea is a labour-intensive crop and whereas on estates these labour costs are a major factor they are negligible for small-holder tea as the farmer and his family cultivate and pluck the crop. Both quality and yields of tea in the Kikuyu Plateau are good, and the tea bush when fully mature yields an average profit of £70–£100 per acre. Because of the low production costs 'small-holder tea' is not only very competitive but can also withstand price fluctuations better than that produced on estates.

Green leaf should reach the factory within six hours of being plucked if good tea is to be manufactured. On a compact estate this is easy but in the Kikuyu Plateau where there are thousands of growers several miles from the factory this is a major problem, especially with a poor road system. The problem is intensified by the fact that the green leaf has to move in a north–south direction, whereas the plateau has been deeply dissected into ridges and valleys running west–east.

Poor communications appear to be the biggest single barrier to further agricultural development in the Kikuyu Plateau as they are restricting marketing at all levels, especially the growing of products for the Kenya market, which is the greatest potential market for the produce of the Kikuyu Plateau.

The farmers of the Kikuyu Plateau have the skill and ability to farm well. There are many, as in most agricultural communities, who are backward and conservative but most are quick to realize the commercial benefits of a new crop or variety and to utilize it. The rapid rise to importance of white maize, wattle, coffee and tea at various times over the last forty years illustrates this admirably. The majority of the people wish to improve their farms and are ready and willing to learn. The agricultural skills of many leave much to be desired but there is considerable latent potential in a people who, using only a digging-stick, grew sufficient food to support an agricultural population density of 100–200 per square mile prior to any contact with more advanced farming techniques.

Consolidation and the profits from coffee have given the farmers a wealth and confidence never before enjoyed. Agricultural finance is being made available on an increasing scale, especially by the commercial banks, from whom the small-holder can now borrow as he can offer his land as security. Prior to consolidation and registration the farmer could not do this as legally he did not 'own' his land.

The basis for further growth has been laid, but the rate of development must be maintained if the ever-growing problem of the landless is to be solved. Already there are signs of illegal subdivision of holdings and although legislation exists prohibiting this, the land will never be held together by legislation alone. The Settlement Schemes are providing temporary relief, but if further agricultural development does not take place the problem of land hunger will be just as intense as ever within a decade. The potential for that growth is there. New cash crops such as tea, fruits, vegetables, vegetable and animal oils and fats are being produced; a good market for these exists but the lack of adequate roads is a major barrier.

The construction of a network of all-weather feeder roads would give a very high return for investment in a relatively short space of time and would stimulate the growth of the whole economy in the plateau to such an extent that it would rapidly become self-generating. The continuing development of the Kikuyu Plateau is vital to Kenya's economic and political future. It is to be hoped that the Kenya Government will be able to find sufficient funds to build the adequate feeder roads necessary for further economic growth.

References

ALLAN, W. 1965. *The African Husbandman*, Edinburgh.
BARLOW, A. R. 1934. Kikuyu land tenure and inheritance. *Report of the Kenya Land Commission: Evidence and Memoranda*, Vol. 1, London.

BEECH, M. H. 1917. Kikuyu system of land tenure. *Jl R. Afr. Soc.*

CAGNOLO, C. 1933. *The Akikuyu*, Nyeri, Kenya.

CAREY JONES, N. S. 1965. The decolonisation of the White Highlands of Kenya. *Geogrl J.* **131**, 186–201.

CLAYTON, E. S. 1964. *Agrarian Development in Peasant Economies*, London.

CORFIELD, F. D. 1960. *The Origins and Growth of Mau Mau*, Nairobi.

DISTRICT AGRICULTURAL OFFICERS. 1900–64. Unpublished annual reports for Kiambu, Fort Hall and Nyeri.

DISTRICT COMMISSIONERS. 1900–64. Unpublished annual reports for Kiambu, Fort Hall and Nyeri.

FAZAN, S. H. 1935. *A Memorandum of the Numbers, Distribution and Rate of Increase of the Kenya Native Population and Some Aspects of the Problems of the Conservation and Development of Native Land*, unpublished paper, Department of Agriculture, Nairobi.

GOVERNMENT OF KENYA, 1964. *Kenya Population Census, 1962. Advance Report of Volumes I and II*, Nairobi.

GOVERNMENT OF KENYA, 1964. *Statistical Abstract*, **1963**, Nairobi.

History of Fort Hall, 1888–1942, unpublished anonymous manuscript, District Commisioner's Office, Fort Hall.

HUMPHREY, N., LAMBERT, H. and WYNN-HARRIS, R. 1945. *The Kikuyu Lands*, Nairobi.

INTERNATIONAL BANK FOR RECONSTRUCTION AND DEVELOPMENT. 1962. *The Economic Development of Kenya*, Nairobi.

KENYATTA, J. 1938. *Facing Mount Kenya*, London.

LAMBERT, H. E. 1956. *Kikuyu Social and Political Institutions*, London.

LEAKEY, L. S. B. 1934. Kikuyu land tenure. *Report of the Kenya Land Commission: Evidence and Memoranda*, Vol. 1, London.

MAINA, J. 1962. *Some Notes on the Use of Manure, Division of Labour and the Use of Ox-cultivation in Fort Hall*, unpublished paper, District Commissioner's Office, Fort Hall.

MEINERTZHAGEN, R. 1957. *Kenya Diary, 1902–06*, London.

MORGAN, W. T. W. 1963. The White Highlands of Kenya. *Geogrl J.* **129**, 140–55.

ROUTLEDGE, W. S. 1910. *With a Prehistoric People: the Akikuyu of British East Africa*, London.

SWYNNERTON, R. J. M. 1954. *A Plan to Intensify the Development of African Agriculture of Kenya*, London.

TAYLOR, D. R. F. 1966. *Fort Hall District, Kenya: A Geographical Consideration of the Problems and Potential of a Developing Area*, unpublished Ph.D. thesis, University of Edinburgh.

18 Agricultural progress in Zambia

G. KAY

Introduction

Zambia has an area of 290,000 square miles, most of which enjoy a tropical climate which is not unfavourable to agriculture. Throughout the greater part of the country the nature of the terrain and soils is such that crop and livestock farming are not only possible but also economically viable; several regions have a high potential for agriculture. In some areas with moderate or good resources tsetse fly at present prohibit domestic livestock, but nowadays tsetse fly cannot be regarded as a major obstacle if the will to develop exists. In short, the physical environment of Zambia is not lacking in agricultural potential and offers very considerable opportunities for the development of a diversified agricultural industry. Furthermore, 80 per cent of the African population live in rural areas and are heavily dependent upon direct exploitation of the land for the greater part of their livelihood. However, in marked contrast to many other countries of tropical Africa, agricultural progress in Zambia has been very limited. In 1965 agriculture (including subsistence production) contributed only 8·9 per cent of the gross domestic product and only 2 per cent of the domestic exports. Zambia's urban population and even some of her rural people consume large quantities of imported foodstuffs which cost £7·1 and £8·3 millions in 1964 and 1965 respectively. Parts of the country are characterized by chronic seasonal food shortages and occasional famine conditions; and protein-calorie malnutrition is widespread among young children in both rural and urban areas. This situation is economically, socially and politically unacceptable, and a massive campaign to develop the rural areas has been launched by the present government since Zambia became independent in October 1964.

This chapter offers an explanatory description of the character and origin of the present state of agriculture in Zambia and attempts to account for regional differences therein. It consists of four sections which deal with successive stages in the development of African agriculture. The first examines indigenous land-use systems and the problems and prospects they present. The second describes early attitudes towards and efforts to improve and change African agricultural land use. The third outlines progress achieved in the post-war years during which serious attempts were made by the colonial government to promote agricultural change. And the fourth

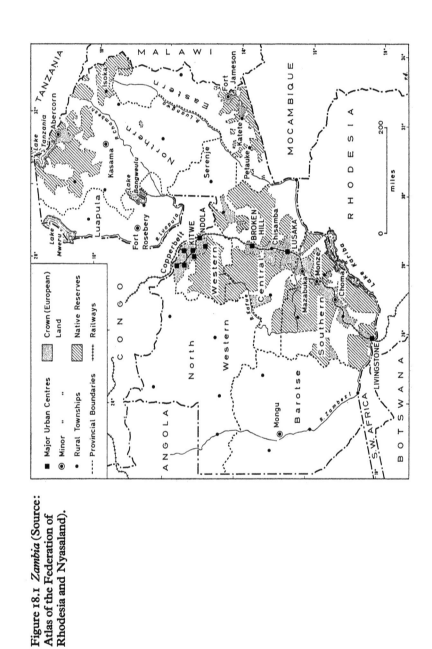

Figure 18.1 *Zambia* (Source: Atlas of the Federation of Rhodesia and Nyasaland).

comments on the post-independence situation and draws attention to some significant changes in policy.

Indigenous systems of land use

The indigenous systems of land use (i.e. those African systems not markedly changed by adoption of modern Western techniques and commercial practices) are extremely diverse. This diversity is largely a reflection of the very wide range of peoples found in Zambia, many of them relatively recent immigrants, and of the limited degree of fusion of the various tribal cultures. It has been increased by adaptation of various systems to different environments so that it is no longer possible to identify a particular system with a particular tribe. For example, Bisa agriculture in the Bangweulu Basin is markedly different from Bisa agriculture on upland plateaux and both are unlike Bisa agriculture in the Luangwa Valley. Furthermore, Bisa agriculture in the distinctive peri-urban areas that have emerged around rural townships differs from all truly rural Bisa systems of land use. On the other hand, selection of agricultural systems with reference to the various physical or natural regions of Zambia is still at a very early stage and no optimum system has been recognized by all tribes in any given region. Finally, it should be noted that indigenous systems of land use are dynamic and have been continuously changing in response to changing circumstances. The formidable diversity of dynamic land-use systems makes the promotion of evolutionary improvements difficult because successful grafting of innovations requires a full knowledge and sympathetic understanding of the old stock. Such knowledge is not available for Zambia as a whole and is not easily acquired.

The diversity of conditions also makes generalization about the indigenous systems of land use somewhat hazardous, but it is possible to identify several characteristics which most, if not all, of these systems have in common, and several regional differences that are relevant to a survey of agricultural change. All of the systems can be fairly described as primitive. They are characterized by a very low level of capital inputs, by the use of no power other than human muscles and fire, and by a dearth of scientific and technical knowledge of the productive processes and modern aids thereto. However, most systems are closely adapted to the physical environment in which they occur and in the absence of abnormal stresses they do no harm to it. They are also closely related to the customary but changing subsistence needs of the people concerned, and they are fully integrated with all aspects of tribal life. It is, in fact, unrealistic in the context of customary tribal life to isolate agriculture from other activities and particularly from other economic activities such as fishing, hunting, trapping, collecting, food processing, manufacturing, building and so on. The rural

African is not a farmer or a fisherman or any such specialist – he is a 'villager' whose polyfunctional life should be viewed as a whole. This absence of specialization adds to the difficulties of effecting evolutionary progress because such development must be on a broad front. Development cannot be pursued separately in different branches of economic activity (which are recognized in the Western world and government head-quarters but not in the village) without setting up tensions and difficulties within the village economy as a whole.

In view of the primitive state of technology and knowledge noted above, it is not surprising that, in spite of high labour inputs, the indigenous systems of land use provide low and uncertain returns *per capita* and per unit area. Surplus production is not unknown. In fact, the villager working for his subsistence requirements endeavours to cultivate sufficient land to provide an adequate supply of foodstuffs during a poor season. Because of marked variations in growing conditions from year to year, he occasionally faces shortages in spite of his caution, but probably in four years out of five he will have adequate supplies and in two of these years he may have a considerable surplus. The market value of this 'normal surplus' often is limited because it consists of commodities grown to satisfy the particular dietetic tastes of the producer and there may not be a market for these items. Also these surpluses are unpredictable and in most areas they do not justify marketing machinery. And in all cases the cash value of the surplus produced by one family from normal subsistence farming is very low be-cause of the small quantities involved. Consequently, even where such surpluses can be sold, poverty is endemic. Furthermore, contact with Western civilization has created new needs and wants, and the total pro-ductive capacity of the rural economy can no longer satisfy even the basic requirements of the people. Cash is necessary to obtain non-customary essentials and indigenous systems of land use afford inadequate opportuni-ties to obtain cash. They fail to provide a standard of life comparable in material terms with that of unskilled labourers in paid employment. Such circumstances preclude the accumulation of capital for investment and there can be no 'ploughing back of profits' until cash incomes exceed the minimum required for consumer goods by the producer and his wide circle of dependants. Also, because of the low levels of income and expenditure, those engaged in indigenous systems of land use contribute little or nothing to national funds on which they impose considerable demands. Alternative sources of both full-time and part-time employment are limited and are subject to ever-increasing pressure from a rapidly growing population with rising aspirations. There is, therefore, an urgent need to increase the total productivity *per capita* of the rural population, and included in the range of commodities there must be items which will procure cash.

The low returns per unit area may appear to be less troublesome in a country where the over-all density of population in the rural areas is only 9·5 persons per square mile. This, however, is not so and there are pressing reasons for increasing productivity of the land. Even in areas of good soils the more intensive indigenous agricultural systems have a low carrying capacity, i.e. the number of persons per unit area a given system can support in perpetuity without adverse effects upon natural resources. For example, the Eastern Province plateau is an area of relatively fertile soils but it can carry no more than 25 persons per square miles under the comparatively intensive use of land by the Chewa, Nsenga and Ngoni. The customary practices of the Lamba support about 18 persons per square mile in the Western Province, and the *citemene* systems commonly practised on the plateaux of northern and central parts of the country support between 4 and 10 persons per square mile. Such low carrying capacities mean that conditions of over-population are easily obtained, while absolute densities remain low. Unless there are compensating changes in man–land relationships, over-population sets in motion degenerative processes which, if not checked or reversed, ultimately must destroy the land.

Land deterioration begins slowly and unobtrusively but gathers momentum as it progresses. At each stage the standard of living of those concerned is affected adversely. Usually the wildlife suffers first from direct and indirect effects of man upon its members and their eco-systems. Simultaneously, the vegetation cover begins to suffer from excessive clearing, burning and, in some areas, grazing. Reduction of the vegetation cover and over-cultivation lead to soil deterioration; organic matter, plant nutrients and soil structure are lost. Subsequently, regeneration of the vegetation becomes more difficult, and increased exposure to the elements leads to sheet erosion of topsoils. Sheet erosion is followed by gullying, and gullies spread into ever more intricate patterns as they eat into the land. Run-off is increased and local water-tables are lowered; surface water supplies are threatened. Eventually the land becomes virtually useless to man and beast and must be returned to nature in the hope that she may heal herself.

Although dynamic, numerous indigenous systems of land use have failed to introduce compensatory changes sufficient to accommodate the rapid general and localized increases of population that have characterized the colonial era. Consequently, the continuation of these systems in overcrowded areas has constituted and still presents a serious threat to the basic means of production. Redistribution of the population has provided and can provide only a palliative, and it is imperative that production per unit area be increased if natural resources are to be conserved.

Indigenous systems of land use, because of their low carrying capacity, foster dispersion of population. Small, scattered settlements make administration and the provision of social services and infrastructure both difficult and costly, prohibitively so in many areas. The problems of this situation are particularly pertinent to an elected, socialist government which has serious political and humanitarian interests in the affairs of the rural population. At present, possibilities of 'regrouping' villages in order to facilitate social development are being investigated, but in the greater part of rural Zambia an increase in land-carrying capacity (i.e. in production per unit area) is a prerequisite for close settlement.

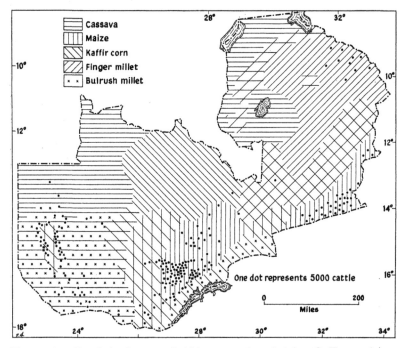

Figure 18.2 *Staple food crops and African-owned cattle* (after G. Kay, 1967).

The above discussion will leave no doubt that by modern standards all the varied indigenous systems of land use make a grossly inefficient use of manpower and natural resources, afford those concerned with a miserable standard of living in material terms, and provide the nation as a whole with little other than a large number of intractable problems. However, some systems are more amenable to development than others and this has been reflected in regional differences in progress made hitherto. The various systems may be grouped according to which of the five staple crops they grow (Fig. 18.2). Two of these crops, cassava and maize, are associated with relatively intensive systems of agriculture based on hoe-cultivation rather than 'slash-and-burn' techniques. Both have fostered close settlement and semi-permanent cultivation of well defined holdings; they give rise to close and lasting ties between cultivators and agricultural lands. There are two main differences between them that are significant for development. First, cassava is generally found in areas of mediocre soils, whereas maize is grown successfully by primitive methods only in relatively good soils. Customary cultivation of maize as the staple crop therefore largely coincides with areas of moderate to high potential. Secondly, cassava is not a popular foodstuff outside the areas where it is grown and, conditioned by long-established practices in southern and central Africa related in part to the production of European farms, the urban market for starchy foods is predominantly for maize flour. The association of maize flour with progressive, modern eating habits is so well established nowadays that large quantities are sold in rural areas where it is not the staple to those Africans who can afford to purchase it. Furthermore, this rural market is for relatively expensive roller-milled (as distinct from the coarser, hammer-milled) flour because this is the standard type sold in towns. In view of the large and growing internal market for maize, those growing it were equipped with a cash crop within the range of their customary subsistence crops. The only other subsistence crop which has found a ready and sizeable market is groundnuts, and it is a major component of subsistence cropping in the Eastern Province which is also a maize-growing region. The other staple crops, finger millet, bulrush millet and Kaffir corn (sorghum) generally are grown by primitive *citemene* or 'slash-and-burn' techniques which involve extensive bush fallowing. They exhibit none of the favourable characteristics associated with maize or cassava.

Indigenous land-use systems also may be grouped according to whether or not cattle are kept (Fig. 18.2). Most tribes of Zambia do not keep cattle. Their present environment would permit most of them to do so, but in many cases this cultural trait may be related to the fact that the tribes concerned migrated from more densely forested regions in the Congo Basin where cattle rearing would be difficult. Whatever its origins, it has unfortunate consequences for agricultural development not only because cattle themselves are valuable assets which utilize otherwise neglected grazing resources but also because considerable reliance has been placed upon cattle as draught animals in efforts to improve agricultural production. It is not a simple matter to introduce cattle to peoples unaccustomed to handling them and this obstacle, in fact, has proved a serious one in many development projects. On the other hand, tribes well acquainted with cattle quickly adopted new uses for them. Figure 18.2 shows that most of the one million or so African-owned cattle in Zambia are found in three

R

areas. Fifty per cent of them are in the maize-growing region of the Southern and Central Provinces; 25 per cent are in Barotse Province, mostly on and around the Zambezi floodplain and Luena Flats; and 15 per cent are on the Eastern Province Plateau – also a maize-growing area. The correlation with maize-growing areas may be no more than coincidence but it is an important association and has greatly assisted development in the two regions so blessed.

Finally, it would be useful to group or rank the various tribes according to their interest in and aptitude for agriculture but this is not possible because of the dearth of suitable data. It is possible, however, to indicate how some of the tribes are preoccupied with non-agricultural activities or have a low regard for cultivation. For example, fishing is a major economic activity in the lower Luapula Valley and Lake Mweru, in the lakes and swamps of the Bangweulu depression, in Lake Tanganyika, Lake Kariba and the Kafue Flats. In the heart of Barotse Province, fishing and fowling, cattle rearing and cultivation together form the basis of the rural economy. Among the Ila, cattle rearing is the prime occupation of the menfolk. Also, in bygone days military tribes such as the Bemba took pride in warlike activities and, although climbing and pollarding trees was accepted as manly sport, agriculture as such was despised and relegated largely to womenfolk and old men. Such attitudes undoubtedly affected behaviour for many years after *pax britannica* terminated actual warfare, and they may still have some significance. On the other hand, there are tribes which appear to have had limited political ambitions, no love of warfare and no major economic activity other than crop and livestock husbandry. Such tribes probably offer better prospects for agricultural development; they include the Tonga-speaking tribes of the Southern Province and the Nyanja-speaking tribes of the Eastern Province, except the Ngoni who were a warfaring tribe. However, the Ngoni have now been greatly influenced by their neighbours and take considerable interest in agriculture.

African agriculture under austere paternalism, 1894–1945

It is against the background of the indigenous systems of land use and in the context of the objectives and abilities of successive colonial governments that changes in African agriculture prior to 1964 are to be understood. The colonial era in Zambia began in 1894 when the British South Africa (B.S.A.) Company was granted powers of direct administration in areas north the Zambezi as well as in Rhodesia. The Company depended for its revenue primarily upon economic progress and therefore was anxious to encourage European settlement and investment as the most expedient means of bringing unused and under-used resources quickly into production. During the 1890s both Zambia and Rhodesia were lands of dreams and hopes, but by 1905, after a flurry of speculations and minor development projects, Zambia had failed in almost every respect to justify the optimism of earlier years; it was being equipped with a railway very largely because of Rhodes' personal wishes. In contrast, development flourished in Rhodesia, and by 1911 Rhodesia had attracted 24,000 Europeans while Zambia had a white population of only 1,500. Apparent differences in potential and alleged differences in the feasibility of European settlement led the B.S.A. Company to adopt markedly different policies in the two countries. Rhodesia was actively developed as a 'white man's country': Zambia was administered as a 'tropical dependency' of Rhodesia. A skeletal administration was established and spontaneous development and settlement were welcomed, but the Company made no serious effort to develop the country. Consequently, throughout Company rule, which ended in 1924, expenditure was kept to a minimum but even so it was never exceeded by revenue collected.

The neglect of Zambia was not entirely due to negative factors. Rhodesia, and also South Africa and the Congo, required labour and because of the rapidity of development and the reluctance of Africans to do unusual work for foreigners in strange surroundings there was a widespread and chronic shortage of workers in these countries – a shortage which continued well into the post-war period. Zambia therefore was soon regarded as a labour pool which was conveniently under the same general administration as that of Rhodesia. The flow of Zambians abroad in search of cash began before the turn of the century and by 1921 there were approximately 50,000 Zambians at work in other countries. During the following decade the emergence of the Copperbelt provided employment opportunities for large numbers within Zambia, and during the years that followed the number of Zambians in employment has repeatedly increased. Nowadays there are about 250,000 men in employment in Zambia and a further 75,000 are at work elsewhere. Labour migration has become an accepted feature of rural life, and wage employment is recognized as the prime means of economic and social advancement.

The stultifying effects of labour migration and of attitudes it has engendered upon prospects for rural development are very considerable. The selective nature of migration has leeched the rural areas of the fittest and most capable workers. For example, 37·5 per cent of all males between twenty-one and forty-five years of age were at work for wages in 1963. At the same time, 75·2 per cent of those with some secondary education and 64·5 per cent of those who had completed seven or eight years at school were in paid employment. The educated African does not wish to dirty his hands, and education generally is viewed as a means of emancipation from physical toil. Agricultural development (which began late) therefore always has had to cope with the anaemic remnants of the rural population. Only recently has the number of Africans with eight or more years of schooling greatly exceeded vacancies for them in paid employment, and it still remains for the present government to change a well established system of values and to make work on the land attractive rather than repulsive to the young man with limited education. It is hampered in this task because it can no longer use previous or existing conditions in the rural areas as a base line from which the prospects of a new life on the land can be 'sold' to the public. The significant criterion nowadays is not the difference between the present and the promised future in agriculture but the gap between rural and urban conditions. This gap is not likely to be closed appreciably in the near future, and therefore agriculture will continue to compete unsuccessfully for the better elements of Zambia's human resources.

The emergence of adverse attitudes towards agriculture has been accompanied by general and local deteriorations in man–land relationships. *Pax britannica* and modern government greatly reduced the effects of Malthusian restrictions on population growth and prevented the hostilities and large-scale population migrations which probably were the main traditional means of readjusting the balance between population and resources. Rapidly and continuously growing populations therefore have had to be accommodated largely within the local regions they occupied at the beginning of the colonial era. The effects of increasing population pressures have acted in concert with new methods of exploitation and changing standards of values to reduce seriously the contribution of non-agricultural sources of foodstuffs. Game-meat and fish everywhere are welcome additions to the diet, but except in highly localized areas they are less plentiful than in the recent past. The collection of insects, fungi, roots, leaves, fruits and honey also has become less rewarding, and the search for many such items is not so favoured as it used to be. Consequently, the burden of feeding the increasing rural population has been thrown more heavily on to agriculture, and pressure on agricultural resources has become critical in many areas. There is ample evidence that the subsistence diet is less adequate than it used to be and, in spite of other pressing demands, much hard-won money is spent in rural areas on basic foodstuffs. A widespread and growing need to improve the diet provided by indigenous agricultural systems has led in recent years to several 'health and nutrition' projects. However, the nutritional defects of indigenous land-use systems should be seen as one aspect only of the general malaise associated with these systems, and development should be directed towards effecting a cure rather than alleviating the symptoms.

The effects of high and increasing population pressures on resources have not been uniform throughout the country. In some areas they were aggravated by the alienation of land for non-African settlement and by the concentration of African population into restricted areas set apart for them. Extensive alienation of land occurred in only three areas: along the line-of-rail; in the North Charterland Concession (i.e. the southern part of the Eastern Province); and on the B.S.A. Company's estates south of Lake Tanganyika (Fig. 18.1). The creation of Native Reserves was restricted to these localities. The first reservations were made on an *ad hoc* basis and it was not until the late twenties that the thirty-nine Native Reserves were legally established (Fig. 18.1). These reserves affected more Africans than subsequent alienations of land merited, but the clearance of large areas scheduled for non-African settlements was mistakenly justified on the grounds that sufficient land had been set aside for the Africans involved. The gross overcrowding that characterized most of the reserves therefore was as much a product of ignorance as of administrative or political necessity. In view of the concentration of local peoples into the Native Reserves it is not surprising that these areas should be amongst the most over-populated in the country. Excessive population pressure has been particularly damaging on the Eastern Province plateau and in the Tonga maize-growing area astride the line-of-rail. The Abercorn–Isoka region, which includes several reserves, also was singled out at an early date as a problem area but this was partly due to its critical position in respect of the headwaters of the Chambezi and Luangwa Rivers. More recently the Kafue headwaters similarly have received special attention from conservation officers because of the national importance of the region. However, as we shall see, the existence at an early date of a dire threat to important natural resources has not been entirely without benefits in the long run.

Perhaps the most significant development in respect of African agriculture during the first forty years of colonial rule was the growth of local markets for foodstuffs, and in particular for maize, pulses and livestock products. The railway linked the more important early markets which were to be found in the towns of Rhodesia, in the mining region of Katanga, and (within Zambia) at Livingstone and Broken Hill. Later the Copperbelt emerged as a major urban region and created a large demand for cheap foodstuffs. The prospects for farmers along the line-of-rail attracted European settlement to the area between Broken Hill and Choma which is characterized by good soils; Chisamba, Lusaka and Mazabuka were the focal points of this early development (Fig. 18.1). The European farming industry deprived Africans of the full stimulus of the markets served by the railway, but these markets have never been monopolized by European producers. In fact, rather than depressing African agriculture, the existence of a European farming industry was instrumental in the development of infrastructure necessary for Africans to take advantage of the market

R 2

opportunities. In particular it brought a marketing organization and a supply of farming equipment (notably ploughs, carts and ancillary items) into the heart of the largest maize-growing, cattle-rearing region of the country. It also demonstrated new methods of cultivation which, during the pioneer struggles by European farmers, were not too remote from those of the African to allow copying.

The outcome of these developments was that Africans, and Tonga in particular, near the line-of-rail began to sell maize, some cattle, and small quantities of other produce. In the first place sales were of surpluses produced by normal subsistence cultivation, but soon enterprising individuals began to plant larger areas specifically to grow maize for sale. This development would have been seriously limited without modern implements, and such implements were not readily and cheaply available away from the line-of-rail until road transport became more efficient in the post-war period. A survey of land usage among the Tonga in 1945 throws light upon the nature of this spontaneous development of agricultural sales (Allan *et al.*, 1948). 85·3 per cent of the families investigated were classed as subsistence cultivators; none of them sold more than 10 bags (each of 200 lb) of maize and about one-third of this group made no sales at all. 14·3 per cent of the families sold between 11 and 100 bags and were described as smallholders, and the remaining 0·4 per cent sold over 100 bags and were described as farmers. Estimates indicated that by the mid-forties subsistence producers contributed 33 per cent of the total maize sales by Africans, while smallholders and farmers contributed 52 per cent and 15 per cent respectively. This trend towards quasi-commercial farming brought much-needed cash into the rural areas concerned and fostered business-like attitudes among the farming élite. However, since it involved extension of the cultivated area rather than increased production per unit area, it lowered the critical density of population or land-carrying capacity and added considerably to the deterioration and destruction of natural resources in already overcrowded areas.

Away from the line-of-rail, widely scattered minor townships, administrative posts and mission stations provided small, local markets, and there was only one large market of regional significance. This was in the Eastern Province – in and around Fort Jameson which was for a time the capital of north-eastern Zambia and for long controlled the eastern gateway to the country. A small community of European farmers eventually established a successful tobacco industry there which began about 1912, reached a peak in 1927 and flourished again briefly in the early fifties. Being in a maize-growing area, the planters found it to their advantage to buy rather than grow foodstuffs for their workers, and they and the townsfolk together created a sizable market. For example, in 1927 from 20,000 to

30,000 bags of maize were sold by Africans to Europeans at an average price of about 7 shillings per bag in the villages. Compared with African sales on the line-of-rail (which averaged over 150,000 bags each year between 1937 and 1940), sales in the Eastern Province were small, and they were mostly from surpluses of normal subsistence cultivation. Nevertheless, this early experience and contact with European farmers introduced a large number of cultivators to practices and possibilities of commercial activity and helped to prepare the way for progress in the Eastern Province.

For the first quarter of this century the changes described above took place without any serious interest in African agriculture by the Government, and official policy (which is of prime importance nowadays) was virtually non-existent. Throughout Company rule 'famine and food supplies' were the only aspects of agriculture the Government expressed real interest in, and its only significant actions were to safeguard against and deal with food shortages. Also Africans were ineffectively harangued from lofty levels about their 'wasteful methods of cultivation' and their destruction of natural resources, but in practice they were left virtually undisturbed to continue their own devices. The Colonial Office took a more serious interest in the agricultural prospects of the country and in 1925 it set up a Department of Agriculture. The first annual report of that department declared that 'the process of gradually building up a native agricultural industry will be slow because of the necessity that it should be on a very sure foundation'. This no doubt comforted many European concerns with vested interests in the *status quo*. The department had to deal with both European and African agriculture, and it was so small and it moved so slowly that its effects were barely felt in the latter field for a full decade. The depression of the early thirties severely restricted the activity of the department; its salaried staff in January 1934 consisted of five men. It is, therefore, not surprising that all its early research and experimental stations were established in or near European farming areas – at Mazabuka (which also accommodated the department's headquarters), at Fort Jameson, and at Lunzuwa near Abercorn and the B.S.A. Company's Tanganyika Estates (Fig. 18.1).

During the mid-thirties official interest in Zambia quickened and, perhaps because of problems of food supply on the Copperbelt or because repeated reports on the destructive effects of over-population at last were taken seriously, this new interest was extended to African agriculture. The Department of Agriculture retained its policy of proceeding with caution, and in 1936 it launched the now famous ecological survey as a stock-taking exercise (Trapnell and Clothier, 1937; Trapnell, 1943). However, in September 1937, Despatch No. 489 from the Secretary of State for the

Colonies urged that 'a programme for the improvement of Native subsistence agriculture' should be put into operation. Such a request required action and a list of priorities was compiled. The Department of Agriculture declared its main objective to be 'the development of production by Europeans and Natives of crops which can be sold, concurrently with the introduction of measures to ensure that the land on which such crops can be grown shall remain permanently productive' (Northern Rhodesia, 1938). Unfortunately, its existing state of knowledge and its limited staff resources restricted action under the first clause of this policy to European farming and to the setting up of experimental production units in African areas under the supervision of established research stations. The location of these stations (noted above) thus became very relevant to African agricultural advancement. In the short term, however, most attention was given to the conservation of resources and the rehabilitation of degraded lands in the worst problem areas of the country. Five such areas were recognized. They were, first, the Native Reserves of the Southern Province where conservation measures were rigorously introduced; secondly, the Native Reserves of Abercorn and Isoka Districts where measures were undertaken to protect the headwaters of the Chambezi and Luangwa Rivers and where attempts were made to rationalize *citemene* agriculture until improved methods of land use could be devised; thirdly, the Native Reserves of the Eastern Province and, fourthly, the Native Reserves of the Central and Western Provinces, where the application of conservation measures was accompanied by massive resettlement schemes to alleviate population pressures within excessively overcrowded areas; and, finally, central Barotse Province was selected for special attention because deterioration of its resources had been accompanied by repeated and severe food shortages.

When, at last, it seemed that development work was to begin, circumstances beyond the control of the colonial government intervened, and inevitably the report of the Agricultural Department for 1939 was gloomy. It noted that:

> ... the year 1939 was to have been one of rapid expansion ... the outbreak of war not only frustrated these plans but necessitated a rapid and drastic modification of existing programmes ... The uncertainty as to the war strength of the department made a revision of policy imperative. It was decided to concentrate all officers, while they remained available, on work which would not be wasted even if it could not be carried through to completion ... In effect this meant the limitation of departmental effort to the reclamation and preservation of native areas in the Southern, Northern and Eastern Provinces (Northern Rhodesia, 1939).

In fact, because of their large populations and the severity of their problems, and because staff there could serve European farming areas too, most attention was given to the reserves and resettlement areas of the Eastern Province and to the Southern Province. During operations to save and safeguard the soils and water supplies, a great deal was learned about the lands and peoples of these areas and a working relationship was established between villagers and government officers. Furthermore, conservation and resettlement had ceased to be regarded as an end in themselves and were conceived as a prelude to the introduction of improved, modern systems of land use; in both regions a limited amount of research was continued to devise such systems. Thus, in several ways, these problem areas were prepared for positive development which was confidently expected to follow the end of the war.

African agriculture under benevolent paternalism, 1946–64

At the end of the Second World War the Colonial Office committed itself to the task of developing African territories for the benefit of all their inhabitants. Colonial governments therefore were required to take a leading part in fostering development in those sectors which were neglected by private enterprise. Inevitably this meant that in Zambia government policy became a major factor (probably the most important single factor) in determining the character and location of changes in African agriculture.

The Department was ill-prepared for its new role as an innovator. It lacked knowledge, staff and resources on which to mount a programme of revolutionary changes. Its main source of countrywide information on natural resources and land-use systems was Trapnell's ecological survey which was not intended as anything more than a preliminary reconnaissance. Its technical and scientific research work was limited in content and extent, and in 1946 'work on native agriculture' was based on only eight stations whose location had been determined, as noted above, by the occurrence of European farming areas and problem areas. Three of them were in the Southern Province, three were in the Eastern Province, one was in the Northern Province and one in the Western Province. In the same year there were only sixteen senior agricultural officers on duty in the whole country and they were responsible for both European and African agriculture. They were to be found in the same provinces and in the same proportions as the agricultural stations noted above. It is, therefore, not surprising that the Department was reluctant to abandon its earlier policy of proceeding with caution and of investigating problems before suggesting solutions. Inevitably its policy statement prepared in 1945 for the Ten-year Development Plan consisted largely of tentative proposals to continue its conservation work and to improve indigenous agricultural systems. Seven general objectives were stated and they may be summarized as follows:

1. The conservation of natural resources.

2. The redistribution of population where necessary in the interests of conservation.

3. The improvement of agricultural methods (by encouraging the use of manure, introducing ploughs and wheeled vehicles, improving weed control, etc.).

4. The improvement of subsistence production (by encouraging selection of seed, introducing new varieties, improving planting techniques, etc.).

5. The introduction and development of cash crops.

6. The introduction on an experimental basis in selected areas of African Farm Settlements.

7. The prosecution of experimental work in connection with the foregoing proposals.

This policy might be described as one of evolutionary change, and it was in harmony with the ideology of indirect rule which had emerged as a powerful influence on colonial governments. It also found support from those who were alarmed by the accelerated rate of 'detribalization' and who regarded the rural areas as a necessary sanctuary for tribal societies. In 1945 a team consisting of two agriculturalists, an ecologist and a social anthropologist assisted by a geologist and a barrister made a survey of *Land holding and land usage among the plateau Tonga of Mazabuka District* (Allan *et al.*, 1948). This team prepared a list of recommendations whereby Tonga land usage might be improved and these were 'designed to fit as far as possible into existing land-holding conceptions and the social system of the Tonga'. Evolutionary progress evidently was in vogue. In November 1947 Allan expressed somewhat different views (Allan, 1949). Commenting on the team study referred to above he claimed that 'such studies, to be of full practical value, must take into account the effects of changing conditions. The industrial and economic changes which are to be expected within the next decade will profoundly affect the whole structure of African life ... ' This might be interpreted as a wistful plea for an adventurous agricultural policy if such were possible, and it may have been prompted by Despatch No. 29 of February 1947 from the Secretary of State for the Colonies.

This Despatch exhorted colonial governments in Africa to 'an endeavour to raise rapidly the standards of living of African populations'. It observed that this objective could be obtained only by 'a revolution in African productivity' and that this revolution would depend upon 'very great changes in the techniques of production'. It also suggested that 'it [would] be necessary to consider very substantial changes in the present customary systems of land tenure'. In short, it called for a far-reaching agricultural revolution. This Despatch and a visitation the previous year by the United Kingdom Groundnut Mission left no doubt as to the urgency the Colonial Office attached to African advancement in the rural economy. The Department of Agriculture, it seemed, was required to review its policy.

Agricultural development in Zambia depends, in the first place, upon providing producers with technical and scientific knowledge whereby productivity may be increased and production methods may be made flexible; with capital whereby the new production units may be suitably equipped; and with managerial ability whereby the stocks of knowledge, capital, land and labour may be put to good use. Secondly, it depends upon the provision of infrastructure including transport and communication facilities and services for processing and marketing agricultural produce. The Department of Agriculture was primarily concerned with the former group of requirements. Its immediate problems were, first, to obtain the necessary knowledge, capital and know-how; secondly, to distribute these assets; and, thirdly, to determine the form or forms of production unit on which they should be employed. These problems are still formidable nowadays; in 1947 they must have been overwhelming. The Director of Agriculture therefore conceded little, and in drafting a reply to the Colonial Secretary's memorandum he suggested that 'the improvement of the traditional agricultural systems by what can best be termed "accelerated evolutionary development" will probably be the most practicable if not the only course'.

However, perhaps in deference to the wishes of the Colonial Office, two 'impact schemes' were adopted, and these became the main vehicles for agricultural development for the next decade or so. This decision did not involve departure from the declared policy of effecting evolutionary change. Indeed, it was decided that the Peasant Farming Scheme 'must do no violence to either traditional social structures or to the accepted systems of land tenure, though its establishment may well help in the evolution of both' (Sec/Nat/214; see Kay, 1965). The object of the impact schemes was to launch progressive individuals (selected by their willingness to volunteer) into modern, commercial agriculture on family holdings. It was hoped that the nucleus of proficient farmers thus created would inspire others to improve their farming and to seek a living from the land. The Peasant Farming Scheme began in 1948 when blocks of farms were laid out near Katete in the Eastern Province and near Serenje in the Central Province. These localities were selected as representative of two major ecological regions. The scheme began without any precedent in the art and science of small-scale farming and it was very much a 'leap into the dark'. However, several important decisions were made which greatly affected the character of the scheme. As its name suggests, the Peasant Farming Scheme was intended to create a group of small-scale farmers, and it was recommended that the size of holdings should be such that it would be possible for 'an average African family' using ox-drawn implements to cultivate a farm 'without requiring any labour outside the immediate family circle'. It also was recommended that each farm should provide the family with their subsistence requirements and a cash income which has been variously stated as £100, £150 and £200 per annum. These recommendations, in fact, have been in conflict with each other, but there can be no doubt that the farms were intended to be small and were to be cultivated by the use of oxen. The economic basis of farming at first was related to local and national market requirements only, and in practice this meant producing saleable quantities of maize, pulses and, if possible, cattle. Those wishing to participate in the scheme registered with the Department of Agriculture and agreed to farm in accordance with instructions. They were provided with some training and then established on a partially cleared farm with all necessary equipment. The cost of clearing the land and of capital assets issued was debited against the Peasant Farmer who was required to repay this loan in ten equal annual instalments; no interest was charged.

On the Eastern Province plateau, the scheme was an immediate success and was enthusiastically taken up by the local populace and by the staff of the Department of Agriculture who pressed for a rapid increase in the number of farms. Success in the Eastern Province was highlighted by poor progress and even failure elsewhere, and especially at Serenje where the experiment was watched closely because it was one of the first pilot trials. The basis of progress in the Eastern Province was threefold. First, aspects of recent history described above had prepared the way for development. Secondly, the agricultural potential of the plateau is moderate to good. And thirdly, because of the character of the indigenous land-use systems, changes required by the scheme were easily accomplished. Maize was the local staple; groundnuts were a major secondary crop; and cattle already were kept. In fact, on the Eastern Province plateau the scheme was an evolutionary measure whereas in most other areas it can be fairly described as a revolutionary project (Kay, 1962). For example, in Serenje District it was recorded that 'the farmers had to be taught everything . . . they had to learn to look after cattle, to work with oxen, to use implements and to carry out each operation' (Northern Rhodesia, 1950). Also they had to learn to grow maize for sale, while they produced millet and sorghum for their own use. Without the natural and historical advantages of the Eastern Province, it is not surprising that the Peasant Farming Scheme floundered in such localities.

The other impact scheme emerged from an experiment which began in the Tonga maize areas of the Southern Province in 1946 and was later fertilized by the Peasant Farming Scheme. In 1946 Tonga farmers (i.e. those with relatively large holdings) were invited to enter into an agreement with the Department of Agriculture whereby they accepted the supervision of the department's staff and adopted improved methods in return for a higher price (22 shillings per bag) than that paid by the Maize Control Board (18 shillings per bag). By 1949 there were 362 farmers in the scheme. The inducements paid to them came from the price stabilization fund which the Maize Control Board had accumulated by paying producers some-what less than their maize actually realized. In 1949 a revised scheme was introduced. The price stabilization fund was renamed the Southern Province African Farmers' Improvement Fund (A.F.I.F.) and was to be replenished annually by a levy on all African sales of maize in the province. Out of this fund individuals who registered as Improved Farmers and faithfully followed instructions were paid a bonus for every acre they so cultivated. In practice, Improved Farmers and Peasant Farmers were set basically similar requirements. They differed in that the latter were financed by government funds and could be established wherever the Government chose, whereas the former were financed by African producers and therefore could be established only where crop sales had reached a sufficient scale to provide the necessary funds. Also, while the Peasant Farmers' only inducement was the supply of capital equipment, loans, training, etc., at very generous rates, the Improved Farmers received a cash bonus as an incentive. In 1951 the African Farmers Improvement Scheme was extended to the 'railway zone' of the Central Province and to the Eastern Province; in the latter area it strengthened the Peasant Farming Scheme because most of those who claimed and qualified for bonuses were, in fact, Peasant Farmers. A.F.I. funds were used to finance capital works as well as to offer inducements to individuals. The following breakdown of expenditure from the Southern Province A.F.I.F. during 1951–3 illustrates the manner in which such funds were used:

	% of total expenditure
1. Expenditure to assist individuals:	
Bonuses and subsidies to Improved Farmers	17
Subsidies on ox-carts	11
2. Capital expenditure to assist local communities:	
Soil conservation works	29
Water conservation and supply schemes	18
Marketing and storage facilities	20
3. Other expenditure (mostly overheads)	5

The greater part of the resources of the Department of Agriculture was devoted to these two schemes, and 'projects' to encourage the production of a particular crop or to increase the take-off of cattle and so on were mostly mounted on or alongside them. Consequently, although no deliberate decision to do so had been made at policy-making levels, development was concentrated into two regions (Fig. 18.3). These were parts of the Eastern Province which had responded well to early opportunities and areas along the line-of-rail in the Southern and Central Provinces which were able to build upon historical, cultural and physical advantages similar to those of the Eastern Province fortified by their position in respect of rail communications and markets. Marketing organizations and other essential items of infrastructure also were largely restricted to these two regions. Elsewhere progress was slow, and serious local problems (such as the occurrence of famines in central Barotse Province) attracted more attention than development prospects.

In June 1957 the Department of Agriculture issued a *Memorandum on African Agricultural Policy* which, for the first time, suggested a regional approach to development as a matter of policy. The most serious bottleneck in development was, and still is, the shortage of scientific and technical staff and of trained extension workers. It was recognized that various categories of staff are complementary to each other and cannot work successfully in isolation, and therefore that staff resources should be viewed not as a given number of individuals but as a number of groups or teams. In order to make optimum use of available field staff such teams must be complete and must be allocated 'according to zones of potential production'; widespread dispersion of staff would stultify the efforts of the entire department. The country, therefore, was divided into three categories and a different policy was adopted in each. In Zones of High Potential intensive total development was advocated. Three such 'zones' were recognized: the maize-growing, cattle-keeping areas of the Southern and Central Provinces; the Petauke, Katete and Kalichero areas of the Eastern Province; and the areas of red loam soils in Ndola Rural District. The variety of criteria used in describing these areas indicates that they were selected after an assessment of a wide range of physical, cultural, historical and economic factors and not with reference to natural resources only. Zones of Medium Potential were more numerous and widespread and consisted of localities worthy of consideration for development in the near future. It was recommended that surveys and experiments and general conservation and extension work be pursued in these areas. The greater part of the country, however, fell into the Zones of Low Potential where work was to be limited to ensuring adequate food supplies and the provision of essential conservation measures.

It is quite clear that this policy of regional development endorsed the actual distribution of activity that had already emerged in an *ad hoc* fashion and put forward rational arguments for continuing with such a distribution of development effort. There was, however, a fourth category of Special Zones which is of particular interest. It consisted of the Gwembe Valley, the Muswishi tsetse area east of Broken Hill, the Ngoni Reserve in the Eastern Province, and the 'Bemba Plateau Intensive Development Area'. The first three were selected for special attention because of the urgency of specific problems – the need to prepare for resettlement in the Gwembe

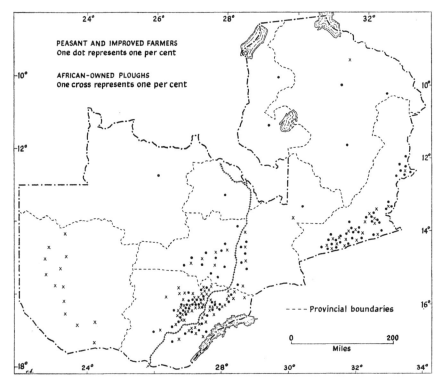

Figure 18.3 *Peasant and Improved Farmers and African-owned ploughs, 1962* (Source: Annual Reports of the Department of Agriculture and of the Secretary for African Affairs).

(Kariba) Valley, the need to protect the Muswishi area from incursions of tsetse fly, and the need to deal with excessive overcrowding in the Ngoni Reserve. The recognition of areas with such problems was in keeping with previous policy. The 'Bemba Plateau' did not fit into this category of problem areas: it suffered not from any specific agricultural problem but from the general malaise found in all rural areas of low potential. It was differentiated from the mass of such areas because it had succeeded in attracting government attention: it was becoming a *political* problem area. In 1947 the Northern Province Development Commission was set up and 'charged with the responsibility of putting into effect, with the utmost expedition, the Government's policy . . . which may be described briefly as implementing, as speedily as possible, a co-ordinated plan of development on whose foundation the future economic, social and political life of the Northern Province can be safely built' (Halcrow, 1961). Between 1957 and 1961 £1,813,627 were spent in the Northern and Luapula Provinces on this crash programme in which agricultural development played a major role. The Northern Province Development Commission marks the introduction of political factors into the location of development effort. It may be significant that the Commission's efforts to increase output (as distinct from its provision of services) were conspicuous by their lack of success.

In 1961 a Working Party was set up to recommend what direction future rural development policy should take. The Working Party was left in no doubt as to the strength of political pressures for changes in the distribution of development effort. The Provincial Commissioner for the Northern Province repeated arguments which had prevailed in 1957:

> The political considerations are vital in areas such as this Province . . .
> In 1957 there was a discontented and disillusioned population and at that time it seemed that it would have soon been necessary to provide major expenditure on security measures . . . The Northern Province Development Plan was timely . . . From the political point of view it is vital to provide funds for development even in comparatively unproductive areas . . . This cannot be sufficiently stressed. (Federation of Rhodesia and Nyasaland, 1961.)

In 1961, however, all the 'forgotten areas' were clamouring for attention and, to counter the 'economic arguments' expressed by the Department of Agriculture in 1956, they all were stressing political, social and humanitarian considerations. The Working Party heard these arguments but tactfully rejected them. It deplored pseudo-development 'which merely injects capital or promotes facilities without raising productivity' and it maintained that 'attention should be directed towards areas with the highest potential'. It went further, and suggested that since farming is essentially a private enterprise care should be given to selecting the most promising individuals for support. This suggestion is worth pursuing more closely than has been done, and criteria for the selection of worthy individuals ought to be given close thought. The Working Party suggested a rule of thumb which is not without merit as an interim measure. It declared that 'the identification of every producer of a cash crop will be the immediate aim and these will then be strengthened in their economy by all necessary means'. The provincial targets set by the Working Party for 1965 give some indication of where it thought the areas of 'highest potential' and the 'producers of cash crops' were likely to be found (Table 18.1).

The Development Plan for the period 1961–5 accepted the recommendations of the Working Party on Rural Economic Development, but during the period covered by the plan political changes were rapid. As independence approached political criteria received a new significance: they could no longer be dealt with lightly. Subsequent shifts in policy and their implications for the future of African agriculture are reflected in three reports dealing with development in independent Zambia.

African agriculture since Independence

The approach of political independence was accompanied by a request in September 1963 from the government to the UN/ECA/FAO to review the economy and recommend lines for further development. The UN Mission appointed to undertake this task took note of the forthcoming political changes and assumed that 'the Government, being elected on a wide franchise, is now in a position to mobilize popular support, especially in the modernization of agriculture' (Northern Rhodesia, 1964). In view of the prevailing attitudes noted above (pp. 503–4), the validity of this assumption is debatable. It is, however, evident that the Government now is subject to pressures from *all* parts of the country in a way that the colonial government never suffered. The Mission was aware of this but, as the following statement shows, it probably failed to recognize the strength of these pressures and hoped that they could be easily countered:

> Some minimum development in every Province is a political requirement. This may appear to cut across the strict application of economic criteria, but the need and opportunities for increase in crop yields throughout the country are great, so policy directed to raising yields will to some extent fulfil this objective. However, naturally the programme must be to some extent concentrated in the districts of highest potential (Northern Rhodesia, 1964).

The Mission also accepted adherence to evolutionary rather than revolutionary methods of change in spite of the relatively slow rate of development implied in the former:

> It was urged on us that as far as possible changes should be achieved without social disruption; this means that existing society should be built upon rather than destroyed (Northern Rhodesia, 1964).

Clearly the UN/ECA/FAO Mission was much influenced by economic criteria and by the recent past – probably by the achievements of the

TABLE 18.1 *Provincial production targets for 1965 set in 1961 by the Working Party on Rural Economic Development*

COMMODITY	PROVINCE							
	Southern	Central	Eastern	Northern	Luapula	North-western	Barotse	Western
Maize (bags)	500,000	300,000	45,000	25,000	10,000	—	15,000	—
Groundnuts (bags)	10,000	10,000	250,000	—	—	50,000	—	—
Tobacco (lb)	400,000	400,000	1,250,000	350,000	150,000	400,000	200,000	100,000
Cattle (head)	30,000	7,500	12,000	4,000	500	1,000	25,000	—
Cotton (lb)	800,000	800,000	800,000	—	—	—	100,000	—
Vegetables (tons)	100	100	—	—	—	—	—	2,400
Wheat (bags)	20,000	—	—	—	—	—	5,000	—
Coffee (tons)	—	—	—	75	25	—	—	—
Poultry (£)	50,000	10,000	—	—	5,000	—	—	10,000

previous decade and by evidence of persons steeped in the policies of these years.

The Transitional Development Plan for the period January 1965 to June 1966 was refreshingly realistic and set the key for future planning. It noted that:

> ... with roughly three-quarters of the country's population (and, of course, of the electorate) still earning what livelihood they can from the land, the over-riding urgency of agricultural development is clear enough ... The moral and political urgency of the problems, the basic promise of Zambia's resources, and the strategic and economic need to achieve greater self-sufficiency and to expand exports – all make boldness essential (Zambia, 1965).

The authors of the Plan, however, were well aware that, as in 1947, Zambia was ill-equipped to engineer a universal agricultural revolution and they therefore advocated 'a mixture of boldness and caution'. The Plan advised that experimentation should be undertaken in the real world, not in the laboratory; and that a wide range of subjects, including many of a controversial nature, should be given a trial. It is implicit in these arguments that widespread development must be attempted and that existing society should no longer be protected from the impact of revolutionary methods. Concern to ensure the gradual evolution of rural society had acted as a check on imaginative economic projects for too long. Typical of the shift of emphasis recommended by the Transitional Development Plan is its plea that the 'oxen-stage of development' should be left out and that agriculture should be mechanized as quickly as possible.

The Plan provided no details of the distribution of proposed investment but an analysis of the allocation of projects (regardless of their size) by provinces does reveal the shift of emphasis from areas of high potential to the country as a whole (Table 18.2). The completion of twenty-two established projects was recommended. The distribution of these reflects earlier policies: all but three of them were in the Southern, Central and Eastern Provinces; of the others, two were in the Northern Province. Sixty-four projects were approved for immediate implementation: half of these were in the three relatively well developed provinces, but it is notable that the Eastern Province received only seven. It can be convincingly argued that the line-of-rail is a region of national importance and that development there is for the benefit of all the peoples of Zambia rather than specifically for the local peoples. It is much more difficult to justify preferential treatment for the Eastern Province. The distribution of Conditionally Approved Projects similarly reflected a wider dispersion of investment than had been the case in earlier years.

TABLE 18.2 *The distribution of rural development projects suggested by the Transitional Development Plan for 1965–66*

PROJECTS	Southern			Central			Eastern			Northern			Luapula			North-Western			Western			Barotse			Total		
	A	B	C	A	B	C	A	B	C	A	B	C	A	B	C	A	B	C	A	B	C	A	B	C	A	B	C
Land settlement and conservation	2	1	2	—	—	—	2	1	—	2	—	—	2	—	—	1	—	—	2	—	—	—	—	—	3	1	12
Crop and livestock development	4	5	5	5	4	4	2	3	3	4	5	—	5	4	—	5	4	—	5	1	—	1	5	—	12	32	31
Marketing and co-operatives	—	4	1	—	6	—	2	1	—	2	—	—	2	—	—	2	—	—	2	—	—	2	—	—	—	22	2
Rural credit	—	2	—	—	2	—	2	—	—	2	—	—	2	—	—	2	—	—	2	—	—	—	—	—	—	—	16
Veterinary	2	5	1	1	1	—	1	2	2	1	2	—	—	—	—	—	1	—	—	1	—	—	—	—	7	9	3
Total	8	15	11	6	10	9	5	7	9	2	7	9	7	7	1	7	8	—	8	3	—	3	8	—	22	64	64

A – 'Carry Overs';   B – Approved Projects;   C – Conditionally Approved Projects*

*(subject to availability of building capacity and materials and/or requisite skilled or professional staff and also subject to successful completion of supporting studies).

Plate 8   PEASANT FARMS IN NORTHERN ZAMBIA
The landscape of the block of Peasant Farms contrasts markedly with that of Bemba *citimene* cultivation
in the surrounding area. (Photograph by Fairey Air Surveys of Rhodesia Ltd, for the Government of
Zambia, Northern Rhodesia.)

The First National Development Plan 1966–70 has been published recently. One of the main objectives of the Plan is 'to provide a radical improvement in the living standards of the whole population', and 'throughout the Plan an effort has been made to localize investment and to increase to a maximum, consistent with their capacity of absorption, investment in the less favoured provinces of the country'. However, it is recognized that 'it would be foolish to exclude the areas of prosperity which will themselves contribute directly and indirectly to raising the prosperity of neighbouring provinces'. The meaning of this policy is made abundantly clear in a section which names the 'less favoured provinces' and 'the areas of prosperity':

> . . . the allocation of investment funds has been orientated, firstly to divert resources to the rural areas rather than to the 'line-of-rail', and secondly to allocate these by Province in such a way as to favour the development of such neglected regions as the Northern, Luapula, Barotse and North-Western Provinces (Zambia, 1966b).

A rather enigmatic phrase follows in which it is claimed that 'this regional orientation constitutes an element of originality in the present Plan'. This chapter will have made it clear that regional allocation of development effort in agriculture was formally adopted in 1956; what is new is the basis of the allocation and the pattern of investment.

Conclusion

This chapter has demonstrated that in Zambia the present state of African agriculture and the character and distribution of innovations can be explained satisfactorily only by reference to diverse human factors. By largely evolutionary measures an agricultural revolution recently has been initiated in parts of the Southern, Central and Eastern Provinces (Plate 8). These areas now market sizable quantities of maize, groundnuts, tobacco, cotton and livestock products. Progress achieved in these areas is spectacular but leaves no room for complacency. Indeed, consolidation is urgently required and it would be unfortunate if progress should slacken (and that of earlier years be undone) because sufficient staff and capital were not forthcoming for these areas. Elsewhere in Zambia unimproved systems of subsistence farming still prevail and it is necessary to account for the lack of progress rather than for improvements. This situation is largely a product of historical, cultural, sociological and political factors. Physical factors (such as climatic and soil conditions) and economic criteria (such as the cost of transport to domestic and overseas markets) cannot be ignored. Indeed, they undoubtedly are of fundamental importance but it would seem that the decisive factors have been human ones.

In recent years, and particularly since independence, political factors

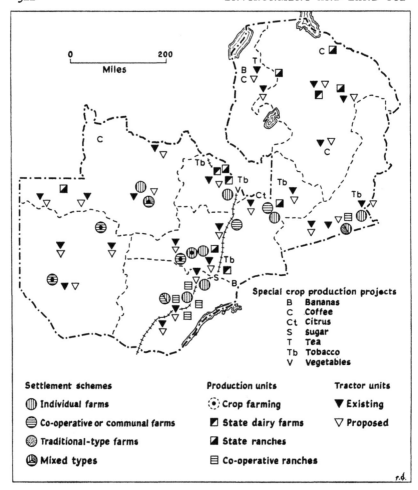

Figure 18.4 *Direct Production Schemes, Settlement Schemes, and Tractor Units, 1966* (Source: First National Development Plan, 1966–70).

have played an increasing role in determining agricultural policies in Zambia. In fact, nowadays political criteria appear to be paramount and current policies are based largely on political and moral realities. The Government is trying to provide all persons in all areas with opportunities for economic and social advancement. In most rural areas such opportunities must be found in agriculture, and an almost desperate search is being made for the means of creating such opportunities. A clear break has been made with the conservative, evolutionary approach of the past. An adventurous policy has been adopted and in addition to normal extension services the Government is experimenting with a variety of projects in widely separated areas (Fig. 18.4). In fact, agriculture in Zambia is in a state of flux and the future character and location of improved commercial farming now is a matter for speculation. Perhaps some of today's problem areas will emerge as the growth areas of tomorrow. However, when the hurly-burly engendered by urgent development policies is done, it seems likely that physical and economic criteria and human factors such as those discussed in this chapter will make themselves felt again. Perhaps then a relatively few highly localized areas will be recognized as 'zones of high potential' and the controlled redistribution of population in relation to resources and economic opportunities may be adopted as a matter of policy.

References

ALLAN, W. 1949. *Studies in African Land Usage in Northern Rhodesia*, Rhodes–Livingstone Pap. 15.
 1965. *The African Husbandman*, Edinburgh.
ALLAN, W., GLUCKMAN, M., PETERS, D. U. and TRAPNELL, C. G. 1948. *Land Holding and Land Usage among the Plateau Tonga of Mazabuka District*, Rhodes–Livingstone Pap. 14.
BARBER, W. J. 1961. *The Economy of British Central Africa*, London.
COSTER, R. N. 1958. *Peasant Farming in the Petauke and Katete Areas of the Eastern Province of Northern Rhodesia*, Agricultural Bulletin 15, Government Printer, Lusaka.
FEDERATION OF RHODESIA AND NYASALAND, 1961. Northern Rhodesia *Report of the Rural Economic Development Working Party*, Government Printer, Lusaka.
 1962. Northern Rhodesia *Draft Development Plan for the Period July 1961 to June 1965*, Government Printer, Lusaka.
GANN, L. H. 1964. *A History of Northern Rhodesia*, London.
HADFIELD, J. 1962. *Report on the Peasant Farming Survey: 1960–61*, Ministry of African Agriculture, Lusaka.
HALCROW, M. 1961. *Report on Intensive Rural Development in the Northern and Luapula Provinces of Northern Rhodesia 1957–61*, Government Printer, Lusaka.
HALL, R. 1965. *Zambia*, London.
JOHNSON, C. E. 1956. *African Farming Improvement in the Plateau Tonga Maize Areas of Northern Rhodesia*, Agricultural Bulletin 11, Government Printer, Lusaka.
KAY, G. 1962. Agricultural change in the Luitikila Basin Development Area, Mpika District, Northern Rhodesia. *Rhodes–Livingstone J.* 31, 21–50.
 1964a. *Chief Kalaba's Village: A Preliminary Survey of Economic Life in an Ushi Village*, Rhodes–Livingstone Pap. 35.
 1964b. Sources and uses of cash in some Ushi villages, Fort Rosebery District, Northern Rhodesia. *Rhodes–Livingstone J.* 35, 14–28.
 1965. *Changing Patterns of Settlement and Land Use in the Eastern Province of Northern Rhodesia*, Occasional Papers in Geography 2, University of Hull Publications.

1967. *A Social Geography of Zambia*, London.

MAKINGS, S. M. 1964. *Problems in African Agricultural Development*, Ministry of African Agriculture, Lusaka.

MORGAN REES, A. M. 1958. *An Economic Survey of Plateau Tonga Improved Farmers*, Agricultural Bulletin 14, Government Printer, Lusaka.

MORGAN REES, A. M. and HOWARD, R. H. 1955. *An Economic Survey of Commercial African Farming among the Sala of Mumbwa District of Northern Rhodesia*, Agricultural Bulletin 10, Government Printer, Lusaka.

NORTHERN RHODESIA, 1938. *Minutes of the Proceedings of the First and Second Meetings of the Native Development Board* (MSS.), Zambia Archives.

1939. *Report of the Department of Agriculture*, Government Printer, Lusaka.

1944-9. *Minutes of the Administrative Conferences of Provincial Commissioners and Heads of Departments* (MSS), Zambia Archives.

1945. *Ten-year Plan of Development – Agriculture*, Acc. 75/1, Report No. 7, Zambia Archives.

1950. *Report of the Department of Agriculture*, Government Printer, Lusaka.

1957. *Board of African Agriculture: Memorandum on African Agricultural Policy*, Acc. 75/1, Zambia Archives.

1964. *Report of the UN/ECA/FAO Economic Survey Mission on the Economic Development of Zambia*, Government Printer, Lusaka.

PETERS, D. U. 1950. *Land Usage in Serenje District*, Rhodes–Livingstone Pap. 19.

1960. Land usage in Barotseland. *Communs Rhodes–Livingstone Inst.* **19**.

SCUDDER, T. 1962. *The Ecology of the Gwembe Tonga*, Manchester University Press for the Rhodes–Livingstone Institute.

TRAPNELL, C. G. 1943. *The Soils, Vegetation and Agricultural Systems of North-eastern Rhodesia*, Government Printer, Lusaka.

TRAPNELL, C. G. and CLOTHIER, J. N. 1937. *The Soils, Vegetation and Agricultural Systems of North-western Rhodesia*, Government Printer, Lusaka.

WHARTON, C. R. 1963. The economic meaning of subsistence. *Malay Econ. Rev.* **8**, 46–58.

ZAMBIA. 1965. *An Outline of the Transitional Development Plan for the Period January 1965 to June 1966*, Government Printer, Lusaka.

1966a. *Economic Report*, Ministry of Finance, Lusaka.

1966b. *First National Development Plan 1966–70*, Office of National Planning and Development, Lusaka.

Appendixes

s

A glossary of terms describing units of the natural environment

Three Major Definitions

1. *Land* (Christian, 1959, p. 591). 'The term land refers to all those physical and biological characteristics of the land surface which affect the possibility of land use, in other words the whole combination of factors which constitute the agro-environment. In practice, it refers more particularly to the obvious inherent features of the land surface, namely to topography, soils, vegetation and climate.'

2. *Landform* (Savigear, 1965, p. 514). 'A landform is a feature of the earth's surface with distinctive form characters which can be attributed to the dominance of particular processes or particular structures in the course of its development and to which the feature can be clearly related. On the other hand, land form (two words) means only the form of the land.'

3. *Soil* defies simple definition but in this context is perhaps best defined as (Soil Survey Staff, 1960, p. 2) 'the upper few feet of the earth's crust having properties differing from the underlying rock material as a result of interactions between climate, living organisms, parent material and relief'.

These definitions are important to the understanding of composite terms listed below:

4. *Ecotype* (Troll, 1963). Although illustrated by Troll it was not specifically defined. It would appear to be similar in concept to the *tessera* (see below).

5. *Land Element* (Brink et al., 1966, p. 9). 'This is the simplest part of the landscape, for practical purposes uniform in lithology, form, soil and vegetation. It is equivalent to Bourne's "site".'

6. *Land Facet* (Brink et al., 1966, p. 9). 'The land facet is a part of a landscape which is reasonably homogeneous and fairly distinct from surrounding terrain.'

The *facet* was defined by Beckett and Webster (1965, p. 8) as 'the largest portion of terrain that can be conveniently treated as one block for purposes of *moderately extensive* land use or construction.'

7. *Land Region* (Brink et al., 1966, p. 15). 'This unit has the small range of surface form and properties expressive of a lithological unit or a close lithological association having everywhere undergone comparable geomorphic evolution.'

8. *Land System* (Christian 1957, p. 76). 'An area or group of areas, throughout which there is a recurring pattern of topography, soils and vegetation.' 'A simple land system is a group of closely related topographic

units, usually small in number, that have arisen as the products of a common geomorphological phenomenon. The topographic units thus constitute a geographically associated series and are directly and consequentially related to one another.'

(Brink *et al.*, 1966, p. 10). 'This is a recurrent pattern of genetically linked land facets. It is defined on its constituent facets and their relationships.'

9. *Land Unit* (Christian, 1959, p. 591). 'A land unit is a particular land form, which at each of its various occurrences has associated with it the same group of soils and vegetation communities. Strict associations are likely to occur only if the various occurrences of the land form have a common genesis.'

(Mabbutt and Stewart, 1963, p. 102). 'In theory, land units are the simplest terms into which the land system can be broken for purposes of correlative description, but in practice land systems are described in terms of combinations of such simple units as are enforced by the scale of work. Such complex units, in so far as they are defined geomorphologically, tend to be elements of larger land forms or to be minor land forms in themselves. In erosional landscapes they are commonly slope segments of characteristic form declivity and position; in depositional landscapes they may be surfaces undergoing particular types of rates of deposition.'

10. *Major Relief Units* (Young, Ch. 13) are 'the main geomorphological divisions of a county'.

11. *Recurrent Landscape Pattern* (Beckett and Webster, 1965, p. 9). 'RLP's are regularly occurring groupings of facets. The facets occurring in one RLP will always exhibit the same spatial inter-relationships.'

12. *Region* (Bourne, 1931, p. 16). 'An association of sites (see below) constitutes a distinct "region".'

13. *Relief Units* (Young, Ch. 13) 'are described in terms of the types and relative extent of landforms within them, and the proportions of slopes at different angles'.

14. *Section* (Linton, 1951, p. 215). 'The section is defined in terms of diversity of rock structure but unity in erosional history.'

15. *Site* (Bourne, 1931, p. 16). 'A site may be defined as an area which appears, for all practical purposes, to provide throughout its extent similar local conditions as to climate, physiography, geology, soil and edaphic factors in general. While a site may be unique, more often the same type of site is to be met with again and again within some readily identifiable area.'

(Clarke, 1957, p. 13). 'A unit of land suitable for a single system of utilisation may be termed a site.'

(Linton, 1951, p. 209). 'Flats and slopes offer a certain limited variety of habitats to growing plants and organisms which ecologists have long recognized under the generic name of sites.'

(Moss, 1968, p. 53). 'A unit of the land surface with a characteristic slope form which constitutes a distinct habitat for plant growth' is a *habitat-site*.

16. *Slope Units* (Young, Ch. 13). 'These are the divisions of an individual slope. They may be rectilinear, and described by their angle, or convex or concave, and described by their curvature and bounding angles.'

17. *Soil Association* (Ellis, 1932, p. 338) is a group of 'topographically related soils on one parent material'.

18. *Soil Catena* (Milne, 1935, p. 197). 'The catena is a grouping of soils which, while they fall wide apart in a natural system of classification on account of fundamental genetic and morphological differences, are yet linked in their occurrence by conditions of topography and are repeated in the same relationship to each other wherever the same conditions are met with.'

19. *Soil Complex* (Soil Survey Staff, p. 304). 'The soil complex is a soil association, the taxonomic members of which cannot be separated individually in a detailed soil survey.'

20. *Soil Series* (Soil Survey Staff, 1951, p. 280). 'The soil series is a group of soils having soil horizons similar in differentiating characteristics and arrangement in the soil profile, except for the texture of the surface soil, and developed from a particular type of parent material.'
(Soil Survey Staff, 1960, p. 15). 'The soil series is a collection of soil individuals essentially uniform in differentiating characteristics and in arrangements of horizons.'

21. *Soil Type* (Soil Survey Staff, 1951, p. 287). 'Is a subdivision of the soil series based on the texture of the surface soil.'

22. *Stow* (Linton, 1951). The stow is not actually defined but refers to a simple unit of terrain exhibiting a characteristic grouping of a few sites. A valley is an obvious example.

23. *Tessera* (Jenny, 1958, p. 5). If an arbitrarily defined spot is selected then, 'there is a soil with organisms below it, and there is plant and animal life above it. The totality constitutes a three dimensional element of landscape, an arbitrary element.' 'The entire landscape can be visualized as being composed of such small landscape elements. This picture is comparable to the elaborate mosaic designs on the walls of Byzantine churches which are made up of little cubes or dice or prisms called tesseras. We shall use the same name, tessera, for a small landscape element.' 'The "thickness" of a tessera is given by the height of the vegetation plus the depth of the soil. The area of a tessera is determined by operational considerations. It is a convenient sampling unit having a specified area.' Tesseras are regarded as 'ecosystem elements' or 'landscape elements' and mosaics of tesseras constitute larger systems akin to the ecosystem.

24. *Tract* (Linton, 1951, p. 211). Tracts were regarded by Unstead (1933) as 'regions of the second order of magnitude' and although not defined by Linton it is clear that they are 'minor regions defined by their characteristic sites'.

25. *Unit Landform* (Lueder, 1959, p. 20). 'A unit landform may be defined as a terrain feature or terrain habit, usually of the third order, created by natural processes in such a way that it may be described and recognized in terms of typical features wherever it may occur, and which, when identified, provides dependable information concerning its own structure, and either composition and texture or uniformity.' Lueder recognized the need for qualifying terms such as 'simple' and 'complex'; 'constructional' and 'tectonic'.

(For books referred to in this Appendix, please see Reference list at the end of Chapter 6.)

Major soil groups in Africa

The following is a brief explanatory glossary of the major soil groups of Africa as defined for mapping purposes by d'Hoore (1964)[1]. The designation of soils within some of the major soil groups cannot always be made with certainty for the specific analytical details required by the definitions are frequently not available. In particular the SiO_2/Al_2O_3 ratio, the percentage of unweathered minerals within the silt to fine sand fractions, and the type of clay mineral that predominates, are rarely to be found in standard analytical descriptions. The interpretation of recorded detail concerning the silt/clay ratio is often complicated by the nature of the parent material. Thus schists and phyllites, even when deeply weathered, maintain a high silt content in the soils.

Ancillary definitive features are usually related to profile morphology. The thickness of the A and B horizons, the presence or absence of free iron and calcium within the profile and their segregation into specific horizons, the presence of a leached A_2 horizon, the presence of structures and the coating of the peds by clay skins, and the amount of humic matter in the top-soil and its distribution throughout the profile are all used. However, they can rarely be used alone to classify a profile, because the A and B notation is not suitable for most tropical soils. Nye (1955) has suggested the use of Cr for the creep horizon and S for the underlying sedentary horizons. The activity of earth-worms and termites, together with the results of colluvial processes, produce what has been called the 'pedisediment'. Many tropical soils have a very distinct textural discontinuity at the Cr/S boundary which is not necessarily related to eluviation and illuviation. Many also have a horizon of loose iron concretions or quartz gravel at this boundary.

Certain kinds of parent material cause particular difficulties. Deep sandy soils derived from weakly cemented sandstones show little further alteration during pedogenesis and distinction between ferrallitic, ferrisolic and ferruginous tropical profiles becomes very difficult. Soils developed from truncated deep weathering profiles over crystalline rocks have been classified generally as ferrisols by d'Hoore (1964). Such soils were described as developing through the erosion of highly leached topsoils of ferrallitic type, but deep weathering is also known beneath ferruginous tropical soils and later erosion is equally possible in these cases. Thus, although the sequence of events may be similar, subsequent pedogenesis would follow separate courses in the ferrallitic and ferruginous tropical soil provinces. Initially similar profiles would be produced in each case, but differences will become more apparent with age. Partial stripping of soil profiles, followed by colluviation as the climate changes also forms profiles that are difficult to classify.

Smyth (1963) has described soils derived from gneiss under semi-deciduous

[1] Additional material for this glossary has been supplied by R. A. Pullan.

rain forest near Ibadan, where the high base saturation throughout the profile and the reserve of weatherable minerals indicates a ferruginous tropical soil. However, the clay is kaolinitic, there is no structural B horizon and clay skins may be found, indicating properties diagnostic of a ferrallitic or ferrisolic soil. On the other hand, Pullan (1964b) has described soils from the Wukari area in Northern Nigeria in which the predominant characteristics are those of ferruginous tropical soils though the base saturation is too low and the soils in which the lateritic ironstone horizon is below 60 in (1·5 m) have a typical ferrisolic profile. Where this horizon is nearer the surface it has prevented eluviation and a zone of accumulation is found, giving the soils a ferruginous tropical profile.

Ferrallitic soils

These have deep, poorly differentiated profiles, usually structureless and and forming a fine porous, friable mass. Clay may be eluviated down the profile and weatherable minerals are rarely found. Though the clay percentage may be high, the cation exchange capacity (CEC) is low and the degree of base saturation is low. Where they are associated with rain forest the topsoil may have a high CEC and be nearly saturated if the humic content is high. Although coarse grained rocks such as granite and sandstone give rise to most characteristic profiles, d'Hoore (1964) also recognizes a type of ferrallitic soil developed on parent materials rich in ferromagnesian minerals and which has a low CEC for its clay content but has a high degree of saturation.

Ferrallitic soils can be divided into yellow and red ferrallites, the former being more highly leached. At altitudes above 1,300 m humic ferrallitic soils have been differentiated by their humic topsoil which exceeds 10 cm in thickness.

The agricultural potential of these soils varies, depending on their topographic position. On well-drained sites they are an excellent growing medium, with ample moisture and room for root development, but they are poor in nutrients. Application of chemical fertilizers to annual crops may be uneconomic as such plants are shallow rooting and the leaching rate is high. The cations are not held in the clay/humus complex; humus being deficient and the kaolinitic clay having a low exchange capacity. However, more efficient use of added nutrients is achieved under deeper rooting, tree crops. Utilization without added nutrients may impoverish these soils if annual cropping under reduced fallow periods is enforced by pressure on land resources. This is especially a problem in areas of nutrient-poor parent materials subject to extreme leaching, such as the sandstones. Analytical data and detailed profile descriptions for various types of these soils and for the ferrisols (below) can be found in Sys (1960).

Ferrisolic soils

Although these soils are regarded as truncated ferrallites by d'Hoore (1964), they nevertheless have a distributional pattern which places them between the ferrallitic soils and the ferruginous tropical soils, and there is no reason why this zone should always be one of accelerated erosion. They are found in areas of lower rainfall than the ferrallites and on certain types of parent material. Whatever their mode of origin, these soils have deep, uniform profiles. They are better structured than the ferrallites in comparable parent materials and they have a higher, but still low, reserve of weatherable minerals and a higher CEC. The degree of saturation is still less than 50 per cent. It is clear that there is no obvious differentiation between ferrallitic and ferrisolic soils in terms of nutrient supply and only a tendency for the latter to have slightly higher reserves. Humic ferrisolic soils are distinguished in the same way and under similar conditions of low temperatures and high rainfall associated with high altitudes, as the humic ferrallitic soils.

The ferrisolic soils are frequently found in areas of moderate annual rainfall and a short dry season so that leaching is not as rapid as in areas of ferrallitic soils, and chemical fertilizers can be applied more economically to annual crops.

Ferruginous tropical soils (fersiallitic soils)

This group of soils contains a wide variety of distinctive units which have been classified by Aubert (1963) into non- or weakly-leached, leached, and those containing concretions or ironstone horizons. The main characteristic of these soils is a marked separation of the free iron oxides into concretions or mottles or leaching of the iron from the soil, possibly to be precipitated in areas of absolute accumulation elsewhere. A comparison of these soils with ferrisolic and ferrallitic soils on similar parent materials shows that tropical ferruginous soils have an appreciable reserve of weatherable minerals. According to d'Hoore (1964) these soils are developed over comparatively shallow weathering profiles, but many of the areas covered by them are known to be deeply weathered and their properties appear to be a result, at least in part, of less active weathering within the profile. Secondary calcium concretions may be associated with such deep, weathered profiles. Other clay minerals are present in addition to kaolinite so that the CEC per 100 g of clay is higher than in either of the major soil groups described above. Rocks of truly basic character frequently give rise to eutrophic brown soils within the zone dominated by ferruginous tropical soils.

The higher nutrient status of the fersiallitic soils is not always apparent, but the release of nutrients from rock weathering ensures that bush fallow can restore them rapidly to the surface. However, these soils are affected seasonally by fire which reduces biological activity and destroys the reserves of nitrogen so that fallowing is no longer effective (Nye and Greenland, 1960). The presence of lateritic ironstone horizons within many of these soils presents specific problems in terms of root penetration, waterlogging and the hardening of such horizons if the topsoil becomes thin. If the soil is removed completely rehabilitation is impossible, and it is essential that the ferruginous tropical soils with lateritic horizons should be conserved against removal of topsoil. Application of specific chemical fertilizers to these soils is known to be economic for various crops.

Various members of this group of soils have been described by Boulet et al. (1964), Fauck (1963) and Maignen (1961a).

Eutrophic brown soils

There is a very close relationship between these soils and areas of shallow soil overlying basalts and other base-rich rocks. They are stony throughout the profile, but weathering produces clay minerals and a high level of base saturation in material which, because of the formation of 2 : 1 lattice clay minerals, has a high CEC. The soils are well structured in the B horizon and permeable, and so form an excellent growing medium if root room is not too restricted by stones. They possess a relatively high organic matter content in the topsoil where they occur in areas of moderate rainfall. Descriptions of these soils have come from Brammer (1962), Dabin et al. (1960), who described them under both rain forest and savanna woodland vegetation in The Ivory Coast, and Laplante (1954), who described them from high mountain grasslands.

These soils are among the most favourable to plant growth provided that they do not dry out quickly owing to excessive drainage resulting from topographic position. On the other hand, their stony character and the broken terrain in which they occur means that they are not suitable for mechanical cultivation, and erosion hazard may be great. Such soils cover only a very small area of the continent.

Reddish brown and brown semi-arid soils

These soils are associated with areas of low rainfall and frequently with unconsolidated, sandy parent materials. They have been described in detail by Maignen (1959b), and by Bocquier and Maignen (1963). They have well-distributed organic matter related to the presence of roots derived from annual herbs, and in particular from grasses, but mineralization is rapid. They frequently have an accumulation of free calcium lower down the profile which is shallow. There is little or no removal of bases so that the saturation percentage is high (over 50 per cent and often as much as 100 per cent) and there may be a reserve of weatherable minerals depending on the character of the parent material. The soil may show a foliated structure near the surface and blocky structure at depth if clay minerals are present. Clays include 2 : 1 lattice varieties. In contrast to soils described previously this group are neutral to slightly alkaline in reaction.

The reddish-brown soils are usually deeper, with a lower organic matter content than the brown soils. They show individualization of iron at depth, and may be slightly leached of bases, giving a slightly acid to neutral reaction.

Accumulation at the surface of fines derived from the Sahara by aeolian activity over a long period of time is an important feature. Under cultivation these soils lose their structure and erosion hazards both from wind and water become serious. Sandy varieties are frequently excessively drained and therefore moisture deficient. The possibility of irrigation for soils of this type must not be ruled out, but elsewhere crop husbandry is hazardous and grazing most suited to the environment.

Vertisols

Vertisolic soils are characterized by the presence of montmorillinitic, 2 : 1 lattice clays in areas of calcium-rich rocks or in topographic depressions into which base-rich waters are draining. The profile is characterized by a self-mulching surface horizon in which large polygonal macrostructure is found. Profiles vary in depth and the lower section is characterized by calcareous concretions. They are poorly drained, frequently flooded during the wet season and they have a very high CEC and high percentage saturation. Leneuf (1954) and Maignen (1961b) have described these soils. They are associated with lagoonal clays near Lake Chad and with basic gneiss on the Accra Plains.

Great difficulty has been experienced in the utilization of these soils owing to their stickiness when wet and their hardness when dry. They are often ignored by shifting agriculturalists, but mechanical cultivation appears to have great possibilities.

Halomorphic soils

These soils are found in areas favourable for the accumulation within the profile of sodium cations, either at the present time or during the Pleistocene. They are frequently difficult to rehabilitate owing to imperfect drainage, a scarcity of suitable water and very high evaporation rates in areas where they occur. In Northern Nigeria irrigation on alluvial soils has frequently increased the salinity hazard.

Hydromorphic soils

Mineral hydromorphic soils represent one of the most neglected soils in Africa. They are characterized by the development of mottled horizons under conditions of imperfect drainage which may be seasonal and imposed by a high, fluctuating water-table or by impermeable, fine-textured horizons. Lateral and vertical variations in texture and nutrient status are important features of these soils, but they present good possibilities for mechanized agriculture and the diversification of crops. In the savannas they provide excellent sites for tree crops such as kola and bananas, if annual fires can be prevented. In West Africa at present they form valuable grazing areas with restricted cultivation of tobacco and sugar cane in the northern savannas and rice in the south.

Organic hydromorphic soils contain within their upper horizons between 20 and 30 per cent organic matter, depending on the nature of the parent material (whether sandy or clayey). Some have a high level of cation saturation.

Weakly developed soils

Juvenile soils occur on recent, sandy riverain or lacustrine deposits, and, in areas of low rainfall and slow pedogenesis, on late Pleistocene materials. Former beach ridges around Lake Chad and the most recent of the fossil dune systems in West Africa are characteristic of the latter type of parent material. These soils may contain reserves of weatherable minerals, but their nutrient status is invariably low as much of the alluvium is derived from highly leached, eroded topsoils. Juvenile soils are also associated with mangrove swamps on fluvio-marine alluvium. They develop a high concentration of sulphides, and as a result a very low pH if improperly managed.

Lithosolic soils are very shallow soils in an immature state of development. Nevertheless, when they are derived from base-rich crystalline rocks they provide excellent growing mediums, though they tend to be excessively well drained. More frequently, however, lithosols are represented either by quartz gravel at the surface with rotting crystalline rock below or by lateritic ironstone. Quartz gravel soils may support good tree growth in areas of adequate rainfall and a short dry season, but they are rarely cultivated and the herbaceous cover is sparse. Over ironstone cuirasses there is frequently a patchy cover of residual colluvium which supports a sparse herbaceous cover, and small shrubs grow within joints. Breakdown of the primary lateritic ironstone produces an ironstone gravel which may support crops. Such areas are cultivated either because of population pressure or because the soils impart a special flavour and texture to such crops as yam (Pullan, 1964b).

Raw mineral soils

Raw mineral soils are associated with desert environments, but they may be found in areas of serious soil erosion. They also form a collar around inselbergs, where they are generally cultivated, because of the appreciable reserves of weatherable minerals which they contain, and because of their free-draining characteristics.

(For books referred to in this Appendix, please see Reference list at the end of Chapter 7.)

Author Index

Subject Index

NOTE: The figures in bold refer to maps.

For Product Safety Concerns and Information please contact our EU
representative GPSR@taylorandfrancis.com
Taylor & Francis Verlag GmbH, Kaufingerstraße 24, 80331 München, Germany

www.ingramcontent.com/pod-product-compliance
Ingram Content Group UK Ltd.
Pitfield, Milton Keynes, MK11 3LW, UK
UKHW020931280425
457818UK00025B/218